Theoretical Physics

Volume 3

Translated from the Russian by
S. Subotić, Belgrade

Translation edited by
J. Schneps, Tufts University, Medford, Mass.
A. J. Manuel, Leeds University

Theoretical Physics

An Advanced Text

Volume 3:

QUANTUM MECHANICS

Benjamin G. Levich, V. A. Myamlin and Yu. A. Vdovin

Institute of Electrochemistry
Academy of Sciences of the USSR, Moscow

1973

North-Holland Publishing Company – Amsterdam • London

Wiley Interscience Division
John Wiley & Sons, Inc. – New York

Library of Congress Catalog Card Number: 68 54501

ISBN North-Holland, complete set: 0 7204 0176 3

Vol. 3: 0 7204 0179 8

Printed in The Netherlands

Title of the Russian edition:
KURS TEORETICHESKOJ FIZIKI

Russian edition published by:
IZDATELSTVO 'NAUKA', GLAVNAJA REDAKCIJA,

FIZIKO-MATEMATIČESKOJ LITERATURY (MOSKVA, 1971)

Publishers:
NORTH-HOLLAND PUBLISHING COMPANY – AMSTERDAM

Sole Distributors for the Western Hemisphere:
WILEY INTERSCIENCE DIVISION
JOHN WILEY & SONS, INC. – NEW YORK
ISBN Wiley Interscience, Vol. 3: 0-471-53115-4

FOREWORD

The first Russian edition of 'Theoretical Physics', which appeared in 1962, has been widely used as a textbook.

Numerous comments from colleagues, lecturers and students have been taken into account in preparing this new edition, which is the first one in English and which will also appear as the second Russian edition.

The material has now been divided into 4 volumes covering the following subjects

Volume 1
> Part I   Theory of the Electromagnetic Field
> Part II   Theory of Relativity

Volume 2
> Part III Statistical Physics
> Part IV   Electromagnetic Processes in Matter

Volume 3
> Part V   Quantum Mechanics

Volume 4
> Part VI   Quantum Statistics and Physical Kinetics

The rapid development of physics and the present wide interest in non-equilibrium and non-stationary processes has compelled us to expand the section on physical kinetics. It has also been transferred to the end of Volume 4 as it is practically impossible to expound this topic without using quantum mechanics.

Part IV — 'Electromagnetic Processes in Matter' — has been substantially revised. Interest in this field has increased recently, mainly in connection with the study of plasmas and plasma-like media, which now have sections devoted to them.

The methods of calculating electrostatic and direct-current fields, and other problems of classical electrodynamics in a medium, are covered very briefly as we have assumed that students will be able to consult the many monographs and handbooks on general physics, electrical- and radio-technology, and the equations of mathematical physics.

As for other modifications and additions, we should draw attention to the introduction of tensor notation, to new ideas in the theories of relativity and electromagnetic fields, the broadening of the introduction to the theory of probability, a brief presentation of the method of correlation functions in statistical physics, the exposition of the thermodynamic theory of ferro-magnetism and the theory of propagation of electromagnetic waves in plasma. A number of paragraphs have been rewritten. We have tried to bring the content of the book even closer to the interests of present-day theoretical physics.

The general level of the book has been preserved and it is still intended to form an introduction to theoretical physics. Problems requiring the use of cumbersome or special mathematical apparatus are still excluded, and the most difficult sections are marked by an asterisk. These may be skipped at will, since there is no reference to them in the main text.

In conclusion we would like to express our gratitude to all those who helped us in preparing this book, in particular to A.M. Brodsky, A.M. Golovin, B.M. Grafov, R.R. Dogonadze, V.S. Krylov and especially V.S. Markin and V.V. Tolmachev. I.V. Savelyev discovered a number of misprints which have now been corrected.

L.D. Konkina helped us in editing the manuscript.

We are grateful to the readers and students who used the first Russian edition of the book for sending us their valuable comments which have been taken into account in this edition.

August 1970

FOREWORD TO VOLUME 3

This volume was written by Benjamin G. Levich in collaboration with Victor A. Myamlin and Yurii A. Vdovin. Dr. Anatol I. Naumov is the author of Chapter 15, *Fundamentals of the theory of elementary particles.* The authors express their deep gratitude to Dr. Naumov for this valuable assistance. The complete volume was written under the supervision of B.G. Levich.

October 1971

FOREWORD TO THE FIRST RUSSIAN EDITION

The continuous development of theoretical physics and the regular expansion of its areas of application create increasing demand for textbooks and manuals.

The rapid development and the complexity of the most recent experimental methods of physical investigation, and the corresponding development and extension of the mathematical apparatus of theoretical physics, have meant that one man usually cannot combine the two methods of investigation. The end of the 19th century and particularly the 20th century therefore saw physicists divided into 'experimentalists' and 'theoreticians', the latter studying physical laws by means of the mathematical methods of theoretical physics.

Obviously, a background in theoretical physics is essential in the education of experimental as well as theoretical physicists.

The experimental and theoretical methods of physical investigation have penetrated into a number of branches of science related to physics (physical chemistry, biophysics, geophysics, astrophysics, and so on) and into technology (metal physics and metallurgical science, thermophysics, electrical technology, radiotechnology, computation, the instrument-making industry etc.). Workers in these branches of science and technology also need a certain minimum knowledge of theoretical physics.

The compilation of a modern textbook on theoretical physics is inevitably associated with certain logical and methodological difficulties. It is impossible at present to divide theoretical physics into classical and quantum parts so that it is also impossible to divide it into separate chapters and sections. For example, the exposition of statistical physics without taking into account the quantum properties of atomic systems is impossible, for it would mean that the general theory remained without practical application. In the theory of electromagnetic processes in matter one has of necessity to make use of the ideas of statistical physics, and so on. It may be that the maximum consistency of composition would be obtained if the book were founded on quantum mechanics but this is completely inadmissible in a book intended as an introductory treatise. Quantum mechanics requires a certain preparedness and the student must be convinced of the necessity of renouncing obvious classical representations. Compromise solutions, which have justified themselves during many years of teaching theoretical physics at the Moscow Engineering-Physical Institute and Moscow State University, are therefore inevitable.

The following general principles have been applied.

(1) The book is written as an introduction to theoretical physics so that aspects requiring the use of cumbersome or special mathematical apparatus have not been included.

(2) As it is to be used for a systematic study of the subject the course is a unique whole and all material necessary for understanding the later sections is contained in the earlier ones.

(3) It would not be feasible to elucidate experimental facts in addition to problems concerning purely theoretical physics. However, physics is a single science, and an attempt to expound the theoretical aspects without taking experiment into account would be quite wrong. The reader is assumed to have some basic experimental knowledge from university courses in general and atomic physics so that we have confined ourselves to references and, in a few instances, to a schematic description of basic experiments.

(4) The acquaintance assumed with general courses in general and atomic physics has allowed us to rely on a certain (very restricted) knowledge of quantum mechanics in our treatment of statistical physics.

(5) Classical mechanics usually forms a separate course so that this topic has been omitted although detailed reference has been made to handbooks of mechanics.

(6) The book similarly does not cover hydrodynamics, aerodynamics, the theory of heat transfer, or problems related to electrical- and radio-technology.

(7) Detailed reference is made to mathematical manuals. The mathematical apparatus utilized, except in the sections marked by an asterisk, is covered by the usual courses in analysis. In the case of quantum mechanics, however, the mathematical apparatus has been included, since it is of a specific character and is not taught in traditional mathematical courses.

(8) As the book is intended as a systematic course in theoretical physics no attempt has been made to achieve the same level of accessibility in all sections. It is a well-known fact that a student's comprehension and assimilation of difficult material increases as a course progresses, and that this is also true for the associated mathematical apparatus. Moreover, experi-

mental physicists will constantly encounter new problems in quantum mechanics which can only be handled using advanced methods of treatment. The section on quantum mechanics (Part V) therefore deals with some topics having a more advanced character than those in other sections. The analysis of applications of the kinetic equations is similarly treated rather extensively.

The uniqueness of the book's objectives has affected the content of individual sections, so that some topics in modern physics have been included at the expense of more traditional material.

Part I contains the foundations of the theory of the electromagnetic field in a vacuum, based on the system of Maxwell-Lorentz equations. A basic knowledge of electromagnetism is assumed. The focus of attention is the theory of radiation and the motion of charged particles in external fields.

In Part II, devoted to the theory of relativity, a four-dimensional form of representation is adopted which not only corresponds to the spirit of the theory but also predominates in contemporary literature. The problems of dynamics in the theory of relativity are treated in some detail. A number of the most recent applications of the theory of relativity, particularly those related to nuclear physics, are covered here for the first time in a textbook.

Part III is a revised version of Levich's 'Introduction to Statistical Physics' and treats statistical physics and the fundamentals of statistical thermo-dynamics. Classical thermodynamics would require too much space, and did not seem indispensable.

Part IV contains the theory of electromagnetic processes in matter. Relatively little attention is paid to problems in theoretical electrical- and radio-technology. The phenomenological theory of electric and magnetic properties of matter is analyzed in some detail, and the notion of the physics of the plasma state of matter is given.

In Part V the basic ideas of present-day relativistic quantum mechanics are included as well as the traditional problems of non-relativistic quantum mechanics. Applications to solid-state theory are considered at length.

Part VI contains the essential concepts of physical kinetics, which are not usually presented in a general course on theoretical physics.

The experience of teaching theoretical physics shows that the greatest difficulties are often encountered not in understanding new physical ideas but in the actual mathematical treatments. All mathematical operations have therefore been performed in sufficient detail.

For convenience we have presented a brief derivation of those formulae of vector analysis which are encountered throughout, as well as the necessary data on Fourier integrals and δ-function theory.

The numbering of formulae and sections starts afresh in each Part and references to appendices have been given Roman numerals.

The author hopes that the readers, after making themselves familiar with the foundations of theoretical physics expounded in this book, will be able to proceed to a more profound study using the many-volume treatise of Landau and Lifshitz. The scientific and educational ideas of their work were of great influence on the author, who is a disciple of Landau.

Parts I–IV and Part VI were written by B.G. Levich. Part V was written by Y.A. Vdovin and V.A. Myamlin under the general scientific guidance of B.G. Levich. Chapter XV of Part V was written by A.I. Naumov.

The author expresses his gratitude to the colleagues who read the book and the manuscripts, and made a number of valuable remarks: B.M. Grafov, R.R. Dogonadze, V.A. Kiryanov, V.S. Krylov, V.S. Markin, V.P. Smilga, Y.A. Chizmadzhev and Y.I. Yalamov.

The creation of a textbook on theoretical physics sufficiently comprehensive in content and clear in presentation is a very complex task. The author is therefore conscious of the fact that shortcomings and errors will be discovered and would be grateful to receive an account of them which can be taken into consideration in the next edition of the book.

1962

Theoretical Physics: Outline of Vols. 1—4

Contents of Volume 3

Part V Quantum mechanics

PART V

QUANTUM MECHANICS

1

The Basic Concepts
of Quantum Mechanics

§1. The physical basis of quantum mechanics

Quantum mechanics, like all physical theories, arose in close connection with the development of a new field of experimental investigations. These investigations, which began with the study of black-body radiation at the turn of the century, were soon extended to the phenomena of the photoelectric effect, and subsequently to atomic systems. In this book we cannot give a chronological account of the history of the development of new concepts about the character of atomic processes which resulted in the creation of contemporary quantum mechanics. We can only point out that this was an agonizing search which required a great effort by some of the most outstanding physicists of our century. The creation of quantum mechanics has undoubtedly been the greatest triumph of contemporary science. The difficulties which the development of quantum mechanics encountered were associated with the fact that the properties of the particles constituting atomic systems are fundamentally different from the properties of macroscopic bodies. The laws of classical mechanics and classical electrodynamics turned out to be inadequate for the description of the behaviour of individual molecules and atoms, as well as elementary particles — electrons, protons, neutrons and so on. Henceforth we shall designate elementary particles, and sometimes individual atoms and molecules, by the term microparticles. As we shall see in what follows, an outstanding feature of microparticles is the fact that their motion does not obey the laws of classical mechanics. To avoid confusion, we shall call particles whose motion does obey the laws of classical mechanics, corpuscles. We have earlier, in particular in the theory of the electromagnetic field and statistical physics, acquainted ourselves with a number of facts which indicate the inadequacy of classical concepts in the realm of atomic processes. Thus in statistical physics we have seen that the energy of individual atoms and molecules, i.e. the basic quantity characterizing their state, assumes discrete values.

A direct proof of the discrete character of the states of atomic systems was provided by the experiments of Franck and Hertz (1913). As is known, in these experiments a beam of electrons of a given energy entered a container filled with a gas. As a function of the accelerating potential, the electron current through the gas displayed a number of sharp minima. The position of these minima is determined by the properties of the atoms of the gas. This dependence of the current on the potential can be interpreted in the following way. In colliding with the atoms, the electrons transfer energy to the former when the energy of the electrons has a value equal to the difference between the energies of two possible states of the atom. The atom then makes transition to a state with a higher energy, an excited state, and a minimum appears in the electron current. If the energies of the electrons have other values, they undergo only elastic collisions. Thus the atom, as a whole, can only obtain definite amounts of energy from outside. This means that the internal energy of the atom has only discrete values or, in other words, the atom possesses a discrete energy spectrum. The discrete character of the energy states is related to the discrete character of atomic transitions. When the atom makes a transition from an excited state to a lower energy state a light quantum is emitted with an energy equal to the difference between the energies of the two states.

The energy of the atom is not the only quantity which can assume only discrete or, as it is said, quantum values. In the experiments of Stern and Gerlach it was shown that the angular momentum of an atom also possesses a discrete spectrum of values. In these experiments a beam of atoms was passed through a magnetic field $\mathscr{H}$ which is nonuniform but constant in direction. Choosing this direction as the z-axis, one can write the expression $\mu_z(\partial \mathscr{H}_z / \partial z)$ for the force acting on the atom, where μ_z is the projection of the magnetic moment of the atom, onto the direction of the field. If it is assumed that the theorem of the proportionality between the magnetic moment and the angular momentum (see Part I, §22) is valid for atoms, then it follows that the mean force is proportional to the value of L_z, where L_z is the projection of the angular momentum of the atom onto the direction of the field (see however ch. 8).

The experiment of Stern and Gerlach showed that the beam of atoms was deflected in the magnetic field, splitting into a number of separate beams. This means that the projection of the angular momentum of the atom onto the direction of the field can assume only discrete values. To every allowed value of L_z there corresponds a definite value of the force and a corresponding magnitude of the deflection in the nonuniform magnetic field. Thus each of the beams produced contains atoms with a given value of the quantity L_z.

The discrete character of the allowed values of the basic quantities which characterize the states of atomic systems profoundly contradicts all the concepts of classical mechanics. It follows from the general propositions of classical mechanics that an infinitesimal force causes an infinitesimal change in the state of a system. Hence all mechanical quantities depending on the state of that system, such as the energy, the momentum and so on, are continuous functions of the state. The discreteness of the states and the discontinuous changes of the states of microparticles directly contradicts this general principle.

The difficulty in understanding the properties of microparticles is increased by the fact that, in addition to the discreteness of certain quantities which characterize the state of particles, in a number of experiments the continuous nature of the same quantities was clearly shown. Thus, for example, the bremsstrahlung of electrons in the nuclear field has a continuous spectrum, which is indicative of a continuous change in the energy of the emitting particles.

It turned out that microparticles combine in a striking way the properties of ordinary particles (corpuscles) and the properties of waves. This fundamental property of microparticles is called the wave–particle duality.

The basic feature of corpuscles as studied in classical mechanics is that they have a definite spatial extent. The idealization of the corpuscle is a material point having no size and moving in a definite trajectory.

The properties of wave motion in classical physics are to a certain extent the opposite of the properties of corpuscular objects. A monochromatic wave, first of all, has an infinite extent in space. Hence it makes no sense to state that 'the monochromatic wave is located at a given point of space'. Also it is meaningless to speak of the trajectory of a monochromatic wave. The localization of a wave in space is inevitably associated with the production of a wave packet (see Part I, §35). The size of the wave packet is smaller, the larger the number of waves of different frequencies which form it. This property of wave motion does not depend at all on their physical nature; it is valid for elastic, electromagnetic and other types of wave. Thus in classical physics localized corpuscles and waves delocalized in space are in a sense opposites.

It turns out that a combination of corpuscular and wave properties, inexplicable form the point of view of the ordinary ideas of classical physics, occurs in microparticles. More precisely, under certain conditions, microparticles behave as corpuscles, while under other conditions the same microparticles display purely wave properties. Finally, in certain experiments both cospuscular and wave properties manifest themselves simultaneoulsy.

The wave--particle duality of the properties of microparticles was first discovered in experiments with light quanta. The wave properties of the electromagnetic field are sufficiently well established. We note only that Newton's corpuscular theory was able to compete succesfully with the wave theory in explaining phenomena such as the rectilinear propagation and refraction of light. However, this theory was completely abandoned after the discovery of interference, diffraction and birefringence.

As to the corpuscular properties of the electromagnetic field, they are manifested in a particularly obvious way in the Compton effect (Part II, § 17; see also ch. 14). Indeed, this effect allows only the corpuscular interpretation. No considerations based on wave concepts can explain the appearance of recoil electrons: the incident electromagnetic wave cannot cause the motion of one of the atomic electrons without perturbing the motion of the remaining electrons. However, as we have seen in § 17 of Part II, the theory based on the concept of the collision of two particles, the incident photon and the atomic electron, describes the process correctly.

The corpuscular nature of light shows up in the same obvious way in the photoelectric effect, in the phenomenon of recoil when atoms are emitting radiation, etc. Thus the wave theory of light, which was successfully applied in considering a wide range of electromagnetic phenomena, turned out to be completely unsuitable for the explanation of a number of processes in which the corpuscular nature of light was manifested.

The situation was characterized briefly by stating that there is a dualism in the properties of the electromagnetic field. Sometimes light manifests its wave nature, and sometimes it behaves as a flux of photons.

The totality of experimental data showed that one must ascribe to every photon an energy E and a momentum p which are respectively equal to

$$E = \hbar \omega ,\tag{1.1}$$

$$p = \frac{E}{c} = \frac{2\pi\hbar}{\lambda} = \frac{\hbar}{\lambdabar} ,\tag{1.2}$$

where $\hbar$ is the Planck constant h divided by 2π and equal to 1.054×10^{-27} erg·sec, and $\hat{\lambda} = \lambda/2\pi$. Further, it turns out that the wave–particle duality appears not only in the case of photons, but for all microparticles.

The corpuscular properties of microparticles were discovered a relatively long time ago. They show up particularly clearly in observations with cloud chambers. As is known, microparticles in passing through a cloud chamber filled with a saturated vapour produce ionization along their path. The ions produced by the microparticles become the centres of condensation which can be observed directly in the form of tracks. Similarly, particles in moving through a thick layer of photographic emulsion leave a photographic image, i.e. a track. All this led to the idea that microparticles move in well defined trajectories and are similar in their properties to ordinary corpuscles. However, the experiments to be described below made it possible to establish that this was not so and that the wave–particle duality is a basic feature of all microparticles. But it should be stressed that the discovery of the wave properties of electrons, protons and other microparticles was preceded by the development of the concepts of quantum mechanics, in which the existence of the wave properties of microparticles was predicted theoretically.

Let us consider the following experiment. Individual electrons which have passed through a fixed accelerating field are let successively through a small opening in an impenetrable screen. After passing through the opening the electrons fall onto a photographic plate giving rise to blackening at the points of impact. If the electrons moved as corpuscles obeying the laws of classical mechanics and did not interact with the edge of the screen, then all of them would fall at the centre of the photographic plate, forming a black spot. In fact, the electrons must interact with the atoms of the screen. Since the latter are in thermal motion, this interaction has a random character. Hence it would be natural to expect the electrons to give rise to a blackening of the photographic plate similar to that caused by a beam of molecules coming out of a small opening. Namely, the number of electrons which are deflected from their rectilinear path and do not fall at the centre of the screen should depend on the magnitude of the deflection according to the normal law of errors. The intensity of the blackening, which is proportional to the number of electrons falling at a given point, should be expressed by the Gaussian distribution.

In fact, nothing of the kind is observed in experiment. If a large number of electrons are succesively let through an opening, then the following is observed:

(1) There are zones on the photographic plate — 'forbidden' zones — in which electrons never arrive. These zones have the character of concentric rings of definite width;

(2) the zones in which electrons do arrive form a system of concentric rings alternating with the 'forbidden' rings.

By carrying out the experiment for a sufficiently long time, i.e. letting through a sufficient number of electrons, one can obtain blackened rings which are identical with those which arise when light is diffracted from a circular opening. Such a diffraction pattern is shown on the right of fig. V.1. In the drawing, white rings correspond to the blackened rings of the photographic plate. The curve of the intensity of electrons as a function of the angle of diffraction ϑ is shown on the left. The same result is also obtained in another arrangement of the experiment. Instead of letting electrons through one by one, a beam of electrons can be directed through the opening of the screen. The beam must be of a sufficiently low intensity, so that the interaction between electrons will be of no importance. When the beam of electrons passes through the opening of the screen, the distribution of the intensity of blackening of the photographic plate immediately appears in the form of a diffraction pattern*. Thus the motion of each individual electron differs fundamentally from that of a classical particle passing through the opening of the screen.

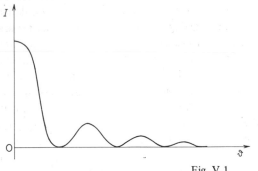

Fig. V.1

At first sight it may seem that the results of the observations described could be interpreted in the following way: for some unknown reason, not all possible trajectories of the motion of the electrons, but only certain allowed ones, can be realized in nature. The sum of these allowed trajectories deter-

---

* From the experimental point of view the second method is simpler. Such an experiment was carried out by Davisson and Germer in 1927, after the creation of the theory of quantum mechanics. The experiment with individual electrons was not carried out until 1948 by V.A. Fabrikant, L.M. Biberman and N. Sushkin.

mines the loci of incidence of electrons on the photographic plate. However, other experiments show this interpretation to be incorrect.

Let us consider an impenetrable screen with two openings. (The experiment discussed below is a schematization of a real experiment in which, instead of diffraction from a screen with two openings, the diffraction of electrons from a crystal lattice was observed.) If in turn each of the openings is covered, while individual electrons are successively let through the other, then after the passage of a large number of electrons the two diffraction patterns described above, with a central spot opposite to each of the openings, will arise on the photographic plate. We now uncover both openings and let electrons through. We assume that each of the electrons moves in a well defined allowed trajectory. Passing through one of the openings, the electron causes blackening at a definite point on the photographic plate. The final diffraction pattern produced by a large number of electrons should be a simple superposition of the intensities of the blackenings arising when electrons are let through one opening. In other words, we should obtain the same diffraction pattern as in the case of the alternative passage of electrons first through one opening and then through the other. In fact, however, the distribution of intensities of blackening is of a completely different character. The blackening of the photographic plate corresponds exactly to the pattern of diffraction from two openings. This means that there are no possible or allowed electron trajectories whatever. The electron, like a wave, possesses interference properties, and it would make no sense to try to establish through which one of the two uncovered openings a given electron 'in reality' passed.

We see that a certain wave motion is associated with the electron; the electron possesses wave properties. It is because of these wave properties that an individual electron passing through an opening can arrive at some regions of the photographic plate but not at others. In the passage through two openings the wave properties of an individual electron become apparent in the fact that its motion is affected by both openings. The allowed and forbidden regions of the photographic plate correspond to the dark and light zones of the diffraction pattern from two openings.

However, it would be incorrect, on the basis of the aforesaid, to try to identify the electron with a particular wave. If this were possible, then the darkening of the photographic plate on which the diffracted wave (the electron) falls would be a pale copy of the darkening produced by many electrons. An individual electron would immediately give the entire diffraction pattern.

We have stressed that experiment shows an individual electron to be incident at a definite point of the plate, like an ordinary corpuscle. The difference between an individual electron and a corpuscle lies in the fact that the loci of incidence on the photographic plate are determined by laws which are completely different from those determining the loci of incidence of a corpuscle. Thus, as is shown by the diffraction experiment, wave properties are inherent in every individual electron, but they manifest themselves in an obvious way only in the repetition of a large number of identical experiments (i.e. the successive passage of a large number of electrons).

We note that, although we have spoken above only of the electron, the same is also valid for other microparticles. Diffraction experiments have been carried out with neutrons, protons and other microparticles.

The quantum-mechanical treatment of the diffraction experiments described will be given in the next section. Here we stress once more that the wave–particle duality which had already been established for light quanta also becomes apparent in diffraction experiments with electrons.

Diffraction experiments make it possible to give an answer to the question: 'what is an electron – a wave or a corpuscle?' Here we use the terms 'wave' and 'corpuscle' in their usual classical meaning. The answer which follows directly from the experiments described is that the electron is neither a wave (otherwise a single electron would give the entire diffraction pattern), nor a corpuscle moving in a definite trajectory (which contradicts the experiment with two openings). The electron is a microparticle possessing specific properties.

§2. The wave function

The fact that the electron possesses wave properties shows that the electron is to be compared with a certain wave field. We shall call the amplitude of this wave field, which depends on the spatial coordinates and time, the wave function $\psi(x, y, z, t)$. For brevity it is sometimes also called the ψ-function.

The physical interpretation of the wave function (which was first given by M. Born) is the following: the quantity $|\psi(x, y, z, t)|^2 dV$ is proportional to the probability that the electron will be found at the instant of time t in a volume element dV in the neighbourhood of the point x, y, z.

Denoting this probability by dW, we have

$$dW \sim |\psi(x, y, z, t)|^2 dV . \tag{2.1}$$

This interpretation is based on the following reasoning. In the experiments with the passage of individual electrons through one or two openings we have seen that the locus of incidence of the electron on the photographic plate is to a certain degree random. The electrons will fall completely randomly at one or other point of the diffraction ring to be formed. Hence the behaviour of an electron must be characterized by a certain probability function. The intensity of blackening of the photographic plate at a given spot is proportional to the number of electrons which fall on this spot. On the other hand, it is clear that this probability function must be connected with the properties of the wave field. Only in this case can the probabilistic character of the blackening of the photographic plate at a given spot be matched with the strict spatial distribution of the bands of blackening. That is, the random character of the incidence of the electron at a given point can be matched with its wave properties only by assuming that the probability of finding the electron at the given point is proportional to the intensity of the wave field $|\psi|^2$. This relation is just that given by formula (2.1).

The physical interpretation of the wave function given by formula (2.1) clearly shows that the wave field $\psi(x, y, z, t)$ is fundamentally different from other wave fields known in classical physics. This is particularly apparent from the fact that only the quantity $|\psi|^2$ has a direct physical meaning. The wave function itself, in general, can be a complex quantity. Furthermore, the wave functions ψ and $A\psi$, where A is an arbitrary constant, correspond to one and the same physical state of the particle, since by virtue of the definition (2.1) the two wave functions lead to one and the same space–time distribution of the probability of finding the particle.

By virtue of the theorem of addition of probabilities (see §2 of Part III) the definition (2.1) can be supplemented by the following normalization condition:

$$\int |\psi(x, y, z, t)|^2 dV = 1 , \tag{2.2}$$

where the integral on the left, taken over all space, is the probability of finding the particle at time t at any point of space. This probability is naturally equal to unity. Wave functions ψ satisfying the normalization condition are said to be normalized. For normalized wave functions the relation (2.1) can be rewritten in the form

$$dW = |\psi(x, y, z, t)|^2 dV = \rho(x, y, z, t) \, dV , \tag{2.3}$$

where $\rho(x, y, z, t)$ is the probability density. The probability $W(V, t)$ of finding the particle in a given finite volume V at an instant of time t, is, according to the theorem of addition of probabilities

$$W(V, t) = \int_V dW = \int_V |\psi(x, y, z, t)|^2 dV. \tag{2.4}$$

The condition (2.2) cannot be satisfied in the case where the integral $\int |\psi|^2 dV$ is divergent. This can occur, in particular, if the square of the modulus of the wave function, $|\psi|^2$, does not tend to zero at infinity. Physically this means that there is a finite probability of finding the particle at every point of space. In §18 it will be shown how the normalization of the wave function is to be carried out in this case.

We note that the wave function normalized by the condition (2.2) is defined with an accuracy to within the factor $e^{i\alpha}$, where α is any real number, in view of the equality $|e^{i\alpha}|^2 = 1$.

In addition to the wave function of a single microparticle it is also necessary to introduce the idea of the wave function of a system of microparticles. Let there be a system of N particles interacting with each other according to an arbitrary law. This system of particles can be represented by the wave function

$$\psi(x_1, y_1, z_1, x_2, y_2, z_2, ..., x_i, y_i, z_i, ..., x_N, y_N, z_N, t) ,$$

where i is the index of the particle.

In the further construction of quantum mechanics we shall proceed from the assumption that there is no difference of principle between the description of an individual microparticle and a system of microparticles and that the interpretation of $\psi(x, y, z, t)$ and $\psi(x_1, y_1, z_1, x_2, y_2, z_2, ..., x_N, y_N, z_N, t)$ must be one and the same. In other words, the physical meaning of the wave function of a system of N particles lies in the fact that the quantity

$$dW \sim |\psi(\mathbf{r}_1, \mathbf{r}_2, ..., \mathbf{r}_N, t)|^2 dV_1 dV_2 ... dV_N \tag{2.5}$$

gives the probability that at a certain instant of time t the first particle be found in the volume element dV_1 surrounding the point $\mathbf{r}_1$, the second particle in the volume element dV_2 surrounding the point $\mathbf{r}_2$, and so on. For brevity, $\mathbf{r}_i$ here denotes the totality of coordinates (x_i, y_i, z_i). We note that on the basis of the theorem of addition of probabilities the quantity

$$dW_1 \sim dV_1 \int \psi(\mathbf{r}_1, \mathbf{r}_2, ..., \mathbf{r}_N, t) \, dV_2 ... dV_N \tag{2.6}$$

represents the probability of finding the first particle in the volume element dV_1 for any distribution of the remaining particles of the system (the integration being carried out with respect to the coordinates of the latter particles). It is obvious that the probability dW_1 given by formula (2.6) must be identical with the definition (2.3). Analogous relations can also be written for other particles of the system. Thus dW gives the probability of finding a given configuration of the system in space.

The normalization condition for the wave function of a system of N particles has the form

$$\int |\psi(\mathbf{r}_1, \mathbf{r}_2, ..., \mathbf{r}_N, t)|^2 dV_1 dV_2 ... dV_N = 1 . \tag{2.7}$$

It is clear that the wave function of a system of N particles is not normalized in the real three-dimensional space but in the $3N$-dimensional configuration space.

In view of the similarity of principle between the wave function of one particle and that of a system of particles, we shall always denote the wave function by the symbol ψ. For the sake of brevity the whole set of coordinates is sometimes denoted by x.

It follows from the aforesaid that the quantity $|\psi|^2$ must be interpreted as a probability not in real space but in configuration space. At the same time, the introduction of the wave function of a system of particles confirms in a particularly obvious way the impossibility of interpreting the wave function as a quantity which describes a wave motion similar to an electromagnetic or an acoustic wave propagating in real space. Indeed, every wave motion in real space is characterized by the set of three variable coordinates and time. However, the wave function of a system of N particles depends on $3N$ coordinates and time. Hence in interpreting the ψ-function as an ordinary wave one would have either to renounce the assumption of the unique meaning of the wave function of one microparticle and of a system of microparticles, or to introduce the hypothesis of the existence of a real multi-dimensional space. Both are in flagrant contradiction with all experimental data.

Let us consider the important particular case of a system of noninteracting particles. Then finding the ith particle in the volume element dV_i, the kth particle in the volume element dV_k and so on, must be independent events. On the basis of the theorem of multiplication of probabilities, formula (2.5) can be written in this case as follows:

$$dW = dW_1 dW_2 ... dW_N =$$
$$= |\psi_1(\mathbf{r}_1, t)|^2 dV_1 |\psi_2(\mathbf{r}_2, t)|^2 dV_2 ... |\psi_N(\mathbf{r}_N, t)|^2 dV_N .$$

This means that the wave function of a system of noninteracting particles is equal to

$$\psi(\mathbf{r}_1, \mathbf{r}_2, ..., \mathbf{r}_N, t) = \psi_1(\mathbf{r}_1, t)\psi_2(\mathbf{r}_2, t) ... \psi_N(\mathbf{r}_N, t) . \tag{2.8}$$

The case where the system consists of identical microparticles (for example, of electrons or protons, and so on) will, in particular, be considered later (see §64).

Before trying to construct the wave function for the simple case of the motion of a microparticle, it is necessary to make the following very important remark. At first sight it might be assumed that it is necessary to introduce into physics new notions in order to describe the states of microparticles, which represent completely new objects as regards their physical nature. It turns out, however, that this is not so. The state and the nature of the motion of microparticles can to a certain degree be characterized by the quantities and terms of classical physics. This is pointed out by the wave–particle duality of microparticles, which consists in the fact that in certain experiments microparticles manifest themselves as objects with a wave nature, while in other experiments they behave as ordinary corpuscles.

When we introduced the statistical interpretation of the wave function we already assumed, to a certain degree, that the concepts of classical mechanics are applicable to microparticles. Indeed, the statement that 'a microparticle can be found in a volume element dV' already implies the assumption that the classical approach is possible for the description of its state by defining its position in space. If the microparticle were in all respects similar to a wave, then the statement of the problem, 'where can the microparticle be found', would make no sense. On the other hand, the presence of the diffraction pattern makes it possible, under certain conditions, to associate the microparticle with another classical notion, a definite wavelength λ, and to speak of the wavelength corresponding to the wave function of the particle.

Classical ideas, such as the position of the particle, the wavelength and so on, can only be applied to microparticles within certain limits. We shall dwell in detail on this in §4. The most important thing is not the fact that in describing microparticles the notions of classical physics are of limited applicability, but the fact that they can and must be used in describing new objects which are so unlike ordinary macroscopic bodies or waves.

We shall assume that the state of an electron moving freely in space can be characterized by an energy E and a momentum $\mathbf{p}$. Then the relation between the energy and momentum is given by the classical formula

$$E = \frac{|\mathbf{p}|^2}{2m}. \tag{2.9}$$

We assume that a beam of electrons which has passed through a strictly defined accelerating potential difference and has acquired a definite energy enters a diffraction arrangement (in practice this arrangement is usually a crystal lattice). Formula (2.9) allows us to speak of a definite momentum of the electron.

On the other hand, knowing the diffraction pattern one can (see §36 of Part IV) find the wavelength λ corresponding to the electron. It turns out that the following relation exists between the quantities λ and p:

$$p = 2\pi\hbar/\lambda = \hbar k .\tag{2.10}$$

The relation (2.10), which was first proposed in 1924 by de Broglie on the basis of theoretical considerations, is called de Broglie's formula. The wave associated with the motion of a microparticle is called the de Broglie wave.

We see that de Broglie's formula is the same as formula (1.2) for light quanta. The frequencies corresponding to the de Broglie waves cannot be determined directly in an experimental way. However, it is natural to assume that the relation between the energy and frequency which holds for light quanta is also applicable to the de Broglie waves, i.e.

$$E = \hbar\omega .\tag{2.11}$$

Based on the relations (2.10) and (2.11), we write the wave function of a free particle in the form of a plane monochromatic wave

$$\psi_p(\mathbf{r}, t) = A e^{i(\mathbf{k}\cdot\mathbf{r} - \omega t)} = A e^{(i/\hbar)(\mathbf{p}\cdot\mathbf{r} - Et)}\tag{2.12}$$

Later it will be explained why $\psi_p(\mathbf{r}, t)$ should be written in the form of an exponential function, and not in the form of a sine or cosine (see §6). The constant A is determined by the normalization condition (see §26).

By definition $\mathbf{k}$ is the wave vector

$$\mathbf{k} = \frac{1}{\hbar}\mathbf{p} .\tag{2.13}$$

By means of formulae (2.9), (2.10) and (2.11) one can find the dispersion law of de Broglie waves

$$\omega = \frac{1}{\hbar}E = \frac{p^2}{2m\hbar} = \frac{\hbar k^2}{2m} .\tag{2.14}$$

The corresponding phase velocity and group velocity are equal to

$$v_{ph} = \frac{\omega}{k} = \frac{\hbar k}{2m} ,\tag{2.15}$$

$$v_{gr} = \frac{d\omega}{dk} = \frac{\hbar k}{m} = \frac{1}{m}p .\tag{2.16}$$

Formula (2.16) shows that the group velocity of de Broglie waves is the same as the ordinary velocity of macroscopic particles. If we took the quantity v_{gr} as the initial expression for the velocity of the particle, then the relation between the energy and frequency (2.11) could be obtained as a consequence of this definition. The phase velocity v_{ph} of de Broglie waves has no direct physical meaning. This becomes particularly clear if one makes use of the relativistic expression for the relation of the energy of the momentum of the particle

$$E = (p^2c^2 + m^2c^4)^{\frac{1}{2}} = (\hbar^2c^2k^2 + m^2c^4)^{\frac{1}{2}} = \hbar\omega .$$

Then

$$v_{ph} = \frac{\omega}{k} = \frac{E}{\hbar k} = \left(c^2 + \frac{m^2c^4}{\hbar^2k^2}\right)^{\frac{1}{2}} > c ,$$

i.e. v_{ph} is larger than the velocity of light.

§3. The principle of superposition. Expansion in plane waves

We have considered above the phenomenon of the diffraction of electrons incident on a screen with a limited number of apertures. In fact, however, what is observed is the diffraction of electrons from a crystal lattice. This case is not only of practical interest, but is also very important from the theoretical standpoint.

We have seen in §36 of Part IV that the selective reflection of X-rays takes place when the Bragg–Woolf condition is fulfilled. If we dispense with unessential details, it turns out that the diffraction of electrons from a crystal lattice is analogous to the diffraction of X-rays. As a result of the scattering of electrons from a single-crystal a number of selective reflections occurs. Each selective reflection corresponds to a definite momentum or, by virtue of (2.10), to a definite wavelength. Thus the crystal lattice is a device which resolves the initial polychromatic beam of electrons into a number of beams each of which corresponds to electrons of a definite wavelength.

We have already pointed out that the experiment with an electron beam is equivalent to the whole set of successive measurements with a large number of electrons which are in identical external conditions. Hence the diffraction grating plays the role of a device which analyzes the initial state of the microparticle, resolving it into a set of individual states with definite values of the momentum. Since to each state with a definite momentum there corresponds a plane wave of the form (2.12), then consequently, the wave function describing the initial state of the electron incident on the grating can in general be written in the form of a superposition of plane waves, i.e.

$$\psi(x, y, z, t) = \int_{-\infty}^{\infty} c(p_x, p_y, p_z)\psi_{\mathbf{p}}(x, y, z, t)\, dp_x\, dp_y\, dp_z . \qquad (3.1)$$

Physically this means that the wave function of an electron in an arbitrary state can be considered as a superposition of the wave functions corresponding to states with a definite value of the momentum.

Hence it is not surprising that one electron (or other microparticle) can be in a definite state without having a definite value of the momentum. Although the concept of momentum, applied to the microparticle from classical mechanics, can be used in quantum mechanics, the state of a microparticle is not defined by the same laws as the state of a particle in classical mechanics. In later sections we shall come back to the discussion of the problem of characterizing the states of microparticles.

Choosing the coefficient A of formula (2.12) in the form (see §26)

$$A = \frac{1}{(2\pi\hbar)^{\frac{3}{2}}}, \qquad (3.2)$$

we have

$$\psi(\mathbf{r}, t) = \int_{-\infty}^{\infty} c(\mathbf{p}) \frac{1}{(2\pi\hbar)^{\frac{3}{2}}} e^{(i/\hbar)(\mathbf{p} \cdot \mathbf{r} - Et)}\, d\mathbf{p} . \qquad (3.3)$$

From the mathematical point of view, formula (3.3) represents the expansion of the function $\psi(\mathbf{r}, t)$ in a Fourier integral. The amplitude $c(\mathbf{p})$ shows the weight with which the state $\psi_{\mathbf{p}}$ is involved in the state described by the wave function $\psi(\mathbf{r}, t)$. According to the Parseval equality (see, in Volume 1, Appendix II eq. II.9), we have

$$\int |\psi|^2 dV = \int |c|^2 d\mathbf{p} . \qquad (3.4)$$

When choosing the coefficient A in the form of (3.2) the equality (3.4) contains no numerical coefficients. It is natural to assume that $|c(\mathbf{p})|^2$ can be related to the density $\rho(\mathbf{p})$ of the probability that the value of the momentum of the particle in the state $\psi(\mathbf{r}, t)$ be equal to $\mathbf{p}$. Namely, it is natural to assume that

$$\rho(\mathbf{p}) = |c(\mathbf{p})|^2 . \qquad (3.5)$$

We write the equality sign and not the proportionality sign $\sim$, because the function ψ is normalized by the condition (2.2). In this case the following equality holds (see (3.4)):

$$\int |c(\mathbf{p})|^2 d\mathbf{p} = 1 .\tag{3.6}$$

We shall return to the discussion of formula (3.5) in §21.

Equation (3.3) is a particular case of one of the most important propositions of quantum mechanics – the principle of superposition. This principle amounts to the following: if the quantum system can be in states described by functions ψ_1, ψ_2, ..., ψ_n, then the linear combination (superposition) of the wave functions ψ_n

$$\psi = \sum_n c_n \psi_n ,\tag{3.7}$$

where c_n are arbitrary constants, is also a wave function describing one of the possible states of the system. The importance of the principle of superposition, in particular, lies in the fact that it restricts the possible equations for the determination of ψ to linear equations (see §6). If the index n characterizing the state runs over a continuous sequence of values, then the summation in formula (3.7) must be replaced by integration. Later we shall come back to the discussion of the notion 'the state of a quantum system' and of the meaning of the coefficients c_n (see §21 and §23).

As an example of the application of the principle of superposition, let us consider a free particle whose momentum has no strictly defined value but can lie in a small interval Δp about the value p_0, namely $p_0 - \Delta p \leqslant p \leqslant p_0 + \Delta p$. For simplicity we consider the case of uniform motion. According to (3.3) the wave function of the electron can be written in the form

$$\psi(x, t) = \int_{p_0 - \Delta p}^{p_0 + \Delta p} c(p) \frac{1}{(2\pi\hbar)^{\frac{1}{2}}} e^{(i/\hbar)(px - Et)} dp =$$

$$= \left(\frac{\hbar}{2\pi}\right)^{\frac{1}{2}} \int_{k_0 - \Delta k}^{k_0 + \Delta k} c'(k) e^{i(kx - \omega t)} dk .\tag{3.8}$$

For uniform motion the coefficient A is equal to $(2\pi\hbar)^{-\frac{1}{2}}$ (see §26). In accordance with the results of §35 of Part I, the expansion of (3.8) is expressed by a formula which is the same as (35.1) of Part I:

$$\psi(x, t) = \left(\frac{\hbar}{2\pi}\right)^{\frac{1}{2}} 2c'(k_0) \frac{\sin\left[\Delta k \left(x - \left(\frac{d\omega}{dk}\right)_0 t\right)\right]}{\left[x - \left(\frac{d\omega}{dk}\right)_0 t\right]} e^{i(k_0 x - \omega_0 t)} .\tag{3.9}$$

Formula (3.9) shows that the superposition of wave functions which correspond to nearby values of the momenta (or the wave numbers) leads to the formation of a wave packet propagating with the group velocity

$$v_{gr} = \left(\frac{d\omega}{dk}\right)_0 = \frac{1}{m}p_0 .$$

It is clear from the form of the wave function (3.9) that the probability of finding the microparticle at a point x at an instant of time t, which is proportional to $|\psi(x, t)|^2$, has a sharp maximum moving with the velocity v_{gr}.

It should be stressed that eq. (3.9) is of an approximate character. Taking account of the subsequent terms of the expansion of the function $\omega(k)$ would lead to an expression for the wave packet whose width would increase with time. Such a wave packet is said to be spreading. The spread of the packet follows directly from the fact that each wave forming the packet moves with its phase velocity $v_{ph} = \omega/k = \hbar k/2m$.

When quantum mechanics first appeared attempts were made to identify the electron with a wave packet made up of de Broglie waves. However, the spread of the wave packet is indicative of the unsoundness of such a treatment. Furthermore, if the electron represented a wave packet, then, as in the case of a single wave, it would be impossible to account for the experiment on the diffraction of individual electrons.

§4. Uncertainty relations and the relationship between quantum mechanics and classical mechanics

We shall use the representation of the wave function in the form of a wave packet for the discussion of a fundamental problem. The question is to what extent and with what degree of accuracy use can be made of the concepts of classical mechanics in their application to microparticles. Here we restrict ourselves to the consideration of the concepts of the momentum and position of a particle in space. In §24 this problem will be studied in full.

We have seen in §35 of Part I that the wave packet possesses a spatial extent given by formula (35.7) of Part I. Applying this to the wave packet (3.9), in which we are interested here, this formula can be written in the form

$$\Delta p_x \Delta x \sim \hbar .$$

Since a spread of the packet takes place, which was not taken into account in deriving this formula, it can be written more correctly as follows:

$$\Delta p_x \Delta x \gtrsim \hbar . \tag{4.1}$$

The numerical factor in formula (4.1) will be defined more precisely in §24. In a similar way one can write the relations for the remaining two coordinate and momentum components:

$$\Delta p_y \Delta y \gtrsim \hbar , \tag{4.2}$$

$$\Delta p_z \Delta z \gtrsim \hbar . \tag{4.3}$$

Formulae (4.1)–(4.3) are called Heisenberg's uncertainty relations. We discuss the meaning of these inequalities by proceeding from the probabilistic interpretation of the wave function. If the width of the wave packet is equal to Δx, then according to what was said in the preceding section the measurements of the coordinates of the electron will show that with a very high probability it will be found in the region of space Δx. In this sense it can be said that the coordinate of the electron is determined to within an accuracy of Δx. However, the electron found in the region Δx is not described by a plane wave and has no definite value of the momentum. To form a wave packet of width Δx it was necessary to form a superposition of plane waves with momenta in the interval $p_0 - \Delta p_x \leqslant p_x \leqslant p_0 + \Delta p_x$, where Δp_x is determined by formula (4.1). This means that the measurements of the momentum of an electron localized in the region Δx will lead to values of the momentum which lie in the interval mentioned. In other words, the uncertainty Δx in the value of the coordinate of an electron localized in the region Δx and the uncertainty Δp_x in the value of its momentum are connected by the relation (4.1). The smaller the width of the packet Δx, the larger Δp_x. Conversely, if the momentum interval Δp_x is defined, then formula (4.1) shows that the particle will be found with a very high probability in a region of space $\Delta x \geqslant \hbar / \Delta p_x$.

It follows from the inequality (4.1) that the values of Δx and Δp_x cannot simultaneously be equal to zero. This means that the x-coordinate and the momentum p_x associated with it cannot simultaneously have sharp values. Thus the classical concepts of the spatial position and the value of the momentum are applicable to microparticles only within definite limits given by Heisenberg's relations. Any attempt to apply simultaneously the concepts of the momentum and the coordinate to a microparticle with an accuracy higher than that given by the uncertainty relations makes no sense. This fact is associated with the very nature of microparticles, with their wave—corpuscle properties.

In this connection the reader should be warned against the erroneous assumption of certain authors that Heisenberg's uncertainty relations give that degree of accuracy with which the coordinates and momentum of microparticles can be determined within the framework of quantum mechanics. In their opinion, for a more accurate simultaneous determination of the coordinates and momenta a further development of theory is necessary.

In reality this is not so. Microparticles are completely new objects, by no means classical, with their characteristic properties and laws of motion. As we have already pointed out, a distinctive feature of microparticles is the dualism of the corpuscular and wave properties which they manifest. It follows from diffraction experiments that particles have no definite trajectory. Hence it is impossible to describe the motion of a particle by giving an accurate value of the coordinate and momentum at every instant of time, as is done in classical mechanics. However, one can indicate, with a certain degree of accuracy, the magnitude of that region of space in which the particle will be found with a very high probability, and the interval of those values of the momentum which it possesses at that time. The value of these quantities is given by Heisenberg's uncertainty relations.

It should be noted that when the particle has a definite value of the momentum, $\Delta p_x = 0$, then according to (4.1) its position is completely indefinite, i.e. $\Delta x \to \infty$. Indeed, a state with definite momentum is described by a plane de Broglie wave. For such a wave the square of the modulus $|\psi_p|^2$ is constant, i.e. the particle can be found with the same probability at any point of space.

On the other hand, if a definite position of the particle at a given instant of time is given, then its momentum is completely indefinite. It may seem that the relation obtained is in contradiction with the existence of distinct tracks of particles in a cloud chamber or on a photographic plate. However, this contradiction is only apparent. Indeed, the track of an electron in a cloud chamber represents liquid drops formed on the ions produced by the electron. The size of the drops gives the degree of accuracy with which the coordinate of the particle can be fixed. Since the size of the drops is of the order of 10^{-4} cm, the uncertainty in the coordinate of the electron is also of the order of 10^{-4} cm. Consequently, the uncertainty in the corresponding momentum component $\Delta p_x \sim \hbar/\Delta x \sim 10^{-23}$ g $\cdot$ cm $\cdot$ sec^{-1}. Since the mass of the electron is equal to $\sim 10^{-27}$ g, then the uncertainty in the velocity component perpendicular to the track will be equal to $\Delta v_x = m^{-1}\Delta p_x = 10^4$ cm $\cdot$ sec^{-1}.

But tracks in a cloud chamber are produced only by fast electrons having a velocity v of the order of $\gtrsim 10^9$ cm/sec. Hence we see that under these conditions $\Delta v_x \ll v$ and one can speak approximately of the motion of the particle along a trajectory in the cloud chamber.

Heisenberg's uncertainty relation written in the form

$$\Delta v_x \Delta x \gtrsim \frac{\hbar}{m} \tag{4.4}$$

shows that the concepts of classical physics turn out to be applicable with a degree of accuracy which is higher, the larger the mass of the particle. In view of the smallness of the quantum constant $\hbar$, the uncertainty in the values of the coordinate and velocity become negligibly small for particles of a macroscopically small but still not atomic size.

Let us, for example, have a body of the size of about 1 micron and with a mass of only 10^{-10} g. Then (4.4) gives $\Delta v_x \Delta x \sim 10^{-17}$ cm^2·sec^{-1}. If, for example, the position of the body is determined with an accuracy of 10^{-6} cm (1/100 of its size), then $\Delta v_x \sim 10^{-11}$ cm·sec^{-1}. The velocity of the Brownian motion of a particle of a mass of 10^{-10} g amounts to $\sim 10^{-4}$ cm·sec^{-1}. We see that the error in the velocity, which is associated with the uncertainty relation, is already negligibly small for such a small body. Even more so, is it of no importance for macroscopic bodies.

The estimates given illustrate a general important proposition of quantum mechanics which is called the correspondence principle: in passing to the limit $\hbar \to 0$, i.e. in assuming that the effects proportional to the quantum constant can be disregarded, the laws and relations of quantum mechanics go over into the corresponding laws and relations of classical mechanics. (For more detail on the transition to classical mechanics see ch. 5.) In particular, for particles of large mass the ratio $\hbar/m$ is so small that in practice the coordinate and velocity have definite values. Such a particle has a trajectory along which it moves in accordance with the laws of classical mechanics. The importance of the correspondence principle lies in the fact that it serves as a method of finding the quantum-mechanical analogues of classical quantities. Quantum mechanics implies classical mechanics as a certain limiting case corresponding to $\hbar \to 0$ (for other conditions of this transition see ch. 5). From the correspondence principle it is possible to establish the relation between certain quantum-mechanical quantities and the concepts of classical mechanics.

Besides the reasoning presented above, the uncertainty relations are often obtained from a discussion of the possible degree of accuracy of the determination of the coordinate and momentum of a microparticle in different experiments which are in principle feasible. We shall not dwell on the analysis of these examples, because a strict derivation of the uncertainty relations will be given in §24.

It should be noted that if the region of possible motion of a microparticle is given, for example the size l of the atom or the nucleus, then the uncertainty relations make it possible to qualitatively estimate the values of its momentum and energy. Indeed, the absolute value of the momentum is of the same order of magnitude as its uncertainty $\Delta p \sim \hbar/l$. Consequently, $p \gtrsim \hbar/l$, and the energy of the particle is

$$E = \frac{p^2}{2m} \gtrsim \frac{\hbar^2}{2ml^2}.$$ (4.5)

We see that the energy increases with decreasing region of localization. For example, for an electron in the atom l is of the order of the size of the atom, i.e. of the order of 10^{-8} cm. Substituting this value into (4.5), we find the energy of the electron in the atom to be $E \gtrsim 10$ eV. This is of the correct order of magnitude.

Further, let us consider a proton or a neutron in a nucleus. The size of the nucleus is of the order of 10^{-12} cm. Setting $l \sim 10^{-12}$ cm and taking into account that the mass of the nucleon is $m \sim 10^{-24}$ g, we estimate the energy E to be $E \gtrsim 1$ MeV. This estimate is also in agreement with experimental data.

§5. The principle of causality in quantum mechanics

We have seen in the preceding section that the concepts of classical physics are applicable to microparticles only within certain limits. The question naturally arises: why can and must we, as a matter of fact, describe the motion of microparticles in terms of classical physics? The necessity of introducing classical concepts into quantum mechanics is associated with the following important fact: the explanation of the properties and laws of motion of micro-objects is possible only by setting them in interaction with macroscopic bodies. A macroscopic body interacting with microparticles is called an apparatus. The process of interaction between the apparatus and a microparticle is called measurement.

Of course, the apparatus in this sense of the word is not necessarily an artificial device for registering the properties of microparticles. The apparatus is any body which can change its state as a result of an interaction with micro-objects and which is described with a sufficient degree of accuracy by the laws of classical physics. The process of interaction of the apparatus with a microparticle (a measurement) is an objective process taking place in space and time. However, since any scientific information can only be based on the fact and character of the interaction mentioned, all characteristics of microparticles must be directly connected with the properties of their interaction with macroscopic bodies. This just means that the description of microparticles must necessarily imply, if only partially, the concepts of classical physics. Of course, there may also exist characteristics and properties of microparticles which manifest themselves in interactions with apparatus but have no classical analogue. We shall see, for example in ch. 8, that such a characteristic is the spin of microparticles.

The interaction between microparticles and macroscopic bodies differs, essentially of course, from the interaction of macroscopic bodies with each other. Namely, for the interaction between two macroscopic bodies, one of which is playing the role of the apparatus, one can always assume that the reaction of the apparatus on the body is as small as one wishes or, if one likes, one can take it accurately into account. Hence it is said that the effect of the apparatus does not change the state of the macroscopic object.

The situation is different in the case of the interaction between physical objects of different natures, i.e. a microparticle and a macroscopic body (the apparatus). Here it is, in principle, impossible to assume that the effect of the apparatus on the microparticle is small and unimportant. Let us consider a simple example. We assume that electrons are let successively through a slit in a screen. The screen with the slit is a macroscopic body (the apparatus) which measures the y-coordinate of an electron with an accuracy Δy, where Δy is the width of the slit.

The state of all the electrons before the interaction was the same. Let, for example, electrons with a definite direction of the momentum $\mathbf{p}$ (e.g. along the x-axis) be incident on the apparatus. Here $p_y = 0$. The state of the apparatus before the interaction is also defined, but in a macroscopic way. In the process of interaction of the apparatus with the electron the latter is localized in the region Δy defined by the size of the opening in the screen. Then the state of the electron essentially changes. The electron passes from a state with the definite momentum component $p_y = 0$ to a state in which the momentum component p_y has a value lying in the interval $\Delta p_y \sim \hbar/\Delta y$. Indeed, as we know, diffraction occurs when electrons pass through a slit and the electrons get a momentum component along the y-axis. If we let electrons successively through the slit and measure the values of their momentum component p_y, then we shall obtain all possible values of p_y which lie in the interval Δp_y.

Thus we see that the effect of the apparatus on the electron changes the state of the latter and in principle cannot be made small. Although before the measurement the micro-object and the apparatus were in a definite state, the result of the interaction with the apparatus is not single-valued: we obtain a state with an indefinite value of the momentum component p_y. We can only find the probability of any one value of this quantity.

As a result of carrying out a successive series of measurements, no matter how large, we would not obtain a more accurate value of p_y, but only a more accurate expression for the distribution of the probabilities of different values of this quantity. If the micro-object were in given external conditions, then it would nevertheless be impossible to predict accurately the result of the measurement. One can speak only of the probability distribution of the results of the measurements. This is not associated with any shortcomings of the theory but with the very nature of microparticles. Hence it follows that the principle of mechanical determinism does not characterize the properties of microparticles.

A given initial value of a certain quantity and a definite law of interaction do not unambiguously determine the measured value of this quantity for the microparticle at subsequent instants of time. Thus the behaviour of an individual microparticle, and not just the behaviour of a set of microparticles, is determined by laws of a statistical type. The law of causality for a microparticle takes the following character. Let the state of a particle be known at the initial instant of time $t = 0$. This means that its wave function $\psi(\mathbf{r}, 0)$ is known. If all interactions which the microparticle undergoes are known, then, as we shall see below (see §6), its wave function can be determined unambiguously at subsequent instants of time $t > 0$. It follows from the meaning of the wave function that we can by this predict the probabilities (see §21) that the quantities characterizing the particle (the coordinate, momentum, energy and others) will have particular values at any instant of time $t > 0$.

The principle of causality formulated in quantum mechanics in this way is of considerably more general character than the dynamical regularity (the Laplace determinism) of classical mechanics*.

---

* The reader may find in the work of V.A. Fok a more detailed consideration of the problems which are touched upon in this section. The interpretation of quantum mechanics may be found in the collection of papers *Filozoficheskie voprosy sovremenoi fiziki* (*Philosophic problems of contemporary physics*) published by the Academy of Sciences of the USSR in 1959. See also N. Bohr, *Atomic physics and human knowledge* (Wiley, New York, 1958).

2

The Schrödinger Equation

§6. The Schrödinger wave equation

In §2 we established the form of the wave function describing the motion of a free particle with a given value of the momentum. This wave function had the form of a plane de Broglie wave. We now turn to the consideration of the motion of particles in external fields of force. For this it is necessary to find the wave function describing the motion of a particle in a given field of force. It turns out that it is possible to establish the form of the differential equation satisfied by the wave function. One can find the wave function itself from the solution of this equation. It should be noted, first of all, that the equation for the wave function must be linear. Indeed, functions satisfying non-linear equations obviously do not meet the requirements of the principle of superposition. Further, it is clear that the wave function we already know, which describes the motion of a free particle, must be the solution of the required differential equation in the particular case where the field is absent. Finding the linear differential equation satisfied by a plane de Broglie wave

$$\psi(x, t) = A e^{(i/\hbar)(\mathbf{p} \cdot \mathbf{r} - Et)} , \qquad (6.1)$$

presents no difficulty. For this we note that

$$\frac{\partial \psi}{\partial t} = -\frac{i}{\hbar} E \psi .$$

Furthermore,

$$\frac{\partial^2 \psi}{\partial x^2} + \frac{\partial^2 \psi}{\partial y^2} + \frac{\partial^2 \psi}{\partial z^2} = -\frac{1}{\hbar^2} (p_x^2 + p_y^2 + p_z^2) \psi .$$

Taking into account that for a free particle

$$\frac{p_x^2 + p_y^2 + p_z^2}{2m} = E , \tag{6.2}$$

we find

$$\frac{\partial \psi}{\partial t} = \frac{i\hbar}{2m} \left(\frac{\partial^2 \psi}{\partial x^2} + \frac{\partial^2 \psi}{\partial y^2} + \frac{\partial^2 \psi}{\partial z^2} \right) .$$

This equation is usually written in the form

$$i\hbar \frac{\partial \psi}{\partial t} = -\frac{\hbar^2}{2m} \nabla^2 \psi . \tag{6.3}$$

This linear differential equation in partial derivatives is called the Schrödinger equation. It does not contain any characteristics of the state of the particle, for example the value of its momentum or energy. It involves only the mass of the particle as well as the universal constant $\hbar$. Equation (6.3) is evidently satisfied not only by a wave function of the form (6.1), which represents the wave function of a particle with a given value of the momentum, but also by any superposition of such wave functions.

The Schrödinger equation possesses the feature that it is an equation of the first order in time and contains the factor i. The latter means that the wave function must be complex.

We note that it seems that a function expressed by a real relation, for example in the form of a travelling wave $\psi = A \cos \hbar^{-1}(\mathbf{p} \cdot \mathbf{r} - Et)$, could be chosen as the wave function of a free particle. However, we could not then construct any equation of the first order in time whose solution would be an arbitrary superposition of such functions. The fact that the Schrödinger equation contains only the first derivative of the wave function with respect to time is closely associated with the expression of the principle of causality in quantum mechanics (see §5). Indeed, if the Schrödinger equation contained, for example, the second derivative of the wave function with respect to time, then the knowledge of the wave function at the initial instant of time would be insufficient for determining the wave function at an arbitrary instant of time t. Namely, it would also be necessary to give the value of the first derivative of the wave function with respect to time for the initial instant of time.

Among the solutions of eq. (6.3) there are solutions depending harmonically on the time

$$\psi(x, t) = \psi(x)\, e^{-(i/\hbar)Et} . \tag{6.4}$$

Substituting (6.4) into (6.3), we obtain an equation for a function which depends only on the coordinates of the particle

$$\nabla^2 \psi(x) + \frac{2mE}{\hbar^2}\, \psi(x) = 0 . \tag{6.5}$$

This equation defines the function $\psi(x)$ for a free particle. Let us generalize eq. (6.5) to the case of a particle moving in a field of force. This generalization is based on the following assumption: the energy E involved in eq. (6.5) represents the kinetic energy of the particle. Indeed, for free motion the kinetic energy is the same as the total energy. If in the required generalization the energy E occurring in eq. (6.5) is assumed to be the total energy, then the wave function describing the motion of electrons in a field of force will not depend on the forces acting on the particle. However, this would make no sense. Thus we arrive at the conclusion that E in eq. (6.5) must be understood to be the kinetic energy of the particle. Denoting the potential energy of the particle by $U(x)$, and the total energy by E, we get

$$\nabla^2 \psi(x) + \frac{2m}{\hbar^2}\, [E - U(x)]\, \psi(x) = 0 . \tag{6.6}$$

Equation (6.6) represents the required generalization of the Schrödinger equation to the case of a particle moving in an arbitrary potential field which does not depend on time. This equation only determines the dependence of the wave function on the coordinates, while the dependence on time is determined as before by the relation (6.4).

Equation (6.6) is called the Schrödinger equation for stationary states. Indeed, the probability density of the measurement of the coordinates of a particle in a state (6.4) does not depend on the time

$$|\psi(x, t)|^2 = |\psi(x, 0)|^2 . \tag{6.7}$$

In §28 it will be shown that the probabilities of the measurement of other physical quantities in a state (6.4) also do not depend on the time.

Substituting the derivative with respect to time $\partial\psi/\partial t$ for the quantity $E\psi$

by means of (6.4), we arrive at the general Schrödinger wave equation

$$i\hbar \frac{\partial \psi}{\partial t} = -\frac{\hbar^2}{2m}\nabla^2 \psi + U\psi , \tag{6.8}$$

where the wave function ψ depends on the coordinates x, y, z and the time t.

Equation (6.8) is the fundamental equation of quantum mechanics. It plays in quantum mechanics the same role as Newton's equations in classical mechanics, and could be called the equation of motion of a quantum particle. To define the law of motion of a particle in quantum mechanics means to define the value of the ψ-function at every instant of time and at every point of space.

It should be noted that the above reasoning is not a derivation of the Schrödinger equation in the strict sense of the word. Like Newton's and Maxwell's equations, the Schrödinger equation appeared, on the one hand, as the generalization of known experimental data and, on the other hand, as a great scientific prediction.

We shall see later how the discreteness of energy levels follows from the Schrödinger equation. Also it will become clear that the Schrödinger equation satisfies the correspondence principle. The validity of the Schrödinger equation and the interpretation of the meaning of the wave function involved in it are confirmed by a vast amount of experimental data from contemporary atomic and nuclear physics. In order to obtain the law of motion of a particle, the wave function $\psi(x, t)$, the initial and boundary conditions must be given in addition to the Schrödinger equation. Since the Schrödinger equation is an equation of the first order in time, it is necessary to know the initial value of the wave function $\psi(x, 0)$.

The set of boundary conditions in general amounts to the requirement of single-valuedness and continuity of the wave function and its first derivatives, and to the fulfillment of certain normalization conditions. The latter is usually a bounding condition on the modulus of the wave function. The whole set of the initial condition and the conditions of single-valuedness, continuity and finiteness of the wave function and of its first derivatives makes it possible, in principle, to find a unique solution of the Schrödinger equation: the wave function $\psi(x, t)$. In other words, if the initial value of the wave function is given, then from the solution of the Schrödinger equation one can determine unambiguously the state of the quantum system for any subsequent instant of time $t > 0$. Namely, for $t > 0$ one can find the wave function of the system $\psi(x, t)$.

We shall see in §23 that by defining $\psi(x, t)$ the quantum particle is characterized as fully as, for example, the particle in classical mechanics is by defining its trajectory. We note in addition that in certain problems of quantum mechanics it is convenient to approximate the potential energy by a discontinuous function. At the point of discontinuity of the potential energy the wave function and its first derivatives must remain continuous. The derivative of the wave function undergoes a jump only at the surface of an infinitely large discontinuity of the potential energy.

The Schrödinger equation, as well as the equations of motion of classical mechanics, allows 'reversal in time'.

Indeed, eq. (6.8) does not change when the transformation $t \rightarrow -t$ and the transition to the complex conjugate function ψ^* are made. Consequently, a process reversed in time is described by the wave function $\psi_{rev}(x, t)$

$$\psi_{rev}(x, t) = \psi^*(x, -t) .\qquad(6.9)$$

We note that for motion in a magnetic field reversal in time takes place only when the direction of the magnetic field is also reversed (see §27). We shall consider the problem of time reversal in more detail in §98.

§7. The probability current density

Generally speaking the wave function describing the motion of a particle changes in space and time. However, this change cannot be arbitrary.

That is, a conservation law holds. To formulate this law, we consider the integral $\int_V |\psi|^2 dV$, which represents the probability of finding the particle in the volume V. Proceeding in the same way as in deriving the law of charge conservation (see §5 of Part I), we find the derivative of the above integral with respect to time. To calculate $\partial \psi/\partial t$ and $\partial \psi^*/\partial t$ we make use of the Schrödinger equation (6.8) and its conjugate equation. Then we obtain

$$\frac{\partial}{\partial t} \int \psi \psi^* dV = \int \left(\frac{\partial \psi}{\partial t} \psi^* + \psi \frac{\partial \psi^*}{\partial t} \right) dV =$$

$$= \frac{\hbar}{2mi} \int (\psi \nabla^2 \psi^* - \psi^* \nabla^2 \psi) dV =$$

$$= \frac{\hbar}{2mi} \int \nabla \cdot (\psi \nabla \psi^* - \psi^* \nabla \psi) dV .\qquad(7.1)$$

Making use of the Gauss-Ostrogradsky theorem, we have

$$\int_V \nabla \cdot (\psi \nabla \psi^* - \psi^* \nabla \psi) dV = \oint_S (\psi \nabla \psi^* - \psi^* \nabla \psi) dS ,$$

where the surface S bounds the volume V. Hence

$$\frac{\partial}{\partial t} \int_V |\psi|^2 dV = \frac{\hbar}{2mi} \oint_S (\psi \nabla \psi^* - \psi^* \nabla \psi)\, dS . \qquad (7.2)$$

We introduce the vector $\mathbf{j}$ defined by the relation

$$\mathbf{j} = \frac{\hbar}{2mi} (\psi^* \nabla \psi - \psi \nabla \psi^*) . \qquad (7.3)$$

Then (7.2) is rewritten in the form

$$-\frac{\partial}{\partial t} \int_V |\psi|^2 dV = \oint_S j_n\, dS . \qquad (7.4)$$

Formula (7.4) shows that the probability density satisfies a conservation law, and the vector $\mathbf{j}$ we introduced has the meaning of the probability current density. Relation (7.4) can be rewritten in differential form as the continuity equation

$$\frac{\partial |\psi|^2}{\partial t} + \nabla \cdot \mathbf{j} = 0 . \qquad (7.5)$$

The integral of the normal component of the vector $\mathbf{j}$ with respect to a particular surface represents the probability that the particle will cross the surface mentioned in unit time.

Let us consider, in particular, free motion. We take the wave function in the form of the plane wave $\psi = A e^{(i/\hbar)(\mathbf{p}\cdot\mathbf{r} - Et)}$. Making use of the relation (7.3), we obtain

$$\mathbf{j} = \frac{1}{m} \mathbf{p}|A|^2 . \qquad (7.6)$$

We now apply the relation (7.4) to all space, i.e. we assume the surface S to be infinitely distant. If ψ is a quadratically integrable function, then the integrand in the integral with respect to the surface decreases more rapidly than r^{-4}, and the surface of integration increases in proportion to r^2. As a result the integral over the surface in.(7.4) reduces to zero. But if ψ does not tend to zero in the way mentioned when $r \to \infty$, as, for example, in the case of a plane wave, then there is a current of particles at infinity. If this current is stationary, then the wave function can be normalized in such a way that the vector $\mathbf{j}$ is the particle-current density vector.

Finally, we note from formula (7.3) that the current density $\mathbf{j}$ obviously reduces to zero if the state of the system is described by a real wave function ψ.

The relation (7.5) written in the form

$$\frac{\partial \rho}{\partial t} + \nabla \cdot \mathbf{j} = 0 , \qquad (7.7)$$

can be interpreted as the law of conservation of the number of particles (see §5 of Part I).

§8. A particle in a one-dimensional rectangular potential well

Before turning to the consideration of real atomic systems we shall discuss the general properties of the solutions of the Schrödinger equation by the use of some simple models. Let us consider first of all the one-dimensional motion of a particle in a potential field defined as follows:

$$U(x) = \begin{cases} 0 & \text{for} \quad 0 < x < l , \\ \infty & \text{for} \quad x \leqslant 0 \quad \text{and} \quad x \geqslant l . \end{cases}$$

We shall call such a potential field an infinitely deep potential well. It is clear that in such a well a particle can move only in the region of space $0 \leqslant x \leqslant l$.

At the boundary of the well the particle is acted upon by arbitrarily large forces which prevent it from getting out, so that the particle behaves as if confined to a region of space bounded by perfectly reflecting walls. It turns out that by such a simple example one can establish a number of properties of quantum-mechanical systems. It is important that these properties are not associated with the model but are of a general character. Furthermore, interest in this problem is also due to the fact that a model of a potential well is often successfully used for a rough description of a number of systems, for example electrons in a metal or nucleons in a nucleus.

The solution of the Schrödinger equation must be written for two regions: outside the potential well and inside it. Since the particle cannot be outside the potential well, its wave function is equal to zero outside the interval $0 \leqslant x \leqslant l$. From the condition of continuity it follows that the wave function is also equal to zero at the points $x = 0$ and $x = l$, i.e. that

$$\psi(0) = \psi(l) = 0 . \qquad (8.1)$$

The requirement (8.1) is the boundary condition for the solution of the Schrödinger equation inside the potential well. In the region $0 < x < l$ the

Schrödinger equation for stationary states (6.6) is of the form

$$-\frac{\hbar^2}{2m}\frac{d^2\psi}{dx^2} = E\psi \; . \tag{8.2}$$

The solution of this equation can evidently be written as

$$\psi = A \sin(kx + \alpha) \; , \tag{8.3}$$

where $k = (2mE/\hbar^2)^{\frac{1}{2}}$. We now make use of the boundary conditions (8.1). From the relation $\psi = 0$ for $x = 0$ it follows that $\alpha = 0$. The condition $\psi(l) = 0$ gives

$$kl = n\pi \; , \tag{8.4}$$

where n is any integer larger than zero. For $n = 0$ we would have $\psi \equiv 0$, which would mean the absence of the particle in all space. The condition (8.4) makes it possible to find the possible values of the energy of the particle

$$E_n = \frac{\pi^2\hbar^2}{2ml^2} n^2 \; . \tag{8.5}$$

We see that the Schrödinger equation has solutions satisfying the boundary conditions only for discrete values of the quantum number n. Thus the energy of the particle in the infinitely deep potential well turns out to be quantized. The discreteness of the energy arose in a natural way, without any subsidiary assumptions. In the case given it turned out to follow directly from the boundary conditions imposed upon the wave function at the limits of the integration range. The state of the particle which has the lowest possible energy will henceforth be called the normal or ground state, while all other states will be called excited states. The energy of the ground state of a particle in an infinitely deep potential well is obtained from formula (8.5) for $n = 1$:

$$E_1 = \frac{\pi^2\hbar^2}{2ml^2} \; . \tag{8.6}$$

We note that the minimum energy value of the particle is consistent with the uncertainty principle. Indeed, the uncertainty in the coordinate of the particle is $\Delta x \sim l$. The uncertainty in the momentum, Δp, is of the order of $\hbar/l$. Since $p \geqslant \Delta p$, the minimum energy of the particle turns out to be equal to

$$\frac{p^2}{2m} \underset{\sim}{\geqslant} \frac{\hbar^2}{2ml^2} \; ,$$

which is in order of magnitude the same as (8.6).

Let us now determine the spacing between neighbouring energy levels

$(\Delta n = 1)$:

$$\Delta E_n = E_{n+1} - E_n = \frac{\pi^2 \hbar^2}{2ml^2}(2n + 1).$$

The spacing between levels increases with decreasing mass of the particle and size of the region of its motion l. Thus, for example, for an electron $(m \sim 10^{-27}\, g)$ confined in a region $l \sim 5 \times 10^{-8}$ cm we find $\Delta E \sim 1$ eV. On the contrary, in the case of a molecule with $m \sim 10^{-23}$ g moving, for example, in a region $l \sim 10$ cm, the spacing between levels amounts to $\Delta E \sim 10^{-20}$ eV. This spacing is so small, for example in comparison with $kT = 0.025$ eV, that in practice the energy of a molecule can be considered to be a continuously varying quantity.

Let us find the ratio $\Delta E_n / E_n$, i.e. the relative spacing between energy levels. We see that $\Delta E_n / E_n \sim n^{-1}$ and tends to zero for very large n. The discreteness of quantum states is no longer significant for large quantum numbers and in fact a transition to a quasi-continuous variation of the energy takes place.

Let us consider in somewhat more detail the properties of the wave function of a particle in a potential well. The wave function corresponding to the nth energy level is of the form

$$\psi_n = A_n \sin \frac{\pi n}{l} x. \tag{8.7}$$

We define the constant A_n by the normalization condition

$$\int_0^1 |\psi_n|^2 \, dx = 1.$$

Then

$$|A_n|^2 \int_0^1 \sin^2 \frac{n\pi}{l} x \, dx = |A_n|^2 \int_0^1 \frac{1}{2}\left(1 - \cos \frac{2n\pi}{l} x\right) dx = |A_n|^2 \frac{l}{2} = 1.$$

Hence

$$A_n = (2/l)^{\frac{1}{2}}. \tag{8.8}$$

Thus the value of the constant does not depend on the quantum number n.

The probability density $|\psi|^2$ of finding the particle at different points inside the well is illustrated in fig. V.2. In classical mechanics a particle moving in a potential well can be found with equal probability at any point

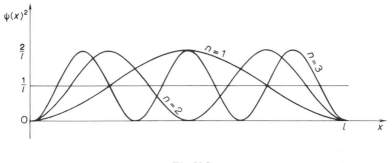

Fig. V.2

inside the well (straight line in fig. V.2). Indeed, the probability dW_{class} of observing the particle in an interval dx is proportional to the time dt of the particle's being in this interval:

$$dW_{class} \sim dt = \frac{1}{v} \, dx \ .$$

Since a particle inside a well is not acted upon by any forces it moves with constant velocity v and, consequently, dW_{class} does not depend on x. As the quantum number n (the energy of the particle) increases the maxima of the probability distribution tend to approach each other. In the limit $n \to \infty$ the probability distribution obtained from the quantum-mechanical calculation leads to the same results as the classical distribution. This follows from the fact that the function $\sin^2 (\pi n x/l)$ rapidly oscillates as x changes, and in integration over any finite interval can be replaced by $\frac{1}{2}$. Thus consideration of the simplest quantum-mechanical system leads us to the following conclusions which, as we shall see later, are of a general character:

(1) the energy of a microparticle moving in a potential well runs over a discrete sequence of values;

(2) even for $E = E_1$ (ground state) the particle is not in a state of complete rest with kinetic energy equal to zero;

(3) the discrete character of the energy levels manifests itself when the mass of the particle and the size of the region in which the motion takes place are small;

(4) for large values of the quantum numbers the quantum-mechanical relations go over into the formulae of classical physics. This statement is a particular case of the correspondence principle which we shall frequently encounter.

Later, in considering the quantum oscillator or atomic systems, we shall see that quantization of states can take place even in systems which are not confined by impenetrable walls. At the same time we shall see that the presence of discrete energy states is not a necessity for quantum-mechanical systems. In certain cases quantum-mechanical systems have a continuous energy spectrum.

§9. A particle in a three-dimensional rectangular potential well

Let us now consider the more complex case of the motion of a particle in a three-dimensional infinitely deep potential well. We shall assume that the region of space in which the particle moves is defined by the inequalities $0 < x < l_1$, $0 < y < l_2$ and $0 < z < l_3$. In this case the wave equation can be written in the form

$$-\frac{\hbar^2}{2m} \left(\frac{\partial^2 \psi}{\partial x^2} + \frac{\partial^2 \psi}{\partial y^2} + \frac{\partial^2 \psi}{\partial z^2} \right) = E\psi \ . \tag{9.1}$$

The boundary conditions are analogous to (8.1) and have the form

$$\psi(0, y, z) = \psi(x, 0, z) = \psi(x, y, 0) =$$
$$= \psi(l_1, y, z) = \psi(x, l_2, z) = \psi(x, y, l_3) = 0 \ . \tag{9.2}$$

We write the solution of eq. (9.1) as follows:

$$\psi = B \sin k_1 x \sin k_2 y \sin k_3 z \ . \tag{9.3}$$

Substituting ψ into the equation, we obtain the relation

$$\frac{\hbar^2}{2m} (k_1^2 + k_2^2 + k_3^2) = E \ . \tag{9.4}$$

From the boundary conditions (9.2) it follows that

$$k_1 l_1 = n_1 \pi \ , \qquad k_2 l_2 = n_2 \pi \ , \qquad k_3 l_3 = n_3 \pi \ , \tag{9.5}$$

where n_1, n_2 and n_3 are integers.

Substituting the values of k_1, k_2 and k_3 into (9.4) and (9.3), we obtain the expressions for the energy and for the wave function

$$E_{n_1, n_2, n_3} = \frac{\pi^2 \hbar^2}{2m} \left(\frac{n_1^2}{l_1^2} + \frac{n_2^2}{l_2^2} + \frac{n_3^2}{l_3^2} \right) , \tag{9.6}$$

$$\psi_{n_1 n_2 n_3} = B \sin \frac{\pi n_1 x}{l_1} \sin \frac{\pi n_2 y}{l_2} \sin \frac{\pi n_3 z}{l_3} \ . \tag{9.7}$$

The constant B is again defined by the normalization condition

$$\int_V |\psi_{n_1 n_2 n_3}|^2 dx\, dy\, dz = 1$$

and is equal to

$$B = \left(\frac{8}{l_1 l_2 l_3}\right)^{\frac{1}{2}} . \tag{9.8}$$

Let us consider, in particular, a particle moving in a potential well of cubic form, i.e. the case $l_1 = l_2 = l_3 = l$. The energy of the particle is equal to

$$E_{n_1,n_2,n_3} = \frac{\pi^2 \hbar^2}{2ml^2}(n_1^2 + n_2^2 + n_3^2) . \tag{9.9}$$

From formula (9.9) it is easily seen that one and the same energy value can be realized by means of different combinations of the numbers n_1, n_2 and n_3. This means that several different quantum states with different wave functions correspond to one and the same energy value. Such energy levels are said to be degenerate, and the number of different states corresponding to a given energy level is called the multiplicity of degeneracy.

Let us consider, for example, the energy level

$$E = \frac{\pi^2 \hbar^2}{2ml^2} \cdot 6 ,$$

where $n_1^2 + n_2^2 + n_3^2 = 6$. Since each of the n's is an integer larger than zero, this equality can be satisfied by the three different combinations of the numbers n_1, n_2, n_3

    (1) $n_1 = 2$, $n_2 = 1$, $n_3 = 1$,

    (2) $n_1 = 1$, $n_2 = 2$, $n_3 = 1$,

    (3) $n_1 = 1$, $n_2 = 1$, $n_3 = 2$.

Thus to the given energy level there correspond 3 different states ψ_{211}, ψ_{121} and ψ_{112}. Consequently, the multiplicity of degeneracy is equal to three. In considering more complex systems, for example atoms, we shall frequently encounter the phenomenon of degeneracy.

§10. The quantum-mechanical oscillator

Turning to more complex quantum-mechanical systems, we shall consider the theory of the linear harmonic oscillator. Such an oscillator represents the quantum analogue of a particle performing small linear oscillations about an equilibrium position. An example of small oscillations in atomic systems are the small oscillations of atoms in a molecule (see §41 of Part III).

Another no less important example is the thermal motion of a crystal, which amounts to a set of linear harmonic oscillators. We shall also deal with the problem of the harmonic oscillator in quantum electrodynamics, where an arbitrary electromagnetic field is represented in the form of the superposition of independent quantum oscillators (see §101).

The above examples show that the theory of the linear harmonic oscillator is one of the important problems of quantum mechanics.

The potential energy of a linear harmonic oscillator is given by the well-known formula $U = \frac{1}{2}m\omega^2 x^2$. Hence the Schrödinger equation (6.6) for the linear harmonic oscillator has the form

$$-\frac{\hbar^2}{2m}\frac{d^2\psi}{dx^2} + \frac{m\omega^2 x^2}{2}\psi = E\psi . \tag{10.1}$$

In solving it, it is convenient to go over to dimensionless variables

$$\xi = \left(\frac{m\omega}{\hbar}\right)^{\frac{1}{2}} x ; \qquad \lambda = \frac{2E}{\hbar\omega} . \tag{10.2}$$

In the new notation the Schrödinger equation takes the form

$$-\frac{d^2\psi}{d\xi^2} + \xi^2\psi = \lambda\psi . \tag{10.3}$$

An important difference of the oscillator from the examples considered before is the fact that in this case the motion of the particle is not restricted by an impenetrable wall. Hence the oscillator has no boundary conditions similar to the conditions (8.1). The only requirement imposed upon the wave function of the oscillator is the requirement that it should be quadratically integrable. We shall see that the Schrödinger equation for the oscillator has a solution satisfying this requirement only for certain definite values of the parameter λ. These values are called the eigenvalues of eq. (10.3).

In order to explain the general character of the solutions of this equation, let us consider the asymptotic behaviour of $\psi(\xi)$ for very large values of the argument $\xi \gg \lambda$.

For $\xi \gg \lambda$ in eq. (10.3) one can drop $\lambda\psi$ as compared with $\xi^2\psi$. We then have, obviously,

$$\frac{d^2\psi}{d\xi^2} - \xi^2\psi = 0 . \tag{10.4}$$

The solution of this equation satisfying the requirement of finiteness for large ξ is the function

$$\psi = A\xi^m e^{-\frac{1}{2}\xi^2} , \tag{10.5}$$

where A is a constant, and m is an arbitrary finite number.

The second independent solution of eq. (10.4), $\psi \sim e^{+\frac{1}{2}\xi^2}$, increases indefinitely for $\xi \to \infty$ and must be dropped.

We try to find a solution of eq. (10.3) in the form

$$\psi = e^{-\frac{1}{2}\xi^2} f(\xi) , \tag{10.6}$$

where $f(\xi)$ is a new unknown function which for $\xi \to \infty$ behaves as ξ^m. Substituting (10.6) into (10.3), we arrive at the following equation for the function f:

$$\frac{d^2 f}{d\xi^2} - 2\xi \frac{df}{d\xi} + (\lambda - 1) f = 0 . \tag{10.7}$$

Since the point $\xi = 0$ is not a singular point of eq. (10.7), we seek a solution of this equation in the form of a power series

$$f(\xi) = \sum_{k=0}^{\infty} a_k \xi^k \tag{10.8}$$

The derivatives $df/d\xi$ and $d^2 f/d\xi^2$ have the form

$$\frac{df}{d\xi} = \sum k a_k \xi^{k-1} , \qquad \frac{d^2 f}{d\xi^2} = \sum k(k-1) a_k \xi^{k-2} . \tag{10.9}$$

Substituting the series (10.9) into eq. (10.7), we obtain

$$\sum k(k-1) a_k \xi^{k-2} - 2\xi \sum k a_k \xi^{k-1} + (\lambda - 1) \sum a_k \xi^k = 0 . \tag{10.10}$$

In order that a power series of the form $\sum_n c_n \xi^n$ be identically equal to zero, it is necessary that all coefficients c_n reduce to zero. Assuming the coefficient of ξ^k to be equal to zero, we obtain the recurrence formula

$$a_{k+2} = \frac{2k + 1 - \lambda}{(k+2)(k+1)} a_k . \tag{10.11}$$

It is easily seen that for $\xi \to \infty$ such a series behaves as e^{ξ^2}, since in this case large k are essential and (10.11) gives $a_{k+2} \approx (2/k) a_k$. In this case the function ψ of (10.6) increases indefinitely. Such a solution must be eliminated.

We obtain a solution satisfying the necessary conditions of finiteness and behaving for $\xi \to \infty$ as (10.5) only in the case where the series (10.8) reduces to a polynomial, i.e. is cut off at a certain term. Thus suppose that $a_n \neq 0$, $a_{n+2} = 0$. Then all subsequent coefficients also vanish, and the function f reduces to a polynomial of the nth degree.

It follows from (10.11) that in this case the condition

$$2n + 1 - \lambda = 0 \tag{10.12}$$

is necessary, where n is an integer, $n \geqslant 0$, since n is the ordinal number of the term at which the series ends.

Substituting this value of λ into (10.2), we obtain

$$E_n = \hbar\omega(n + \tfrac{1}{2}) . \tag{10.13}$$

Hence it is seen that the energy of the oscillator can only take on discrete values, and the energy levels are equally spaced at intervals of $\hbar\omega$.

We write the wave function corresponding to the nth excited energy level in the form

$$\psi_n(\xi) = A_n e^{-\frac{1}{2}\xi^2} f_n(\xi) , \tag{10.14}$$

where $f_n(\xi)$ is a polynomial of the nth power with coefficients which are defined by the relation (10.11), and A_n is a factor determined by the normalization condition. The polynomials $f_n(\xi)$ are called the Chebyshev–Hermite polynomials and are denoted by $H_n(\xi)$. The Chebyshev–Hermite polynomials are often written in the form

$$H_n(\xi) = (-1)^n e^{\xi^2} \frac{d^n e^{-\xi^2}}{d\xi^n} . \tag{10.15}$$

They satisfy the differential equation

$$\frac{d^2 H_n}{d\xi^2} - 2\xi \frac{dH_n}{d\xi} + 2nH_n = 0 , \tag{10.16}$$

which is obtained from (10.7) taking into account condition (10.12).

We give the first four Chebyshev–Hermite polynomials:

$$H_0(\xi) = 1 , \quad H_1(\xi) = 2\xi , \quad H_2(\xi) = 4\xi^2 - 2 ,$$
$$H_3(\xi) = 8\xi^3 - 12\xi , \quad H_4(\xi) = 16\xi^4 - 48\xi^2 + 12 . \tag{10.17}$$

Knowing the general form of the Chebyshev–Hermite polynomials, one can calculate the normalization integral. One then obtains* for A_n

$$A_n = \left(\frac{m\omega}{\hbar\pi}\right)^{\frac{1}{4}} \left(\frac{1}{n!2^n}\right)^{\frac{1}{2}}. \tag{10.18}$$

The form of the wave functions for different quantum numbers n is shown in fig. V.3. We note that the wave function corresponding to the ground state $n = 0$ of the oscillator is nowhere equal to zero. The wave function corresponding to the level $n = 1$ is equal to zero once, at $x = 0$, while $\psi_2(x)$ $(n = 2)$ is equal to zero twice, and so on. The points at which the wave function is equal to zero are called the nodes of the wave function. It is easily seen that the number of nodes of the wave function is equal to the quantum number n.

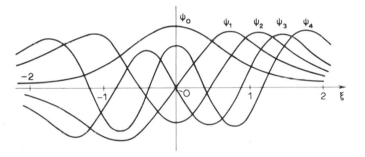

Fig. V.3

This statement is not specific to the oscillator. It can be stated** that in general in the one-dimensional case the number of nodes of the wave function is determined by the quantum number n. The probability of finding the particle at the point x in the interval dx is equal to

$$W_n(x)\, dx = |\psi_n(x)|^2 dx .$$

These probabilities for different n are shown in fig. V.4. Let us compare the expressions obtained with the probability of finding the particle at a given point as calculated by means of classical mechanics. The latter is defined as the ratio of the time dt spent in the neighbourhood of the given point to the period of the motion. The classical probability turns out to be highest in the

* See, for example, L.D.Landau and E.M.Lifshitz, *Quantum mechanics, Non-relativistic theory* (Pergamon Press, Oxford, 1965).

** R.Courant and D.Hilbert, *Methods of mathematical physics,* Vol. 1 (Interscience, New York, 1953).

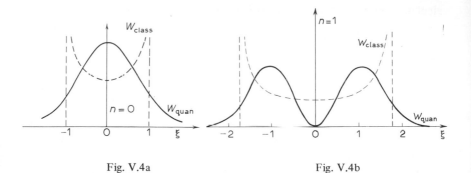

Fig. V.4a   Fig. V.4b

neighbourhood of the turning points $x = \pm x_0$, at which the velocity of motion reduces to zero. On the other hand, in the neighbourhood of the point $x = 0$ the particle has its largest velocity and the probability of finding it is a minimum.

It is seen from the curves of fig. V.4 that the probability of finding a quantum particle differs from zero even in the region outside the turning points, which is unattainable in classical mechanics. For large quantum numbers (fig. V.5) the quantum probability distribution approaches the classical one, in agreement with the correspondence principle.

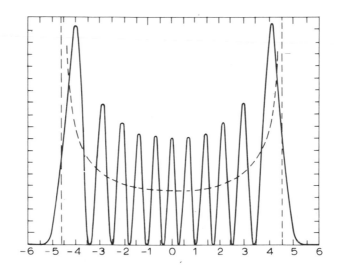

Fig. V.5

In conclusion we note that the lowest possible energy value of the oscillator, equal to $\frac{1}{2}\hbar\omega$, is different from zero. This means that the quantum oscillator can never be in a state of absolute rest. This fact in its turn is associated with the uncertainty principle. In order of magnitude the energy of the oscillator is

$$E \geqslant \frac{\Delta p^2}{2m} + \frac{m\omega^2}{2}\Delta x^2 \geqslant \frac{\Delta p^2}{2m} + \frac{m\omega^2}{2}\left(\frac{\hbar}{\Delta p}\right)^2 .$$

Considering this quantity as a function of Δp, it is easily established that it has a minimum for $\Delta p \sim (m\omega h)^{\frac{1}{2}}$ and is of the order of magnitude of $\hbar\omega$. Experimentally the zero-point energy E_0 is observed in the scattering of light by a crystal at a temperature close to absolute zero. At absolute zero the crystal is in the ground (lowest energy) state. Nevertheless, atoms perform zero-point oscillations which cause scattering of the light.

§11. The three-dimensional oscillator

Let us now consider the motion of a spatial three-dimensional oscillator. For generality we shall assume that the natural frequencies are different in three mutually perpendicular directions and equal respectively to ω_1, ω_2 and ω_3. Then the potential energy is expressed by the formula

$$U = \frac{m\omega_1^2}{2}x^2 + \frac{m\omega_2^2}{2}y^2 + \frac{m\omega_3^2}{2}z^2 . \tag{11.1}$$

The Schrödinger equation correspondingly has the form

$$-\frac{\hbar^2}{2m}\nabla^2\psi + \frac{m}{2}(\omega_1^2 x^2 + \omega_2^2 y^2 + \omega_3^2 z^2)\psi = E\psi . \tag{11.2}$$

We try to find the solution of eq. (11.2) in the form of a product of functions each of which depends on only one coordinate

$$\psi(x, y, z) = \psi_1(x)\,\psi_2(y)\,\psi_3(z) . \tag{11.3}$$

Substituting (11.3) into (11.2) and separating the variables we obtain

$$-\frac{\hbar^2}{2m}\frac{d^2\psi_i}{dx_i^2} + \frac{m\omega_i^2 x_i^2}{2}\psi_i(x_i) = E_i\psi_i(x_i) , \qquad i = 1, 2, 3 , \tag{11.4}$$

where $x_1 = x$, $x_2 = y$, $x_3 = z$; $E_1 + E_2 + E_3 = E$.

Thus the problem is reduced to the one-dimensional case. In correspondence with this, making use of (10.13), (10.14) and (10.18) we can write

$$\psi_{n_1 n_2 n_3}(x_1, x_2, x_3) = \left(\frac{m^3 \omega_1 \omega_2 \omega_3}{h^3 \pi^3}\right)^{\frac{1}{4}} \left(\frac{2^{-(n_1+n_2+n_3)}}{n_1! \, n_2! \, n_3!}\right)^{\frac{1}{2}} \times$$

$$\times \exp\left[-\tfrac{1}{2}(\xi_1^2 + \xi_2^2 + \xi_3^2)\right] H_{n_1}(\xi_1) H_{n_2}(\xi_2) H_{n_3}(\xi_3) , \quad (11.5)$$

where $\xi_i = (m\omega_i/h)^{\frac{1}{2}} x_i$ $(i = 1, 2, 3)$.

The total energy of the oscillator is equal to

$$E = \hbar\omega_1(n_1 + \tfrac{1}{2}) + \hbar\omega_2(n_2 + \tfrac{1}{2}) + \hbar\omega_3(n_3 + \tfrac{1}{2}) . \quad (11.6)$$

In particular, for an isotropic oscillator, which has $\omega_1 = \omega_2 = \omega_3 = \omega$, the total energy is of the form

$$E_n = \hbar\omega(n + \tfrac{3}{2}) , \quad \text{where} \quad n = n_1 + n_2 + n_3 , \quad (11.7)$$

i.e. E_n depends on the sum of the quantum numbers n_1, n_2 and n_3. This means that a given energy value (given n) can be obtained from different combinations of n_1, n_2 and n_3. Hence it follows that all the energy levels, except the ground level $n = 0$, are degenerate. The multiplicity of degeneracy is easily calculated. For this we fix, in addition to n, the quantum number n_1. Then the number of possible combinations of n_1, n_2 and n_3 will be equal to the number of possible values of n_2, i.e. equal to $n-n_1 + 1$, since n_2 can vary from zero up to $n-n_1$. Summing the expression obtained over all possible values of the number n_1, we find the total number of combinations of the three quantum numbers n_1, n_2 and n_3 which add up to the given number n, i.e. the multiplicity of degeneracy of the nth energy level:

$$\sum_{n_1=0}^{n} (n-n_1 + 1) = \tfrac{1}{2}(n + 1)(n + 2) . \quad (11.8)$$

§12. Reflection from and penetration through a potential barrier

Among the other relatively simple problems of quantum mechanics we shall consider the motion of particles in a field of force which has the form of a potential barrier. This means that the forces act on the particle in a certain limited region of space. Outside this region the particle moves as a free particle. We shall see that the study of the motion of particles in a field having the form of a barrier of the simplest shape will allow a number of important and, in principle new, properties of quantum particles to be exhibited. We shall

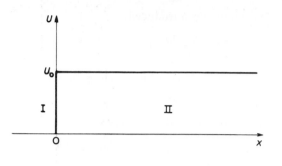

Fig. V.6

begin our consideration with the simple rectangular one-dimensional barrier of infinite extent shown in fig. V.6. In classical mechanics any particle moving from left to right with an energy smaller than the barrier height U_0 is completely reflected from the potential wall. The region $x > 0$ is inaccessible to it, since in this region the total energy of the particle would be less than the potential energy. This would mean that the kinetic energy must be negative, which is evidently impossible. If, on the contrary, E is larger than U_0, then according to the laws of classical mechanics the particle passes freely above the barrier, moving in the region $x > 0$ with a lower kinetic energy equal to $E - U_0$.

Let us now consider the motion of a particle in the same situation according to the laws of quantum mechanics. For this we write the Schrödinger equation for the stationary states of the particle in the field of a barrier of infinite extent

$$-\frac{\hbar^2}{2m} \frac{d^2\psi}{dx^2} + U(x)\psi = E\psi \, , \tag{12.1}$$

where U is the potential energy whose graph is shown in fig. V.6. The solutions of eq. (12.1) are conveniently considered in two different regions. Region I ranges from $x = -\infty$ to $x = 0$, and region II from $x = 0$ to $x = \infty$.

We write the Schrödinger equation for each of the regions mentioned:

$$\frac{d^2\psi}{dx^2} + k^2\psi = 0 \, , \qquad x < 0 \, ,$$

$$\frac{d^2\psi}{dx^2} + k'^2\psi = 0 \, , \qquad x > 0 \, , \tag{12.2}$$

where the following notation is introduced:

$$k^2 = \frac{2mE}{\hbar^2}, \qquad k'^2 = \frac{2m}{\hbar^2}(E - U_0).$$ (12.3)

The solutions of these equations are respectively written in the form

$$\psi(x) = A_1 e^{ikx} + B_1 e^{-ikx}, \qquad x < 0,$$

$$\psi(x) = A_2 e^{ik'x} + B_2 e^{-ik'x}, \qquad x > 0.$$ (12.4)

In these formulae a term of the form e^{ikx} represents a plane wave propagating in the positive direction of the x-axis, and e^{-ikx} represents a plane wave propagating in the opposite direction. The amplitudes A_1, B_1, A_2 and B_2 are integration constants. We define a flux of particles incident on the barrier. Let j_0 be the incident particle flux density. Then, according to (7.6),

$$j_0 = \frac{\hbar k}{m}|A_1|^2.$$

For simplicity we choose the flux such that we can set $A_1 = 1$.

In order to define the other constants we consider the behaviour of the wave function at the boundary of regions I and II at the point $x = 0$. By virtue of the general conditions imposed upon the wave function and its derivative (see §6), they must remain continuous even at the point of discontinuity of the potential energy. Hence for $x = 0$ the following equalities must hold:

$$\psi(+0) = \psi(-0),$$ (12.5)

$$\psi'(+0) = \psi'(-0).$$ (12.6)

From relations (12.5) and (12.6) one can also define the two integration constants A_2 and B_1. As to the constant B_2, we must set $B_2 = 0$. As a matter of fact, we only define the particle flux propagating in the positive direction of the x-axis. For $E > U_0$ (i.e. for real k') the term of the wave function proportional to $e^{-ik'x}$ represents a plane wave propagating in the opposite direction. The reflected wave propagates in region I in the negative direction of the x-axis. Obviously, the reflected wave is not present in region II and, consequently, the wave propagating from the right to the left is absent in this region. Hence we have to set the amplitude B_2 of this wave equal to zero. But if $E < U_0$ (k' is a purely imaginary quantity), then the function $e^{-ik'x}$ increases exponentially as $x \to -\infty$, which contradicts the condition of finiteness of the wave function. By virtue of this the coefficient B_2 must also be equal to zero for imaginary values of k', i.e. for $E < U_0$.

Let us consider the case where the total energy of the particle is larger than the potential barrier height, $E > U_0$, in more detail.

Taking into account (12.4), from relation (12.5) and (12.6) we have

$$1 + B_1 = A_2 , \qquad k(1 - B_1) = k'A_2 .$$

From these equations we find the amplitudes A_2 and B_1:

$$B_1 = \frac{k - k'}{k + k'} , \qquad A_2 = \frac{2k}{k + k'} . \tag{12.7}$$

We see that B_1, the amplitude of the reflected wave, is different from zero, although $E > U_0$. This fact is due to the wave properties of particles. The wave is in part reflected, and in part passes into region II. The ratio of the flux density of reflected particles j_R to the flux density of incident particles j_0 will be called the reflection coefficient R. Correspondingly, the ratio of the flux density of transmitted particles j_D to the flux density of incident particles will be called the transmission coefficient D. Taking into account (12.4), we find

$$j_R = \frac{\hbar k}{m} |B_1|^2 , \qquad j_D = \frac{\hbar k'}{m} |A_2|^2 .$$

Since $j_0 = \hbar k / m$, we obtain

$$R = \left(\frac{k - k'}{k + k'} \right)^2 , \qquad D = \frac{4kk'}{(k + k')^2} . \tag{12.8}$$

We see that the following relation is automatically fulfilled:

$$R + D = 1 . \tag{12.9}$$

This relation expresses the law of conservation of the number of particles.

We note that the expressions (12.8) turn out to be symmetric with respect to k and k', i.e. for particles of a given energy E the reflection coefficient (as well as the transmission coefficient) turns out to be independent of the direction of motion of the particles. Particles moving from the left to the right, i.e. against the action of the force at the point $x = 0$, have the same probability of being reflected at this point as particles of the same energy moving from the right to the left, in the direction of the action of the force at the point $x = 0$. This fact is also due to the wave character of the process and has a corresponding optical analogy.

Let us now consider the case $E < U_0$. Then k' is a purely imaginary quantity which is conveniently written in the form $k' = i\kappa$, where

$$\kappa = \frac{1}{\hbar} [2m(U_0-E)]^{\frac{1}{2}} .$$ (12.10)

The amplitude of the reflected wave B_1 turns out to be a complex quantity, and the reflection coefficient R is equal to

$$R = |B_1|^2 = \left|\frac{k - i\kappa}{k + i\kappa}\right|^2 = 1 , \qquad D = 1-R = 0 .$$ (12.11)

The reflected wave is written in the form

$$\psi_R = \frac{k - i\kappa}{k + i\kappa} e^{-ikx} = e^{-i(kx+\delta)} ,$$ (12.12)

i.e. the reflection leads to a shift of the phase of the wave. From (12.12) it follows that this shift is equal to

$$\delta = \text{arc tan} \frac{2k\kappa}{k^2-\kappa^2} .$$ (12.13)

Although the reflection is complete, nevertheless the wave function in region II is different from zero and has the form

$$\psi(x) = A_2 e^{-\kappa x} = \frac{2k}{k + i\kappa} e^{-\kappa x} \qquad (x > 0) .$$ (12.14)

Correspondingly the probability density for finding the particle at point x in the region $x > 0$ is equal to

$$|\psi(x)|^2 = \frac{4k^2}{k^2 + \kappa^2} e^{-2\kappa x} .$$ (12.15)

We see that the behaviour of quantum particles differs essentially from that of classical particles. For a particle moving according to the laws of classical mechanics the region $x > 0$ for $E < U_0$ was forbidden. On the contrary, a particle moving according to the laws of quantum mechanics can, with a certain probability, penetrate this region. The penetration of particles into the region of forbidden energies represents a specific quantum effect which is called the tunnel effect. As is seen from the formula, the effective depth of penetration into region II, i.e. the distance δx from the boundary of region II at which the probability of finding the particle is still considerably different from zero, is of the order of magnitude of $\delta x \sim \kappa^{-1}$. For $x \gg \delta x$ the probability density (12.15) turns out to be exponentially small.

Let us estimate the effective depth of penetration for an electron, assuming that $U_0-E \sim 1$ eV $= 1.6 \times 10^{-12}$ erg. For δx we evidently have

$$\delta x \sim \frac{\hbar}{[2m(U_0-E)]^{\frac{1}{2}}} \sim \frac{10^{-27}}{[2 \times 10^{-27} \times 1.6 \times 10^{-12}]^{\frac{1}{2}}} \sim 10^{-8} \text{ cm} .$$

This estimate shows that the effect can be significant only in the realm of microscopic dimensions. Thus, as was to be expected, the tunnel effect cannot be observed in the motion of macroscopic bodies for which the laws of classical mechanics are valid.

In order to actually observe the particle in region II, we have to localize it there in a certain small interval $\Delta x \lesssim \delta x$. By localizing the particle we change its state (its energy), since by virtue of the uncertainty relation $\Delta p \gtrsim \hbar/\Delta x \gtrsim \gtrsim \hbar\kappa$. The particle which we shall detect somewhere in region II will no longer possess the initial energy E. The uncertainty in the momentum is related to the uncertainty in the kinetic energy of the particle:

$$\Delta T \gtrsim \frac{\Delta p^2}{2m} \gtrsim \frac{\hbar^2 \kappa^2}{2m} .$$

Substituting here the expression for κ (12.10), we obtain

$$\Delta T \gtrsim U_0-E .$$

Thus the uncertainty in the energy of a particle localized in the region behind the barrier is larger than the energy that it lacks to reach the barrier height.

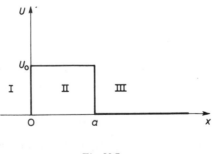

Fig. V.7

Let us consider briefly a barrier of finite extent, as shown in fig. V.7. Particles are incident on the barrier, moving in the positive direction of the x-axis. We can immediately write the wave function for the three different regions:

$$\psi(x) = e^{ikx} + B_1 e^{-ikx} , \qquad\qquad x < 0 , \qquad U = 0 , \qquad\qquad \text{I} \qquad (12.16)$$

$$\psi(x) = A_2 e^{ik'x} + B_2 e^{-ik'x} , \qquad 0 \leqslant x \leqslant a , \qquad U = U_0 \qquad \text{II} \qquad (12.17)$$

$$\psi(x) = A_3 e^{ikx} , \qquad\qquad\qquad x > a , \qquad U = 0 . \qquad\qquad \text{III} \qquad (12.18)$$

In this case (as in the case of a barrier of infinite extent) we again set the amplitude of the incident wave equal to unity. Since in region III there is no reflected wave, only the wave propagating in the positive direction of the x-axis is taken into account. We write the conditions of continuity of the wave function and of its first derivative at the boundaries of the regions, analogous to (12.5) and (12.6):

$$\psi(+0) = \psi(-0) , \qquad \psi'(+0) = \psi'(-0) ,$$
$$\psi(a+0) = \psi(a-0) , \qquad \psi'(a+0) = \psi'(a-0) . \qquad (12.19)$$

Substituting (12.16), (12.17) and (12.18) into these relations, we obtain a system of equations with respect to B_1, A_2, B_2 and A_3

$$1 + B_1 = A_2 + B_2 , \qquad\qquad A_2 e^{ik'a} + B_2 e^{-ik'a} = A_3 e^{ika} ,$$
$$\qquad\qquad\qquad (12.20)$$
$$k(1 - B_1) = k'(A_2 - B_2) , \qquad A_2 e^{ik'a} - B_2 e^{-ik'a} = \frac{k}{k'} A_3 e^{ika} .$$

Let us consider immediately the most interesting case, $E < U_0$. If the motion of particles proceeded according to the laws of classical mechanics, then the barrier would be completely impenetrable for them, and at the point $x = 0$ the particles would undergo total reflection from the barrier. The situation is different in the case of microparticles, whose motion is described by the laws of quantum mechanics. Solving the system (12.20) with respect to A_3 and taking into account that $k' = i\kappa$, where κ is defined by (12.10), we obtain

$$A_3 = \frac{4ik\kappa e^{-ika}}{(k + i\kappa)^2 e^{\kappa a} - (k - i\kappa)^2 e^{-\kappa a}} . \qquad (12.21)$$

The amplitude of the plane wave turns out to be different from zero in the region behind the barrier, although the energy of the particle is smaller than the barrier height $E < U_0$. This means that a microparticle can, with a certain probability, pass through the potential barrier by means of the tunnel effect.

The tunnel transmission of particles, which earlier seemed to be paradoxical, is at present not only observed experimentally, but plays a fundamental role in a number of fields of physics, in particular in nuclear physics. It suffices to say that tunnel passage through a barrier is associated with the α-decay of radioactive nuclei and with the phenomenon of spontaneous fission of uranium nuclei. The tunnel effect is also related to the phenomenon of emission of electrons from cold metals in a strong electric field and to a number of other processes.

Let us define the coefficient of transmission D of microparticles through a barrier

$$D = \frac{j_D}{j_0} = |A_3|^2 = \frac{4k^2\kappa^2}{(k^2 + \kappa^2)^2 \sinh^2 \kappa a + 4k^2\kappa^2}. \tag{12.22}$$

If $\kappa a \gg 1$, then $\sinh \kappa a \approx \frac{1}{2}e^{\kappa a}$ and the expression (12.22) is simplified:

$$D \approx \frac{16k^2\kappa^2}{(k^2 + \kappa^2)^2} e^{-2\kappa a}. \tag{12.23}$$

The basic dependence of the transmission coefficient on the width and height of the barrier is defined by the exponential function $e^{-2\kappa a}$. Denoting the factor multiplying this by D_0, we have

$$D = D_0 \exp\left[-\frac{2}{\hbar}[2m(U_0 - E)]^{\frac{1}{2}}a\right]. \tag{12.24}$$

We see that the probability of passing through a barrier is not too low if

$$\frac{2}{\hbar}[2m(U_0 - E)]^{\frac{1}{2}}a \lesssim 1. \tag{12.25}$$

The condition (12.25) can evidently be fulfilled only in the realm of micro-phenomena. If we substitute into (12.25) values on a nuclear scale $a \sim 10^{-13}$ cm, $m \sim 10^{-24}$ g (nucleon mass), $U_0 - E \sim 10$ MeV (10^{-5} erg), then carrying out the estimation we find that $D \sim e^{-1}$. Thus the particle can with considerable probability pass through a barrier whose height exceeds its energy by 5–10 MeV. A completely different result is obtained for the same particle and the same barrier height, if the spatial extent of the barrier amounts to $a \sim 1$ cm. Then $D \sim 10^{-13}$. This means that in the realm of macroscopic phenomena the effect of tunnel transmission is virtually absent. The probability of tunnel passage through a barrier of arbitrary form will be considered in §42.

§13. One-dimensional motion*

In this chapter we have considered a number of simple problems of quantum mechanics regarding one-dimensional motion. The three-dimensional problems considered also reduce to one-dimensional problems, because the Schrödinger equation with the potential $U(x, y, z) = U_1(x) + U_2(y) + U_3(z)$ reduces to one-dimensional equations with the potentials U_1, U_2 and U_3 respectively. The results obtained allow one to draw some general conclusions on the properties of the one-dimensional motion of particles.

For one-dimensional motion the energy levels can belong to a discrete spectrum (see §8–11) as well as to a continuous spectrum (§12). As we have seen, there correspond to the states of the discrete spectrum quadratically integrable wave functions, i.e. wave functions for which the normalization condition can be written in the form $\int |\psi_n(x)|^2 dx = 1$. This condition indicates that the motion is finite, i.e. that the probability of observing the particle at arbitrarily large distances is negligible.

On the contrary, if the particle can go off to arbitrarily large distances, i.e. if the motion is infinite, its wave function is not quadratically integrable. It can be shown that in this case the energy has a complex spectrum. Suppose that the potential energy $U(x)$ of the particle somehow varies from the value $U(\infty)$ for $x \to \infty$, which we shall choose as the zero energy, to the value $U(-\infty) = U_0$ for $x \to -\infty$. We shall assume for definiteness that U_0 is positive, $U_0 > 0$. The function $U(x)$ may vary quite arbitrarily. We suppose only that it has a minimum $U_{min} < 0$. Then for an energy E such that $U_{min} < E < 0$ the particle cannot go off to infinity. For these energy values the motion is finite, and the spectrum is discrete. The energy levels of a discrete spectrum are not degenerate. This statement is easily proved by assuming the contrary. Indeed, if it is assumed that ψ_1 and ψ_2 are two solutions of the Schrödinger equation corresponding to one and the same energy value E, then they satisfy the relation

$$\frac{1}{\psi_1} \frac{d^2\psi_1}{dx^2} = \frac{2m}{\hbar^2} (U-E) = \frac{1}{\psi_2} \frac{d^2\psi_2}{dx^2},$$

i.e.

$$\frac{1}{\psi_1} \frac{d^2\psi_1}{dx^2} = \frac{1}{\psi_2} \frac{d^2\psi_2}{dx^2}.$$

* For a detailed treatment see the book of L.D.Landau and E.M.Lifshitz, *Quantum mechanics* (Pergamon Press, Oxford, 1965).

Integrating this relation with respect to x, we obtain

$$\psi_2 \frac{d\psi_1}{dx} - \psi_1 \frac{d\psi_2}{dx} = \text{const} . \tag{13.1}$$

But at infinity $\psi_1 = \psi_2 = 0$. Hence the constant on the right-hand side of relation (13.1) is equal to zero and, consequently,

$$\psi_2 \frac{d\psi_1}{dx} = \psi_1 \frac{d\psi_2}{dx} .$$

Integrating once more with respect to x, we obtain $\psi_2 = \text{const}.\psi_1$. This means that the two functions describe one and the same state, i.e. degeneracy is absent.

In the range $0 < E < U_0$ the particle can move arbitrarily distant in the direction of positive x. Therefore the motion is infinite, and the energy spectrum is continuous. The wave functions in this case are also not degenerate. In fact, the preceding proof is again valid, since the wave functions reduce to zero when $x \to -\infty$. The asymptotic expressions for the wave functions for $x \to \pm \infty$ are easily obtained from the Schrödinger equation

$$\frac{d^2\psi}{dx^2} + \frac{2m}{\hbar^2} [E - U(x)] \psi = 0$$

if one substitutes here $U = 0$ for $x \to \infty$ and $U = U_0$ for $x \to -\infty$. Correspondingly we obtain

$$\psi(x) = A \sin (kx + \alpha) \quad \text{for} \quad x \to \infty \tag{13.2}$$

$$\psi(x) = B e^{\kappa x} \qquad \text{for} \quad x \to -\infty \tag{13.3}$$

where

$$k = \frac{1}{\hbar}(2mE)^{\frac{1}{2}} \quad \text{and} \quad \kappa = \frac{1}{\hbar}[2m(U_0 - E)]^{\frac{1}{2}} ,$$

i.e. the solution has the form of a standing plane wave when $x \to \infty$, and is exponentially damped when $x \to -\infty$.

In the energy range $E > U_0$ the motion is infinite in both directions. The energy spectrum is continuous. Since the Schrödinger equation is of second order, it has two linearly independent solutions. In this energy range both solutions satisfy the necessary requirements. Therefore the energy levels are two-fold degenerate. The asymptotic expression for the wave function is of the form

$$\psi = A_1 e^{ikx} + A_2 e^{-ikx} , \tag{13.4}$$

where one term corresponds to the particle moving in the positive direction of the x-axis, and the other corresponds to the particle moving in the negative direction of the x-axis.

We now assume that the field increases indefinitely, i.e. that $|U(x)| \to \infty$ as $x \to \pm\infty$. As a simple and at the same time important example we shall consider the problem of the motion of a particle in a uniform external field

$$U(x) = -fx .$$

We choose the x-axis to be in the direction of the field; f denotes the force acting on the particle: $f = -dU/dx$. The potential energy is measured from its value at $x = 0$, hence $U(0) = 0$. The Schrödinger equation for the motion in such a field is of the form

$$\frac{d^2\psi}{dx^2} + \frac{2m}{\hbar^2}(E + fx)\psi = 0 . \tag{13.5}$$

We now introduce in place of x the new variable

$$\eta = \left(\frac{2mf}{\hbar^2}\right)^{\frac{1}{3}} \left(x + \frac{E}{f}\right) .$$

Correspondingly eq. (13.5) will have the form

$$\frac{d^2\psi}{d\eta^2} + \eta\psi = 0 . \tag{13.6}$$

The solution of eq. (13.6) can be expressed in terms of Bessel functions, but the expression of the solution of (13.6) in terms of the so-called Airy function is more convenient. Namely, the solution of eq. (13.6), finite for all values of η, has the form

$$\psi(\eta) = C\Phi(-\eta) , \tag{13.7}$$

where $\Phi(\eta)$ denotes the Airy function

$$\Phi(\eta) = \pi^{-\frac{1}{2}} \int\limits_0^\infty \cos\left(\tfrac{1}{3}u^3 + u\eta\right) du , \tag{13.8}$$

and C is the normalization constant.

Thus the Schrödinger equation (13.5) has a solution satisfying the necessary requirements for any energy value E. Consequently, for motion in a uniform field the energy spectrum of the particle is continuous, which corresponds to infinite motion. In the given case $U \to -\infty$ as $x \to \infty$, i.e. the motion in the positive direction of the x-axis is infinite.

The wave function (13.7) has a rather simple form for $\eta \to \pm \infty$. Making use of the known asymptotic expressions for the Airy function, we have

$$\psi(\eta) = \frac{C}{2|\eta|^{\frac{1}{4}}} \exp\left(-\tfrac{2}{3}|\eta|^{\frac{3}{2}}\right) \quad \text{for} \quad \eta \to -\infty \tag{13.9}$$

and

$$\psi(\eta) = \frac{C}{\eta^{\frac{1}{4}}} \sin\left(\tfrac{2}{3}\eta^{\frac{3}{2}} + \tfrac{1}{4}\pi\right) \quad \text{for} \quad \eta \to \infty. \tag{13.10}$$

The constant C is defined by the normalization condition. Of course, the integral of the square of the modulus of the wave function (13.7) over all space diverges, which corresponds to infinite motion. Normalization rules for such cases will be discussed in § 18.

§ 14. The Schrödinger equation for a system of particles

In the preceding sections we have considered the laws of motion of a single particle in an external field. But this greatly restricted the range of problems we could consider. As a matter of fact, even the simplest system, the hydrogen atom, represents, strictly speaking, a system of two particles. This holds all the more for systems such as many-electron atoms, molecules, atomic nuclei, matter in the solid state and so on. Generalizing the results obtained in § 6, we formulate the fundamental equation of quantum mechanics: the Schrödinger equation for a system of N particles. It has the form

$$i\hbar \frac{\partial \psi}{\partial t} = \sum_{i=1}^{N} \left(-\frac{\hbar^2}{2m_i}\right) \nabla_i^2 \psi + \sum_{i=1}^{N} U_i(\mathbf{r}_i)\psi + U_{\text{int}}(\mathbf{r}_1, ..., \mathbf{r}_N)\psi. \tag{14.1}$$

Here the Laplacian ∇_i^2

$$\nabla_i^2 = \frac{\partial^2}{\partial x_i^2} + \frac{\partial^2}{\partial y_i^2} + \frac{\partial^2}{\partial z_i^2}$$

acts on the coordinates of the ith particle. $U_i(\mathbf{r}_i)$ is the potential energy of the ith particle in the external field, U_{int} is the potential energy of the interaction of the particles with each other, and m_i is the mass of the ith particle. The summation is carried out over all particles of the system. The wave function describing a system of particles, in accordance with §2, depends on the coordinates of all the particles and the time, $\psi(\mathbf{r}_1, \mathbf{r}_2, ..., \mathbf{r}_N, t)$.

The Schrödinger equation for stationary states has the form

$$\sum_{i=1}^{N} \left(-\frac{\hbar^2}{2m_i}\right) \nabla_i^2 \psi + \sum_{i=1}^{N} U_i(\mathbf{r}_i)\psi + U_{int}(\mathbf{r}_1, ..., \mathbf{r}_N)\psi = E\psi . \tag{14.2}$$

As the simplest example of the integration of eq. (14.2) let us consider a system of particles which do not interact with each other, i.e. let us assume that the energy of interaction is equal to zero, $U_{int} = 0$. In this case the Schrödinger equation can be rewritten in the form

$$\sum_{i=1}^{N} \left(-\frac{\hbar^2}{2m_i}\nabla_i^2 + U_i(\mathbf{r}_i)\right) = E\psi . \tag{14.3}$$

where the terms in each bracket depend only on the coordinates of the corresponding particle. We seek a wave function ψ in the form of a product of functions which depend on the coordinate of individual particles.

$$\psi = \psi_1(\mathbf{r}_1)\psi_2(\mathbf{r}_2) ... \psi_N(\mathbf{r}_N) . \tag{14.4}$$

On substituting into the Schrödinger equation, we obtain

$$\sum_{i=1}^{N} \psi_1(\mathbf{r}_1) ... \psi_{i-1}(\mathbf{r}_{i-1})\psi_{i+1}(\mathbf{r}_{i+1}) ...$$

$$... \psi_N(\mathbf{r}_N) \left(-\frac{\hbar^2}{2m_i}\nabla_i^2 + U_i(\mathbf{r}_i)\right) \psi_i(\mathbf{r}_i) = E\psi .$$

Dividing the right and left sides of the equation by ψ, we find

$$\sum_{i=1}^{N} \frac{1}{\psi_i(\mathbf{r}_i)} \left(-\frac{\hbar^2}{2m_i}\nabla_i^2 + U_i(\mathbf{r}_i)\right) \psi_i(\mathbf{r}_i) = E .$$

There is a constant quantity on the right-hand side of the equation. The left-hand side is made up of a sum of terms each of which is a function of its independent variable. In order that the equation hold for all values of the independent variables, the following conditions must be fulfilled:

$$-\frac{\hbar^2}{2m_i}\nabla_i^2 \psi_i(\mathbf{r}_i) + U_i(\mathbf{r}_i)\psi_i(\mathbf{r}_i) = E_i\psi_i(\mathbf{r}_i) , \qquad \sum_{i=1}^{N} E_i = E ,$$

where the E_i are constants, which, as is easily seen, represent the energies of individual particles.

Thus, if the left-hand side of the Schrödinger equation can be written in the form of the sum (14.3), then the wave function of the system resolves into a product of wave functions, while the energy of the system appears as the sum of the energies of the individual particles.

These results have a simple physical meaning. We have assumed that the energy of interaction between the particles is equal to zero. It is therefore natural that the total energy of the entire system is made up of the sum of the energies of the individual particles, since the motion of each of the particles is independent of the motion of the other particles. The probability of observing a given set of coordinates of the particles is written in the form

$$dW(\mathbf{r}_1, ..., \mathbf{r}_N) = |\psi_1(\mathbf{r}_1)|^2 |\psi_2(\mathbf{r}_2)|^2 ... |\psi_N(\mathbf{r}_N)|^2 \, dV_1 ... dV_N .$$

The above result is in complete agreement with the theorem of multiplication of probabilities of independent events.

Further, let us consider in more detail a system of two particles with masses m_1 and m_2. We assume that the potential energy of interaction depends only on the distance between the particles, and that there is no external field. In this case the Schrödinger equation for stationary states has the form

$$-\frac{\hbar^2}{2m_1} \nabla_1^2 \psi(\mathbf{r}_1, \mathbf{r}_2) - \frac{\hbar^2}{2m_2} \nabla_2^2 \psi(\mathbf{r}_1, \mathbf{r}_2) +$$

$$+ U(|\mathbf{r}_1 - \mathbf{r}_2|)\psi(\mathbf{r}_1, \mathbf{r}_2) = E\psi(\mathbf{r}_1, \mathbf{r}_2) . \tag{14.5}$$

We transform this equation by introducing new coordinates $\mathbf{R}$ and $\mathbf{r}$ which are defined by the relation

$$\mathbf{R} = \frac{m_1 \mathbf{r}_1 + m_2 \mathbf{r}_2}{m_1 + m_2}, \qquad \mathbf{r} = \mathbf{r}_1 - \mathbf{r}_2 . \tag{14.6}$$

We note that the new variables are completely analogous to the coordinates of the centre of mass and those of relative motion in classical mechanics. As a result of somewhat lengthy but not complicated transformations the Schrödinger equation takes the form

$$-\frac{\hbar^2}{2M} \nabla_R^2 \psi - \frac{\hbar^2}{2\mu} \nabla_r^2 \psi + U(r)\psi = E\psi . \tag{14.7}$$

Here M and μ are the total and reduced masses of the system

$$M = m_1 + m_2 , \qquad \mu = \frac{m_1 m_2}{m_1 + m_2} . \tag{14.8}$$

We see that the left-hand side of the Schrödinger equation is resolved into a sum of two terms and has a form similar to eq. (14.3). In this case the Schrödinger equation can be written in the form

$$\psi(\mathbf{r}_1, \mathbf{r}_2) = \varphi(\mathbf{R})\psi_0(\mathbf{r}) . \tag{14.9}$$

Substituting (14.9) into (14.7) and repeating the transformations which were carried out before, we obtain

$$-\frac{\hbar^2}{2M}\nabla_{\mathbf{R}}^2 \varphi = E_{\mathbf{R}}\varphi , \tag{14.10}$$

$$-\frac{\hbar^2}{2\mu}\nabla_{\mathbf{r}}^2 \psi_0 + U(r)\psi_0 = E_{\mathbf{r}}\psi_0 , \tag{14.11}$$

$$E_{\mathbf{R}} + E_{\mathbf{r}} = E . \tag{14.12}$$

Equation (14.10) is the Schrödinger equation for a free particle with mass M. Its solution is the function

$$\varphi(\mathbf{R}) = A e^{(i/\hbar)(\mathbf{P} \cdot \mathbf{R})} . \tag{14.13}$$

The quantity $E_{\mathbf{R}} = |\mathbf{P}|^2/2M$ represents the kinetic energy of motion of the system as a whole.

Thus corresponding to (14.9) and (14.13) the solution of the Schrödinger equation can be written in the form

$$\psi(\mathbf{r}_1, \mathbf{r}_2) = A e^{(i/\hbar)(\mathbf{P} \cdot \mathbf{R})} \psi_0(\mathbf{r}) . \tag{14.14}$$

From the formulae obtained it is seen that the centre of mass of the system moves in space as a free particle, while the relative motion of the particles proceeds independently of the motion of the centre of mass and is described by the function ψ_0 satisfying eq. (14.11). The total energy of the system is made up of the energies of the relative motion and the motion of the centre of mass. Hence we see that in quantum mechanics, as in classical physics, the problem of the motion of two particles whose potential energy of interaction U depends only on the distance between them, $U(|\mathbf{r}_1 - \mathbf{r}_2|)$, reduces to the problem of the motion of a single particle with reduced mass μ in an external field U.

3

The Mathematical Apparatus
of Quantum Mechanics

§15. Linear operators

We have seen above that in solving the Schrödinger equation one can find the wave functions and the energy of a system. The latter in certain cases (e.g. a particle in a potential well) has a discrete sequence of values, while in other cases (e.g. a free particle; a particle passing through a barrier) it has a continuous sequence of values.

Knowing the wave function ψ, we could obtain the probability of finding the particle at a given point of space, as well as the mean values of the quantities depending on the coordinates. Then, as was shown in §4, the coordinates and the corresponding momentum components of the particle have no definite simultaneous values. However, the mathematical apparatus we used was inadequate to solve a number of important problems. As examples we put these questions: Which quantities cannot simultaneously have definite values? How are the mean values of the quantities which are not functions of the coordinates to be found? What characteristics of a quantum-mechanical system must be given in order that its state is completely defined?

The peculiarity of the problems of quantum mechanics required the development and application of a special mathematical technique.

The mathematical technique of quantum mechanics must correspond to the physical statement of its problems. It turned out that the corresponding mathematical technique – the theory of linear operators – had already been worked out in mathematics. We shall first consider the basic concepts of this theory, and in what follows we shall show how the theory of linear operators can be associated with the problems of quantum mechanics.

We shall understand by an operator a rule by which a function $\varphi(x_1, x_2, x_3, ...)$ of the variables $x_1, x_2, x_3, ...$ is related to another function $\chi(x_1, x_2, x_3, ...)$ of the same variables.

In what follows we shall denote operators by means of letters with the sign $\hat{}$, for example $\hat{F}$. By means of the symbol $\hat{F}$ the rule of the transition from the function φ to the function χ can be written in the form

$$\chi = \hat{F}\varphi . \tag{15.1}$$

Let us consider some simple operators.

The operator $\hat{F}$ can, for example, denote differentiation with respect to any variable

$$\chi(x_1, x_2, ...) = \frac{\partial}{\partial x_i} \varphi(x_1, x_2, ...) .$$

This operator is written symbolically as follows:

$$\hat{F} = \frac{\partial}{\partial x_i} .$$

The differential operators will be encountered particularly often. The operator $\hat{F}$ can also denote multiplication by any quantity, raising to a power, and so on.

We define the operator of the independent variable x as the multiplication by this variable: $\chi = x\varphi$.

An integral relation between the functions φ and χ can also be represented in operator form:

$$\chi(x) = \int K(x, \xi)\varphi(\xi)\, d\xi = \hat{F}\varphi . \tag{15.2}$$

The function $K(x, \xi)$ is called the kernel of the integral operator $\hat{F}$.

We note that earlier we have also used differential operators: the operator $\mathbf{V} = \mathbf{i}(\partial/\partial x) + \mathbf{j}(\partial/\partial y) + \mathbf{k}(\partial/\partial z)$, the Laplacian $\mathbf{V}^2 = \partial^2/\partial x^2 + \partial^2/\partial y^2 + \partial^2/\partial z^2$ and others.

We now define a linear operator $\hat{F}$ as an operator for which the equalities

$$\hat{F}(\varphi_1 + \varphi_2) = \hat{F}\varphi_1 + \hat{F}\varphi_2 , \tag{15.3}$$

$$\hat{F}C\varphi = C\hat{F}\varphi \tag{15.4}$$

are fulfilled, where C is an arbitrary constant. Hence it follows that

$$\hat{F}(C_1\varphi_1 + C_2\varphi_2) = C_1\hat{F}\varphi_1 + C_2\hat{F}\varphi_2 , \tag{15.5}$$

where C_1, C_2 are arbitrary constants. It is obvious that the operators mentioned above are linear operators.

For reasons which will be clear from what follows we shall, in quantum mechanics, deal only with linear operators.

By combining two given operators $\hat{F}$ and $\hat{R}$ one can define their sum and product. We shall understand the sum of the operators $\hat{F}$ and $\hat{R}$ to be the operator $\hat{G}$ defined by the relation

$$\hat{G} = \hat{F} + \hat{R} , \qquad \hat{G}\varphi = \hat{F}\varphi + \hat{R}\varphi . \tag{15.6}$$

We shall understand the product of two operators $\hat{F}$ and $\hat{R}$ to be the operator $\hat{L} = \hat{F}\hat{R}$ consisting of the consecutive application of the operators $\hat{R}$ and $\hat{F}$,

$$\hat{L}\varphi = \hat{F}(\hat{R}\varphi) . \tag{15.7}$$

If one first applies the operator $\hat{F}$ and then the operator $\hat{R}$, their product will be the operator $\hat{L}' = \hat{R}\hat{F}$,

$$\hat{L}'\varphi = \hat{R}(\hat{F}\varphi) . \tag{15.8}$$

We note that the operators $\hat{L}$ and $\hat{L}'$, generally speaking, are not identical to each other, i.e. the product of the operators depends critically on the order of the factors. Corresponding to this, the algebra of operators is an algebra of non-commuting quantities. Two operators are said to commute with each other if the product of the operators does not depend on the order of the factors; otherwise the operators are said to be non-commuting. As an example let us find the product of the operator of differentiation with respect to x and the operator of multiplication by x for both orders of the factors, i.e. let us assume that $\hat{F} = x$, $\hat{R} = d/dx$. In correspondence with (15.7) the operator of the product $\hat{L}$ will be $\hat{L} = x(\partial/\partial x)$. We now find the operator $\hat{L}' = \hat{R}\hat{F}$

$$\hat{L}'\psi = \frac{\partial}{\partial x}(x\psi) = \left(1 + x\frac{\partial}{\partial x}\right)\psi .$$

We see that in this case the operator $\hat{L}'$ is equal to

$$\hat{L}' = 1 + x\frac{\partial}{\partial x}$$

and is not identical to the operator $\hat{L}$. Thus the operators $\hat{L}$ and $\hat{L}'$ do not commute. Making use of the expressions obtained for $\hat{L}$ and $\hat{L}'$, we can write that

$$\frac{\partial}{\partial x}x - x\frac{\partial}{\partial x} = 1 .$$

It is natural to call the right-hand side of this operator relation the unit operator. If we took the operator of multiplication by any other independent variable, say y, as the operator $\hat{F}$, then it would turn out that the operators $\partial/\partial x$ and y commute

$$y\frac{\partial}{\partial x} - \frac{\partial}{\partial x}y = 0 . \tag{15.9}$$

For certain operators it turns out that the following relation holds:

$$\hat{F}\hat{R} = -\hat{R}\hat{F} . \tag{15.10}$$

In this case the operators $\hat{F}$ and $\hat{R}$ are said to anticommute. We shall call the operator $\hat{F}\hat{R} - \hat{R}\hat{F}$ the commutator of the operators $\hat{F}$ and $\hat{R}$, and shall denote it by brackets, i.e.

$$\hat{F}\hat{R} - \hat{R}\hat{F} = \{\hat{F}, \hat{R}\} . \tag{15.11}$$

An operator $\hat{F}$ can be contrasted with the inverse operator $\hat{F}^{-1}$. The inverse operator is defined by the relations

$$\hat{F}^{-1}\hat{F}\psi = \psi , \qquad \hat{F}\hat{F}^{-1}\psi = \psi$$

or

$$\hat{F}\hat{F}^{-1} = \hat{F}^{-1}\hat{F} = 1 . \tag{15.12}$$

If $\hat{F}$ is a differential operator, then the inverse operator $\hat{F}^{-1}$ has the form of an integral operator. Indeed, suppose that the relation

$$\hat{F}\psi(x) = \varphi(x) . \tag{15.13}$$

holds. Then, acting on the right-hand and left-hand sides of this equality with the operator $\hat{F}^{-1}$, we obtain

$$\psi(x) = \hat{F}^{-1}\varphi(x) . \tag{15.14}$$

On the other hand, the relation (15.14) can be written in the form

$$\psi(x) = \int G(x, x')\varphi(x')\,dx' , \tag{15.15}$$

where the function $G(x, x')$, called the Green's function of eq. (15.13), satisfies the relation

$$\hat{F}G(x, x') = \delta(x-x') . \tag{15.16}$$

Indeed, if we act with the operator $\hat{F}$ on the right-hand and left-hand sides of the equality (15.15), then under the condition (15.16) we again arrive at the relation (15.13). Comparing (15.14) and (15.15), we see that the Green's function $G(x, x')$ is the kernel of the integral operator $\hat{F}^{-1}$. The Green's function $G(x, x')$ is not determined unambiguously from eq. (15.16). For a single-valued determination it is necessary to give in addition certain conditions of the nature of boundary conditions.

The relation (15.13) can be considered as an equation relating the function $\psi(x)$ to a given function $\varphi(x)$, and whose solution is given by formula (15.15). It should only be borne in mind that in order to obtain a general solution we have to add the general solution $\psi_0(x)$ of the homogeneous equation $\hat{F}\psi_0(x) = 0$ to (15.15). We then have

$$\psi(x) = \psi_0(x) + \int G(x, x')\varphi(x')\,dx' . \tag{15.17}$$

We shall need the above relation in what follows.

§16. Eigenvalues and eigenfunctions of operators

Let us consider the operator relation

$$\hat{F}\psi = F\psi . \tag{16.1}$$

Relation (16.1) means that if the operator $\hat{F}$ is applied to the function ψ, then one again obtains the function ψ multiplied by a certain constant F. It is obvious that for a given form of the operator $\hat{F}$ the relation (16.1) cannot be satisfied by every function ψ. In other words, relation (16.1) is an equation. The form of the function ψ can be obtained by solving eq. (16.1). If the operator $\hat{F}$ is a linear differential operator, then eq. (16.1) will be a differential equation. Since from the form of the equation it is immediately clear that $\psi = 0$ is its trivial solution, (16.1) represents a linear homogeneous differential equation. The study of such linear homogeneous equations is the most important problem of the theory of operators.

In what follows we shall not be interested in arbitrary operators $\hat{F}$ and arbitrary functions ψ, but only in functions which satisfy certain definite conditions:

(1) the function ψ must exist over the entire range of the independent varia-

bles. For example, in the case of Cartesian coordinates, in the range $-\infty < x < \infty$, $-\infty < y < \infty$, $-\infty < z < \infty$;

(2) in the region of existence the function ψ must be finite and continuous, with the exception, in some cases, of singular points;

(3) the function ψ must be single-valued.

We shall call the set of conditions (1)–(3) the standard conditions. It turns out, generally speaking, that eq. (16.1) has solutions which differ from the trivial one, and which satisfy the standard conditions, not for all values of the parameter F but only for certain selected values of it. The selected values of F for which the non-trivial solutions of eq. (16.1) exist are called the eigenvalues of the operator $\hat{F}$, and the corresponding solutions of eq. (16.1) are called the eigenfunctions of the operator $\hat{F}$.

We shall first of all present the problems on eigenfunctions and eigenvalues with which we are already acquainted.

(1) In considering the problem of the motion of a particle in a potential well we have solved eq. (16.1) with the differential operator $\hat{F} = -d^2/dx^2$. The boundary conditions led to the eigenvalues (8.5) and to the eigenfunctions (8.7) of the operator $\hat{F}$.

(2) If, for the same form of the operator, we require no reduction of ψ to zero at the boundaries of the interval $(0, l)$, then the solutions of (8.2) will have the form

$$\psi = A e^{ikx} + B e^{-ikx} .$$

If $k^2 > 0$, then for all values of x the function ψ is finite, so that the solution satisfies the standard conditions. For negative k^2

$$\psi = A e^{-\kappa x} + B e^{\kappa x}, \quad \text{where} \quad k = i\kappa ,$$

there are no solutions which satisfy the standard conditions.

(3) In the problem of the oscillator we have considered the solution of eq. (16.1) for the operator (see (10.3))

$$\hat{F} = -\frac{d^2}{dx^2} + x^2 .$$

The problem has the solution $F = 2E/\hbar\omega = 2n + 1$.

It is clear from the examples given that the whole set of eigenvalues of an operator, which we shall call its spectrum, can be discrete (example 1 and 3) as well as continuous (example 2). It can be proved that the eigenfunctions which correspond to the discrete spectrum of the eigenvalues are quadratically integrable, i.e. the integral $\int |\psi|^2 dV$ converges. The eigenfunctions corresponding to the continuous spectrum of the eigenvalues are not quadratically integrable. If to each eigenvalue of the operator there belongs one and only one eigenfunction ψ, the spectrum is said to be non-degenerate. On the contrary, if to one eigenvalue F there correspond several, for example s, different eigenfunctions, then the given eigenvalue is said to be degenerate with degeneracy s.

The examples given above are important because they illuminate our interest in the theory of operators. The problem of finding the solutions of the Schrödinger equation is a particular case of the problem of the eigenfunctions of operators of a particular form.

Before passing over from this heuristic reasoning to the establishment of a more complete relationship between the concepts of quantum mechanics and the theory of linear operators, it is necessary in addition to consider important properties of a particular class of operators.

§17. Hermitian operators

The eigenvalues F in the operator equation (16.1) can, generally speaking, be complex. However, we shall be solely interested in equations which lead only to real eigenvalues. It turns out that there is a class of operators which can possess only real eigenvalues. Such operators are called Hermitian or self-adjoint. Each linear operator $\hat{F}$ can be compared to a certain operator $\hat{F}^\dagger$ which we shall call the adjoint operator or the Hermitian conjugate. The adjoint operator is defined by the condition

$$\int \psi_1^* \hat{F} \psi_2 \, dV = \int \psi_2 (\hat{F}^\dagger \psi_1)^* \, dV . \tag{17.1}$$

Here, as always, the asterisk denotes complex-conjugate quantities. The integration in (17.1) is carried out over the entire region of variation of the independent variables. We have denoted by dV a volume element of this region.

The functions ψ_1, ψ_2 must satisfy the necessary requirements for the convergence of the integrals in (17.1). Furthermore, they must satisfy certain boundary conditions which usually amount to the requirement that the functions ψ_1 and ψ_2 reduce to zero at infinity. But in other respects the functions ψ_1 and ψ_2 are rather arbitrary. If the operator $\hat{F}$ coincides with its adjoint operator $\hat{F}^\dagger = \hat{F}$, then such an operator is said to be Hermitian or self-adjoint. In this case relation (17.1) has the form

$$\int \psi_1^* \hat{F} \psi_2 \, dV = \int \psi_2 \hat{F}^* \psi_1^* \, dV . \tag{17.2}$$

Here we have denoted by $\hat{F}^*$ the operator defined by the relation $\hat{F}^*\psi^* = (\hat{F}\psi)^*$.

As an example let us find the adjoint operator of the differential operator $\hat{F} = d/dx$. Assuming that the functions ψ_1, ψ_2 reduce to zero at infinity and carrying out the integration by parts in (17.1), we obtain

$$\int_{-\infty}^{\infty} \psi_1^* \frac{d}{dx} \psi_2 dx = - \int_{-\infty}^{\infty} \psi_2 \frac{d}{dx} \psi_1^* dx .$$

Comparing with (17.1), we find the operator $\hat{F}^\dagger$: $\hat{F}^\dagger = - d/dx$. We see that the operator $\hat{F}^\dagger$ in the given case does not coincide with the operator $\hat{F}$, i.e. the differential operator is not self-adjoint. If, however, the operator $i(d/dx)$ is taken as the operator $\hat{F}$, then it is easily seen that such an operator is Hermitian. Indeed, in this case we have, on integrating by parts

$$i \int_{-\infty}^{\infty} \psi_1^* \frac{d}{dx} \psi_2 dx = - i \int_{-\infty}^{\infty} \psi_2 \frac{d}{dx} \psi_1^* dx ,$$

and relation (17.2) is now valid. Hence it follows that the operator $\hat{F} = i(d/dx)$ is Hermitian.

We also define the operator $\tilde{\hat{F}}$, which is called the transpose of the operator $\hat{F}$

$$\int \psi_1^* \hat{F} \psi_2 dV = \int \psi_2 \tilde{\hat{F}} \psi_1^* dV . \tag{17.3}$$

Comparing (17.3) with (17.1), we obtain $\hat{F}^\dagger = \tilde{\hat{F}}^*$.

Further we find the operator $\hat{L}^\dagger$ adjoint of the operator $\hat{L}$ which is the product of two operators $\hat{L} = \hat{F}\hat{R}$. From the definition (17.1) we have

$$\int \psi_1^* \hat{F}(\hat{R}\psi_2) dV = \int (\hat{R}\psi_2)(\hat{F}^\dagger \psi_1)^* dV .$$

We exchange the functions $(\hat{R}\psi_2)$ and $(\hat{F}^\dagger \psi_1)$. Then we get

$$\int \psi_1^* \hat{F}\hat{R} \psi_2 dV = \int (\hat{F}^\dagger \psi_1)^* \hat{R} \psi_2 dV .$$

Further, we again use the relation (17.1)

$$\int \psi_1^* \hat{F}\hat{R} \psi_2 dV = \int \psi_2 (\hat{R}^\dagger (\hat{F}^\dagger \psi_1))^* dV .$$

From this expression we obtain the operator $\hat{L}^\dagger = (\hat{F}\hat{R})^\dagger$

$$(\hat{F}\hat{R})^\dagger = \hat{R}^\dagger \hat{F}^\dagger . \tag{17.4}$$

We see that the adjoint operator of the product is equal to the product of the adjoint operators taken, however, in the reverse order. Thus if the operators $\hat{F}$ and $\hat{R}$ are self-adjoint, i.e. $\hat{F}^\dagger = \hat{F}$, $\hat{R}^\dagger = \hat{R}$, then their product will be a self-adjoint operator only in the case where they commute. Indeed, under these conditions we have

$$(\hat{F}\hat{R})^\dagger = \hat{R}\hat{F} = \hat{F}\hat{R} \ . \tag{17.5}$$

Since each operator undoubtedly commutes with itself, it follows from (17.5) that if the operator $\hat{F}$ is Hermitian, then so also will be the operator $\hat{F}^2 = \hat{F}\hat{F}$ as well as, in general, the operator $\hat{F}^n = \underbrace{\hat{F}\cdot\hat{F}...\hat{F}}_{n}$, where n is a positive integer.

We now pass on to the proof of the basic theorem on the reality of the eigenvalues of Hermitian operators. For this we once again write eq. (16.1), assuming for concreteness that the operator $\hat{F}$ possesses a discrete spectrum $\hat{F}\psi_n = F_n\psi_n$. Multiplying the equation from the left by ψ_n^* and integrating, we obtain

$$F_n = \frac{\int \psi_n^* \hat{F}\psi_n dV}{\int |\psi_n|^2 dV} \ .$$

If the operator $\hat{F}$ is Hermitian, then it is easily seen that the eigenvalues F_n, determined in (16.1), are real. Indeed, taking into account the Hermitian property (17.2), we find

$$F_n^* = \frac{\int \psi_n \hat{F}^* \psi_n^* dV}{\int |\psi_n|^2 dV} = F_n \ .$$

Thus we have proved that Hermitian (self-adjoint) operators have real eigenvalues only.

§18. The orthogonality and normalization of the eigenfunctions of Hermitian operators

The eigenfunctions of a linear Hermitian operator $\hat{F}$, which correspond to different eigenvalues F_n and F_m, are mutually orthogonal, i.e. satisfy the relation

$$\int \psi_m^* \psi_n dV = 0 \qquad \text{(for } m \neq n) \ . \tag{18.1}$$

The functions ψ_n and ψ_m^* satisfy eq. (16.1),

$$\hat{F}\psi_n = F_n\psi_n \ , \qquad \hat{F}^* \psi_m^* = F_m \psi_m^* \ . \tag{18.2}$$

Since the operator F is Hermitian, we have:

$$\int \psi_m^* \hat{F} \psi_n dV = \int \psi_n \hat{F}^* \psi_m^* dV .$$ (18.3)

Making use of eqs. (18.2), we rewrite eq. (18.3) in the form

$$F_n \int \psi_m^* \psi_n dV = F_m \int \psi_m^* \psi_n dV .$$

Hence it follows that

$$(F_m - F_n) \int \psi_m^* \psi_n dV = 0 .$$ (18.4)

Since by assumption $F_m \neq F_n$, then

$$\int \psi_m^* \psi_n dV = 0 ,$$ (18.5)

which proves our statement.

Because the eigenfunctions satisfy a homogeneous linear equation, they are determined to within an arbitrary constant.

Keeping in mind what is to follow, we shall normalize the eigenfunctions of a discrete spectrum by the condition

$$\int \psi_n^* \psi_n dV = 1 .$$ (18.6)

The eigenfunctions satisfying relation (18.6) are said to be normalized to unity. We combine formulae (18.1) and (18.6) into one formula

$$\int \psi_m^* \psi_n dV = \delta_{nm} ,$$ (18.7)

where δ_{nm} is the Kronecker symbol:

$$\delta_{nm} = \begin{cases} 1 & \text{if } n = m , \\ 0 & \text{if } n \neq m . \end{cases}$$

We now consider the case of degenerate states, where several eigenfunctions $\psi_{n1}, \psi_{n2}, ..., \psi_{ns}$, belong to the same eigenvalue F_n.

One can take as the solution of eq. (18.2) corresponding to the eigenvalue F_n arbitrary linear combinations of these functions

$$\psi_{nk}' = \sum_{r=1}^{s} a_{kr} \psi_{nr} .$$ (18.8)

By appropriate choice of the coefficients a_{kr} one can obtain mutual orthogonality of the eigenfunctions ψ_{nk}' which belong to one and the same eigenvalue F_n. Imposing also the normalization condition, we obtain

$$\int \psi_{nk}'^* \psi_{nl}' dV = \delta_{kl} .$$ (18.9)

The condition (18.9) still does not completely determine the values of the coefficients a_{kr}. Indeed, if the functions ψ_{nk} are already orthogonal to each other and if we have carried out transformation (18.8), then the orthogonality will be preserved if

$$\sum_{r=1}^{s} a_{kr}^* a_{lr} = \delta_{kl} . \tag{18.10}$$

Thus there is still a certain arbitrariness in the choice of the coefficients a_{kl}.

Finally, we consider the wave functions of a continuous spectrum. For the wave functions of continuous spectrum $\psi_F(x)$ the condition of orthogonality is proved analogously to (18.3)–(18.5):

$$\int \psi_F^*(x)\,\psi_{F'}(x)\,\mathrm{d}V = 0 . \tag{18.11}$$

On the other hand, the condition of normalization can no longer be written in the form of (18.6), because the wave functions of the continuous spectrum are not quadratically integrable. For these functions the integral $\int |\psi_F|^2 \mathrm{d}V$ diverges. This divergence is associated with the fact that the eigenfunctions of the continuous spectrum do not reduce to zero at infinity. The eigenfunctions of the continuous spectrum are conveniently normalized to the Dirac δ-function (see Vol. 1, Appendix III), since the conditions of orthogonality and normalization can be expressed analogously to (18.7),

$$\int \psi_F^*(x)\,\psi_{F'}(x)\mathrm{d}V = \delta(F - F') . \tag{18.12}$$

The normalization to the δ-function, of course, is not the only possible one. Later we shall encounter other methods of normalizing the eigenfunctions of the continuous spectrum (see, for example, §26).

§19. Expansion in terms of eigenfunctions

In the preceding section we have proved that the system of eigenfunctions of an arbitrary linear self-adjoint operator is a system of orthogonal functions. It turns out that such a system of functions is complete. An arbitrary continuous function, determined in the same region of variation of the independent variables and satisfying a wide class of conditions, can be expanded in this set of eigenfunctions*.

We shall first give here the conditions for the completeness of a system of

* V.I.Smirnov, *A course of higher mathematics* (Pergamon Press, Oxford, 1964).

eigenfunctions for the case of an operator $\hat{F}$ which possesses a discrete spectrum. We write the expansion of a function ψ in a series in terms of the eigenfunctions ψ_n, assuming the latter to be normalized to unity, in the form

$$\psi(x) = \sum_n c_n \psi_n(x) . \tag{19.1}$$

The amplitudes c_n can be determined by making use of the orthogonality of the eigenfunctions. Multiplying (19.1) by $\psi_m^*(x)$ and integrating over all the regions of variation of the independent variables, we obtain

$$\int \psi_m^*(x) \psi(x) \, \mathrm{d}V = \sum_n c_n \int \psi_m^*(x) \psi_n(x) \, \mathrm{d}V .$$

Here we have changed the order of summation and integration. By virtue of the orthogonality of the eigenfunctions (18.7), of all terms of the sum on the right-hand side of the equation only the term with $n = m$ is different from zero. Consequently we have

$$c_m = \int \psi_m^*(x) \psi(x) \, \mathrm{d}V . \tag{19.2}$$

Substituting this expression into (19.1) and again changing the order of summation and integration, we obtain

$$\psi(x) = \int \psi(x') \left(\sum_n \psi_n^*(x') \psi_n(x) \right) \mathrm{d}V' . \tag{19.3}$$

For this expression to hold for an arbitrary continuous function $\psi(x)$ it is necessary that the equality

$$\sum_n \psi_n^*(x') \psi_n(x) = \delta(x - x') \tag{19.4}$$

be fulfilled. The relation (19.4) expresses the condition for the completeness of a system of eigenfunctions $\psi_n(x)$. If the operator $\hat{F}$ possesses a continuous spectrum, then the expansion of the function $\psi(x)$ in terms of its eigenfunctions will be no longer a sum but an integral:

$$\psi(x) = \int c(F) \psi_F(x) \, \mathrm{d}F . \tag{19.5}$$

The amplitudes $c(F)$ are found in the same way as in the case of a discrete spectrum. Multiplying the left-hand and right-hand sides of eq. (19.5) by the function $\psi_{F'}^*(x)$ and integrating over the entire region of variation of the independent variables, we find

$$\int \psi_{F'}^*(x) \psi(x) \, \mathrm{d}V = \int c(F) \, \mathrm{d}F \int \psi_{F'}^*(x) \psi_F(x) \, \mathrm{d}V .$$

Assuming that the eigenfunctions $\psi_F(x)$ are normalized to the δ-function, we obtain finally

$$c(F) = \int \psi_F^*(x)\,\psi(x)\,\mathrm{d}V \ . \tag{19.6}$$

We have already encountered in §3 a particular case of such an expansion (expansion in terms of plane waves). The condition for completeness in the case of a continuous spectrum is written analogously to (19.4)

$$\int \psi_F^*(x')\,\psi_F(x)\,\mathrm{d}F = \delta(x - x') \ . \tag{19.7}$$

§20. Quantum-mechanical variables and operators

We can now turn to the discussion of the basic postulate of quantum mechanics which establishes the connection between real physical quantities which characterize the properties of quantum-mechanical systems and the mathematical apparatus of quantum mechanics.

In classical mechanics the state of a system is determined by the whole set of coordinates and momenta (or variables expressed in terms of the latter) involved in the equations of motion. All variables characterizing the state of a system are called mechanical variables. In quantum mechanics, variables which play an analogous role will be called quantum-mechanical variables. They are also often said to be physical or dynamical variables.

In the examples considered above certain properties of quantum systems have been elucidated. These are, in the first place:

(1) the existence of an uncertainty relation between the values of canonically conjugate physical variables (such as, for example, the coordinate and momentum);

(2) the existence of a discrete spectrum and a continuous spectrum of values of physical variables (for example, the energy of a quantum oscillator and of a free particle);

(3) the existence of a superposition of quantum states (for example, the superposition of states of a free particle);

(4) the continuous transition from the concepts of quantum mechanics to those of classical mechanics in passing to systems in which the Planck constant can be assumed to be an infinitesimal quantity, while quantum numbers can be assumed to be infinitely large (the correspondence principle).

The first and second of these properties just correspond to the properties of linear operators — their non-commutativity and the existence of a spectrum of eigenvalues. Hence it is natural to make the following basic assumption: 'To each quantum-mechanical variable F there corresponds a certain linear

Hermitian operator $\hat{F}$. The spectrum of eigenvalues of the operator $\hat{F}$ represents the spectrum of the possible (measured) values of this variable'.

The eigenfunction $\psi_F(x)$ of the operator $\hat{F}$ represents the wave function of the system in the state in which the variable represented by the operator $\hat{F}$ has the given definite value F.

The requirement of hermiticity of the operator is connected, obviously, with the reality of the values of real physical quantities, whereas the requirement of linearity is associated with the principle of superposition. It is clear that this statement will assume a concrete meaning only after being supplemented by the indication of how the operator corresponding to a given quantum-mechanical quantity can be found. If such a recipe were known, then the postulate formulated would make it possible to determine the spectrum of the possible values of this quantity. The validity of the basic postulate can be established only by the agreement between the inferences of quantum mechanics and experiment.

For the determination of the form of the linear operators which correspond to definite quantum-mechanical variables — quantum-mechanical operators — it is necessary to make use of the correspondence principle. Namely, it is natural to assume that between quantum-mechanical operators describing the motion of particles in quantum mechanics there are the same relations as between their 'originals', the variables of classical mechanics. Thus, for example, the total energy operator $\hat{H}$ is connected with the kinetic energy operator $\hat{T}$ and the potential energy operator $\hat{U}$ by the relation

$$\hat{H} = \hat{T} + \hat{U} . \tag{20.1}$$

In its turn the operator $\hat{T}$ is equal to

$$\hat{T} = \frac{|\hat{\mathbf{p}}|^2}{2m} , \tag{20.2}$$

where $\hat{\mathbf{p}}$ is the momentum operator, and so on.

We have, in essence, already made use of these relations in the preceding chapter in obtaining the Schrödinger equation. If quantum-mechanical operators are connected with each other by the ordinary relations of classical mechanics, then it is sufficient to obtain the expression for one operator in order to construct subsequently the total system of operators of quantum mechanics. The limiting transition to classical mechanics as $\hbar \to 0$ will automatically be ensured, provided the initial operator is correctly chosen, taking into account this condition. Such an approach appears to be quite reasonable, although not strict. In what follows another, more consistent method for the construction of operators will be presented.

One can choose as initial operators the coordinate operator and the momentum operator.

The coordinate operator $\hat{\mathbf{r}}$ amounts to multiplication by this variable, as does every operator corresponding to an independent variable, i.e.

$$\hat{x} = x \; ; \qquad \hat{y} = y \; ; \qquad \hat{z} = z \; . \tag{20.3}$$

To establish the form of the momentum operator $\hat{\mathbf{p}}$, use can be made of the fact that the free particle is described by the Schrödinger equation (6.5),

$$-\frac{\hbar^2}{2m} \nabla^2 \psi = E\psi \; .$$

On the other hand, by virtue of what was said above, this equation can be written as

$$\frac{1}{2m} (\hat{p}_x^2 + \hat{p}_y^2 + \hat{p}_z^2) \, \psi = E\psi \; .$$

Hence it follows that the operators $\hat{p}_x, \hat{p}_y, \hat{p}_z$ can be chosen in the form

$$\hat{p}_x = \frac{\hbar}{i} \frac{\partial}{\partial x} \; ; \qquad \hat{p}_y = \frac{\hbar}{i} \frac{\partial}{\partial y} \; ; \qquad \hat{p}_z = \frac{\hbar}{i} \frac{\partial}{\partial z} \; ; \qquad \hat{\mathbf{p}} = \frac{\hbar}{i} \nabla \; . \tag{20.4}$$

Thus the momentum component operator amounts to differentiation with respect to the corresponding coordinate. The factor i ensures the hermiticity of the operator $\hat{\mathbf{p}}$. Before considering the construction of operators which correspond to quantum-mechanical variables by a more consistent method we shall consider two questions of principle: the question of the meaning of the eigenfunctions of operators, and the question of the possibility of simultaneous measurement of two quantum-mechanical quantities.

§ 21. The wave function and the probability of the results of measurements

Let $\hat{F}$ represent a certain quantum-mechanical operator for which one can write

$$\hat{F}\psi_n = F_n \psi_n \; .$$

For definiteness we assume that the operator $\hat{F}$ has a discrete spectrum of eigenvalues F_n and that to each of these there corresponds an eigenfunction ψ_n (the spectrum is non-degenerate). Since the eigenfunctions ψ_n form a complete system of functions, the wave function ψ can be expanded in a series

$$\psi = \sum_n c_n \psi_n .$$ (21.1)

On the basis of the principle of superposition we can conclude that the state of the system described by the wave function ψ can be written in the form of a superposition of the states with definite values F_n of the physical quantity F.

The amplitude c_m in the expansion (21.1) shows the weight with which the state ψ_m is represented in the state ψ. In other words, the amplitude c_m characterizes the probability that a value equal to F_m will be found when measurements of the quantity F are carried out on the system in the state with wave function ψ. In quantum mechanics it is assumed that this probability is equal to the square of the modulus of the amplitude in the expansion, $|c_m|^2$. Thus, if we want to find the probability of finding the value F_m for the physical quantity F when measurements are carried out on the system in the state ψ, the wave function must be expanded in terms of the eigenfunctions of the operator $\hat{F}$. The square of the modulus of the corresponding amplitude in the expansion, $|c_m|^2$, gives the probability sought. If the quantity F changes continuously (continuous spectrum), then one can speak of the probability that in a measurement one will obtain a value of F lying in the interval between F and $F+dF$. The corresponding probability is given by the expression

$$dW = |c(F)|^2 dF .$$ (21.2)

Thus, in expanding ψ in terms of plane waves (see §3) the square of the modulus of the corresponding amplitude in the expansion gives the probability that in a measurement a certain given value of the momentum will be obtained.

The probabilities of the measurements of given values of the quantity F, which are defined in the way shown above, satisfy the relations

$$\sum_n |c_n|^2 = 1 , \qquad \int |c(F)|^2 dF = 1$$ (21.3)

(under the condition that the wave function ψ is quadratically integrable, while the eigenfunctions of the operator $\hat{F}$ are normalized by the condition (18.7) of (18.12)).

Let us, as an example, prove the last of these relations. Making use of

(19.5) and (19.6), we obtain

$$\int c^*(F)\, c(F)\, dF = \int c^*(F)\, dF \int \psi_F^*(x)\, \psi(x)\, dV =$$

$$= \int \psi(x)\, dV \int c^*(F)\, \psi_F^*(x)\, dF = \int \psi(x)\, \psi^*(x)\, dV = 1 \ . \quad (21.4)$$

The whole set of amplitudes c_n (or $c(F)$ in the case of a continuous spectrum) determines the wave function ψ completely. Hence the definition of the amplitudes in the expansion of the wave function in terms of the eigenfunctions of an arbitrary operator is equivalent to the definition of the wave function itself.

In this connection the following terminology is often used. The wave function $\psi(x)$ is said to be a wave function given in the coordinate representation (x-representation); the whole set of all amplitudes $c(F)$ is called the wave function in the F-representation. In this sense the relations (19.5) and (19.6) must be considered as completely symmetric. The relation (19.5) expresses the expansion of the wave function ψ, taken in the coordinate representation, in terms of the eigenfunctions $\psi_F(x)$ of the operator $\hat{F}$ which is also taken in the x-representation. The amplitudes of the expansion $c(F)$ represent the wave function in the F-representation. On the other hand, the relation (19.6) expresses the expansion of the wave function $c(F)$, taken in the F-representation, in terms of the functions $\psi_F^*(x)$ which have the meaning of the eigenfunctions of the coordinate operator taken in the F-representation (see (48.19)). The amplitudes of the expansion $\psi(x)$ represent the wave function in the x-representation. We shall say also that a certain operator $\hat{D}$ is given in the F-representation if it acts on a function given in the F-representation, for example, $\hat{D}c(F) = b(F)$. From this point of view the statement formulated above that $|c(F)|^2 dF$ is equal to the probability of observing the system in the state with a given value of F becomes almost obvious. Indeed, $|\psi(x)|^2 dx$ is the probability that the coordinate of the particle lies in the interval dx. In view of the equivalence of the x-representation and F-representation, $|c(F)|^2 dF$ is naturally interpreted as the probability that a measurement of F will lead to a value which lies in the interval between F and $F+dF$.

§22. Mean values

Let us assume that the state of a system is described by a wave function $\psi(x)$ which is not an eigenfunction of the operator $\hat{F}$ corresponding to the quantum-mechanical quantity F. As we have already explained above, this

means that in the given state the quantity F has no definite value. In measurements carried out on the system one can obtain, with a certain probability,
any eigenvalue F_n. In this connection it is natural to try to find the mean
value of the quantity F in the given state. We understand the mean, as always,
to be the mathematical expectation (the arithmetic mean) of the given quantity.

Let us consider an ensemble, i.e. a large number of completely identical
samples of a system. Each of these systems is described by one and the same
wave function ψ. We carry out the measurement of the quantity F in each of
the systems. The mean value obtained from the entirety of these measurements will be called the mean value of the quantity F. According to the general formulae of the theory of probability (see ch. 1 of Part III) we can write

$$\bar{F} = \sum_n W_n F_n , \qquad (22.1)$$

where W_n is the probability of obtaining the eigenvalue F_n in measuring the
quantity F. Making use of the expressions for the probabilities W_n which we
have obtained in the preceding section, we have for the case of the discrete
spectrum

$$\bar{F} = \sum_n |c_n|^2 F_n \qquad (22.2)$$

or, if the operator $\hat{F}$ possesses a continuous spectrum,

$$\bar{F} = \int |c(F)|^2 F dF . \qquad (22.3)$$

These formulae can be transformed in such a way that, instead of the
amplitudes of the expansion of the wave function in terms of the eigenfunctions of the operator $\hat{F}$, they contain directly the wave function $\psi(x)$ (i.e.
they can be transformed into the coordinate representation or any other
representation). For concreteness we assume that the operator $\hat{F}$ possesses a
discrete spectrum (in the case of a continuous spectrum the transformation
formulae are derived in an analogous way). Making use of the expression
(19.2) for the amplitudes, we obtain

$$\bar{F} = \sum_n c_n^* c_n F_n = \sum_n c_n F_n \int \psi_n(x) \psi^*(x) dV .$$

Since the eigenfunctions $\psi_n(x)$ satisfy the equation

$$\hat{F} \psi_n(x) = F_n \psi_n(x) ,$$

the last relation can be rewritten in the form

$$\bar{F} = \sum_n c_n \int \psi^* \hat{F} \psi_n \, dV = \int dV \psi^* \hat{F} \sum_n c_n \psi_n \, .$$

Taking into account (21.1), we obtain finally

$$\bar{F} = \int \psi^* \hat{F} \psi \, dV \, . \tag{22.4}$$

We note that this expression must be written in a somewhat more general form, if the wave function ψ is not normalized to unity. In this case

$$\bar{F} = \frac{\int \psi^* \hat{F} \psi \, dV}{\int \psi^* \psi \, dV} \, . \tag{22.5}$$

If the wave function ψ is an eigenfunction of the operator $\hat{F}$

$$\hat{F} \psi = F_m \psi \, ,$$

then the quantity F has a definite value equal to the eigenvalue F_m. In this case, as was to be expected, the mean value of the quantity F is the same as this eigenvalue, $\bar{F} = F_m$.

The relation (22.4) can be the starting point in choosing the operator which corresponds to a given physical quantity. It then follows immediately that in the coordinate representation the coordinate operator amounts to multiplication by this coordinate.

Indeed, proceeding from the physical meaning of the wave function, we can write the expression for the mean value of the coordinate in the form

$$\bar{x} = \int |\psi|^2 x \, dV = \int \psi^* x \psi \, dV \, . \tag{22.6}$$

Comparing this expression with (22.4), we see that the coordinate operator $\hat{x}$ is multiplication by the coordinate x. In an analogous way, if we have an arbitrary function of coordinates $U(x, y, z)$, then its mean value is given by the expression

$$\overline{U(x, y, z)} = \int |\psi|^2 U(x, y, z) \, dV = \int \psi^* U \psi \, dV \, . \tag{22.7}$$

It follows from this that the operator of an arbitrary function of coordinates, taken in the coordinate representation, is multiplication by the function. This, of course, corresponds to the statement we made earlier. In general the operator corresponding to a physical quantity F in its own F-representation is multiplication by the quantity F. This general statement is easily explained in the same way as we have done in the example of the coordinate. The mean value of the quantity F, obtained by means of the function $c(F)$, i.e. by means of the wave function in the F-representation, is given by the formula

$$\bar{F} = \int |c(F)|^2 F \, dF = \int c^*(F) F c(F) \, dF \; .$$

On the other hand, the general expression for the mean in terms of the operator $\hat{F}$, taken in the F-representation, must have the form

$$\bar{F} = \int c^*(F) \hat{F} c(F) \, dF \; .$$

Comparing these expressions, we see that the operator $\hat{F}$ in its own representation is just multiplication by F.

§23. Commutation of operators

One of the most important problems arising in quantum mechanics is that of the possibility of the simultaneous measurement of values of physical quantities of a given quantum-mechanical system.

In order that two quantities F and R may have sharp values in a state described by a wave function $\psi_n(x)$, it is obvious that this wave function must be an eigenfunction of both the operators $\hat{F}$ and $\hat{R}$, i.e. that the following two equations must simultaneously be satisfied:

$$\begin{aligned}
\hat{F}\psi_n(x) &= F\psi_n(x) \; , \\
\hat{R}\psi_n(x) &= R\psi_n(x) \; .
\end{aligned} \tag{23.1}$$

We operate on the first equation with the operator $\hat{R}$, and on the second with the operator $\hat{F}$:

$$\begin{aligned}
\hat{R}\hat{F}\psi_n &= \hat{R}F\psi_n = FR\psi_n \; , \\
\hat{F}\hat{R}\psi_n &= \hat{F}R\psi_n = RF\psi_n \; ,
\end{aligned}$$

The right-hand sides of these equations are equal and, consequently, the left-hand sides are also equal, i.e.

$$\hat{R}\hat{F}\psi_n = \hat{F}\hat{R}\psi_n$$

or

$$(\hat{R}\hat{F} - \hat{F}\hat{R}) \psi_n = 0 \; . \tag{23.2}$$

If the general eigenfunctions ψ_n form a complete system of functions, then an arbitrary wave function ψ can be expanded in this system of functions. Operating on the function with the commutator $\hat{R}\hat{F} - \hat{F}\hat{R}$, we obtain, obviously,

$$(\hat{R}\hat{F} - \hat{F}\hat{R})\psi = \sum c_n (\hat{R}\hat{F} - \hat{F}\hat{R})\psi_n = 0 . \qquad (23.3)$$

The last equation can be written symbolically in the form

$$\hat{R}\hat{F} - \hat{F}\hat{R} = 0 . \qquad (23.4)$$

Thus we have proved that if two quantum-mechanical quantities can have sharp values simultaneously, then the corresponding operators must commute. Of course, if these quantities simultaneously have sharp values only in certain particular states (so that the common eigenfunctions ψ_n do not form a complete system of functions), then the corresponding operators do not commute (see, for example, § 30).

The converse of the theorem can also be proved: if two operators $\hat{F}$ and $\hat{R}$ commute, they will have common eigenfunctions. To prove this we operate on the equation for the eigenfunctions of the operator $\hat{F}$ with the operator $\hat{R}$. We make use of the fact that the operators $\hat{F}$ and $\hat{R}$ commute with each other. Then we obtain

$$\hat{F}(\hat{R}\psi) = F(\hat{R}\psi) .$$

We see that the function $\psi' = \hat{R}\psi$ is also an eigenfunction of the operator $\hat{F}$ corresponding to the eigenvalue F. If there is no degeneracy, then the function ψ' describes the same state as the function ψ, and, consequently, can differ from ψ only by the constant factor R, i.e.

$$\hat{R}\psi = R\psi .$$

Thus we have proved that the function ψ will simultaneously be an eigenfunction of the operators $\hat{F}$ and $\hat{R}$. The proof is easily generalized to the case where there is degeneracy. However, in this case not every eigenfunction of the operator $\hat{F}$ or $\hat{R}$ will simultaneously be an eigenfunction of both operators. Nevertheless, for commuting operators one can always construct a complete system of common eigenfunctions.

Summing up what has been said above, we can say that if to two quantum-mechanical quantities there correspond commuting operators, then these quantities will simultaneously have sharp values; but if the operators do not commute, then these quantities, generally speaking, cannot simultaneously have sharp values, with the exception of certain particular cases (see § 30).

Let us illustrate this by a concrete example. The coordinate operator $\hat{x}$ and the operator corresponding to the momentum component, $\hat{p}_x$, can be chosen as an example of non-commuting operators. The corresponding quantities x and p_x, as we know, do not simultaneously (i.e. in one and the same state) have sharp values. On the contrary, the coordinate operator and a momentum component operator corresponding to an orthogonal coordinate, for example $\hat{x}$ and $\hat{p}_y$, commute with each other. The corresponding quantities x and p_y are simultaneously measurable. The coordinate in a given direction and momentum components in a direction orthogonal to the given direction can simultaneously have sharp values.

We can now define more precisely the concept 'state of a system' in quantum mechanics. A state of a system is given if the wave function describing this system is given. However, in no circumstances can we measure directly the wave function itself. Only the square of its modulus, interpreted as a probability, has a physical meaning. A way out of this apparent contradiction is as follows. When we say that a state of a system is given, then this means that the value of a definite set of quantum-mechanical quantities is given. This set of quantities, the definition of which completely determines the state of the system, is called a complete set of quantum-mechanical quantities. In classical physics, in order to define the state of a system at an arbitrary instant of time, we have to give the values of all the generalized momenta and generalized coordinates at that instant of time. If the classical system has n degrees of freedom, we have to give the values of $2n$ variables. For a microsystem, i.e. a system described by quantum mechanics, it is obvious that the complete set cannot include both the momenta and coordinates of the particles, because these quantities have not simultaneously sharp values. To define the state of a system in quantum mechanics it is sufficient to give only the coordinates of the particle or only its momenta or, in general, any set of independent quantities which are simultaneously measurable and whose number is equal to the number of degrees of freedom of the system. Then the wave function describing the given state of the system will be an eigenfunction of the operators of the quantities which enter into the complete set corresponding to the given eigenvalues.

For example, if the system possesses three degrees of freedom, then the momentum components p_x, p_y, p_z can be chosen as the quantities which form the complete set. The corresponding wave function has the form (2.12).

States characterized by a complete set of quantities defined at a given instant of time are said to be completely specified or 'pure' states. These states are unambiguously described by a corresponding wave function. At a given instant of time this wave function is chosen as the eigenfunction of the operators of all the quantities which enter into the complete set. We note that we shall also obtain a 'pure' state in the case where the wave function corresponding to this state is represented by a certain superposition of eigenfunctions, for example by a superposition of plane waves (3.3). We obtain information on the development of the system in time by solving the Schrödinger equation (6.8) with given initial conditions, thus defining the wave function at subsequent instants of time.

It should be noted that as well as 'pure' states one sometimes has to deal with so-called 'mixed' states (see §89). In these states the wave function of the system is not an eigenfunction. One can speak only of the probability P_n of the realization of one or other 'pure' state φ_n.

If we are interested in the probability of measuring the F_mth value of the quantity F, then in the 'pure' state $\psi(x)$ this probability is determined by the square of the modulus $|c_m|^2$ of the corresponding amplitude in the expansion of the function ψ in terms of the eigenfunctions $\psi_k(x)$ of the operator $\hat{F}$

$$\psi(x) = \sum_k c_k \psi_k(x) \, ,$$

$$W(F_m) = |c_m|^2 = \left| \int \psi(x) \, \psi_m^*(x) \, dV \right|^2 . \tag{23.5}$$

If the system is in a 'mixed' state, then in order to obtain the required probability we have to expand the wave function $\varphi_n(x)$ in terms of the functions $\psi_k(x)$

$$\varphi_n(x) = \sum_k c_{kn} \psi_k(x) \, ,$$

where

$$c_{kn} = \int \varphi_n(x) \, \psi_k^*(x) \, dV . \tag{23.6}$$

The probability of observing the F_mth value of the quantity F in the state φ_n is given by the square of the modulus of the expansion amplitude c_{mn}. In its turn the state $\varphi_n(x)$ is realized with a probability P_n. Thus, finally, according to the theorem of the multiplication of probabilities we obtain

$$W'(F_m) = \sum_n P_n |c_{mn}|^2 . \tag{23.7}$$

In order to compare the results obtained, we write the wave function ψ in the form of a superposition of the functions $\varphi_n(x)$

$$\psi(x) = \sum_n b_n \varphi_n(x) \, .$$

Then, as is easily seen from (23.5) and (23.6),

$$c_m = \sum_n b_n c_{mn} . \tag{23.8}$$

Substituting this value into the expression (23.5), we get

$$W(F_m) = \left| \sum_n b_n c_{mn} \right|^2 = \sum_n |b_n|^2 |c_{mn}|^2 + \frac{1}{2} \sum_{n \neq k} \sum b_n b_k c_{mn} c_{mk} , \qquad (23.9)$$

where

$$\sum_n |b_n|^2 = \sum_n P_{\hat{n}} = 1 .$$

We see that the expression obtained differs from the result given by formula (23.7) by the presence of the double sum which expresses interference between the states. In the case of 'mixed' states there is no such interference.

The above reasoning can be generalized directly to the case of a continuous spectrum.

§24. Heisenberg inequalities

We have in the preceding section found the conditions under which the simultaneous measurement of two physical quantities is possible. We now assume that two physical quantities F and R do not simultaneously have sharp values. Then the operators $\hat{F}$ and $\hat{R}$ corresponding to these quantities do not commute with each other. We assume that the following relation holds:

$$\hat{F}\hat{R} - \hat{R}\hat{F} = i\hat{B} , \qquad (24.1)$$

where $\hat{B}$ is a certain Hermitian operator.

It is of interest to determine in a general form the minimum possible value of the product of fluctuations of given quantities. We choose the mean square deviations (dispersions) $\overline{\Delta F^2}$ and $\overline{\Delta R^2}$, where

$$\Delta F = F - \overline{F} ,$$

$$\Delta R = R - \overline{R} ,$$

as the measure characterizing the deviations of the individual results of measuring the quantities F and R from their mean values. Correspondingly, for the mean square deviations we have

$$\overline{\Delta F^2} = \overline{(F - \overline{F})^2} = \overline{F^2} = \overline{F}^2 .$$

$$\overline{\Delta R^2} = \overline{(R - \overline{R})^2} = \overline{R^2} - \overline{R}^2 . \qquad (24.2)$$

Without affecting the generality of the argument we can set $\overline{F} = 0$ and

$\bar{R} = 0$ (in other words, we can understand F and R to be the deviation of these quantities from their mean value).

Let us consider the integral

$$J(\alpha) = \int |(\alpha\hat{F} - i\hat{R})\psi|^2 dV . \tag{24.3}$$

Here ψ is the wave function, the integration is carried out over the entire region of variation of the independent variables, and α is an arbitrary real parameter. The integral (24.3) is not negative: $J(\alpha) \geqslant 0$. We rewrite it in the form

$$J(\alpha) = \int (\alpha\hat{F} - i\hat{R})\psi \cdot (\alpha\hat{F}^* + i\hat{R}^*)\psi^* dV .$$

Making use of the self-conjugate property of the operators $\hat{F}$ and $\hat{R}$, we obtain

$$J(\alpha) = \int \psi^*(\alpha\hat{F} + i\hat{R})(\alpha\hat{F} - i\hat{R})\psi \, dV =$$
$$= \int \psi^*(\alpha^2\hat{F}^2 - \alpha i(\hat{F}\hat{R} - \hat{R}\hat{F}) + \hat{R}^2)\psi \, dV .$$

Taking into account (24.1) and using expression (22.4) for the mean value, we have

$$J(\alpha) = \alpha^2\overline{F^2} + \alpha\bar{B} + \overline{R^2} = \alpha^2\overline{\Delta F^2} + \alpha\bar{B} + \overline{\Delta R^2} .$$

The condition for this trinomial quadratic in α to be negative can be written in the form

$$4\,\overline{\Delta F^2}\,\overline{\Delta R^2} \geqslant \bar{B}^2 \tag{24.4}$$

or

$$(\overline{\Delta F^2 \Delta R^2})^{\frac{1}{2}} \geqslant \tfrac{1}{2}|\bar{B}| . \tag{24.5}$$

Formula (24.5) gives the relation which we have sought between the uncertainties ΔF and ΔR. It establishes the minimum possible value of the product of these errors.

Let us consider a particular case, taking as the quantities F and R respectively p_x and x. Then it follows from (20.3) and (20.4) that $\hat{B} = -\hbar$, and we have

$$(\overline{\Delta p_x^2})^{\frac{1}{2}} (\overline{\Delta x^2})^{\frac{1}{2}} \geqslant \tfrac{1}{2}\hbar . \tag{24.6}$$

Thus the uncertainty relation (24.5) is of a general character. The uncertainty relation for the coordinate and momentum is a particular case of the relation (24.5).

Conjugate quantum-mechanical quantities cannot be measured simultaneously. The minimum uncertainties in their values in a simultaneous measurement are connected with the quantity $\bar{B}$. On the contrary, mutually commuting quantum-mechanical quantities, for which $\hat{B} = 0$, can be measured simultaneously with an arbitrary degree of accuracy.

§25. Poisson brackets

In §20 we have considered one of the possible methods of finding the operators describing physical quantities. This problem was considered in a more consistent way by Dirac. He assumed that in quantum mechanics, as in classical mechanics, the concept of Poisson brackets can be introduced*. Thus, if to two classical quantities F, R there corresponds the Poisson bracket

$$[F, R] = -\sum_i \left(\frac{\partial F}{\partial p_i} \frac{\partial R}{\partial q_i} - \frac{\partial F}{\partial q_i} \frac{\partial R}{\partial p_i} \right) ,$$

then to the operators $\hat{F}$, $\hat{R}$ describing these quantities there corresponds the quantum Poisson bracket $[\hat{F}, \hat{R}]$. Further it was assumed that the properties of the quantum Poisson brackets correspond exactly to those of the classical Poisson brackets except that for the quantum brackets the consecutive order of the two factors is important. We write down the properties of the Poisson brackets:

$$[\hat{F}, \hat{R}] = - [\hat{R}, \hat{F}] , \tag{25.1}$$

$$[\hat{F}, C] = 0 , \tag{25.2}$$

where C is a constant.

$$[\hat{F}_1 + \hat{F}_2, \hat{R}] = [\hat{F}_1, \hat{R}] + [\hat{F}_2, \hat{R}] , \tag{25.3}$$

$$[\hat{F}, \hat{R}_1 + \hat{R}_2] = [\hat{F}, \hat{R}_1] + [\hat{F}, \hat{R}_2] , \tag{25.4}$$

$$[\hat{F}_1\hat{F}_2, \hat{R}] = [\hat{F}_1, \hat{R}]\hat{F}_2 + \hat{F}_1[\hat{F}_2, \hat{R}] , \tag{25.5}$$

$$[\hat{F}, \hat{R}_1\hat{R}_2] = [\hat{F}, \hat{R}_1]\hat{R}_2 + \hat{R}_1[\hat{F}, \hat{R}_2] , \tag{25.6}$$

$$[\hat{F}_1, [\hat{F}_2, \hat{F}_3]] + [\hat{F}_3, [\hat{F}_1, \hat{F}_2]] + [\hat{F}_2, [\hat{F}_3, \hat{F}_1]] = 0 . \tag{25.7}$$

The choice of Poisson brackets as the basis for the construction of a

---

* For a discussion of Poisson brackets in classical mechanics see L.D.Landau and E.M.Lifshitz, *Mechanics* (Pergamon Press, Oxford, 1960); H.Goldstein, *Classical mechanics* (Addison-Wesley, Cambridge, Mass., 1950).

system of quantum-mechanical operators is associated with the fact that, as we shall see, they can be expressed directly in terms of the commutators of the corresponding operators. The last combination of operators is the basis for their physical interpretation.

Let us consider the Poisson bracket $[\hat{F}_1\hat{F}_2, \hat{R}_1\hat{R}_2]$, for the calculation of which we can use expressions (25.5) and (25.6). Correspondingly we obtain

$$[\hat{F}_1\hat{F}_2, \hat{R}_1\hat{R}_2] = [\hat{F}_1, \hat{R}_1\hat{R}_2]\hat{F}_2 + \hat{F}_1[\hat{F}_2, \hat{R}_1\hat{R}_2] =$$
$$= [\hat{F}_1, \hat{R}_1]\hat{R}_2\hat{F}_2 + \hat{R}_1[\hat{F}_1, \hat{R}_2]\hat{F}_2 + \hat{F}_1[\hat{F}_2, \hat{R}_1]\hat{R}_2 + \hat{F}_1\hat{R}_1[\hat{F}_2, \hat{R}_2]$$

and

$$[\hat{F}_1\hat{F}_2, \hat{R}_1\hat{R}_2] = [\hat{F}_1\hat{F}_2, \hat{R}_1]\hat{R}_2 + \hat{R}_1[\hat{F}_1\hat{F}_2, \hat{R}_2] =$$
$$= \hat{F}_1[\hat{F}_2, \hat{R}_1]\hat{R}_2 + [\hat{F}_1, \hat{R}_1]\hat{F}_2\hat{R}_2 + \hat{R}_1\hat{F}_1[\hat{F}_2, \hat{R}_2] + \hat{R}_1[\hat{F}_1, \hat{R}_2]\hat{F}_2 .$$

Equating these two results, we find

$$(\hat{F}_1\hat{R}_1 - \hat{R}_1\hat{F}_1)\,[\hat{F}_2, \hat{R}_2] = [\hat{F}_1, \hat{R}_1]\,(\hat{F}_2\hat{R}_2 - \hat{R}_2\hat{F}_2) .$$

Since the last equality must be satisfied identically, we have

$$[\hat{F}_1, \hat{R}_1] = iC(\hat{F}_1\hat{R}_1 - \hat{R}_1\hat{F}_1) ,$$
$$[\hat{F}_2, \hat{R}_2] = iC(\hat{F}_2\hat{R}_2 - \hat{R}_2\hat{F}_2) ,$$

where C is a real constant.

The reality of C follows from the fact that the Poisson bracket of two real variables must also be real. Thus, if $\hat{F}^\dagger = \hat{F}$, $\hat{R}^\dagger = \hat{R}$, then also $[\hat{F}, \hat{R}]^\dagger = [\hat{F}, \hat{R}]$. However,

$$[\hat{F}, \hat{R}]^\dagger = -iC^*(\hat{F}\hat{R} - \hat{R}\hat{F})^\dagger = -iC^*(\hat{R}\hat{F} - \hat{F}\hat{R}) = C^*C^{-1}[\hat{F}, \hat{R}]$$

and hence $C^* = C$. It follows from the classical theory of Poisson brackets that the constant C has the dimensionality $\text{erg}^{-1} \cdot \text{sec}^{-1}$. Its numerical value can be determined only by comparing the inferences of the theory with experimental data. It turns out to be equal to $\hbar^{-1}$. Finally we have

$$(\hat{F}\hat{R} - \hat{R}\hat{F}) = \frac{\hbar}{i}\,[\hat{F}, \hat{R}] . \tag{25.8}$$

In the transition to classical mechanics (see ch. 5), i.e. for $\hbar \to 0$, the commutator $\{\hat{F}\hat{R}\}$ reduces to zero, as was to be expected. It is natural to assume if only in the simplest cases, that the Poisson brackets themselves have the same values as classical brackets. For canonically conjugated variable coordinates and momenta in classical mechanics we have:

$$[p_i, p_k] = 0 ,$$

$$[x_i, x_k] = 0 , \tag{25.9}$$

$$[p_i, x_k] = \delta_{ik} \qquad (i, k = 1, 2, 3) .$$

Here and in what follows we make use of the notation

$$x_1 = x , \qquad p_1 = p_x ,$$

$$x_2 = y , \qquad p_2 = p_y ,$$

$$x_3 = z , \qquad p_3 = p_z .$$

The same expressions can be written for the quantum operators of the coordinate and of the momentum components. Hence the commutators of the corresponding values assume the form

$$\hat{x}_i \hat{x}_k - \hat{x}_k \hat{x}_i = 0 ,$$

$$\hat{p}_i \hat{p}_k - \hat{p}_k \hat{p}_i = 0 , \tag{25.10}$$

$$\hat{p}_i \hat{x}_k - \hat{x}_k \hat{p}_i = \frac{\hbar}{i} \delta_{ik} \qquad (i, k = 1, 2, 3) .$$

We shall make use of these equalities for the determination of the coordinate operator and the momentum operator.

§26. Coordinate and momentum operators and eigenfunctions

We begin with the establishment of the form of operators in the coordinate representation*. In this representation the wave function characterizing the state of the particle depends on its coordinates $\psi(x, y, z)$. The coordinates x, y, z are independent variables. Hence the corresponding operators, in correspondence with the conclusions of §20 and §22, reduce to multiplication by these coordinates

$$\hat{x}_i = x_i \qquad (i = 1, 2, 3) . \tag{26.1}$$

* See V.A.Fok, *Nachala kvantovoi mekhaniki* (*Principles of quantum mechanics*) (KUBUCH, 1932) p. 32.

We rewrite the commutation relations (25.10) in the form

$$(\hat{p}_k x_k - x_k \hat{p}_k)\psi = \frac{\hbar}{i}\psi \qquad (k = 1, 2, 3), \qquad (26.2)$$

$$(\hat{p}_i x_k - x_k \hat{p}_i)\psi = 0 \qquad (i \neq k), \qquad (26.3)$$

$$(\hat{p}_i \hat{p}_k - \hat{p}_k \hat{p}_i)\psi = 0 \qquad (i, k = 1, 2, 3). \qquad (26.4)$$

Equations (26.2)–(26.4) can be satisfied by an arbitrary function ψ, setting

$$\hat{p}_k = \frac{\hbar}{i}\frac{\partial}{\partial x_k} + \frac{\partial \alpha(x_1, x_2, x_3)}{\partial x_k}, \qquad (26.5)$$

where $\alpha(x_1, x_2, x_3)$ is an arbitrary real function. The reality of α is required for the hermiticity of the operator $\hat{p}_k$. The function α can, without loss of generality of the result, be set equal to zero. Indeed, the action of the operator (26.5) on an arbitrary function ψ transforms it into the function

$$\psi' = \frac{\hbar}{i}\frac{\partial \psi}{\partial x_k} + \frac{\partial \alpha}{\partial x_k}\psi .$$

On the other hand, if we act on the function $e^{(i/\hbar)\alpha}\psi$ with the operator $(\hbar/i)(\partial/\partial x_k)$, we then obtain the function $e^{(i/\hbar)\alpha}\psi'$. Consequently, the transition from the operator (26.5) to the operator $(\hbar/i)(\partial/\partial x_k)$ is equivalent to the transition from the wave function ψ to the function $e^{(i/\hbar)\alpha}\psi$

$$\psi \to e^{(i/\hbar)\alpha}\psi ,$$

$$\frac{\hbar}{i}\frac{\partial}{\partial x_k} + \frac{\partial \alpha}{\partial x_k} \to \frac{\hbar}{i}\frac{\partial}{\partial x_k} = e^{(i/\hbar)\alpha}\left(\frac{\hbar}{i}\frac{\partial}{\partial x_k} + \frac{\partial \alpha}{\partial x_k}\right)e^{-(i/\hbar)\alpha}, \qquad (26.6)$$

since

$$e^{(i/\hbar)\alpha}\left(\frac{\hbar}{i}\frac{\partial}{\partial x_k} + \frac{\partial \alpha}{\partial x_k}\right)e^{-(i/\hbar)\alpha}e^{(i/\hbar)\alpha}\psi = e^{(i/\hbar)\alpha}\psi' = \frac{\hbar}{i}\frac{\partial}{\partial x_k}(e^{(i/\hbar)\alpha}\psi) .$$

Wave functions, and operators, connected with each other by the transformations (26.6) have identical physical properties. This will be proved in its most general form in §46.

The operators (26.5) and $\hat{p}_k = (\hbar/i)(\partial/\partial x_k)$ $(k = x, y, z)$ have the same spectrum of eigenvalues. Hence without loss of generality we can use, instead of the operators (26.5), the operators for the momentum components which have, in the coordinate representation, the form (see §20)

$$\hat{p}_x = \frac{\hbar}{i}\frac{\partial}{\partial x}; \qquad \hat{p}_y = \frac{\hbar}{i}\frac{\partial}{\partial y}; \qquad \hat{p}_z = \frac{\hbar}{i}\frac{\partial}{\partial z} \qquad (26.7)$$

or in vector form

$$\hat{\mathbf{p}} = \frac{\hbar}{i}\mathbf{V} , \tag{26.8}$$

where $\mathbf{V}$ is the gradient operator.

Let us now use the momentum representation, in which the wave function depends on three momentum components: p_x, p_y, p_z. The corresponding operators reduce to multiplication by the quantities p_x, p_y, p_z. The coordinate operators in this representation are found on the basis of the same commutation relations, and turn out to be equal to

$$\hat{x} = i\hbar \frac{\partial}{\partial p_x} ; \qquad \hat{y} = i\hbar \frac{\partial}{\partial p_y} ; \qquad \hat{z} = i\hbar \frac{\partial}{\partial p_z} , \tag{26.9}$$

or

$$\hat{\mathbf{r}} = i\hbar \left(\mathbf{i}\frac{\partial}{\partial p_x} + \mathbf{j}\frac{\partial}{\partial p_y} + \mathbf{k}\frac{\partial}{\partial p_z} \right) \equiv i\hbar \frac{\partial}{\partial \mathbf{p}} .$$

Making use of (26.7) it is easy to establish the commutation relations of the operator $\hat{\mathbf{p}}$ and an arbitrary function $U(x, y, z)$

$$\hat{\mathbf{p}}U - U\hat{\mathbf{p}} = \frac{\hbar}{i}\mathbf{V}U . \tag{26.10}$$

The commutation of the operator $\hat{\mathbf{r}}$ with an arbitrary function $f(p_x, p_y, p_z)$ is calculated in an analogous way

$$\hat{\mathbf{r}}f - f\hat{\mathbf{r}} = i\hbar \frac{\partial f}{\partial \mathbf{p}} . \tag{26.11}$$

The equations for the eigenfunctions and eigenvalues of the operators $\hat{p}_x, \hat{p}_y, \hat{p}_z$ are of the form

$$\frac{\hbar}{i}\frac{\partial \psi_{p_x}}{\partial x} = p_x \psi_{p_x} , \qquad \frac{\hbar}{i}\frac{\partial \psi_{p_y}}{\partial y} = p_y \psi_{p_y} , \qquad \frac{\hbar}{i}\frac{\partial \psi_{p_z}}{\partial z} = p_z \psi_{p_z} . \tag{26.12}$$

We write down the solution of the first equation

$$\psi_{p_x} = a(y, z)\,e^{(i/\hbar)p_x x} ,$$

where $a(y, z)$ is an arbitrary function. Analogous solutions are also obtained for the functions ψ_{p_y} and ψ_{p_z}. The functions $\psi_{p_x}, \psi_{p_y}, \psi_{p_z}$ satisfy the necessary requirements, in particular the condition of finiteness (see §16) for real values of p_x, p_y, p_z Thus the momentum operator has a continuous spectrum of eigenvalues. The wave function

$$\psi_{\mathbf{p}} = A e^{(i/\hbar)(\mathbf{p}\cdot\mathbf{r})} ,\tag{26.13}$$

where A is a constant, is an eigenfunction of the operators $\hat{p}_x, \hat{p}_y, \hat{p}_z$ and describes the state with given momentum $\mathbf{p}$. A freely moving particle can be in such a state. This conclusion is in complete agreement with the result of §2.

The constant A is defined by the normalization condition. Since the momentum operator has a continuous spectrum, its eigenfunctions are conveniently normalized to the δ-function. Let us first find the normalization coefficient A in the case of one-dimensional motion.

Setting $\int \psi_{p_x}^* \psi_{p_x'} \, dx = \delta(p_x - p_x')$ and taking into account (III.5), we obtain $A = (2\pi\hbar)^{-\frac{1}{2}}$, so that finally

$$\psi_{p_x} = (2\pi\hbar)^{-\frac{1}{2}} e^{(i/\hbar)p_x x} .\tag{26.14}$$

In the three-dimensional case, for the wave function (26.13) we have correspondingly

$$\psi_{\mathbf{p}} = (2\pi\hbar)^{-\frac{3}{2}} e^{(i/\hbar)(\mathbf{p}\cdot\mathbf{r})} .\tag{26.15}$$

Another method of normalizing plane waves, called normalization in a 'box', sometimes turns out to be more convenient. We define the wave function in an arbitrarily large but finite volume V. As the normalization volume we choose a cube with edge L and centre at the origin. We require that at the walls of the cube the wave functions (26.12) satisfy the condition of periodicity, i.e. that at corresponding points of opposite faces the wave functions take on the same values. Under these conditions the momentum vector no longer changes continuously, but runs over a discrete set of values

$$p_x = \frac{2\pi\hbar}{L} n_x ; \qquad p_y = \frac{2\pi\hbar}{L} n_y ; \qquad p_z = \frac{2\pi\hbar}{L} n_z ,\tag{26.16}$$

where n_x, n_y, n_z are positive or negative integers, including zero. Choosing the edge of the cube, L, to be sufficiently large, the spacing between neighbouring eigenvalues of the momentum vector can be made as small as one pleases. The normalization coefficient defined by the condition

$$|A|^2 \int_{L^3} |e^{(i/\hbar)\mathbf{p}\cdot\mathbf{r}}|^2 dV = 1$$

is equal to $A = L^{-\frac{3}{2}}$. Correspondingly the wave function for such a normalization has the form

$$\psi_{\mathbf{p}} = L^{-\frac{3}{2}} e^{(i/\hbar)(\mathbf{p} \cdot \mathbf{r})} = V^{-\frac{1}{2}} e^{(i/\hbar)(\mathbf{p} \cdot \mathbf{r})} . \tag{26.17}$$

In §12 and 13 we normalized wave functions of the form (26.13), defining the probability current density $\mathbf{j}_0$. As a matter of fact, in this state, according to (7.6),

$$\mathbf{j}_0 = |A|^2 \frac{\mathbf{p}}{m} . \tag{26.18}$$

Setting, for example, $A = 1$, we obtain

$$\mathbf{j}_0 = \frac{\mathbf{p}}{m} = v , \tag{26.19}$$

i.e. for such a normalization the probability current density is numerically equal to the velocity of the particle. But if $A = v^{-\frac{1}{2}}$, then this corresponds to a normalization to unit probability current density.

It is easily seen that the operators $\hat{p}_x, \hat{p}_y, \hat{p}_z$ are related in a simple way to the operators of an infinitesimal translation along the x-, y- and z-axes respectively. Indeed, let us shift our system or, what is equivalent, the origin, a distance Δx along the x-axis. Then the old and new coordinates are connected by the relation

$$x' = x - \Delta x ,$$

$$y' = y ,$$

$$z' = z .$$

We express the function $\psi(x, y, z)$ in terms of the new coordinates x', y', z'. Confining ourselves to the first term of a series expansion we obtain

$$\psi(x, y, z) = \psi(x' + \Delta x, y', z') = \psi(x', y', z') + \frac{\partial \psi}{\partial x'} \Delta x =$$

$$= \left(1 + \Delta x \frac{\partial}{\partial x'}\right) \psi(x', y', z') .$$

It is natural to call the operator $1 + \Delta x (\partial/\partial x')$ the operator of displacement by a distance Δx along the x-axis. We denote this operator by $\hat{R}_{x'}$, so that

$$\psi(x, y, z) = \hat{R}_{x'} \psi(x', y', z') . \tag{26.20}$$

We see that the displacement operator $\hat{R}_x$ is connected with the operator of the corresponding momentum component $\hat{p}_x$

$$\hat{R}_x = 1 + \frac{i}{\hbar} \Delta x \hat{p}_x .$$ (26.21)

The form of the momentum operator $\hat{p}_x$ could also be obtained proceeding from the expression for the operator $\hat{R}_x$. *

We write down the equation for the eigenfunctions and eigenvalues of the coordinate operator in the coordinate representation

$$\hat{x} \psi_{x_0}(x) = x_0 \psi_{x_0}(x) .$$ (26.22)

Here x_0 is a particular value of the coordinate x. The operator $\hat{x}$ in its own representation reduces to multiplication by x. From eq. (26.22) it follows that

$$\psi_{x_0}(x) = 0 \quad \text{for} \quad x \neq x_0 .$$

Furthermore, the functions $\psi_{x_0}(x)$ must satisfy the orthogonality and normalization conditions

$$\int \psi_{x_0}^*(x) \, \psi_{x_0'}(x) \, dx = \delta(x_0 - x_0') .$$

From these relations it follows that the function $\psi_{x_0}(x)$ is of the form (see Appendix III)

$$\psi_{x_0}(x) = \delta(x - x_0) .$$ (26.23)

The eigenfunctions of the operators $\hat{y}$ and $\hat{z}$ are written in an analogous way. Since the coordinate operators $\hat{x}, \hat{y}, \hat{z}$ commute with each other, their values are simultaneously measurable. Correspondingly, if the system has three degrees of freedom, the three coordinate projections x, y, z can be chosen as quantities forming a complete set. The wave function describing the state with three defined coordinates x_0, y_0, z_0 is of the form

$$\psi_{\mathbf{r}_0}(\mathbf{r}) = \delta(x - x_0) \delta(y - y_0) \delta(z - z_0) = \delta(\mathbf{r} - \mathbf{r}_0) .$$ (26.24)

The eigenfunction of the momentum operator in the momentum representation is written in an analogous way.

---

* See L.D.Landau and E.M.Lifshitz, *Quantum mechanics* (Pergamon Press, Oxford, 1965). An analogous result also applies to the operators of the y- and z-momentum components.

§27. The Hamiltonian operator

The most important operator of quantum mechanics is the total energy operator $\hat{H}$. As in classical mechanics, it is made up of the kinetic energy operator and the potential energy operator. We construct, first of all, the operator of the kinetic energy of a particle. In the non-relativistic approximation, in which we are now interested, the kinetic energy is connected with the momentum of the particle by the usual relation

$$T = \frac{\mathbf{p}^2}{2m} = \frac{p_x^2 + p_y^2 + p_z^2}{2m} .$$

(27.1)

Replacing the momentum $\mathbf{p}$ of the particle in this relation by the operator $\hat{\mathbf{p}}$, we obtain the operator $\hat{T}$ which we shall call the kinetic energy operator (see also §20):

$$\hat{T} = \frac{1}{2m}\hat{\mathbf{p}}^2 = \frac{1}{2m}(\hat{p}_x^2 + \hat{p}_y^2 + \hat{p}_z^2) \equiv -\frac{\hbar^2}{2m}\nabla^2 .$$

(27.2)

It is obvious that the kinetic energy operator commutes with the momentum operator.

We now consider the total energy operator $\hat{H}$. Since the potential energy depends only on the coordinates x, y, z, the corresponding operator in the coordinate representation is simply the function $U(x, y, z)$. Consequently we have

$$\hat{H} = -\frac{\hbar^2}{2m}\nabla^2 + U(x, y, z) .$$

(27.3)

Since the total energy operator in formula (27.3) is expressed in terms of the momentum operator (but not the velocity operator), it represents the quantum-mechanical Hamiltonian operator which is usually just called the Hamiltonian. The expression for the Hamiltonian can easily be generalized to the case where the particle moves in non-stationary external fields. Then

$$\hat{H} = -\frac{\hbar^2}{2m}\nabla^2 + U(\mathbf{r}, t) ,$$

(27.4)

where $U(\mathbf{r}, t)$ is the so-called force function, connected with the force which acts on the particle by the relation

$$\mathbf{f} = -\overline{\nabla}U .$$

The formulae found for the Hamiltonian operator are inapplicable in the case of the motion of a particle in a field of force which depends on its velocity. An example of such a case is the motion of a charged particle in a magnetic field.

To obtain the Hamiltonian operator in this case we make use of general rules. We write down the Hamiltonian function of classical mechanics for particles moving in an electromagnetic field. According to (41.4) of Part I, we have

$$H = \frac{1}{2m} \left(\mathbf{p} - \frac{e}{c} \mathbf{A} \right)^2 + e\varphi \, , \tag{27.5}$$

where the vector $\mathbf{p}$ is the generalized momentum of the particle, $\mathbf{A}$ and φ are the vector and scalar potentials, and e is the charge of the particle. According to the general rule, we replace the Hamiltonian function in formula (27.5) by the Hamiltonian operator, and the generalized momentum by the momentum operator. The vector potential and scalar potential, which depend only on the coordinates and time, can be left unchanged, since in the coordinate representation the application of the corresponding operators amounts to multiplication by these functions. We then find

$$\hat{H} = \frac{1}{2m} \left(\hat{\mathbf{p}} - \frac{e}{c} \mathbf{A} \right)^2 + e\varphi \, . \tag{27.6}$$

By means of the Hamiltonian operator found above the basic equation of quantum mechanics, the Schrödinger equation, can be written in the form

$$i\hbar \frac{\partial \psi}{\partial t} = \hat{H}\psi \, . \tag{27.7}$$

The operator form of notation of the Schrödinger equation has a very general character and is suitable for the description of the motion of a particle in arbitrary stationary or non-stationary fields. In particular, in such a form it is valid for the case of the motion of a particle in an electromagnetic field. Just as for the classical Hamiltonian function, the Hamiltonian can be transformed to an arbitrary curvilinear system of coordinates. For this one need only transform the differential Laplacian operator ∇^2 to this system. Depending on the symmetry of the force field it is convenient to choose a system of curvilinear coordinates in which the expression for the potential energy of the particle assumes the simplest form. In particular, as we shall see in §35, it is often convenient to write the Hamiltonian operator in spherical coordinates.

The Hamiltonian operator of a system of particles can be constructed according to the same scheme which has already been successfully applied to the case of one particle. Namely, one has to write the classical expression for the Hamiltonian function and then replace all quantities involved in it by their quantum-mechanical operators.

The classical expression for the Hamiltonian of a system of N particles has the form

$$H = \sum_{k=1}^{N} \frac{\mathbf{p}_k^2}{2m_k} + \sum_{k=1}^{N} U_k(\mathbf{r}_k) + U_{\text{int}} , \qquad (27.8)$$

where $\mathbf{p}_k$, m_k and $U_k(\mathbf{r}_k)$ are respectively the momentum, mass and potential energy of the kth particle in the external field; U_{int} is the potential energy of interaction of the particles.

We obtain the Hamiltonian operator if we replace the momenta of the particles by the corresponding operators $\hat{\mathbf{p}}_k$, where the index k denotes differentiation with respect to the coordinates of the kth particle. After this replacement we obtain the Schrödinger equation. It has the form

$$i\hbar \frac{\partial \psi}{\partial t} = \hat{H} \psi , \qquad \hat{H} = \sum_{k=1}^{N} - \frac{\hbar^2}{2m_k} \nabla_k^2 + \sum_{k=1}^{N} U_k(\mathbf{r}_k) + U_{\text{int}} . \quad (27.9)$$

The expression for the Hamiltonian operator for a system of charged particles in an external electromagnetic field is generalized directly from (27.6).

§28. Stationary states

We assume that the Hamiltonian of the system does not depend explicitly on the time. In this case it is possible to separate the variables in the Schrödinger equation (27.7). We have already made use of this in §6. However, we can now analyze more profoundly the solution obtained.

We seek the solution of the Schrödinger equation (27.7) in the form

$$\psi(x, t) = \chi(t) \psi(x) , \qquad (28.1)$$

where we understand x to be the entire set of coordinates on which the wave function depends.

Substituting this expression into (27.7), we obtain

$$i\hbar \frac{d\chi(t)}{dt} \psi(x) = \chi(t) \hat{H} \psi(x) .$$

Separating variables in this equation gives:

$$i\hbar \frac{d\chi}{dt} \frac{1}{\chi} = \frac{\hat{H}\psi(x)}{\psi(x)}.$$

The expression on the left-hand side of the equation can depend only on the time t, while the expression on the right-hand side can depend only on the coordinates of the system. It follows from the equality of these expressions that each of them is equal to one and the same constant which we shall denote by E. We then obtain

$$\chi(t) = Ce^{-(i/\hbar)Et}, \qquad \hat{H}\psi(x) = E\psi(x),$$

where C is an arbitrary constant.

We see that the constant E has the meaning of an eigenvalue of the operator $\hat{H}$, i.e. determines the possible values of the energy of the system, and the function $\psi(x)$ describes a state with given energy.

The Hamiltonian operator can possess a discrete as well as a continuous spectrum, as we have seen in the examples already discussed. One also often encounters a mixed spectrum, i.e. a discrete spectrum in one energy interval, and a continuous spectrum in another.

Assuming for definiteness that the operator $\hat{H}$ possesses a discrete spectrum, we write the wave functions (28.1)

$$\psi_n(x, t) = \psi_n(x)e^{-(i/\hbar)E_n t}. \tag{28.2}$$

The states of a system described by wave functions of the type (28.2) are said to be stationary. The wave functions of stationary states depend harmonically on the time with frequencies $\omega_n = E_n/\hbar$. As we have already noted in §6, in a stationary state the density of the probability of finding a particle at a given point of space does not depend on the time. Indeed,

$$W_n(x, t) = |\psi_n(x, t)|^2.$$

Substituting the expression (28.2) for the wave function, we find

$$W_n(x, t) = |\psi_n(x)|^2 = W_n(x, 0). \tag{28.3}$$

This statement can easily be generalized. The probability of observing the eigenvalue F_k in the stationary state $\psi_n(x, t)$ does not depend on the time. According to our general rules (see §21), in order to obtain the probability to be determined we must expand the wave function $\psi_n(x, t)$ in terms of the eigenfunctions ψ_{F_k} of the operator $\hat{F}$ and take the square of the modulus of the corresponding expansion amplitude c_k. According to formula (19.2)

$$c_k(t) = \int \psi_n(x, t)\, \psi_{F_k}^*(x)\, dV = e^{-(i/\hbar)E_n t} \int \psi_n(x)\, \psi_{F_k}^*(x)\, dV \ .$$

The corresponding probability $W(F_k, t)$ is equal to

$$W(F_k, t) = |c_k(t)|^2 = \left| \int \psi_n(x)\, \psi_{F_k}^*(x)\, dV \right|^2 = W(F_k, 0) \ . \qquad (28.4)$$

An arbitrary solution of the Schrödinger equation $\psi(x, t)$ can be expanded in terms of the wave functions (28.2). This function $\psi(x, t)$ describes a state in which the energy of the system has no sharp value.

§29. Integral form of the Schrödinger equation

The Schrödinger differential equation can also be represented by an integral equation. In a number of cases this latter formulation has a number of advantages both from the theoretical standpoint and from the point of view of purely mathematical convenience. The theoretical advantage of the integral representation of the equations of quantum mechanics is closely associated with the development of Feynman's ideas in quantum field theory (see ch. 14).

In §58 we shall dwell in detail on the advantages of approximate methods of solving the Schrödinger equation in integral form.

Let us consider a particle with Hamiltonian $\hat{H}$ depending, in general, on the time. At the initial instant of time let the wave function

$$\psi_0 = \psi(\mathbf{r}_1, t_1) \qquad (29.1)$$

be defined. The wave function satisfies the Schrödinger equation

$$i\hbar \frac{\partial \psi}{\partial t} = \hat{H}\psi \ . \qquad (29.2)$$

The wave function of the particle satisfying eq. (29.2) for the boundary condition (29.1) at the instant of time $t_2 > t_1$ can be written in the form

$$\psi(\mathbf{r}_2, t_2) = \int K(\mathbf{r}_2, t_2; \mathbf{r}_1, t_1)\, \psi(\mathbf{r}_1, t_1)\, d\mathbf{r}_1 \ . \qquad (29.3)$$

The function $K(\mathbf{r}_2, t_2; \mathbf{r}_1, t_1)$ is the Green's function of eq. (29.2) (see §19). The interpretation of formula (29.3) is obvious: the Green's function represents the amplitude for the transition of the particle from the initial state with wave function $\psi(\mathbf{r}_1, t_1)$ into the state with wave function $\psi(\mathbf{r}_2, t_2)$, where $t_2 > t_1$.

Since formula (29.3) defines the wave function only for $t_2 > t_1$, the Green's function can be predetermined by the requirement

$$K(\mathbf{r}_2, t_2; \mathbf{r}_1, t_1) = 0 \quad \text{for} \quad t_2 < t_1 \,.$$

For the relation (29.3) to be equivalent to (29.2) and (29.1), the Green's function must satisfy the equation

$$\left[i\hbar \frac{\partial}{\partial t_2} - \hat{H}(\mathbf{r}_2, t_2) \right] K(\mathbf{r}_2, t_2; \mathbf{r}_1, t_1) = i\hbar \delta(\mathbf{r}_2 - \mathbf{r}_1) \delta(t_2 - t_1) \,. \tag{29.4}$$

Indeed, for $t_2 > t_1$ the Green's function satisfies the equation

$$\left[i\hbar \frac{\partial}{\partial t_2} - \hat{H}(\mathbf{r}_2, t_2) \right] K(\mathbf{r}_2, t_2; \mathbf{r}_1, t_1) = 0 \,. \tag{29.5}$$

Operating on both sides of (29.3) with the operator $i\hbar(\partial/\partial t_2) - \hat{H}$ for $t_2 > t_1$, and taking into account eq. (29.5), we arrive at an identity. It is easily seen that the Green's function defined by eq. (29.4) satisfies the initial condition (29.1) for $t_2 = t_1$, if (29.4) is integrated over the infinitesimal interval $2\Delta t \to 0$ about the instant t_1. We then have

$$\int_{t_1 - \Delta t}^{t_1 + \Delta t} \left[i\hbar \frac{\partial}{\partial t_2} - \hat{H}(\mathbf{r}_2, t_2) \right] K(\mathbf{r}_2, t_2; \mathbf{r}_1, t_1) \, dt_2 =$$

$$= i\hbar \delta(\mathbf{r}_2 - \mathbf{r}_1) \int_{t_1 - \Delta t}^{t_1 + \Delta t} \delta(t_2 - t_1) \, dt_2 = i\hbar \delta(\mathbf{r}_2 - \mathbf{r}_1) \,.$$

Evidently we have

$$\lim_{\Delta t \to 0} \int_{t_1 - \Delta t}^{t_1 + \Delta t} \hat{H}(\mathbf{r}_2, t_2) K(\mathbf{r}_2, t_2; \mathbf{r}_1, t_1) \, dt_2 = 0 \,,$$

$$\lim_{\Delta t \to 0} \int_{t_1 - \Delta t}^{t_1 + \Delta t} i\hbar \frac{\partial}{\partial t_2} K(\mathbf{r}_2, t_2; \mathbf{r}_1, t_1) \, dt_2 = i\hbar K(\mathbf{r}_2, t_1; \mathbf{r}_1, t_1) \,.$$

Hence

$$K(\mathbf{r}_2, t_1; \mathbf{r}_1, t_1) = \delta(\mathbf{r}_2 - \mathbf{r}_1) \,. \tag{29.6}$$

Thus if the Green's function satisfies eq. (29.4), then (29.3) represents the solution of the Schrödinger equation with the corresponding initial conditions (the Cauchy problem).

In other words, if the transition amplitude is known, then the wave function is also known. On the other hand, the transition amplitude has certain important features which make it in many respects more convenient, and as completely characteristic of the system as the wave function.

Let us first consider the case where the Hamiltonian $\hat{H}_0$ of the system does not depend explicitly on the time. Then one can find the general relation between the wave functions of the stationary states $\psi_n(\mathbf{r}, t)$, the energy levels of the system, and the transition amplitude. We shall denote the latter by K_0 for a system with Hamiltonian $\hat{H}_0$. The amplitude must satisfy the equation

$$\left[i\hbar \frac{\partial}{\partial t_2} - \hat{H}_0(\mathbf{r}_2) \right] K_0(2, 1) = i\hbar \delta(\mathbf{r}_2 - \mathbf{r}_1) \delta(t_2 - t_1) . \qquad (29.7)$$

If $\hat{H}_0$ does not depend explicitly on the time, then the wave function satisfying eq. (29.2) can be written in the form

$$\psi_n = U_n(\mathbf{r}) \exp\left[-(i/\hbar) E_n t \right] ,$$

where the ψ_n form a complete set of orthonormal functions. By virtue of this property of the ψ_n, one can always write the expansion

$$K_0(2, 1) \equiv K_0(\mathbf{r}_2, t_2; \mathbf{r}_1, t_1) = \sum C_n(\mathbf{r}_1, t_1) \psi_n(\mathbf{r}_2, t_2) \theta(t_2 - t_1) , \quad (29.8)$$

where $\theta(t_2 - t_1)$ is the step function

$$\theta(x) = \begin{cases} 1 & (x > 0) \\ 0 & (x < 0) . \end{cases} \qquad (29.9)$$

The behaviour of K_0 for $t_2 < t_1$ is taken into account by means of the θ-function. Substituting the expansions (29.8) into (29.4), we find

$$\left[i\hbar \frac{\partial}{\partial t_2} - \hat{H}_0(\mathbf{r}_2) \right] \sum C_n(\mathbf{r}_1, t_1) \psi_n(\mathbf{r}_2, t_2) \theta(t_2 - t_1) =$$

$$= i\hbar \sum C_n \psi_n \frac{d\theta(t_2 - t_1)}{dt_2} + \sum C_n \theta(t_2 - t_1) \left[i\hbar \frac{\partial}{\partial t_2} - \hat{H}_0(\mathbf{r}_2) \right] \psi_n =$$

$$= i\hbar \sum C_n \psi_n \delta(t_2 - t_1) .$$

On the other hand using (29.7),

$$\left[i\hbar \frac{\partial}{\partial t_2} - \hat{H}_0(\mathbf{r}_2) \right] \sum C_n \psi_n \theta(t_2 - t_1) = i\hbar \delta(\mathbf{r}_2 - \mathbf{r}_1) \delta(t_2 - t_1) .$$

Hence $\Sigma C_n \psi_n = \delta(\mathbf{r}_2 - \mathbf{r}_1)$. But for ψ_n the normalization condition holds:

$$\sum \psi_n^*(\mathbf{r}_1) \psi_n(\mathbf{r}_1) = \delta(\mathbf{r}_2 - \mathbf{r}_1) .$$

Hence for the coefficient C_n we find $C_n = \psi_n^*$, and, finally, we obtain

$$K_0(2, 1) = \theta(t_2 - t_1) \sum_n U_n(\mathbf{r}_2) U_n(\mathbf{r}_1) \exp\left[-(i/\hbar) E_n(t_2 - t_1)\right] . \qquad (29.10)$$

The summation goes over into integration for a continuous spectrum. As an example, let us find the explicit expression for the Green's function of a free particle. In this case

$$\psi = (2\pi\hbar)^{-\frac{3}{2}} \exp(i\mathbf{p} \cdot \mathbf{r}) \exp\left(-\frac{ip^2 t}{2m\hbar}\right) ,$$

so that

$$K_0(2, 1) = \theta(t_2 - t_1)(2\pi\hbar)^{-3} \int \exp[-i\mathbf{p} \cdot (\mathbf{r}_1 - \mathbf{r}_2)] \exp\left(-\frac{ip^2(t_2 - t_1)}{2m\hbar}\right) d\mathbf{p} =$$

$$= \frac{\theta(t_2 - t_1)}{[2\pi i\hbar(t_2 - t_1)/m]^{\frac{3}{2}}} \exp\left(\frac{im\hbar|\mathbf{r}_2 - \mathbf{r}_1|^2}{2(t_2 - t_1)}\right).$$

The transition amplitude $K_0(\mathbf{r}_2, t_2; \mathbf{r}_1, t_1)$, or more briefly $K_0(2, 1)$, has the following important properties:
(1) The transition amplitude depends only on the difference $t_2 - t_1$, as is seen from formula (29.10): $K_0(\mathbf{r}_2, t_2; \mathbf{r}_1, t_1) = K_0(\mathbf{r}_2, \mathbf{r}_1; t_2 - t_1)$.
(2) Owing to this the transition amplitude $K_0(\mathbf{r}_3, t_3; \mathbf{r}_1, t_1) = K_0(\mathbf{r}_3, \mathbf{r}_1; t_3 - t_1)$ can be written in the form

$$K_0(\mathbf{r}_3, \mathbf{r}_1; t_3 - t_1) = \int K_0(\mathbf{r}_3, \mathbf{r}_2; t_3 - t_2) K_0(\mathbf{r}_2, \mathbf{r}_1; t_2 - t_1) d\mathbf{r}_2 . \qquad (29.11)$$

This means that the transition can be considered as a set of successive transitions $(1 \to 2), (2 \to 3)$ for all possible states 2.

Formula (29.11) expresses the principle of superposition. Its proof is elementary:

$$\psi(\mathbf{r}_3, t_3) = \int K_0(\mathbf{r}_3, \mathbf{r}_2; t_3 - t_2) \psi(\mathbf{r}_2, t_2) d\mathbf{r}_2 =$$

$$= \int\int K_0(\mathbf{r}_3, \mathbf{r}_2; t_3 - t_2) K_0(\mathbf{r}_2, \mathbf{r}_1; t_2 - t_1) \psi(\mathbf{r}_1, t_1) d\mathbf{r}_2 d\mathbf{r}_1 ,$$

$$\psi(\mathbf{r}_3, t_3) = \int K_0(\mathbf{r}_3, \mathbf{r}_1; t_3 - t_1) \psi(\mathbf{r}_1, t_1) d\mathbf{r}_1 .$$

Comparison of these formulae gives (29.11) immediately.
(3) The Fourier component of the transition amplitude K_0 defines the spectrum of eigenvalues of the energy of the system. Let us find the Fourier

component of the function $K_0(\mathbf{r}_2, \mathbf{r}_1; t_2 - t_1)$

$$
\begin{aligned}
K_0(\mathbf{r}_2, \mathbf{r}_1; \omega) &= \int K_0(\mathbf{r}_2, \mathbf{r}_1; t_2 - t_1) \exp[i\omega(t_2 - t_1)] \, d(t_2 - t_1) = \\
&= \sum U_n(\mathbf{r}_2) U_n^*(\mathbf{r}_1) \int \exp[-(i/\hbar) E_n(t_2 - t_1)] \exp[i\omega(t_2 - t_1)] \times \\
&\quad \times \theta(t_2 - t_1) \, d(t_2 - t_1) = \\
&= \sum U_n(\mathbf{r}_2) U_n^*(\mathbf{r}_1) I \, ,
\end{aligned}
$$

where

$$
I = \int \theta(t_2 - t_1) \exp[-(i/\hbar) E_n(t_2 - t_1)] \exp[i\omega(t_2 - t_1)] \, d(t_2 - t_1) \, .
$$

The θ-function can be written in the form of a contour integral

$$
\theta(x) = \lim_{\gamma \to 0} \frac{1}{2\pi i} \int_{\infty}^{\infty} \frac{e^{i\alpha x}}{\alpha - i\gamma} \, d\alpha \, . \tag{29.12}
$$

The contour of integration for $x > 0$ is shown in fig. V.8.

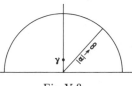

Fig. V.8

Then for I we find

$$
\begin{aligned}
I &= \lim_{\gamma \to 0} \frac{1}{2\pi i} \int \frac{d\alpha}{\alpha - i\gamma} \int \exp[-i(\omega_n - \omega - \alpha)\tau] \, d\tau = \\
&= \lim_{\gamma \to 0} \frac{1}{i} \int \frac{d\alpha}{\alpha - i\gamma} \delta(\omega_n - \omega - \alpha) = \lim_{\gamma \to 0} \frac{i}{\omega - \omega_n - i\gamma} \, . \tag{29.13}
\end{aligned}
$$

Hence for $K_0(\mathbf{r}_2, \mathbf{r}_1; \omega)$ we obtain

$$
K_0(\mathbf{r}_2, \mathbf{r}_1; \omega) = \lim_{\gamma \to 0} \sum U_n(\mathbf{r}_2) U_n^*(\mathbf{r}_1) \frac{1}{\omega - (E_n/\hbar) - i\gamma} \, . \tag{29.14}
$$

We see that the poles of the Fourier component of the transition amplitude, $\omega = E_n/\hbar$, correspond to the energy eigenvalues. Thus knowing the transition amplitude K_0 one can find directly the spectrum of energies E_n.

Let us turn to the general case of a Hamiltonian depending on time. It can usually be written in the form of a sum $\hat{H} = \hat{H}_0(\mathbf{r}) + U(\mathbf{r}, t)$, where $\hat{H}_0$ does not depend on time. $U(\mathbf{r}, t)$ often represents a variable external field acting on the particle. In this case the Green's function K satisfies the equation

$$\left[i\hbar \frac{\partial}{\partial t_2} - \hat{H}_0 - U \right] K(2, 1) = i\hbar\delta(\mathbf{r}_2 - \mathbf{r}_1)\delta(t_2 - t_1) \qquad (29.15)$$

and reduces to zero for $t_2 < t_1$,

$$K(2, 1) = 0 \quad \text{for} \quad t_2 < t_1 . \qquad (29.16)$$

The differential equation (29.15) for the Green's function can be compared to the integral equation

$$K(2, 1) = K_0(2, 1) - \frac{i}{\hbar} \int K_0(2, 3) U(3) K(3, 1) \, d^4x_3 \qquad (29.17)$$

where $d^4x = dx\,dy\,dz\,dt$. In this integral equation $K_0(2, 1)$ is considered to be a known function, and $K_0(2, 3)U(3)$ is its kernel. This can easily be seen by acting on eq. (29.17) with the operator $[i\hbar(\partial/\partial t_2) - \hat{H}_0(\mathbf{r}_2)]$. Then, taking into account (29.7), we obtain

$$\left[i\hbar \frac{\partial}{\partial t_2} - \hat{H}_0(\mathbf{r}_2) \right] K(2, 1) = i\hbar\delta(t_2 - t_1)\delta(\mathbf{r}_2 - \mathbf{r}_1) + U(2)K(2, 1) .$$

Thus we again arrive at eq. (29.15).

The initial condition (29.16) is contained in (29.17) since $K_0(2, 1) = 0$ for $t_2 < t_1$. The integral form of the equation for the transition amplitude (29.7) is especially convenient because it allows one to obtain $K(2, 1)$ in the form of a series of successive approximations (see §58).

§30. The eigenvalues and eigenfunctions of the angular momentum operator and of the operator of the square of the angular momentum

Let us now form some operators which play important roles in our subsequent considerations — the operators of the angular momentum components and of the square of the angular momentum. Replacing mechanical quantities in the classical definition of angular momentum by quantum-mechanical operators according to general rule, we find

$$\hat{l}_x = y\hat{p}_z - z\hat{p}_y = \frac{\hbar}{i} \left(y \frac{\partial}{\partial z} - z \frac{\partial}{\partial y} \right) ,$$

$$\hat{l}_y = z\hat{p}_x - x\hat{p}_z = \frac{\hbar}{i} \left(z \frac{\partial}{\partial x} - x \frac{\partial}{\partial z} \right) , \qquad (30.1)$$

$$\hat{l}_z = x\hat{p}_y - y\hat{p}_x = \frac{\hbar}{i} \left(x \frac{\partial}{\partial y} - y \frac{\partial}{\partial x} \right) .$$

We shall call the set of operators $\hat{l}_x$, $\hat{l}_y$ and $\hat{l}_z$ the angular momentum operator $\hat{\mathbf{l}}$. This quantity possesses all the properties of angular momentum. In particular, as we shall show below, it obeys the same conservation laws as angular momentum in classical mechanics.

Further, we construct the operator of the square of the angular momentum

$$\hat{\mathbf{l}}^2 = \hat{l}_x^2 + \hat{l}_y^2 + \hat{l}_z^2 . \tag{30.2}$$

Let us consider the commutation relations for these operators. We first of all note that not all of the operators of the different angular momentum components commute with each other. Let us calculate, for example, the commutator $\hat{l}_x\hat{l}_y - \hat{l}_y\hat{l}_x$. Making use of expressions (30.1) we have

$$\hat{l}_x\hat{l}_y = -\hbar^2\left(y\frac{\partial}{\partial z} - z\frac{\partial}{\partial y} \right)\left(z\frac{\partial}{\partial x} - x\frac{\partial}{\partial z} \right) =$$

$$= -\hbar^2\left(y\frac{\partial}{\partial x} + yz\frac{\partial^2}{\partial z\partial x} - xy\frac{\partial^2}{\partial z^2} - z^2\frac{\partial^2}{\partial y\partial x} + xz\frac{\partial^2}{\partial y\partial z} \right) .$$

On the other hand, interchanging the operators we find

$$\hat{l}_y\hat{l}_x = -\hbar^2\left(z\frac{\partial}{\partial x} - x\frac{\partial}{\partial z} \right)\left(y\frac{\partial}{\partial z} - z\frac{\partial}{\partial y} \right) =$$

$$= -\hbar^2\left(zy\frac{\partial^2}{\partial x\partial z} - z^2\frac{\partial^2}{\partial x\partial y} - xy\frac{\partial^2}{\partial z^2} + x\frac{\partial}{\partial y} + xz\frac{\partial^2}{\partial z\partial y} \right) .$$

Subtracting the lower equation from the upper, we obtain finally

$$\hat{l}_x\hat{l}_y - \hat{l}_y\hat{l}_x = \hbar^2\left(x\frac{\partial}{\partial y} - y\frac{\partial}{\partial x} \right) = i\hbar\hat{l}_z . \tag{30.3}$$

Carrying out the cyclic permutation of the coordinates x, y, z we obtain two more relations:

$$\hat{l}_y\hat{l}_z - \hat{l}_z\hat{l}_y = i\hbar\hat{l}_x , \qquad \hat{l}_z\hat{l}_x - \hat{l}_x\hat{l}_z = i\hbar\hat{l}_y . \tag{30.3'}$$

From the relations (30.3) it follows that the components of the angular momentum of a particle l_x, l_y, l_z cannot simultaneously have sharp values. An exception to this is the state when the angular momentum is equal to zero, so that then $\bar{l}_x = \bar{l}_y = \bar{l}_z = 0$. On the other hand the angular momentum component operators $\hat{l}_x$, $\hat{l}_y$ and $\hat{l}_z$ do commute with the operator of the square of the angular momentum $\hat{l}^2$, i.e. the following relations hold:

$$\hat{l}_x\hat{l}^2 - \hat{l}^2\hat{l}_x = 0 ,$$
$$\hat{l}_y\hat{l}^2 - \hat{l}^2\hat{l}_y = 0 , \qquad\qquad (30.4)$$
$$\hat{l}_z\hat{l}^2 - \hat{l}^2\hat{l}_z = 0 .$$

These relations are easily proved by means of (30.3). Let us prove, for example, the first of these. From relation (30.3), multiplying on the right and on the left by $\hat{l}_y$, we have

$$\hat{l}_x\hat{l}_y^2 = \hat{l}_y\hat{l}_x\hat{l}_y + i\hbar\hat{l}_z\hat{l}_y ,$$
$$\hat{l}_y^2\hat{l}_x = \hat{l}_y\hat{l}_x\hat{l}_y - i\hbar\hat{l}_y\hat{l}_z .$$

Subtracting the second relation from the first, we obtain

$$\hat{l}_x\hat{l}_y^2 - \hat{l}_y^2\hat{l}_x = i\hbar(\hat{l}_z\hat{l}_y + \hat{l}_y\hat{l}_z) .$$

Analogously,

$$\hat{l}_x\hat{l}_z^2 - \hat{l}_z^2\hat{l}_x = - i\hbar(\hat{l}_z\hat{l}_y + \hat{l}_y\hat{l}_z) .$$

Also taking into account the fact that $\hat{l}_x\hat{l}_x^2 - \hat{l}_x^2\hat{l}_x = 0$ and adding up the equalities obtained, we find

$$\hat{l}_x\hat{l}^2 - \hat{l}^2\hat{l}_x = 0 .$$

The two remaining relations (30.4) are proved in the same way. From these relations it follows that the square of the total angular momentum and one of its projections onto an arbitrary axis can simultaneously have definite values.

We note that commutation relations analogous to (30.3), (30.3') also hold for the angular momentum operator and the coordinate operator, and the angular momentum operator and momentum operator. Omitting the simple proof, we write down the two relations

$$\hat{l}_x\hat{y} - \hat{y}\hat{l}_x = i\hbar\hat{z} ,$$
$$\hat{l}_x\hat{p}_y - \hat{p}_y\hat{l}_x = i\hbar\hat{p}_z . \qquad\qquad (30.5)$$

The remaining four relations are obtained by the cyclic permutation of the indices. Relations (30.3), (30.4) and (30.5) are the same as the corresponding classical expressions if, of course, we pass from commutators to the classical Poisson brackets.

Further, let us determine the possible values of the angular momentum projection onto an arbitrarily chosen direction in space and the possible values of the square of the angular momentum (i.e. the eigenvalues of the corresponding operators). In solving the equations for the eigenfunctions and eigenvalues it is convenient to use spherical coordinates.

We carry out the transition from the Cartesian coordinates x, y, z to the variables r, ϑ, φ in formulae (30.1) and (30.2) according to the ordinary rules of replacement of variables. Omitting these elementary calculations, we simply given the result

$$\hat{l}_z = \frac{\hbar}{i} \frac{\partial}{\partial \varphi}, \tag{30.6}$$

$$\hat{l}_x = \frac{\hbar}{i} \left(\sin \varphi \frac{\partial}{\partial \vartheta} + \cotan \vartheta \cos \varphi \frac{\partial}{\partial \varphi} \right), \tag{30.7}$$

$$\hat{l}_y = \frac{\hbar}{i} \left(\cos \varphi \frac{\partial}{\partial \vartheta} - \cotan \vartheta \sin \varphi \frac{\partial}{\partial \varphi} \right), \tag{30.8}$$

$$\hat{l}^2 = -\hbar^2 \left[\frac{1}{\sin \vartheta} \frac{\partial}{\partial \vartheta} \left(\sin \vartheta \frac{\partial}{\partial \vartheta} \right) + \frac{1}{\sin^2 \vartheta} \frac{\partial^2}{\partial \varphi^2} \right] = -\hbar^2 \nabla^2_{\vartheta, \varphi}, \tag{30.9}$$

where $\nabla^2_{\vartheta, \varphi}$ is the angular part of the Laplacian in spherical coordinates.

Choosing an arbitrary direction in space as the z-axis, we define the eigenfunctions and eigenvalues of the operator of the component of angular momentum in this direction. The equation for the eigenfunctions and eigenvalues of the operator $\hat{l}_z$ is of the form

$$\frac{\hbar}{i} \frac{\partial \psi}{\partial \varphi} = l_z \psi. \tag{30.10}$$

The solution of this equation is

$$\psi = \psi(r, \vartheta) \exp(i l_z \varphi / \hbar) \tag{30.11}$$

where $\psi(r, \vartheta)$ is an arbitrary function.

The wave function which is the solution of eq. (30.10) must satisfy the condition of single-valuedness. Since φ is a cyclic variable varying from 0 to 2π, the condition of single-valuedness is written in the form $\psi(\varphi) = \psi(\varphi + 2\pi)$ or

$$\exp[(i/\hbar) l_z \varphi] = \exp[(i/\hbar) l_z (\varphi + 2\pi)].$$

This last condition is fulfilled if $l_z = m\hbar$, where m is a positive or negative integer (including zero). In what follows it will be called the magnetic quantum number.

Since the z-axis is not specified by any physical condition, the same result also holds for the operators $\hat{l}_x$ and $\hat{l}_y$.

Thus the angular momentum component along an arbitrarily chosen direction in space takes on integer (in units of $\hbar$) values. For a sharp value of the projection l_z the two other projections have no well defined value. This means that if in a state with given l_z the values of the projections l_x and l_y are measured, then any possible value may be found for them.

The eigenfunction of the operator $\hat{l}_z$ depending on the angle φ and normalized to unity by the condition

$$\int_0^{2\pi} \psi_m^*(\varphi)\,\psi_{m'}(\varphi)\,\mathrm{d}\varphi = \delta_{mm'}\,,$$

has the form

$$\psi_m(\varphi) = (2\pi)^{-\frac{1}{2}}\,\mathrm{e}^{im\varphi}\,. \tag{30.12}$$

Let us now determine the eigenvalues and eigenfunctions of the operator of the square of the angular momentum, $\hat{l}^2$

$$\hat{l}^2\psi = l^2\psi\,. \tag{30.13}$$

Substituting into (30.13) the expression for $\hat{l}^2$ given by formula (30.9), we obtain the equation

$$\frac{1}{\sin\vartheta}\frac{\partial}{\partial\vartheta}\left(\sin\vartheta\frac{\partial\psi}{\partial\vartheta}\right) + \frac{1}{\sin^2\vartheta}\frac{\partial^2\psi}{\partial\varphi^2} + \frac{l^2}{\hbar^2}\psi = 0\,. \tag{30.14}$$

The equation for the eigenfunctions of the operator $\hat{l}^2$ is the well-known equation for the spherical harmonic functions*.

Equation (30.14) only has solutions satisfying the standard conditions formulated in §16 for the values $l^2/\hbar^2 = l(l+1)$, where l is a positive integer (including zero). The quantum number l is called the orbital angular momentum quantum number. Thus the operator of the square of the angular momentum has a discrete spectrum of eigenvalues

$$l^2 = \hbar^2 l(l+1)\,. \tag{30.15}$$

The solution of eq. (30.14) for the eigenfunctions of the operator of the

---

* See, for example, V.I.Smirnov, *A course of higher mathematics,* Vol. III (Pergamon Press, Oxford, 1964) and V.A.Fok, *Nachala kvantovoi mekhaniki (Principles of quantum mechanics)* (KUBUCH, 1932) p. 118.

square of the angular momentum is of the form

$$\psi_{lm}(\vartheta, \varphi) = Y_{lm}(\vartheta, \varphi) =$$

$$= (-1)^k \left(\frac{(l - |m|)! \, (2l + 1)}{(l + |m|)! \, 4\pi} \right)^{\frac{1}{2}} P_l^m(\cos \vartheta) \, e^{im\varphi}, \quad (30.16)$$

where m is an integer taking on the values $m = 0, \pm 1, \pm 2, \dots \pm l$; $k = m$ for $m \geqslant 0$ and $k = 0$ for $m < 0$.

We denoted by P_l^m the associated Legendre polynomial

$$P_l^m = (1 - \xi^2)^{\frac{1}{2}|m|} \frac{d^{|m|}}{d\xi^{|m|}} P_l(\xi) = \frac{1}{2^l l!} (1 - \xi^2)^{\frac{1}{2}|m|} \frac{d^{|m|+l}}{d\xi^{|m|+l}} (\xi^2 - 1)^l. \quad (30.17)$$

The constant factor in formula (30.16) is defined by the condition of normalization of the function Y_{lm} to unity.

$$\int\limits_0^\pi \int\limits_0^{2\pi} Y_{lm}^*(\vartheta, \varphi) \, Y_{l'm'}(\vartheta, \varphi) \sin \vartheta \, d\vartheta \, d\varphi = \delta_{ll'} \delta_{mm'}. \quad (30.18)$$

From formulae (30.15) and (30.16) it follows that to each eigenvalue of the square of the angular momentum there correspond $2l+1$ eigenfunctions Y_{lm} (differing in the number m). Thus the eigenvalues of the square of the angular momentum are degenerate. The meaning of this degeneracy, and consequently of the number m, is easily understood. We act on the wave function Y_{lm} with the operator $\hat{l}_z$. We then obtain

$$\hat{l}_z Y_{lm}(\vartheta, \varphi) = \hbar m Y_{lm}(\vartheta, \varphi). \quad (30.19)$$

We see that the wave function Y_{lm} is simultaneously an eigenfunction of the operators $\hat{l}_z$ and $\hat{l}^2$. Hence it is clear that the quantum number m involved in (30.16) characterizes the value of the angular momentum projection onto the z-axis in a given state, and the wave function Y_{lm} describes a state with a given total angular momentum and a given projection on the z-axis.

Summarizing we can say that the value of the total angular momentum is defined according to formula (30.15) by the orbital angular momentum quantum number l running over a sequence of integer values. For a fixed value of the square of the angular momentum the projection of the angular momentum onto an arbitrarily oriented z-axis can take on $2l+1$ values from $-l$ up to $+l$ (in units of $\hbar$). Any other values of this angular momentum projection for a given l are impossible. Since the z-axis is oriented quite arbitrarily, it is natural that the angular momentum projections onto the x-axis and y-axis for a given l also take on values form $-l$ up to $+l$. For $l = 0$ the angular momentum projection onto any axis is also equal to zero. This is the only state in which the angular momentum projections onto different axes simultaneously have sharp values. In this case the function Y_{lm} ($l = 0$) reduces to a constant which is an eigenfunction of all the operators $\hat{l}_x, \hat{l}_y, \hat{l}_z$.

We note that the eigenvalue of the square of the total angular momentum $1^2 = \hbar^2 l(l+1)$ is always larger than the square of the maximum projection of the angular momentum which is equal to $\hbar^2 l^2$. If these quantities were the same, then this would mean that in a state in which the angular momentum projection onto a certain axis has its maximum value the other two projections would be equal to zero. This is, however, impossible, since for a sharp value of one of the angular momentum projections the other two cannot have well defined values, not even zero.

Finally, we shall show that the angular momentum operator is related to the operator of an infinitesimal rotation of the system about the origin. Let us rotate the coordinate system through a small angle $\delta\varphi$ about, for example, the z-axis. The old and new coordinates of a point are connected by the relation

$$x' = x + y\,\delta\varphi\,, \qquad x = x' - y'\,\delta\varphi\,,$$

$$y' = -x\,\delta\varphi + y\,, \qquad y = x'\,\delta\varphi + y'\,,$$

$$z' = z\,, \qquad z = z'\,.$$

Consequently, upon rotation the wave function $\psi(x, y, z)$ expressed in terms of the new variables has the form

$$\psi(x, y, z) = \psi(x' - y'\,\delta\varphi, \quad y' + x'\,\delta\varphi, \quad z') =$$

$$= \psi(x', y', z') - y'\,\delta\varphi\,\frac{\partial\psi}{\partial x'} + x'\,\delta\varphi\,\frac{\partial\psi}{\partial y'} =$$

$$= \left[1 + \delta\varphi\left(x'\,\frac{\partial}{\partial y'} - y'\,\frac{\partial}{\partial x'}\right)\right]\psi(x', y', z) =$$

$$= \left(1 + \frac{i}{\hbar}\,\delta\varphi\hat{l}_z\right)\psi(x', y', z') = \hat{W}_z\cdot\psi(x', y', z')\,.$$

It is natural to call the operator $\hat{W}$ the rotation operator. We found the operator $\hat{W}_z$ of the rotation through a small angle $\delta\varphi$ about the z-axis to be connected with the operator $\hat{l}_z$ by the relation

$$\hat{W}_z = 1 + \frac{i}{\hbar}\,\delta\varphi\hat{l}_z\,. \qquad\qquad (30.20)$$

Such a relation also holds, of course, for any other axis.

§31. Differentiation of operators with respect to time

We now construct the operator $\hat{\dot{F}}$ corresponding to the derivative with respect to time of the quantum-mechanical quantity described by the operator $\hat{F}$. It is clear that the ordinary definition of the derivative of a function is inapplicable to the quantum-mechanical quantity described by the operator $\hat{F}$. To define the notion of the derivative we again make use of the analogy with classical mechanics. As is known, in classical mechanics the derivative with respect to time of a mechanical quantity F can be expressed in terms of the classical Poisson bracket

$$\frac{\mathrm{d}F}{\mathrm{d}t} = \frac{\partial F}{\partial t} + [H, F] \ ,$$

where H is the Hamiltonian.

Passing from classical quantities to quantum-mechanical operators and from the classical Poisson bracket to the quantum one, we obtain the expression for the operator $\hat{\dot{F}}$

$$\hat{\dot{F}} = \frac{\partial \hat{F}}{\partial t} + [\hat{H}, \hat{F}] \ . \tag{31.1}$$

If the operator $\hat{F}$ does not depend explicitly on time, then the operator $\hat{\dot{F}}$ has the form

$$\hat{\dot{F}} = [\hat{H}, \hat{F}] = \frac{\mathrm{i}}{\hbar} (\hat{H}\hat{F} - \hat{F}\hat{H}) \ . \tag{31.2}$$

The expressions for the derivative of the sum $\hat{\dot{F}}$ and product $\hat{\dot{L}}$ of two operators $\hat{D}$ and $\hat{R}$

$$\hat{\dot{F}} = \hat{\dot{D}} + \hat{\dot{R}} \ , \tag{31.3}$$

$$\hat{\dot{L}} = \hat{\dot{D}}\hat{R} + \hat{D}\hat{\dot{R}} \tag{31.4}$$

follow immediately from the properties of the quantum Poisson brackets.

By means of formula (31.1) for the derivative of the quantum operator one can find the expression for the derivative with respect to time of the mean value of the quantity F.

Differentiating the expression (22.4) for the mean, we find

$$\dot{\bar{F}} = \int \frac{\partial \psi^*}{\partial t} \hat{F} \psi \, \mathrm{d}V + \int \psi^* \frac{\partial \hat{F}}{\partial t} \psi \, \mathrm{d}V + \int \psi^* \hat{F} \frac{\partial \psi}{\partial t} \, \mathrm{d}V \ .$$

We express the derivatives $\partial \psi / \partial t$ and $\partial \psi^* / \partial t$ in terms of the wave functions by means of the Schrödinger equation and the equation which is conjugate to it. We then have

$$\dot{\bar{F}} = \int \psi^* \frac{\partial \hat{F}}{\partial t} \psi \, dV - \frac{i}{\hbar} \int \psi^* \hat{F}(\hat{H}\psi) \, dV + \frac{i}{\hbar} \int (\hat{H}^* \psi^*) \hat{F}\psi \, dV \, ,$$

$$\int (\hat{H}^* \psi^*) \hat{F}\psi \, dV = \int (\hat{F}\psi) \hat{H}^* \psi^* \, dV \, ,$$

because the integral does not change when the integrands are exchanged. It follows from the Hermitian property of the operator $\hat{H}$ that

$$\int (\hat{F}\psi) \hat{H}^* \psi^* \, dV = \int \psi^* \hat{H}\hat{F}\psi \, dV \, .$$

We finally obtain

$$\dot{\bar{F}} = \int \psi^* \left(\frac{\partial \hat{F}}{\partial t} + \frac{i}{\hbar}(\hat{H}\hat{F} - \hat{F}\hat{H}) \right) \psi \, dV \, . \tag{31.5}$$

Comparing the above expression with the definition of the mean of the derivative $\dot{F}$, we arrive at the important equality $\dot{\bar{F}} = \bar{\dot{F}}$.

As an example let us define the operators $\dot{\hat{x}}$ and $\dot{\hat{p}}_x$. Since the coordinate operator and momentum operator do not depend explicitly on the time, we have

$$\dot{\hat{x}} = [\hat{H}, \hat{x}] \, , \qquad \dot{\hat{p}}_x = [\hat{H}, \hat{p}_x] \, . \tag{31.6}$$

In such a form the operator equations (31.6) are analogous to the classical Hamilton equations. We evaluate the commutators on the right-hand sides of the equations (31.6), assuming that the Hamiltonian has the form

$$\hat{H} = \frac{1}{2m}(\hat{p}_x^2 + \hat{p}_y^2 + \hat{p}_z^2) + U(x, y, z, t) \, .$$

Taking into account that the coordinate and momentum operators are

$$\hat{x} = x \, , \qquad \hat{p}_x = \frac{\hbar}{i} \frac{\partial}{\partial x} \, ,$$

we obtain

$$[\hat{H}, \hat{x}] = \frac{i}{2m\hbar}(\hat{p}_x^2 x - x\hat{p}_x^2) \, ,$$

since x and $U(x, y, z, t)$ commute.

Calculating the commutator of the operators $\hat{p}_x^2$ and x, we find

$$\hat{p}_x^2 x - x\hat{p}_x^2 = -\hbar^2 \left(\frac{\partial^2}{\partial x^2} x - x \frac{\partial^2}{\partial x^2} \right) =$$

$$= -\hbar^2 \left(x \frac{\partial^2}{\partial x^2} + 2 \frac{\partial}{\partial x} - x \frac{\partial}{\partial x^2} \right) =$$

$$= -2\hbar^2 \frac{\partial}{\partial x} = -2i\hbar \, \hat{p}_x \, ,$$

We finally obtain

$$\hat{\dot{x}} = [\hat{H}, \hat{x}] = \frac{1}{m} \hat{p}_x \, . \tag{31.7}$$

We see that the velocity operator $\hat{\dot{x}}$ is connected with the momentum operator $\hat{p}_x$ by the same relation as in classical mechanics. We find the operator $\hat{\dot{p}}_x$ by

$$[\hat{H}, \hat{p}_x] = \frac{i}{\hbar} (U\hat{p}_x - \hat{p}_x U) = -\frac{\partial U}{\partial x} \, .$$

Thus we have

$$\hat{\dot{p}}_x = -\frac{\partial U}{\partial x} \, . \tag{31.8}$$

We have obtained the operator equation of motion in the form of Newton's equation. Equations (31.7) and (31.8) can also be written for the mean values of the corresponding quantities

$$\bar{\dot{x}} = \dot{\bar{x}} = \frac{1}{m} \bar{p}_x \, , \qquad \bar{\dot{p}}_x = \dot{\bar{p}}_x = -\frac{\overline{\partial U}}{\partial x} \, . \tag{31.9}$$

These last relations are called the Ehrenfest theorems. Expressing $\bar{\dot{p}}_x$ in terms of $\ddot{x}$, we find

$$m\ddot{\bar{x}} = -\frac{\overline{\partial U}}{\partial x} \, . \tag{31.10}$$

In such a form this equation is very close in appearance to Newton's equation of classical mechanics.

§ 32. Constants of the motion

Suppose the operator $\hat{F}$ does not depend explicitly on the time and commutes with the Hamiltonian $\hat{H}$. In this case, according to (31.2), the operator corresponding to the derivative with respect to time is equal to zero, and from relation (31.5) it follows that the mean value of the quantity F does not change with time

$$\dot{\bar{F}} = 0 .$$

(32.1)

The probability that, in measuring F, we shall obtain a possible value F_n is also constant in time. Indeed, this probability is given by the square of the modulus of the coefficient of expansion $|c_n(t)|^2$ of the wave function $\psi(x, t)$ describing the state of the system at the instant of time t in terms of the eigenfunctions of the operator $\hat{F}$. Since, however, the operator $\hat{F}$ commutes with the operator $\hat{H}$, both operators have the common eigenfunctions

$$\psi_n(x, t) = \psi_n(x) \exp[(-i/\hbar)E_n t]$$

(see §23). The expansion of $\psi(x, t)$ in terms of the eigenfunctions of the operator $\hat{F}$ can be written in the form

$$\psi(x, t) = \sum_n c_n(0) \exp[(-i/\hbar)E_n t] \, \psi_n(x) = \sum_n c_n(t) \, \psi_n(x) .$$

(32.2)

Consequently,

$$|c_n(t)|^2 = |c_n(0)|^2 = \text{const.}$$

In quantum mechanics such quantities, as in classical mechanics, are usually called constants of the motion. From the above it is clear that a quantum-mechanical quantity is a constant of the motion if: (1) its operator does not depend explicitly on the time, (2) this operator commutes with the Hamiltonian.

Knowing the operators corresponding to different quantum-mechanical quantities and the Hamiltonian, one can find the conservation laws.

Finding conservation laws in quantum mechanics is as important for the study of the motion of a system as in classical mechanics. As in classical mechanics*, the laws of conservation of momentum and angular momentum are closely associated with the properties of homogeneity and isotropy of space. Thus from the isotropy of space it follows that the Hamiltonian of a

---

* See, for example, L.D.Landau and E.M.Lifshitz, *Mechanics* (Pergamon Press, Oxford, 1960).

closed system or of a system in a centrally symmetric force field must not change when an arbitrary infinitesimal rotation is made. Mathematically this is expressed by the fact that the Hamiltonian $\hat{H}$ must commute with the rotation operator $\hat{W}$. But, as we know (see §30), the operator corresponding to rotation through a small angle about a certain axis (for example the z-axis) is related in a simple way to the operator of the component of angular momentum along this axis. Therefore a consequence of the commutation of the operator $\hat{W}_z$ with the Hamiltonian $\hat{H}$ is the commutation of the operator $\hat{l}_z$ with the Hamiltonian, hence the law of conservation of this quantity follows. The fact that we have considered rotation only through a small angle is not important, since a rotation through a finite angle can be resolved into a succession of small rotations.

Thus we see that the conservation of angular momentum is associated with the isotropy of space.

Similarly it is easily seen that momentum conservation is associated with the homogeneity of space. Indeed, from the homogeneity of space it follows that the displacement operator must not change the Hamiltonian of a closed system, i.e. it must commute with the Hamiltonian. However, since the displacement operator $\hat{R}$ is related to the operator of the corresponding momentum component (see §26), we arrive immediately at the momentum conservation law.

The law of conservation of energy in a closed system or a system in a stationary external field can be associated with the arbitrariness of the choice of the zero of time (homogeneity in time). This means that the laws of motion of the system must not depend on the choice of the zero of time.

We introduce the operator corresponding to the translation over a small time interval δt, $\hat{\mathcal{V}}(\delta t)$, defined by the relation

$$\hat{\mathcal{V}}(\delta t)\,\psi(x, t) = \psi(x, t + \delta t)\,. \tag{32.3}$$

Expanding the function $\psi(x, t+\delta t)$ in a series in terms of the small interval δt and confining ourselves to terms of the first order of small quantities, we obtain

$$\hat{\mathcal{V}}(\delta t)\,\psi(x, t) = \left(1 + \delta t \frac{\partial}{\partial t}\right)\psi(x, t)\,.$$

Hence it follows that the operator $\hat{\mathcal{V}}(\delta t)$ is of the form

$$\hat{\mathcal{V}}(\delta t) = 1 + \delta t \frac{\partial}{\partial t}\,. \tag{32.4}$$

The requirement of the independence of the laws of motion of the system of the choice of zero time is expressed by the commutation of the operator $\hat{\mathcal{V}}(\delta t)$ with the Hamiltonian of the system

$$\hat{\mathcal{V}}(\delta t)\hat{H} = \hat{H}\hat{\mathcal{V}}(\delta t) . \tag{32.5}$$

Using the expression (32.4) for $\hat{\mathcal{V}}(\delta t)$, we can rewrite relation (32.5) in the form

$$\frac{\partial \hat{H}}{\partial t} = 0 . \tag{32.6}$$

But eq. (32.6) just expresses the energy conservation law. Indeed, the operator $\hat{H}$ commutes with itself, and the condition $\dot{\hat{H}} = 0$, denoting the energy conservation law, amounts to (32.6).

To the existence of a constant of the motion there corresponds a simple property of the wave function. If the operator $\hat{I}$ corresponds to a certain conserved quantity, then the Schrödinger equation will be satisfied not only by the wave function ψ but also by the wave function

$$\psi' = e^{i\alpha \hat{I}}\psi , \tag{32.7}$$

where α is an arbitrary real number. By definition

$$e^{i\alpha \hat{I}} = 1 + i\alpha\hat{I} + \frac{(i\alpha)^2}{2!}\hat{I}^2 + \dots .$$

Substituting ψ' into the Schrödinger equation, we find

$$i\hbar \frac{\partial \psi'}{\partial t} = i\hbar \frac{\partial}{\partial t}(e^{i\alpha I}\psi) = \hat{H}e^{i\alpha \hat{I}}\psi . \tag{32.8}$$

But, since $\hat{I}$, as the operator of a conserved quantity, satisfies the commutation condition $\hat{I}\hat{H} - \hat{H}\hat{I} = 0$, $\partial \hat{I}/\partial t = 0$, we have

$$\frac{\partial}{\partial t}(e^{i\alpha \hat{I}}\psi) = e^{i\alpha \hat{I}}\frac{\partial \psi}{\partial t} , \qquad \hat{H}e^{i\alpha \hat{I}}\psi = e^{i\alpha \hat{I}}\hat{H}\psi ,$$

and eq. (32.8) is satisfied directly.

Let us consider some simple examples. We begin with the case of a free particle. Then the Hamiltonian is of the form

$$\hat{H} = \frac{1}{2m}(\hat{p}_x^2 + \hat{p}_y^2 + \hat{p}_z^2) .$$

Evidently, $[\hat{H}, \hat{p}_x] = [\hat{H}, \hat{p}_y] = [\hat{H}, \hat{p}_z] = 0$. Consequently,

$$\dot{\hat{p}}_x = \dot{\hat{p}}_y = \dot{\hat{p}}_z = 0 . \tag{32.9}$$

If at a certain initial instant the free particle was in a state with definite momentum, then this value of the momentum is conserved in time.

As another example let us consider a particle moving in the field produced by an infinite uniform plane (xy-plane). The potential energy of a particle in such a field depends only on the distance from the plane $U = U(|z|)$, so that the Hamiltonian is of the form

$$\hat{H} = -\frac{\hbar^2}{2m}\nabla^2 + U(|z|) .$$

The operators $\hat{p}_x, \hat{p}_y, \hat{l}_z$ commute with such a Hamiltonian. This means that in the case of motion in the field of a uniform plane (xy) the components of the momentum of the particle, p_x and p_y, and the z-component of the angular momentum, l_z, are conserved.

§33. Parity

The conservation laws considered above — the laws of conservation of energy, momentum and angular momentum — are the quantum-mechanical analogues of the conservation laws of classical mechanics. It turns out that in quantum mechanics there are also conservation laws which have no classical analogue. One such law is closely associated with the properties of space and is of a very general character. Namely, the Hamiltonian of a closed system must not change under the following transformations of the coordinates:
(1) translation of the origin by an arbitrary segment;
(2) rotation through an arbitrary angle;
(3) inversion, i.e. the substitution $x_i \rightarrow -x_i$, in which the signs of all coordinates are changed.

As we have seen in the preceding section, the first two transformations are associated with the laws of conservation of momentum and angular momentum. In quantum mechanics it turns out that inversion is associated with still another general conservation law. As for the translation and rotation operators, which have been introduced earlier, one can also introduce the corresponding inversion operator $\hat{I}$

$$\hat{I}\psi(\mathbf{r}, t) = a\psi(-\mathbf{r}, t) , \tag{33.1}$$

where a is a constant.

When the inversion operator $\hat{I}$ is applied twice we arrive at the initial state. Hence it follows that $a^2 = 1$, i.e. $a = \pm 1$. Thus, in general, the following is fulfilled:

$$\hat{I}\psi(\mathbf{r}, t) = \pm \psi(-\mathbf{r}, t) , \qquad (33.2)$$

i.e. the wave function itself, and not only the argument $\mathbf{r}$ on which it depends, can change sign directly under inversion. Whether the transformation of the wave function under inversion will have $a = +1$ or $a = -1$ depends on the intrinsic properties of the particles described by this wave function.

The particles which are described by wave functions satisfying the condition

$$\hat{I}\psi(\mathbf{r}, t) = \psi(-\mathbf{r}, t)$$

are said to possess even intrinsic parity. On the contrary, particles which are described by wave functions satisfying the condition

$$\hat{I}\psi(\mathbf{r}, t) = -\psi(-\mathbf{r}, t)$$

have odd intrinsic parity.

We assume that the Hamiltonian of a closed system has the form

$$\hat{H} = \sum_i -\frac{\hbar^2}{2m_i} \nabla_i^2 + \frac{1}{2} \sum_{i \neq k} U_{ik}(|\mathbf{r}_i - \mathbf{r}_k|) .$$

It is easily seen that this Hamiltonian does not change under the substitution $\mathbf{r}_i \rightarrow -\mathbf{r}_i$, i.e. it satisfies the condition $\hat{I}\hat{H}\psi = \hat{H}\hat{I}\psi$. This means that the operator $\hat{I}$ commutes with the Hamiltonian

$$\hat{I}\hat{H} = \hat{H}\hat{I} . \qquad (33.3)$$

We determine the eigenvalues λ of the inversion operator

$$\hat{I}\psi_\lambda(x) = \lambda\psi_\lambda(x) . \qquad (33.4)$$

We apply the inversion operator to this equation once more. Since under the two-fold inversion we come back to the initial value of the coordinates, this transformation is simply

$$\hat{I}^2\psi_\lambda = \psi_\lambda = \lambda\hat{I}\psi_\lambda = \lambda^2\psi_\lambda . \qquad (33.5)$$

Whence we find that the eigenvalues λ are equal to ± 1. A state with $\lambda = +1$ is said to have even parity or to be even. On the other hand, a state with $\lambda = -1$ had odd parity or is odd. If the parity operator commutes with the Hamiltonian operator, then the parity conservation law holds. The parity conservation law, like other conservation laws, imposes definite restrictions upon possible changes of the states of a system. Namely, if the system was in an even state, then it will remain in such a state, and not pass over into an odd state. Naturally, the situation is analogous in the case of a system in an odd state.

Let us determine the parity of the state of a particle with angular momentum l. The fact that the angular momentum and parity can be determined simultaneously follows from the commutation of the corresponding operators:

$$\{\hat{I}, \hat{l}_x\} = 0 ; \qquad \{\hat{I}, \hat{l}_y\} = 0 ; \qquad \{\hat{I}, \hat{l}_z\} = 0 ; \qquad \{\hat{I}, \hat{\mathbf{l}}^2\} = 0 . \qquad (33.6)$$

From the expressions for the angular momentum operators $\hat{l}_x, \hat{l}_y, \hat{l}_z$ it is clear that they do not change under inversion. In the spherical system of coordinates the inversion has the form

$$r \to r ; \qquad \vartheta \to \pi - \vartheta ; \qquad \varphi \to \varphi + \pi . \qquad (33.7)$$

The dependence of the wave function of a particle with definite angular momentum l on the angles ϑ, φ is given by the spherical function $Y_m(\vartheta, \varphi)$ (see §30). Under inversion (33.7) we have $\cos \vartheta \to -\cos \vartheta$ and $e^{im\varphi} \to \to (-1)^m e^{im\varphi}$. From formula (30.17) it is easily established how the associated Legendre polynomial $P_l^m(\xi)$ is transformed under a change of sign of its argument. Since $P_l(-\xi) = (-1)^l P_l(\xi)$, we obtain that $P_l^m(-\xi) = (-1)^{l+m} P_l^m(\xi)$. Taking into account the factor $(-1)^m$ which is given by the function $e^{im\varphi}$, we find that under inversion the wave function on the whole is multiplied by the factor $(-1)^l$. Taking into account also the factor $a = \pm 1$ associated with the intrinsic properties of particles, we get

$$\lambda = (-1)^l a . \qquad (33.8)$$

Thus the states with even l have even parity if $a = 1$, and odd parity if $a = -1$. The states with odd l have, correspondingly, odd parity if $a = 1$, and even parity if $a = -1$. If we have a system of non-interacting particles, then the parity of the system is determined by the product of the parities of the individual particles. Indeed, in §14 we have seen that the wave function of a system of non-interacting particles can be written in the form of the product of the wave functions of the individual particles. Hence it follows immediately that under inversion the parities of individual particles are multiplied. If each of the particles is in a state with definite angular momentum (motion in a central field), then the parity of the entire system can be written in the form

$$\lambda = (-1)^{\Sigma_k l_k} \prod_k a_k , \qquad (33.9)$$

where the second factor is determined by the product of the intrinsic parities of the particles.

In addition to other conservation laws the parity conservation law is one of the most general laws of nature. The impossibility of transitions of a closed quantum-mechanical system from states with one parity into states with another parity – so-called forbidden transitions – is confirmed by a vast amount of experimental data in atomic as well as nuclear physics. However, it has been established (see §122) that the parity conservation law is not a universal physical law. The parity conservation law is violated in certain processes involving elementary particles.

§34. The uncertainty relation for time and energy

The relation between the uncertainty in the energy ΔE and a time interval Δt can be derived from the general apparatus of quantum mechanics, as was shown by Mandelshtam and Tamm*. Indeed, the total energy of a closed system can have no definite values which are constant in time. As we have explained in §32, its mean value and the probabilities of observing one or other possible value are constant in time. In other words, the form of the energy distribution function is conserved in time.

Knowing the distribution function one can define the value of the root-mean-square deviation of the energy ΔE, which is naturally also conserved in time in the usual way. The energy will have a definite value ($\Delta E = 0$) only if the system is in a stationary state. A characteristic indication of a stationary state is the constancy in time of the physical quantities of a given system.

Let us assume that a closed system is in a state with indefinite energy E at the initial instant of time. Further, let R be a quantity whose operator $\hat{R}$ does not depend explicitly on time. For the given quantity one can, in the usual way, define its root-mean-square deviation ΔR and the mean value $\bar{R}$. Making use of (24.5) and (31.5), we write the relations

$$\Delta E\, \Delta R \geqslant \tfrac{1}{2}|\overline{(\hat{H}\hat{R} - \hat{R}\hat{H})}|\,, \tag{34.1}$$

$$\hbar\dot{\bar{R}} = i\overline{(\hat{H}\hat{R} - \hat{R}\hat{H})}\,. \tag{34.2}$$

* L.I.Mandelshtam and I.E.Tamm, Izvestiya Akad. Nauk SSSR, physical series, 9 (1945) 122.

Substituting (34.2) into (34.1), we obtain correspondingly

$$\Delta E \,\Delta R \geqslant \tfrac{1}{2}\hbar |\dot{\bar{R}}| . \tag{34.3}$$

This relation connects the uncertainty in the energy ΔE, the uncertainty ΔR in the value of R, and the rate of change of the mean value of the quantity R. Relation (34.3) can be rewritten in a somewhat more convenient form if one introduces the interval Δt — the time for which the mean value of R changes by an amount of the order of magnitude of its root-mean-square deviation ΔR

$$\Delta t = \frac{\Delta R}{\dot{\bar{R}}} . \tag{34.4}$$

Then we have

$$\Delta E \,\Delta t \geqslant \tfrac{1}{2}\hbar . \tag{34.5}$$

In particular, it follows from (34.3) that for the value of $\bar{R}$ to change with time, R must possess a dispersion different from zero.

Thus we see that there is a definite relation between the dispersion of the total energy of the system and the rate of change of arbitrary quantities characterizing the system under consideration.

As a simple example let us consider a one-dimensional wave packet. We take the coordinate x as the quantity R, $R = x$. Then ΔR is the width of the packet, and Δt is the time of flight of the packet past a certain point of space. The relation (34.5) shows that the transit time essentially depends on the dispersion of the total energy ΔE.

From the inequality (34.5) then also follows a definite relation between the lifetime of a given state and the uncertainty in the energy, ΔE, of this state. Thus, assuming Δt to be equal to τ, the half-life, we obtain that in order of magnitude

$$\Gamma \sim \frac{\hbar}{2} \cdot \frac{1}{\tau} , \tag{34.6}$$

where Γ is the uncertainty in the energy of the initial state and gives the width of the corresponding spectral line. The problem of the connection of the decay law with the energy distribution function is considered in more detail in a work of Krylov and Fok*. In this work it is also shown that relation (34.5) cannot be applied to the measurement processes, because it is derived by making use of the Schrödinger equation. This follows, for example

* I.S.Krylov and V.A.Fok, J. Exp. Theor. Phys. (USSR), 17 (1947) 93.

from the fact that a given object during the process of measurement is no longer a closed quantum-mechanical system.

For the measurement processes the corresponding inequality must be formulated in the form of a certain physical principle (stated by Bohr)

$$\Delta(E - E')\Delta t > \hbar , \qquad (34.7)$$

where E and E' are the values of the energy of the object before and after the measurement process, and $\Delta(E - E')$ is the absolute value of the uncertainty in the measurement of the energy of the object, i.e. the corresponding error of the measurement if it was carried out during a time Δt.

Relation (34.7) is very important in the analysis of the results of measurements, i.e. for an experimental check of the results given by quantum mechanics. We shall illustrate it by the simple example of a free particle. For the measurement of the quantities E, p, v (energy, momentum, velocity) of the particle it is necessary to consider the collision of this particle with another system (apparatus). Assuming for simplicity that the motion is one-dimensional, we write the momentum conservation law

$$p + k - p' - k' = 0 . \qquad (34.8)$$

Here we denote by k and k' the momentum of the apparatus before and after the collision. We shall denote by primes the quantities referring to the systems after the collision. It can be assumed that the momentum of the apparatus before and after the collision is accurately measured. Then from relation (34.8) there follows the equality of the errors in the measurement of the momentum of the particle before and after the collision

$$\Delta p = \Delta p' . \qquad (34.9)$$

The error in the measurement of the energy can be expressed in terms of the error in the measurement of the momentum, since

$$\Delta E = \frac{\partial E}{\partial p} \Delta p = v \Delta p ,$$

$$\Delta E' = \frac{\partial E'}{\partial p'} \Delta p' = v' \Delta p' .$$

In view of the equality of the errors Δp and $\Delta p'$, we have

$$\Delta(E - E') = |v - v'| \Delta p . \qquad (34.10)$$

We multiply (34.10) by the time of measurement Δt. We then obtain

$$\Delta(E - E')\Delta t = |v - v'| \Delta p \Delta t . \qquad (34.11)$$

But the quantity $|v - v'| \Delta t$ represents an additional error in the coordinate, which appeared during the time of measurement Δt. The total uncertainty in the coordinate Δx can be written in the form

$$\Delta x = (\Delta x)_0 + |v - v'| \Delta t ,$$

where $(\Delta x)_0$ is the uncertainty in the coordinate of the particle which existed before the collision being considered. In particular, $(\Delta x)_0$ can be made arbitrarily small. The fact that the value of $(\Delta p)_0$ will then be large is of no importance, because $(\Delta p)_0$ is in no way connected with the error Δp considered.

Heisenberg's uncertainty relation $\Delta p \, \Delta x \gtrsim \hbar$ must be fulfilled irrespective of the value of $(\Delta x)_0$. Consequently

$$|v - v'| \Delta t \, \Delta p > \hbar . \tag{34.12}$$

Comparing with (34.11), we arrive at the inequality

$$\Delta(E - E') \Delta t > \hbar \tag{34.13}$$

in accordance with (34.7). We see that the error in the measurement of the energy tends to zero provided the measurement process lasts a sufficiently long time (in the limit $\Delta t \to \infty$).

We note in addition that, as follows from (34.12), the measurement of the momentum for a given value of the error Δp leads to a change in the velocity of the particle,

$$|v - v'| > \frac{\hbar}{\Delta p \, \Delta t} ,$$

and, consequently, to a change in the momentum. Only if the measurement is carried out during an infinitely long time ($\Delta t \to \infty$) does the momentum not change. Of course, a measurement of the momentum over a long period of time can make sense only if the particle is free. Thus we see that the process of measurement of the momentum in small time intervals is irreproducible. The measurement brings the micro-object into a completely new state (see §5).

4

Motion in a Centrally Symmetric Field

§35. The Schrödinger equation

We can now apply the mathematical apparatus of quantum mechanics, developed in the preceding chapter to the study of the properties of real systems. It is natural to consider, first of all, the hydrogen atom, the simplest atomic system. In the hydrogen atom the potential energy of interaction of the electron with the nucleus depends only on the distance between them, $|\mathbf{r}_1 - \mathbf{r}_2|$. The problem of the motion of two particles with the interaction law $U(|\mathbf{r}_1 - \mathbf{r}_2|)$ amounts, as we have explained in §14, to the problem of the motion of one particle with reduced mass μ in a field $U(r)$. In view of the large difference in the masses, the reduced mass μ is very close to the mass of the electron. If also the size of the proton is neglected, then the hydrogen atom represents an electron moving in the Coulomb field of a motionless centre. Such a field is a particular case of a centrally symmetric field in which the potential energy depends only on the distance from the force centre. We shall first consider the motion of an electron in a centrally symmetric field of the most general form, after which we shall pass over to the case of the Coulomb field.

The Schrödinger equation for the stationary states of a particle moving in a force field with potential energy $U(r)$ has the form

$$\nabla^2 \psi + \frac{2\mu}{\hbar^2} [E - U(r)] \psi = 0 . \tag{35.1}$$

In the case of a centrally symmetric potential field it is convenient to transform the Schrödinger equation to spherical coordinates, since the potential energy depends only on the distance from the origin r. Expressing the Laplacian operator in spherical coordinates, we have

$$\frac{1}{r^2} \frac{\partial}{\partial r} \left(r^2 \frac{\partial \psi}{\partial r} \right) + \frac{1}{r^2} \nabla^2_{\vartheta\varphi} \psi + \frac{2\mu}{\hbar^2} [E - U(r)] \psi = 0 . \tag{35.2}$$

This equation is conveniently transformed by introducing into it explicitly the operator of the square of the angular momentum $\hat{l}^2$. Substituting its value according to formula (30.9), we have

$$\frac{1}{r^2} \frac{\partial}{\partial r} \left(r^2 \frac{\partial \psi}{\partial r} \right) - \frac{\hat{l}^2}{\hbar^2 r^2} \psi + \frac{2\mu}{\hbar^2} [E - U(r)] \psi = 0 . \tag{35.3}$$

We shall first of all show that in the case of motion in a centrally symmetric field two more conservation laws are satisfied, in addition to the energy conservation law: the total angular momentum conservation law and the law of conservation of the z-component of the angular momentum where the z-axis is arbitrarily oriented in space. When we speak here of the conservation of total angular momentum we mean the quantity described by the operator $\hat{l}^2$ (the square of the angular momentum). For this, according to general rules, we consider the conditions for the commutation of the operators $\hat{l}^2$ and $\hat{l}_z$ with the Hamiltonian. It is obvious that in our case the Hamiltonian $\hat{H}$ can be written in the form

$$\hat{H} = -\frac{\hbar^2}{2\mu} \frac{1}{r^2} \frac{\partial}{\partial r} \left(r^2 \frac{\partial}{\partial r} \right) + \frac{\hat{l}^2}{2\mu r^2} + U(r) . \tag{35.4}$$

The operator $\hat{l}^2$ involves only the angular variables ϑ, φ, and the differential operators with respect to these variables. Hence the operator $\hat{l}^2$ commutes with any operator of differentiation with respect to r, as well as with the operator of the coordinate r itself

$$\hat{H}\hat{l}^2 - \hat{l}^2\hat{H} = 0 . \tag{35.5}$$

An analogous relation also holds for the operator $\hat{l}_z$ in view of the fact that, as we have seen in §30, it commutes with the operator $\hat{l}^2$ (30.4):

$$\hat{H}\hat{l}_z - \hat{l}_z\hat{H} = 0 . \tag{35.6}$$

Since, in motion in a centrally symmetric field, three quantities are conserved — the energy, the square of the angular momentum l^2, and the projection l_z of the angular momentum onto an arbitrary axis — we shall consider states with given values of the three quantities.

It should be noted that in motion in a centrally symmetric field the laws of conservation of energy, of total angular momentum and of the z-component of angular momentum also hold in classical mechanics.

We have considered previously the states of a system with given values of the total angular momentum and its projection onto the z-axis. The eigenvalues of the operators $\hat{l}^2$ and $\hat{l}_z$ were characterized by the azimuthal and magnetic quantum numbers l and m, while the spherical functions $Y_{lm}(\vartheta, \varphi)$ with the indices l, m were the eigenfunctions of these operators.

Equation (35.3) allows one to separate the variables. Its angular part is the same as eq. (30.14). It describes the motion with given values l and m. Hence it is natural to seek the solution of (35.3) in the form

$$\psi(r, \vartheta, \varphi) = R(r) Y_{lm}(\vartheta, \varphi) . \qquad (35.7)$$

Substituting expression (35.7) into eq. (35.3) and taking into account that $\hat{l}^2 Y_{lm} = \hbar^2 l(l + 1) Y_{lm}$, we arrive at the following equation for the radial part of the wave function $R(r)$:

$$\frac{1}{r^2} \frac{d}{dr} \left(r^2 \frac{dR}{dr} \right) + \frac{2\mu}{\hbar^2} \left(E - U(r) - \frac{\hbar^2}{2\mu} \frac{l(l + 1)}{r^2} \right) R = 0 . \qquad (35.8)$$

We see that the expression for the radial component R of the wave function ψ depends essentially on the form of the potential energy $U(r)$. At the same time the angular part $Y_{lm}(\vartheta, \varphi)$ of the wave function is determined only by the value of the angular momentum of the particle (the number l) and its z-component (the number m). States with a given angular momentum are denoted by small letters:

$$l = 0 \quad 1 \quad 2 \quad 3 \quad 4 \quad 5 \quad 6 \quad 7$$

$$s \quad p \quad d \quad f \quad g \quad h \quad i \quad k$$

Also the parity of the state is determined by the value of the quantum number l. In §33 we have shown that in a state with given total angular momentum and given z-component of angular momentum, the parity is equal to $(-1)^l$, i.e. under inversion the spherical function Y_{lm} goes over into $(-1)^l Y_{lm}$. Since the radial wave function, which depends on the absolute value of the radius vector, does not change under inversion, the transformation law mentioned also refers to the total wave function

$$\psi(r, \vartheta, \varphi) \to (-1)^l \psi(r, \vartheta, \varphi) .$$

Thus the states s, d, g, ... are even, while p, f, h, ... are odd (for even intrinsic parity).

The probability that an electron in the state $\psi(r, \vartheta, \varphi) = R(r) Y_{lm}(\vartheta, \varphi)$ will be observed in an infinitesimal volume element with coordinates r, ϑ, φ, is given by the formula

$$dW(r, \vartheta, \varphi) = |\psi(r, \vartheta, \varphi)|^2 r^2 \, dr \, d\Omega , \qquad (35.9)$$

where $d\Omega = \sin \vartheta \, d\vartheta \, d\varphi$. If this expression is integrated with respect to all values of the angles ϑ, φ, then we shall obtain the probability of observing the electron in a spherical layer between r and $r + dr$

$$dW(r) = |R(r)|^2 r^2 \, dr . \qquad (35.10)$$

Integrating (35.9) with respect to all values of the radius r from 0 to ∞, we find the probability $dW(\vartheta, \varphi)$ of observing the electron in the solid angle $d\Omega$ in the direction defined by the angles ϑ, φ

$$dW_{lm}(\vartheta, \varphi) = |Y_{lm}|^2 \, d\Omega . \qquad (35.11)$$

It follows from the definition of the spherical function (30.16) that the last expression does not depend on the angle φ. This means that in the plane perpendicular to the z-axis the distribution of the probability of finding the particle is completely symmetric. It should be noted that we understand the z-axis to be an arbitrarily chosen direction in space; the projection of the angular momentum onto this direction is fixed. Thus it follows from (35.11) that

$$dW_{lm} \sim |P_l^m(\cos \theta)|^2 \, d\Omega . \qquad (35.12)$$

The probability distribution (35.12) is determined by the two quantum numbers l and m, i.e. it depends on the value of the total **angular momentum** and its projection on the z-axis.

The state with $l = 0$ (s-state) possesses spherical symmetry, because for $l = 0$ (consequently also $m = 0$) $P_0^0 = \text{const}$

$$dW_{00} = \frac{1}{4\pi} \, d\Omega . \qquad (35.13)$$

In the p-state ($l = 1$) the probability distribution is given by the following expressions:

$$dW_{1,\pm1} = \frac{3}{8\pi}\sin^2\vartheta\,d\Omega\,, \qquad dW_{10} = \frac{3}{4\pi}\cos^2\vartheta\,d\Omega\,. \qquad (35.14)$$

The distributions (35.12) for different l and m are presented graphically in fig. V.9 in the form of polar diagrams. The probability $dW_{lm}/d\Omega$ is plotted on the radius vector drawn at the angle ϑ to the z-axis.

Let us consider eq. (35.8) for the radial component of the wave function in more detail. First of all, it follows from this equation that the energy of the particle does not depend on the z-component of the angular momentum. This is associated with the fact that in a spherically symmetric field all directions are equivalent. Thus the isotropy of space leads to degeneracy of the levels of the system in which the energy does not depend on the quantum number m. It should be noted that degeneracy is always due to definite symmetry properties of the system considered.

Instead of the function R it is convenient to introduce the function $\chi(r)$:

$$R(r) = \frac{1}{r}\chi(r)\,. \qquad (35.15)$$

For $\chi(r)$ we find

$$\frac{d^2\chi}{dr^2} + \frac{2\mu}{\hbar^2}\left(E - U(r) - \frac{\hbar^2}{2\mu}\frac{l(l+1)}{r^2}\right)\chi = 0\,. \qquad (35.16)$$

The condition of the finiteness of the wave function for $r = 0$ leads to the requirement

$$\chi(0) = 0\,. \qquad (35.17)$$

The equation for the radial function (35.8) amounts to the equation of a one-dimensional motion with an effective potential energy equal to

$$U_{\text{eff}}(r) = U(r) + \frac{\hbar^2}{2\mu}\frac{l(l+1)}{r^2}\,. \qquad (35.18)$$

As in classical mechanics, the quantity $\hbar^2 l(l+1)/2\mu r^2$ is called the centrifugal energy.

Without fixing the detailed form of the potential energy $U(r)$, one can nevertheless make definite conclusions about the behaviour of the wave function near the origin and at very large distances from the force centre.

Let us first study the region of small distances $r \to 0$. We assume that near the origin the potential energy of interaction $U(r)$ changes so slowly that the

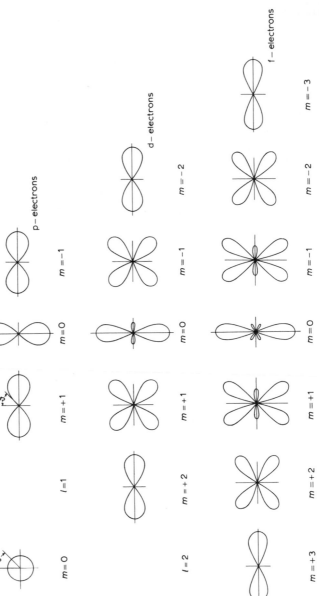

Fig. V.9

following condition holds:

$$\lim_{r \to 0} r^2 U(r) = 0 .$$ (35.19)

This condition means that $|U(r)|$ for $r \to 0$ increases more slowly than $1/r^2$. It is fulfilled, in particular, for the electron in the Coulomb field of the nucleus. Then in eq. (35.16) for $r \to 0$ the terms $E\chi$ and $U(r)\chi$ can be disregarded in comparison with the term $\hbar^2 l(l + 1)/2\mu r^2 \chi$, and we obtain

$$\frac{d^2\chi}{dr^2} - \frac{l(l + 1)}{r^2} \chi = 0 .$$

We seek the solution of the last equation in the form $\chi = Ar^\gamma$. Substituting this expression into the equation, we have

$$\gamma(\gamma - 1) = l(l + 1) .$$ (35.20)

Equation (35.20) has two roots: $\gamma_1 = l + 1$; $\gamma_2 = -l$. We must discard the second root, since it corresponds to a function R which increases indefinitely for $r \to 0$. Thus we find that at small distances $\chi(r) \sim r^{l+1}$, and the radial part of the wave function is expressed by the formula

$$R(r) = Ar^l .$$ (35.21)

The probability of finding the particle at a given distance r from the centre, independent of the angles ϑ and φ, is given by the square of the modulus of the radial function, i.e. by the quantity $|R|^2 r^2 \, dr$.

It follows from (35.21) that for small r this probability is proportional to $r^{2l+2} dr$ and is smaller the larger l becomes. The centrifugal force acts as if to throw the particle out from the centre.

Further, we study the asymptotic behaviour of the wave function at large distances from the origin. At such distances the force acting on the particle tends to zero, and consequently, the potential energy $U(r)$ tends to a constant. If not specified otherwise, we shall choose this constant as the zero of potential energy, i.e. we shall assume that $\lim_{r \to \infty} U(r) = 0$. Then in eq. (35.16), for large r, the terms $U\chi$ and $\hbar^2 l(l + 1)/2\mu r^2\chi$ can be disregarded* in comparison with the term $E\chi$. In this case eq. (35.16) assumes the

---

* From a more detailed analysis it follows that this is legitimate if the potential energy at infinity decreases according to the law $1/r^n$, where $n > 1$. See, for example, L.D. Landau and E.M. Lifshitz, *Quantum mechanics* (Pergamon Press, Oxford, 1965); V.A. Fok, *Nachala kvantovoi mekhaniki* (*Principles of quantum mechanics*) (KUBUCH, 1932) p. 126.

form

$$\frac{d^2\chi}{dr^2} + k^2\chi = 0 , \qquad k = \left(\frac{2\mu E}{\hbar^2}\right)^{\frac{1}{2}} . \tag{35.22}$$

The solution of this last equation can obviously be written as

$$\chi = A_1 e^{ikr} + A_2 e^{-ikr} , \tag{35.23}$$

where A_1 and A_2 are integration constants.

Let us consider, first of all, the solutions corresponding to positive values of the energy. For $E > 0$ the quantity k defined by formula (35.22) is real. The radial part of the wave function (35.15) amounts to the sum of the two functions

$$R(r) = A_1 \frac{e^{ikr}}{r} + A_2 \frac{e^{-ikr}}{r} . \tag{35.24}$$

Since both terms are restricted in modulus, neither of the constants A_1 and A_2 can be equal to zero. At a large distance from the force centre the radial function represents the superposition of a converging and a diverging spherical wave.

A definite conclusion can be also made about the energy spectrum of a particle for an arbitrary form of the energy of interaction $U(r)$. Indeed, the function (35.23) does not reduce to zero at infinity, which corresponds to an infinite motion, i.e. a motion in which the particle or the system goes off to infinity. The integral of the square of the modulus of function (35.24), taken over all space, diverges. But, as we have noted in §16, such functions correspond to a continuous spectrum. Consequently, for $E > 0$ the energy spectrum is continuous. If the radial component of the current density is equal to zero, then function (35.24) must be real. Correspondingly we assume that

$$A_1 = \frac{1}{2i} A' e^{i\alpha} , \qquad A_2 = -\frac{1}{2i} A' e^{-i\alpha} , \tag{35.25}$$

A' and α being real.

Then corresponding to (35.24) the radial function R assumes the form

$$R = A' \frac{\sin (kr + \alpha)}{r} , \tag{35.26}$$

where the phase α depends on k, l, as well as on the actual form of the function $U(r)$. In the following section we shall show that for a free particle

$(U \equiv 0)$

$$\alpha = -\tfrac{1}{2}l\pi .$$

In accordance with this we assume that

$$\alpha = -\tfrac{1}{2}l\pi + \delta_l , \tag{35.27}$$

where the phases δ_l are directly connected with the action of the force field on the particle and reduce to zero for free motion.

We now consider the region of negative energies, $E < 0$. Since the kinetic energy of the particle is always positive, the total energy can be negative only in the case of attraction of the particle towards the centre. If $E < 0$, the quantity k has purely imaginary values, i.e. $k = i\kappa$, where $\kappa = (-2\mu E/\hbar^2)^{\frac{1}{2}}$. The radial function (35.24) is written in the form

$$R = A_1 \frac{e^{-\kappa r}}{r} + A_2 \frac{e^{\kappa r}}{r} . \tag{35.28}$$

In order to satisfy the requirement of the finiteness of the wave function for $r \to \infty$, we have to assume that the constant A_2 is equal to zero

$$R = A_1 \frac{e^{-\kappa r}}{r} . \tag{35.29}$$

Then the radial wave function R tends to zero as $r \to \infty$. This means that the probability of finding the particle at an infinitely large distance from the force centre is equal to zero. Consequently the motion of the particle is finite. We see that there is a similarity between the conclusions of quantum and classical mechanics: for a positive total energy $(E > U(\infty))$ the particles go off to infinity, while for a negative total energy they perform a finite motion.

Let us now consider the energy spectrum for $E < 0$. As we have explained, a finite motion corresponds to these energies and the corresponding wave functions (35.29) are quadratically integrable. Such wave functions, as was pointed out in §16, belong to a discrete spectrum. Consequently, for $E < 0$ we have a discrete energy spectrum.

The general solution of the Schrödinger equation (35.2) can be written in the form of a superposition of the wave functions (35.7)

$$\psi(r, \vartheta, \varphi) = \sum_{l,m} B_{lm} R_l(r) Y_{lm}(\vartheta, \varphi) . \tag{35.30}$$

For a solution which does not depend on the angle φ we obtain a simpler expression (superposition of states with $m = 0$)

$$\psi(r, \vartheta) = \sum_l c_l R_l(r) P_l(\cos \vartheta) . \tag{35.31}$$

§36. The free motion of a particle with given angular momentum

So far we have represented a freely moving particle by a plane wave $e^{i(\mathbf{k \cdot r} - \omega t)}$, where $\mathbf{k}$ is the wave vector of the particle $\mathbf{k} = \mathbf{p}/\hbar$, and $\omega = E/\hbar$. This wave function describes a stationary state with a definite value of the momentum and energy $E = p^2/2m$ of the particle. For what follows, we need to find the wave functions of the stationary states of a freely moving particle, in which, in addition to a definite value of the energy E, the values of the angular momentum and the z-component of the angular momentum are also given. In classical mechanics a free particle moving with a definite momentum also possesses automatically a definite angular momentum. In quantum mechanics the situation is fundamentally altered. In a state with a given momentum the angular momentum is an indefinite quantity. On the other hand, in a state where the angular momentum and its projection onto the z-axis are given the direction of the momentum is indefinite. This is associated with the fact that the corresponding quantities cannot simultaneously have sharp values.

In order to find the wave function required, let us consider the motion of a free particle in spherical coordinates. Setting $U(r) \equiv 0$ in the Schrödinger equation (35.3), we have

$$-\frac{\hbar^2}{2m} \frac{1}{r^2} \frac{\partial}{\partial r} \left(r^2 \frac{\partial \psi}{\partial r} \right) + \frac{\hat{l}^2}{2mr^2} \psi = E\psi . \tag{36.1}$$

We seek the wave function of the free particle in the form

$$\psi_{klm}(r, \vartheta, \varphi) = R_{kl}(r) Y_{lm}(\vartheta, \varphi) . \tag{36.2}$$

In this case the radial function R_{kl} must satisfy eq. (35.8) in which one must set $U \equiv 0$

$$\frac{1}{r^2} \frac{d}{dr} \left(r^2 \frac{dR_{kl}}{dr} \right) + \left(k^2 - \frac{l(l+1)}{r^2} \right) R_{kl} = 0 . \tag{36.3}$$

Here we have expressed the energy E in terms of the wave number k. For

$l = 0$ the equation is rewritten in the form

$$\frac{1}{r^2} \frac{d}{dr} \left(r^2 \frac{dR_{k0}}{dr} \right) + k^2 R_{k0} = 0 . \tag{36.4}$$

The solution of the above equation which does not go to infinity at the origin is the function

$$R_{k0} = A \frac{\sin kr}{r} . \tag{36.5}$$

To find the solution of eq. (36.3) for $l \neq 0$ we introduce a new function given by the formula

$$R_{kl} = s^{-\frac{1}{2}} Z \tag{36.6}$$

where $s = kr$. For such a substitution eq. (36.3) is easily transformed to the form

$$\frac{d^2 Z}{ds^2} + \frac{1}{s} \frac{dZ}{ds} + \left(1 - \frac{(l + \frac{1}{2})^2}{s^2} \right) Z = 0 . \tag{36.7}$$

The solution of eq. (36.7) satisfying the condition of finiteness of the wave function at the origin is a Bessel function of half-integer order

$$Z(s) = C J_{l+\frac{1}{2}}(s) . \tag{36.8}$$

Correspondingly for the radial function we have

$$R_{kl} = (kr)^{-\frac{1}{2}} C J_{l+\frac{1}{2}}(kr) . \tag{36.9}$$

At a large distance from the origin ($r \to \infty$) one can make use of the known asymptotic expression for the Bessel function and obtain the asymptotic value $R_{kl}(r)$

$$R_{kl}(r) = C(2/\pi)^{\frac{1}{2}} \frac{\sin (kr - \frac{1}{2} l\pi)}{kr} . \tag{36.10}$$

The constant C is determined by the normalization condition. At small distances from the force centre ($r \to 0$) the radial function (36.9) assumes the form

$$R_{kl} \sim r^l \tag{36.11}$$

in accordance with the general expression (35.21).

§37. The spherical well

As a simple and at the same time important example we shall consider the motion of a particle in a centrally symmetric field defined by the expression

$$U(r) = \begin{cases} -U_0 & (r \leqslant a), \\ 0 & (r > a). \end{cases}$$

A field of this type is called a spherically symmetric potential well. The potential well shown in fig. V.10 represents an idealized model of a system in which the interaction with the centre is realized by so-called short-range forces. Short-range forces are understood to be forces which decrease with distance so rapidly that they can be assumed to be practically equal to zero at distances exceeding a certain distance, a, called the range of the short-range force. The importance of the consideration of systems with short-range forces is clear, for example, from the fact that the forces of interaction between nucleons, nuclear forces, are of such a type.

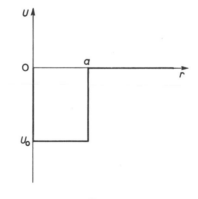

Fig. V.10

The idealization of a system by means of the model of a spherical potential well amounts to the assumptions of total isotropy of the forces and the constancy of the potential energy for $r < a$.

For simplicity let us consider the motion of a particle with angular momentum $l = 0$. It is obvious that two different modes of motion are possible. For $E < 0$ the total energy of the particle is smaller than the potential energy at infinity, which corresponds to a finite motion. On the contrary, for $E > 0$ there is an infinite motion. To the first case, to which we now confine ourselves, there corresponds a discrete energy spectrum, while to the second case there corresponds a continuous spectrum.

The wave function of a particle with $l = 0$ depends only on the coordinate r, and not on the angles ϑ and φ. Upon the substitution $\chi(r) = rR(r)$ the Schrödinger equation will have the form (35.16)

$$\frac{d^2\chi}{dr^2} + \frac{2m}{\hbar^2}(E + U_0)\chi = 0 \qquad (r \leqslant a), \tag{37.1}$$

$$\frac{d^2\chi}{dr^2} + \frac{2m}{\hbar^2}E\chi = 0 \qquad (r > a). \tag{37.2}$$

We write the solution of eq. (37.1) in the form

$$\chi(r) = A \sin \kappa r + B \cos \kappa r, \tag{37.3}$$

where

$$\kappa = \left(\frac{2m}{\hbar^2}(U_0 - |E|)\right)^{\frac{1}{2}}.$$

For the wave function R to be finite at the origin it is necessary to set $\chi(0) = 0$. Consequently, inside the well the solution of eq. (37.1) has the form

$$\chi(r) = A \sin \kappa r. \tag{37.4}$$

The solution outside the well which reduces to zero at infinity is expressed by the formula

$$\chi(r) = Be^{-\kappa' r}, \tag{37.5}$$

where κ' denotes the quantity $\kappa' = [(2m/\hbar^2)|E|]^{\frac{1}{2}}$.

It follows from the continuity of the wave function that solution (37.4) must go over continuously into solution (37.5) at the surface of the sphere $r = a$. The derivative of the wave function also must be continuous at this surface. Hence we can equate to each other the logarithmic derivative of the functions (37.4) and (37.5) for $r = a$. We then obtain

$$\kappa \cotan \kappa a = -\kappa'. \tag{37.6}$$

This relation can be rewritten in the form

$$\sin \kappa a = \pm \left[\frac{\kappa'^2}{\kappa^2} + 1\right]^{-\frac{1}{2}} \tag{37.7}$$

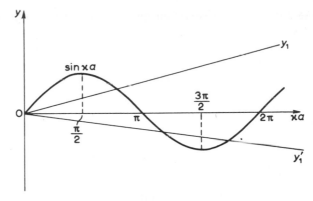

Fig. V.11

or, taking into account the expressions for κ and κ', we have

$$\sin \kappa a = \pm \left(\frac{\hbar^2}{2mU_0 a^2} \right)^{\frac{1}{2}} \kappa a . \tag{37.8}$$

The roots of eq. (37.8) determine the energy levels of the particle in the well. Equation (37.8) is conveniently solved graphically. Namely, the roots of eq. (37.8) are the intersections of the straight lines

$$y_1 = \left(\frac{\hbar^2}{2mU_0 a^2} \right)^{\frac{1}{2}} \kappa a \quad \text{and} \quad y_1' = - \left(\frac{\hbar^2}{2mU_0 a^2} \right)^{\frac{1}{2}} \kappa a$$

with the curve $\sin \kappa a$ (see fig. V.11). Only crossover points for which $\cotan \kappa a$ has negative values may be chosen in correspondence with (37.6). From the graph in fig. V.11 it is seen that the roots of eq. (37.8) do not always exist. In order that a bound state (energy level) may exist the well must be sufficiently deep. Let us determine the minimum depth $U_{0\,\text{min}}$ corresponding to the appearance of the first energy level. As is seen from fig. V.11, the first level will appear when the straight line passes through the peak of the sinusoidal curve at $\kappa a = \frac{1}{2}\pi$. The tangent of the slope angle is equal to $2/\pi$. Consequently, the minimum potential energy $U_{0\,\text{min}}$ for which there is a bound state of the particle in the spherical well is determined by the condition

$$\left(\frac{\hbar^2}{2mU_{0\,\text{min}} a^2} \right)^{\frac{1}{2}} = \frac{2}{\pi} ,$$

whence

$$U_{0\,\text{min}} = \frac{\pi^2 \hbar^2}{8ma^2} . \tag{37.9}$$

We find the first energy level in the potential well of minimum depth $U_{0\,\text{min}}$ from the condition $\kappa a = \frac{1}{2}\pi$ or

$$\left(\frac{2ma^2}{\hbar^2}(U_{0\,\text{min}} - |E_1|)\right)^{\frac{1}{2}} = \frac{\pi}{2}.$$

Taking into account the value of $U_{0\,\text{min}}$ we find that $E_1 = 0$, i.e. the energy of the particle in the first level is equal to zero and there are no other levels in the well. Also, the energy of the first level decreases with increasing depth of the well and becomes negative. In the graph this corresponds to a decrease in the slope of the straight line with respect to the abscissa. For a certain slope another root will appear in addition to the root corresponding to the first level. This new root corresponds to the appearance of a second energy level in the well. The number of crossover points in the graph increases with increasing depth of the well, which corresponds to an increase in the number of allowed energy levels of the particle in the potential well.

In conclusion we stress that the absence of bound states for a particle in a potential well of depth $U_0 < U_{0\,\text{min}}$ represents a specific quantum-mechanical effect which has no analogue in classical physics. Indeed, however small the depth of the well in classical physics, a particle which falls into it with an initial kinetic energy less than the depth of the well will be confined in it. In quantum mechanics this proposition, in general, does not hold.

§38. Motion in a Coulomb field

As we have already pointed out, the most important example of the motion of a particle in a centrally symmetric field is the motion of an electron in the Coulomb field of the atomic nucleus. The simplest atomic system of such a kind, consisting of a nucleus and an electron, is the hydrogen atom, and also the ion of any atom in which only one electron remains. The meso-hydrogen atom, consisting of a proton and a negatively charged meson, is another example.

The problem of the motion of two bodies, a nucleus and an electron, reduces to the problem of the motion of one particle with reduced mass μ in the Coulomb field (see §14).

It is clear that the theory of the hydrogen atom and hydrogen-like systems is extremely important, since these systems are the simplest atomic systems. Furthermore it turns out that in the case of the motion of a particle in the Coulomb field of a nucleus one can obtain a complete analytical solution of the Schrödinger equation. This makes it possible to follow the appearance of general quantum-mechanical regularities in atomic systems.

The potential energy of an electron moving in the field of a nucleus with charge Ze is given by the formula

$$U(r) = -\frac{Ze^2}{r} . \tag{38.1}$$

We write the Schrödinger equation for the radial wave function (35.8)

$$\frac{d^2R}{dr^2} + \frac{2}{r}\frac{dR}{dr} - \frac{l(l+1)}{r^2}R + \frac{2\mu}{\hbar^2}\left(E + \frac{Ze^2}{r}\right)R = 0 . \tag{38.2}$$

We are at first interested in states belonging to a discrete energy spectrum. In correspondence with §16 these states correspond to a finite motion of the electron and, consequently, their energy is negative, $E < 0$ (see §35).

In solving eq. (38.2) it is convenient to use dimensionless quantities. This will make all formulae less cumbersome.

We choose as basic quantities the charge of the electron e, its reduced mass μ, and the Planck constant $\hbar$. From these quantities one can make a combination having the dimensionality of a length

$$a = \frac{\hbar^2}{\mu e^2} . \tag{38.3}$$

As we shall see below, this length is a characteristic atomic dimension. If the reduced mass μ is set equal to the mass of the electron m, then $a = 0.529\times10^{-8}$ cm.

The system of units based on the quantities e, μ and a is called the Coulomb system.

The quantity $e^2/\hbar$, equal to 1/137 of the velocity of light, will be the unit of velocity, while the quantity

$$E_0 = \frac{\mu e^4}{\hbar^2} = \frac{e^2}{a} \tag{38.4}$$

will be the unit of energy. For $\mu = m$, $E_0 = 4.30\times10^{-11}$ erg = 27.07 eV. We introduce into eq. (38.2) the dimensionless variable ρ and the energy ϵ

$$\rho = r/a , \qquad \epsilon = -E/E_0 . \tag{38.5}$$

Then this equation is rewritten in the form

$$\frac{d^2R}{d\rho^2} + \frac{2}{\rho}\frac{dR}{d\rho} + \left(-2\epsilon - \frac{l(l+1)}{\rho^2} + \frac{2Z}{\rho}\right)R = 0 . \tag{38.6}$$

At small distances the function R behaves, according to (35.21), as ρ^l. At large distances this function has the form $R \sim \exp\left[-(2\epsilon)^{\frac{1}{2}}\rho\right]$ (see (35.29)). Corresponding to this we shall seek the solution of eq. (38.6) in the form

$$R(\rho) = \rho^l e^{-\beta\rho} v(\rho) . \tag{38.7}$$

where $\beta = (2\epsilon)^{\frac{1}{2}}$. Substituting expression (38.7) into (38.6), we obtain after simple calculations

$$\rho \frac{d^2 v}{d\rho^2} + 2 \frac{dv}{d\rho} (l - \beta\rho + 1) + 2v(Z - \beta - \beta l) = 0 . \tag{38.8}$$

We introduce a new variable

$$\xi = 2\beta\rho . \tag{38.9}$$

Denoting differentiation with respect to this new variable by a prime, we have

$$\xi v'' + v'(2l + 2 - \xi) + v\left(\frac{Z}{\beta} - l - 1\right) = 0 . \tag{38.10}$$

The radial wave function R must remain finite over the entire region of variation of the variable ξ, for $\xi \to \infty$ as well as for $\xi \to 0$.

We seek the solution of eq. (38.10) in the form of a series

$$v(\xi) = \sum_{k=0}^{\infty} a_k \xi^k . \tag{38.11}$$

Substituting (38.11) into eq. (38.10) and gathering terms with the same powers of ξ, we obtain

$$\sum_k \xi^k \left[(k + 1)(2l + 2 + k)a_{k+1} + \left(\frac{Z}{\beta} - l - 1 - k\right) a_k \right] = 0 . \tag{38.12}$$

Equation (38.12) will be satisfied for arbitrary values of ξ if the coefficients of all powers of ξ are equal to zero. Hence, equating the square bracket to zero we arrive at the following recurrence formula:

$$a_{k+1} = \frac{k + l + 1 - (Z/\beta)}{(k + 1)(2l + 2 + k)} a_k . \tag{38.13}$$

We note that the function v defined by the series (38.11), with coefficients a_k which satisfy (38.13), can be expressed in terms of the confluent hypergeometric function*

$$v = AF\left(1 + l - \frac{Z}{\beta}, 2l + 2, \xi\right).$$ (38.14)

It is easily shown, by analogy with what was done in §10, that the series (38.11) diverges as e^{ξ} for $\xi \to \infty$. This means that if the wave function were expressed by the series (38.11) it would not satisfy the condition of being finite at arbitrarily large distances from the force centre. In order to define the function which possesses the necessary properties and is a solution of eq. (38.10) we must, as was done in solving the problem of the oscillator, cut off the series at a certain term, i.e. reduce it to a polynomial. If for a certain value of the number $k = n_r$ the coefficient a_{n_r+1} reduces to zero, then according to (38.13) all subsequent coefficients a_{n_r+2}, a_{n_r+3} and so on also reduce to zero. In this case the infinite series reduces to a polynomial of the n_rth degree. For large values of ξ the function $v(\xi)$ will increase according to the power law $v(\xi) \sim \xi^{n_r}$, while the wave function will tend to zero at infinity on account of the exponential factor. For $\xi \to 0$ the polynomial $v(\xi)$ tends to the constant quantity a_0 and the wave function (38.7) correspondingly reduces to zero or tends to a constant. Thus we see that the wave function will satisfy the standard boundary conditions.

Let us now consider the conditions under which the coefficient of the series a_{n_r+1} reduces to zero. For this it is necessary, according to (38.13), that

$$n_r + l + 1 - \frac{Z}{\beta} = 0.$$ (38.15)

Since n_r is an integer (including zero), the sum $(n_r + l + 1)$ is also an integer. We denote it by n; $n = n_r + l + 1$. The integer n is called the principal quantum number, and n_r is called the radial quantum number. For a fixed value of the angular momentum quantum number l we have

$$n \geqslant l + 1.$$

It is obvious that relation (38.15) determines the ordering of the energy levels of the system. Taking into account the value of β, we find

$$\epsilon = \frac{Z^2}{2n^2}.$$ (38.16)

---

* V.I.Smirnov, *A course of higher mathematics* (Pergamon Press, Oxford, 1964).

Passing over from atomic units to ordinary units, (38.4) and (38.5), we obtain

$$E_n = -\frac{\mu e^4 Z^2}{2\hbar^2 n^2} = -13.5 \frac{Z^2}{n^2} \, \text{eV} \, . \tag{38.17}$$

This formula, first obtained by N. Bohr before the appearance of modern quantum mechanics, determines the discrete energy levels in the hydrogen atom and hydrogen-like ions. We see that the energy levels depend only on the principal quantum number n. The lowest energy level (the ground state) of the particle in the Coulomb field corresponds to the value $n = 1$. The spacing between levels decreases with increasing n, the levels coming nearer to each other. As $n \to \infty$, $\Delta E \to 0$ and the discrete spectrum goes over into a continuous one.

The radial function R_{nl} is given by the formula

$$R_{nl} = \text{const} \, \xi^l e^{-\xi/2} v(\xi) \, , \tag{38.18}$$

where the polynomial $v(\xi)$, with coefficients determined by the recurrence formula (38.13), coincides except for a constant factor with the generalized Laguerre polynomial*. Hence in our case the radial function assumes the form

$$R_{nl} = A_{nl} \xi^l e^{-\xi/2} L_{n+l}^{2l+1}(\xi) \, . \tag{38.19}$$

The generalized Laguerre polynomial $L_n^m(\xi)$ is expressed in terms of the derivatives of the Laguerre polynomials which are determined by the relation

$$L_n(\xi) = e^\xi \frac{d^n}{d\xi^n} (e^{-\xi} \xi^n) \, , \tag{38.20}$$

so that

$$L_n^m(\xi) = \frac{d^m}{d\xi^m} L_n(\xi) \, . \tag{38.21}$$

The coefficients A_{nl} of (38.19) are determined from the normalization condition**. The radial wave functions belonging, for example, to the two lowest energy levels have the form

$$R_{10}(\rho) = 2 \left(\frac{Z^3}{a^3} \right)^{\frac{1}{2}} e^{-Z\rho} \, , \tag{38.22}$$

---

* See the reference on p. 138.

** The calculation of the normalization integral is carried out, for example, in the book of L.D.Landau and E.M.Lifshitz, *Quantum mechanics* (Pergamon Press, Oxford, 1965).

$$R_{20}(\rho) = \left(\frac{Z^3}{2a^3}\right)^{\frac{1}{2}} e^{-\frac{1}{2}Z\rho}(1 - \tfrac{1}{2}Z\rho), \tag{38.23}$$

$$R_{21}(\rho) = \left(\frac{Z^3}{6a^3}\right)^{\frac{1}{2}} e^{-\frac{1}{2}Z\rho}\,\tfrac{1}{2}Z\rho. \tag{38.24}$$

Here the variable ρ (see (38.9)) is again introduced instead of ξ. We stress that the wave function is determined by the whole set of values of the three quantum numbers n, l, and m, whereas the energy levels (38.17) depend only on the principal quantum number n. Thus the energy levels of the hydrogen atom are degenerate. We have seen in §35 that degeneracy in the magnetic quantum number m is a general property of motion in a centrally symmetric field. However, in a Coulomb field the energy levels turn out to also be degenerate in the angular momentum quantum number l. This degeneracy is characteristic only for motion in a Coulomb field. A slight change in the law of force and the energy becomes dependent on the angular momentum quantum number. Hence the degeneracy characteristic of the Coulomb field is called an accidental degeneracy. Let us find the multiplicity of the degeneracy of the nth energy level. Since for a given n the angular momentum quantum number runs over all integers from 0 up to $n-1$ and, in its turn, to each l there correspond $2l+1$ possible values of the quantum number m, the degeneracy is equal to

$$\sum_{l=0}^{n-1} (2l+1) = n^2. \tag{38.25}$$

To each energy level E_n there belong n^2 different wave functions.

Let us consider in more detail the energy levels of the hydrogen atom. They are given by formula (38.17), in which one must set $Z = 1$. The energy of the ground state determines the ionization potential of the hydrogen atom. According to the quantum theory of light emission, the differences between the energy states determine the frequency of electromagnetic waves emitted by the atom (see §103):

$$\hbar\omega = E_m - E_n. \tag{38.26}$$

The quantity $E_n/\hbar$ is called the spectral term. The differences between these spectral terms determine the frequencies of radiation. Substituting expression (38.17) into formula (38.26), we obtain

$$\nu = \frac{\omega}{2\pi} = R\left(\frac{1}{n^2} - \frac{1}{m^2}\right), \qquad m > n. \tag{38.27}$$

The quantity R is called the Rydberg constant

$$R = \frac{e^4 \mu}{4\pi\hbar^3} = 3.27 \times 10^{15} \text{ sec}^{-1} . \tag{38.38}$$

All frequencies referring to transitions to one and the same lower level form a spectral series. Thus if we set $n = 1$ in formula (38.27), we obtain the Lyman series. It lies in the ultraviolet part of the spectrum. The transitions to the level $n = 2$ lie in the visible part of the spectrum. The whole set of these spectral lines forms the Balmer series. The spectral series corresponding to transitions to the levels $n = 3$ and so on lie in the infrared region of the spectrum. For hydrogen-like ions the corresponding spectral lines are shifted toward shorter wavelengths, because the frequencies increase by a factor of Z^2.

Further, we find the probability (35.10) of observing the electron in different quantum states at a given distance r from the nucleus. The ground state of the electron in the hydrogen atom is described by the wave function $\psi_{100} = R_{10} Y_{00}$. For $l = 0, m = 0$ the angular part of the wave function reduces to a constant (see §30), i.e. the state is spherically symmetric. The probability of observing the electron in the ground state ψ_{100} at a given distance from the nucleus is given by the expression

$$dW_{10} = |\psi_{100}|^2 4\pi r^2 \, dr .$$

Making use of (38.22), we obtain

$$dW_{10} = \frac{4}{a^3} e^{-2r/a} r^2 \, dr . \tag{38.29}$$

We see that the probability is different from zero over all space, although it decreases rapidly with increasing r. A simple calculation shows that the curve dW_{10}/dr has a maximum at the distance $r = a$, where the quantity a is determined by formula (38.3) and is called the Bohr radius. The form of the function $|R_{nl}|^2 r^2$ for different n and l is shown in fig. V.12. The distance from the centre $\rho = r/a$ is measured along the abscissa, and the probability density $a^3 |R_{nl}|^2 \rho^2$ is measured along the ordinate. We note that the number of zeros of the radial wave function R_{nl} is equal to the value of the radial quantum number n_r. At large distances the radial wave function has the form

$$R_{nl}(r) \sim e^{-Zr/na} \left(\frac{2Zr}{na}\right)^{n-1} + \dots . \tag{38.30}$$

The probability density calculated by means of this function rapidly decreases at distances above the order of magnitude of na/Z. Hence it is seen that the

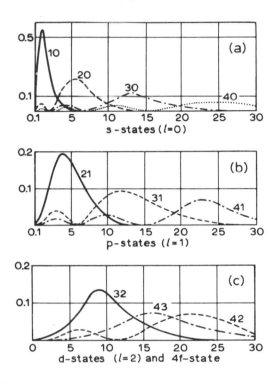

Fig. V.12

quantity na/Z characterizes the size of the atom, because the probability of observing the electron at larger distances is very small.

Up to now we have considered the bound states of an electron in the Coulomb field of the nucleus. Other negatively charged particles, for example π-mesons and muons, can also be in a bound state in the Coulomb field. As we have already mentioned, such systems are called mesic atoms.

As the simplest example let us consider the mesic atom of hydrogen or, as it is called, mesohydrogen. The energy levels of mesohydrogen and the wave functions of the meson are given by formulae (38.17)–(38.19) in which, however, the reduced mass μ of the electron must be replaced by the reduced mass μ' of the meson. The effective size of the atom of mesohydrogen is determined by the value of $a' = (\mu/\mu')a$, which is substantially smaller than the effective size of the hydrogen atom. In particular, the mass of the π^--meson is equal to 273 electron masses and, correspondingly, $a' \approx 0.2 \times 10^{-10}$ cm. In the mesohydrogen atom the π-meson is situated at considerably smaller distances from the nucleus than the electron. The presence of the nuclear interaction of the π-meson with the nucleus leads to a displacement of the energy levels (38.17) obtained for the pure Coulomb field. Experimental investigation of this displacement allows one to draw certain conclusions on the character of the nuclear interaction of π-mesons and nucleons. It should be noted that the lifetime of mesic atoms is restricted by the lifetime of the mesons themselves. As is known, mesons are unstable particles undergoing decay with a mean lifetime τ which is characteristic of the given kind of meson.

Up to now we have restricted ourselves to the consideration of the discrete energy spectrum, i.e. we have considered the energy to be negative.

Let us now consider the continuous energy spectrum $E > 0$, $\epsilon = -E/E_0 < 0$ (38.5). We introduce the following notation taking into account (38.7), (38.9) and (38.15):

$$\beta = (2\epsilon)^{\frac{1}{2}} = i(2E/E_0)^{\frac{1}{2}} = ik, \qquad n = Z/\beta = -iZ/k, \qquad \xi = 2ik\rho. \tag{38.31}$$

Making use of (38.7), (38.14) and (38.31), we write the radial wave function of the continuous spectrum in the form

$$R_{kl} = \frac{C_k}{(2l+1)!} (2k\rho)^l e^{-ik\rho} F\left(1 + l + i\frac{Z}{k}; \quad 2l+2, \quad 2ik\rho\right), \tag{38.32}$$

Here C_k is a normalization factor.

If the functions R_{kl} are normalized to the δ-function in k, then this factor is equal to

$$C_k = \left(\frac{2}{\pi}\right)^{\frac{1}{2}} \frac{k}{Z} e^{\pi Z/2k} \left|\Gamma\left(l+1-\frac{iZ}{k}\right)\right|. \tag{38.33}$$

The asymptotic expression of the radial wave function for large ρ is determined by the formula*

$$R_{kl} \approx \left(\frac{2}{\pi}\right)^{\frac{1}{2}} \frac{1}{Z\rho} \sin\left(k\rho + \frac{Z}{k}\ln 2k\rho - \frac{\pi}{2}l + \delta_l\right), \tag{38.34}$$

where

$$\delta_l = \arg\Gamma\left(l+1-i\frac{Z}{k}\right)$$

* See L.D.Landau and E.M.Lifshitz, *Quantum mechanics* (Pergamon Press, Oxford, 1965); V.A.Fok, *Nachala kvantovoi mekhaniki* (*Principles of quantum mechanics*), (KUBUCH, 1932) p. 155.

(Γ is the gamma-function of a complex variable. Its argument is equal to δ_l).

   The expression of the wave function (38.34) (Coulomb field) differs from the general asymptotic expression of the radial wave function in a centrally symmetric field (35.26) by the presence of the slowly increasing logarithmic term in the argument of the sine.

5

The Quasi-classical Approximation

§39. The limiting transition to classical mechanics

We have more than once referred to the existence of the correspondence principle and the rules for the transition of the relations of quantum mechanics into the formulae of classical mechanics for $\hbar \to 0$. We shall now define more precisely the conditions of this transition and we shall at the same time obtain an important approximate method of solving the Schrödinger equation* (the WKB method).

If one sets $\hbar = 0$ in the Schrödinger equation

$$i\hbar \frac{\partial \psi}{\partial t} = \left(-\frac{\hbar^2}{2m} \nabla^2 + U \right) \psi \, , \tag{39.1}$$

it becomes meaningless. Hence to carry out the limiting transition mentioned above we write the wave function ψ in the form

$$\psi = e^{(i/\hbar)S} \, . \tag{39.2}$$

---

* The Wentzel–Kramers–Brillouin method. G.Wentzel, Z. Phys. 38 (1926) 518; L.Brillouin, Comptes Rendus 183 (1926) 24, J. de Physique 7 (1926) 353; H.A.Kramers, Z. Phys. 39 (1926) 828, J.Jeffreys, Proc. London Math. Soc. (2) 23 (1923) 428.

Substituting this expression into eq. (39.1), we obtain an equation for the function S:

$$-\frac{\partial S}{\partial t} = \frac{1}{2m}(\nabla S)^2 - \frac{i\hbar}{2m}\nabla^2 S + U \,, \tag{39.3}$$

We now formally expand the function S in powers of $\hbar/i$

$$S = S_0 + \left(\frac{\hbar}{i}\right)S_1 + \left(\frac{\hbar}{i}\right)^2 S_2 + \dots \,. \tag{39.4}$$

We substitute the expansion (39.4) into eq. (39.3) and equate the coefficients of the same powers of $\hbar$. We obtain two equations

$$-\frac{\partial S_0}{\partial t} = \frac{1}{2m}(\nabla S_0)^2 + U \,, \tag{39.6}$$

$$-\frac{\partial S_1}{\partial t} = \frac{1}{m}\nabla S_0 \nabla S_1 + \frac{1}{2m}\nabla^2 S_0 \,. \tag{39.6}$$

to within terms proportional to the first power of $\hbar$. Equation (39.5) is the same as the Hamilton–Jacobi equation of classical mechanics* for the action function S_0. This means that in the zeroth order approximation the motion of the particle follows the classical trajectory. To elucidate the meaning of eq. (39.6) we write the expression for the probability density of finding the particle at a given point of space in the form

$$\rho = |\psi|^2 = e^{2S_1} \,. \tag{39.7}$$

Multiplying (39.6) by ρ and taking into account that

$$\frac{\partial \rho}{\partial t} = 2\frac{\partial S_1}{\partial t}\rho \,; \qquad \nabla\rho = 2\nabla S_1 \rho \,,$$

we obtain

$$-\frac{\partial \rho}{\partial t} = \frac{1}{m}(\nabla S_0 \nabla \rho + \rho \nabla^2 S_0) = \nabla\cdot\left(\frac{1}{m}\rho\nabla S_0\right) \,. \tag{39.8}$$

Equation (39.8), equivalent to eq. (39.6), represents a continuity equation. It shows that the probability density moves in space with the same velocity $\mathbf{v} = m^{-1}\nabla S_0$ and on the same trajectory as the particle would move in classi-

---

* For the Hamilton–Jacobi equation in classical mechanics see L.D.Landau and E.M.Lifshitz, *Mechanics* (Pergamon Press, Oxford, 1960); H.Goldstein, *Classical mechanics* (Addison-Wesley, Cambridge, Mass., 1950).

cal mechanics. We note that, since the velocity is directed along the normal to the surfaces $S_0 = $ const, the trajectories of the classical particle are orthogonal to the surfaces $S_0 = $ const. In the quasi-classical approximation it is natural to call the surfaces $S = $ const the equi-phase surfaces of the wave function.

We now find the wave function of the stationary states of the particle in the quasi-classical approximation, confining ourselves to one-dimensional motion, so that $\psi = \psi(x, t)$. Because of the stationary state we have

$$\psi(x, t) = e^{-(i/\hbar)Et} \psi(x) . \tag{39.9}$$

In correspondence with this, in formulae (39.2) and (39.4) we set

$$S_0(x, t) = -Et + S_0'(x) , \tag{39.10}$$

while the functions $S_1, S_2, \ldots$ and so on can be assumed to be independent of time.

From eq. (39.5) we obtain

$$E = \frac{1}{2m} \left(\frac{dS_0'}{dx} \right)^2 + U(x) , \tag{39.11}$$

whence

$$S_0'(x) = \pm \int [2m(E - U(x))]^{\frac{1}{2}} \, dx = \pm \int^x p(x) \, dx , \tag{39.12}$$

where

$$p(x) = [2m(E - U(x))]^{\frac{1}{2}} .$$

As was to be expected, we have obtained the ordinary formulae of classical mechanics.

We can now determine the function S_1 from eq. (39.6). Taking into account that it is constant in time, we find

$$\frac{dS_0'}{dx} \frac{dS_1}{dx} + \frac{1}{2} \frac{d^2 S_0'}{dx^2} = 0 \tag{39.13}$$

or

$$\frac{dS_1}{dx} = -\frac{1}{2} \frac{d^2 S_0'/dx^2}{dS_0'/dx} = -\frac{1}{2p} \frac{dp}{dx} . \tag{39.14}$$

On integrating we obtain

$$S_1 = -\tfrac{1}{2} \ln p \tag{39.15}$$

(we shall take the integration constant into account directly in the expression for the wave function).

From the definition (39.2) and the expressions (39.10) and (39.15) we easily find the wave function of the particle with an accuracy to within terms up to the first order in powers of $\hbar/i$, for $E > U$ and $E < U$ respectively:

$$\psi(x) = \frac{C_1}{[p(x)]^{\frac{1}{2}}} \exp\left(\frac{i}{\hbar} \int p(x)\,dx\right) + \frac{C_2}{[p(x)]^{\frac{1}{2}}} \exp\left(-\frac{i}{\hbar} \int p(x)\,dx\right), \quad (39.16)$$

$$\psi(x) = \frac{C_1'}{[|p(x)|]^{\frac{1}{2}}} \exp\left(\frac{1}{\hbar} \int |p(x)|\,dx\right) + \frac{C_2'}{[|p(x)|]^{\frac{1}{2}}} \exp\left(-\frac{1}{\hbar} \int |p(x)|\,dx\right), \tag{39.16'}$$

The character of the wave function obtained depends critically on the sign of the difference $(E - U)$. If $E > U$, then the momentum is real. This corresponds to a motion of the particle in the region allowed by classical mechanics. In this case the wave function has the character of an oscillatory function. The period of oscillation is smaller, the larger the value of the momentum p. The factor $p^{-\frac{1}{2}} \sim v^{-\frac{1}{2}}$ has a simple meaning. The probability of finding the particle in a region from x to $x+dx$ is proportional to the time during which the particle is in this region; $|\psi(x)|^2\,dx \sim v^{-1}\,dx \sim dt$, i.e. the same result is obtained as in classical mechanics. The wave function in the region of forbidden energies, for $E < U$, has a completely different character. Here the momentum becomes imaginary, and the wave function goes over into a sum of exponential expressions. At the point $E = U$ (called the turning point) $p = 0$ and the expression obtained for the wave function is meaningless.

As is clear from what follows, the quasi-classical approximation becomes inapplicable near the turning point. However, without knowing the wave function at the turning point one cannot close the wave function at the boundary of the allowed and forbidden regions. In other words, one cannot determine the constants figuring in the oscillating and exponential expressions, and without this the quasi-classical wave functions have no practical validity. However, before considering the calculation of the wave function at the turning point it should be explained why the quasi-classical solution is meaningless at this point. For this we estimate the limits of applicability of the expressions (39.16) and (39.16'). First of all we note that in substituting $S = S_0$ into eq. (39.3) we have dropped the term $(i\hbar/2m)\nabla^2 S_0$ as negligible. For this to be correct the following inequality must be satisfied

$$\left|\frac{i\hbar}{2m} \nabla^2 S_0\right| \ll \frac{1}{2m} (\nabla S_0)^2, \tag{39.17}$$

or, taking into account that $\nabla S_0 = \mathbf{p}$,

$$\hbar|\nabla \mathbf{p}| \ll \mathbf{p}^2 . \tag{39.18}$$

For one measurement the above inequality can be rewritten in the form

$$\hbar\left|\frac{dp}{dx}\right| \ll p^2 . \tag{39.19}$$

Introducing the wave number k instead of the momentum p and the corresponding wavelength $\lambdabar = \hbar/p = k^{-1}$, we have

$$\frac{d\lambdabar}{dx} \ll 1 . \tag{39.20}$$

Thus we see that the Schrödinger equation reduces to eqs. (39.5) and (39.6) when the condition (39.20) is satisfied. Namely, for the applicability of the quasi-classical approximation it is necessary that the de Broglie wavelength should change sufficiently slowly from point to point in space. In other words, the relative change of the wave number over the extent of a wavelength must be small in comparison with unity $\lambdabar k^{-1} |dk/dx| \ll 1$. We note also that the relative change in the derivative of the wave number k over the extent of a de Broglie wavelength must be small. Indeed, in obtaining eqs. (39.5) and (39.6) we have also made use of the condition

$$\hbar|\nabla^2 S_1| \ll |\nabla^2 S_0| . \tag{39.21}$$

Taking into account that the problem is one-dimensional and introducing the wave number k, we rewrite this inequality in the form

$$\left|\frac{d^2 k}{dx^2}\right| \lambdabar \ll \left|\frac{dk}{dx}\right| . \tag{39.22}$$

For the motion of a particle in a potential field $U(x)$ it is convenient to express the wave number in terms of the potential energy according to the formula

$$k = \frac{p}{\hbar} = \left(\frac{2m}{\hbar^2}(E - U)\right)^{\frac{1}{2}}$$

and to write the condition (39.20) in the form

$$\left|\frac{m\hbar}{p^3}\frac{dU}{dx}\right| \ll 1 \tag{39.23}$$

or

$$\frac{m}{\hbar^2} \left| \lambdabar^3 \frac{dU}{dx} \right| \ll 1 \, .$$ (39.24)

Making use of (39.22) one can also write the corresponding inequality for the second derivative of the function U. Hence it is seen that the quasi-classical approximation is valid:

(1) when the de Broglie wavelength is sufficiently small (i.e. when the particle is moving sufficiently quickly;

(2) for a sufficiently slow change in the potential energy from point to point, when no considerable change in the momentum of the particle takes place over a length of the order of magnitude of $\lambdabar$.

It becomes clear from (39.23) why the quasi-classical expression for the wave function makes no sense at the turning point. Near the turning point the momentum of the particle becomes small and the quasi-classical approximation becomes inapplicable.

The formulations 'sufficiently small' and 'sufficiently slow' in (1) and (2) stress the fact that, since the criteria (39.20)–(39.24) involve the mass of the particle and the actual characteristics of the field – the quantity dU/dx – then for different fields and different particles the quasi-classical approximation will be valid for motions with different energies. For qualitative estimates one can rewrite (39.20) in a simplified form. Namely, assuming that the change in the wavelength takes place in a region of action of the field having extent a, one can write instead of (39.20) $\lambdabar \ll a$ or

$$E \gg \frac{\hbar^2}{2ma^2} \, .$$ (39.25)

For α-particles ($m = 6.7 \times 10^{-24}$ g) with energy $E = 1$ MeV passing through an atomic shell ($a \sim 10^{-8}$ cm) the inequality (39.25) is fulfilled to a good approximation. On the contrary, for the same α-particles with an energy of 10 MeV undergoing direct collision with a nucleus ($a \sim 10^{-13}$ cm) the quasi-classical consideration is inapplicable. In the region of substantially larger energies the application of the quasi-classical approximation turns out to be possible in considering certain processes connected with nuclear collisions.

§40. The solution of the Schrödinger equation near a turning point

Let us now return to the consideration of the behaviour of the wave function near a turning point.

The idea here is as follows: since near a turning point the quasi-classical approximation turns out to be unsatisfactory, it is necessary to find a solution of the Schrödinger equation without making use of this approximation. The possibility of obtaining such a solution for an arbitrary form of the potential energy is associated with the fact that the expression for the potential energy near a turning point makes possible an essential simplification (see below). If the solution sought is found, then one has to determine its asymptotic behaviour at large distances from the turning point in both directions, in those regions where the quasi-classical approximation is already valid. Requiring that the quasi-classical solution be the same as this asymptotic expression, we shall be able to determine the corresponding constants.

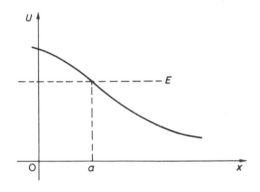

Fig. V.13

To carry out this programme, we note that one can expand the potential energy $U(x)$ in the vicinity of the turning point (fig. V.13) in a series with respect to the small displacement $\xi = x - a$ and retain the linear term of this expansion. In this case we assume that at the turning point the curve $U(x)$ is smooth, as is shown in fig. V.13. We shall also assume that the region $x > a$ extends to infinity. Near the point $x = a$ we can write

$$U(x) = U(a) + \frac{dU}{dx}\bigg|_{x=a} (x - a) + \dots . \tag{40.1}$$

The potential energy at the point a is the same as the total energy of the particle $U(a) = E$. We denote by f the force acting on the particle at the turning point, $f = - dU/dx|_{x=a}$, and introduce the new variable $\xi = x - a = = (E - U)/f$. We write the Schrödinger equation near the point $x = a$ as follows:

$$\frac{d^2\psi}{d\xi^2} + f\frac{2m}{\hbar^2}\xi\psi = 0 .$$ (40.2)

The Schrödinger equation was considered in such a form in §13. Equation (40.2) is the same as eq. (13.5) for $E = 0$. Consequently, the wave function satisfying eq. (40.2) and finite for $\xi \to \pm\infty$ is expressed in terms of the Airy function. We shall use immediately the asymptotic expressions (13.9) and (13.10). This means that we consider values of ξ which are sufficiently large for the asymptotic expressions to be used, and which are at the same time such that expansion (40.1) is still applicable. Such a region, as a rule, exists for fields satisfying the quasi-classical conditions.

Correspondingly, in the region $x \gg a$ the solution of eq. (40.2) can be written in the form

$$\psi = \frac{2C}{(2mf\xi)^{\frac{1}{4}}} \sin\left(\frac{2}{3\hbar}(2mf)^{\frac{1}{2}}\xi^{\frac{3}{2}} + \tfrac{1}{4}\pi\right)$$ (40.3)

where C is the normalization constant.

For motion in the field (40.1) the momentum p is of the form

$$p = [2m(E - U)]^{\frac{1}{2}} = (2mf\xi)^{\frac{1}{2}} .$$ (40.4)

We express the action in terms of the variable ξ

$$\int_a^x p\,dx = \int_0^\xi p\,d\xi = (2mf)^{\frac{1}{2}}\int_0^\xi \xi^{\frac{1}{2}}\,d\xi = \tfrac{2}{3}(2mf\xi^3)^{\frac{1}{2}} .$$ (40.5)

Making use of (40.5) we can write the wave function (40.3) in the form

$$\psi(x) = 2Cp^{-\frac{1}{2}}\sin\left(\hbar^{-1}\int_a^x p\,dx + \tfrac{1}{4}\pi\right) = 2Cp^{-\frac{1}{2}}\cos\left(\hbar^{-1}\int_a^x p\,dx - \tfrac{1}{4}\pi\right).$$
(40.6)

We see that the function (40.6) has the quasi-classical form (see (39.16)).

We now find the function in the region $x \ll a$. Again making use of the asymptotic expression (13.9) and expressions (40.4), and (40.5), we have

$$\psi = \frac{C}{(2mf\xi)^{\frac{1}{4}}}\exp\left[-\tfrac{2}{3}\hbar^{-1}(2mf\xi^3)^{\frac{1}{2}}\right] = \frac{C}{|p|^{\frac{1}{2}}}\exp\left[-\hbar^{-1}\int_x^a |p|\,dx\right],$$ (40.7)

where C is the same normalization constant as in formula (40.6). Thus we have obtained the expression for the quasi-classical wave function valid on both the left and right side of the turning point $x = a$.

We finally write the expressions for the quasi-classical wave function:

$$\psi(x) = \begin{vmatrix} C|p|^{-\frac{1}{2}} \exp\left(-\hbar^{-1} \int\limits_{x}^{a} |p|\,dx \right) & (x < a), \\ \\ 2Cp^{-\frac{1}{2}} \cos\left(\hbar^{-1} \int\limits_{a}^{x} p\,dx - \tfrac{1}{4}\pi \right) & (x > a). \end{vmatrix} \qquad (40.8)$$

The constant C is defined by the normalization condition.

Analogously, if the allowed region lies on the left of the turning point b, i.e. $U(x) < E$ for $x < b$ and $U(x) > E$ for $x > b$ (see, for example, fig. V.14 for $x > a$), then the wave function is written in the form

$$\psi(x) = \begin{vmatrix} C'|p|^{-\frac{1}{2}} \exp\left(-\hbar^{-1} \int\limits_{b}^{x} |p|\,dx \right) & (x > b), \\ \\ 2C'p^{-\frac{1}{2}} \cos\left(\hbar^{-1} \int\limits_{x}^{b} p\,dx - \tfrac{1}{4}\pi \right) & (x < b). \end{vmatrix} \qquad (40.9)$$

Thus we have found in the quasi-classical approximation the function $\psi(x)$ satisfying the Schrödinger equation. The solution obtained is still not complete, since a linear equation of second order has two linearly independent solutions. In the case of a wave function depending on one independent variable the other solution of the Schrödinger equation can easily be obtained. That is, if ψ_1 and ψ_2 are two linearly independent functions satisfying the one-dimensional Schrödinger equation corresponding to an energy E, then they are always connected by the relation

$$\frac{1}{\psi_1}\frac{d^2\psi_1}{dx^2} = \frac{1}{\psi_2}\frac{d^2\psi_2}{dx^2} = \frac{2m}{\hbar^2}(U - E). \qquad (40.10)$$

Integrating, we obtain

$$\psi_2\frac{d\psi_1}{dx} - \psi_1\frac{d\psi_2}{dx} = \text{const}. \qquad (40.11)$$

We seek the solution linearly independent of (40.8) in the form

$$\psi(x) = \begin{vmatrix} B_1 |p|^{-\frac{1}{2}} \exp\left(\hbar^{-1} \int_x^a |p|\, dx\right) & (x < a), \\[2em] B_2 p^{-\frac{1}{2}} \cos\left(\hbar^{-1} \int_a^x p\, dx + \alpha\right) & (x > a). \end{vmatrix} \qquad (40.12)$$

The expressions (40.8) and (40.12) are to be substituted into relation (40.11). Then, by virtue of inequality (39.20), it is sufficient to restrict oneself to differentiation with respect to the arguments of the exponential and trigonometric functions. Equating the expressions obtained for $x < a$ and $x > a$, we find $B_1 = B_2 \sin\left(\frac{1}{4}\pi + \alpha\right)$. Finally we set $B_1 = B_2$, $\alpha = \frac{1}{4}\pi$. Thus the solution linearly independent of (40.8) can be chosen in the form

$$\psi(x) = \begin{vmatrix} B |p|^{-\frac{1}{2}} \exp\left(\hbar^{-1} \int_x^a |p|\, dx\right) & (x < a), \\[2em] B p^{-\frac{1}{2}} \cos\left(\hbar^{-1} \int_a^x p\, dx + \frac{1}{4}\pi\right) & (x > a). \end{vmatrix} \qquad (40.13)$$

Correspondingly, the solution linearly independent of (40.9) is written as

$$\psi(x) = \begin{vmatrix} B' |p|^{-\frac{1}{2}} \exp\left(\hbar^{-1} \int_b^x |p|\, dx\right) & (x > b), \\[2em] B' p^{-\frac{1}{2}} \cos\left(\hbar^{-1} \int_x^b p\, dx + \frac{1}{4}\pi\right) & (x < b). \end{vmatrix} \qquad (40.14)$$

The expressions obtained will be unsuitable in the case where at the turning point, for example at point b, the potential energy becomes infinite in a discontinuous way. In this case $\psi = 0$ in the region $x \geqslant b$. The phase of the wave function for $x < b$ can be determined if the conditions of applicability of the quasi-classical approximation (39.20) remain valid up to the point $x = b$. Then, taking into account that $\psi(b) = 0$, we obtain

$$\psi(x) = A p^{-\frac{1}{2}} \sin\left(\hbar^{-1} \int_x^b p\, dx\right). \qquad (40.15)$$

§41. Motion in a potential well in the quasi-classical approximation

We apply the results obtained to the motion of a particle in a potential well. We then find an approximate formula for the energy spectrum. Its comparison with accurate formulae will clearly allow us to judge the degree of accuracy and the merits of the quasi-classical approximation. At the same time the solution of the problem posed is of great interest in another respect. It makes it possible to elucidate the connection between quantum mechanics and the old Bohr theory.

Let us consider, first of all, a potential well with infinitely high walls (see §8). The wave function in the quasi-classical approximation is given by a formula of the type (40.15). Namely,

$$\psi(x) = Ap^{-\frac{1}{2}} \sin \left(\hbar^{-1} \int_q^x p\,dx \right) . \tag{41.1}$$

In the potential well there will be two turning points a and b at which the wave function must reduce to zero. Thus at the two turning points the condition $\psi = 0$ or

$$Ap^{-\frac{1}{2}} \sin \left(\hbar^{-1} \int_a^b p\,dx \right) = 0 \tag{41.2}$$

must be fulfilled. This condition is fulfilled if

$$\hbar^{-1} \int_a^b p\,dx = n\pi , \tag{41.3}$$

where the n are the positive integers beginning with unity. Since the momentum is constant and equal to $(2mE)^{\frac{1}{2}}$, we find

$$E = \frac{\hbar^2\pi^2}{2ml^2} n^2 , \tag{41.4}$$

where $l = b - a$ is the width of the well.

We see that in the simplest case of an infinitely deep potential well the quasi-classical approximation leads to an accurate expression for the energy spectrum (see §8).

Let us now consider the general case of a potential well as shown in fig. V.14. We assume that the forbidden region extends infinitely to the right and left of the turning points. Then the quasi-classical wave function will not

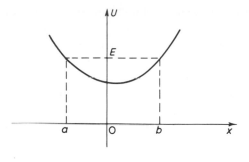

Fig. V.14

contain any exponentially increasing terms and will be given by formulae of the type (40.8) or (40.9). The two wave functions (40.8) and (40.9) describing the motion of a particle in a well must be identical.

$$2Cp^{-\frac{1}{2}}\cos\left(\hbar^{-1}\int_a^x p\,dx - \tfrac{1}{4}\pi\right) = 2C'p^{-\frac{1}{2}}\cos\left(\hbar^{-1}\int_x^b p\,dx - \tfrac{1}{4}\pi\right)$$

$$(a < x < b).\tag{41.5}$$

This is possible only in the case where the sum of the two phases is equal to an integer multiple of π

$$\hbar^{-1}\int_a^b p\,dx - \tfrac{1}{2}\pi = n\pi,\tag{41.6}$$

where n is an integer. Then

$$C' = (-1)^n C.\tag{41.7}$$

If one introduces the integral with respect to the period of the classical motion of a particle from a to b and conversely $\int_a^b p\,dx = \tfrac{1}{2}\oint p\,dx$, then from (41.6) we obtain

$$\oint p\,dx = 2\pi\hbar(n + \tfrac{1}{2}).\tag{41.8}$$

The above expression is none other than the Bohr quantization rule, from which the stationary states of a particle in the quasi-classical case are determined. Thus Bohr's theory with its inconsistent imposition of quantization conditions upon purely classical quantities, turns out to be completely valid within the limits of the quasi-classical approximation. We note that the number n is equal to the number of roots of the quasi-classical wave function between the turning points a and b, because as x changes from a to b the phase of the wave function increases from $-\frac{1}{4}\pi$ to $(n+\frac{1}{2})\pi-\frac{1}{4}\pi$ and, consequently, the cosine reduces to zero n times. The larger the quantum number n, i.e. the smaller the de Broglie wavelength, the better the conditions of applicability of the quasi-classical approximation (39.20). Consequently, we expect that the energy levels obtained from condition (41.8) coincide, for large values of n, with their exact values as calculated from the solution of the Schrödinger equation. However, in some cases, such as, for example, the harmonic oscillator, formula (41.8) gives the correct value of the energy level for any value of n. The integral on the left-hand side of eq. (41.8) represents the area bounded in its phase plane by the classical phase trajectory of a particle with energy E. According to (41.8), this area is equal to $2\pi\hbar n$ for $n \gg 1$. Since an energy level of the system corresponds to each node of the wave function, the number n gives the number of states with energies less than or equal to E. Thus to each quantum state in the phase plane there corresponds an area equal to $2\pi\hbar$. The number of states corresponding to an area $\Delta p \Delta x$ in the phase plane will thus be equal to

$$\frac{\Delta p \, \Delta x}{2\pi\hbar} . \tag{41.9}$$

Generalizing this formula to the three-dimensional case it is obvious that we shall obtain the number of states corresponding to the volume $\Delta x \, \Delta y \, \Delta z \, \Delta p_x \, \Delta p_y \, \Delta p_z$ in phase space to be

$$\frac{\Delta x \, \Delta y \, \Delta z \, \Delta p_x \, \Delta p_y \, \Delta p_z}{(2\pi\hbar)^3} . \tag{41.10}$$

This formula was the basis of our exposition of statistical physics. We see that the quasi-classical approximation is like a bridge connecting classical and quantum mechanics. It enables one to understand the meaning of Bohr's theory and the correspondence principle, and it makes it possible to eliminate all apparent contradictions between the different aspects of the behaviour of real particles. By means of the quasi-classical approximation we can find directly the conditions and the degree of accuracy with which one can pass over to the classical description of the motion of particles in many problems. At the same time it gives a relatively simple method of describing quantum systems approximately, such as, in particular, finding energy levels for particles with a high energy.

§42. Potential barrier penetration

In §13 we considered the passage of a microparticle through a rectangular potential barrier. In this section we shall obtain more general formulae for the case of the passage of particles through potential barriers of arbitrary form (fig. V.15). We assume that the energies E of the particles are sufficiently large, and that the curve of the potential energy is sufficiently smooth that, with the exception only of small regions around the turning points a and b, the conditions of applicability of the quasi-classical approximation are everywhere fulfilled.

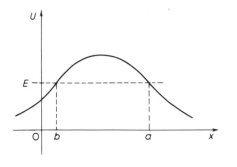

Fig. V.15

Let a particle moving from left to right along the x-axis fall onto a barrier. Then in the region behind the turning point a, i.e. for $x > a$, there must be only a wave propagating in the positive direction of the x-axis (in this region there is no reflected wave). The quasi-classical wave function for $x > a$ can be written in the form of a superposition of the expressions (40.8) and (40.13)

$$\psi(x) = 2Cp^{-\frac{1}{2}} \cos \left(\hbar^{-1} \int_a^x p\,dx - \tfrac{1}{4}\pi \right) + Bp^{-\frac{1}{2}} \cos \left(\hbar^{-1} \int_a^x p\,dx + \tfrac{1}{4}\pi \right). \tag{42.1}$$

Since at large distances from the point a the momentum p changes little, each term of (42.1) is a superposition of two plane waves propagating in opposite directions. It is easily seen that the superposition (42.1) describes a wave propagating from left to right only under the condition $B = -2C\mathrm{i}$. In this case

$$\psi = 2Cp^{-\frac{1}{2}} \exp \left(\mathrm{i}\hbar^{-1} \int_a^x p\,dx - \tfrac{1}{4}\mathrm{i}\pi \right) \qquad (x > a). \tag{42.2}$$

Let us find the quasi-classical wave function in the region $b < x < a$. Taking the superposition (40.8) and (40.13) and taking into account the relation between B and C, we obtain

$$\psi(x) = C|p|^{-\frac{1}{2}} \left[\exp \left(-\hbar^{-1} \int_x^a |p| \, dx \right) + \right.$$

$$\left. + 2i^{-1} \exp \left(\hbar^{-1} \int_x^a |p| \, dx \right) \right] \qquad (b < x < a) \qquad (42.3)$$

This relation is conveniently rewritten in the form

$$\psi(x) = C|p|^{-\frac{1}{2}} \left[e^{-L} \exp \left(\hbar^{-1} \int_b^x |p| \, dx \right) - \right.$$

$$\left. - 2ie^{L} \exp \left(-\hbar^{-1} \int_b^x |p| \, dx \right) \right] \qquad (b < x < a) \qquad (42.4)$$

where

$$L = \hbar^{-1} \int_b^a |p| \, dx \ .$$

Making use now of the relations (40.9) and (40.14), one easily obtains the expression for the wave function in the region in front of the barrier

$$\psi(x) = Cp^{-\frac{1}{2}} \left[4i^{-1} e^{L} \cos \left(\hbar^{-1} \int_x^b p \, dx - \tfrac{1}{4}\pi \right) + \right.$$

$$\left. + e^{-L} \cos \left(\hbar^{-1} \int_x^b p \, dx + \tfrac{1}{4}\pi \right) \right] =$$

$$= 2i^{-1} Cp^{-\frac{1}{2}} \left[(e^{L} - \tfrac{1}{4}e^{-L}) \exp \left(i\hbar^{-1} \int_x^b p \, dx - \tfrac{1}{4}i\pi \right) + \right.$$

$$\left. + (e^{L} + \tfrac{1}{4}e^{-L}) \exp \left(-i\hbar^{-1} \int_x^b p \, dx + \tfrac{1}{4}i\pi \right) \right] \qquad (x < b) . \qquad (42.5)$$

Thus we have written the wave function in the region in front of the barrier in the form of a superposition of incident and reflected waves. We now determine the transmission coefficient, D, of particles through the barrier. Making use of (42.2), we calculate the current density of particles transmitted through the barrier j_{tr}

$$j_{tr} = 4|C|^2 m^{-1} .\tag{42.6}$$

The current density of incident particles, in correspondence with (42.5), is equal to

$$j_{inc} = 4|C|^2 m^{-1}(e^L + \tfrac{1}{4}e^{-L})^2 .\tag{42.7}$$

Consequently, we obtain the following expression for the transmission coefficient D:

$$D = \frac{j_{tr}}{j_{inc}} = \frac{e^{-2L}}{(1 + \tfrac{1}{4}e^{-2L})^2} .\tag{42.8}$$

For a sufficiently wide potential barrier $e_b^{-2L} \ll 1$, we obtain

$$D = e^{-2L} = \exp\left(-2\hbar^{-1} \int_a^b p\,dx \right) .\tag{42.9}$$

We note that if the potential energy on one side of the barrier changes rapidly enough, so that the quasi-classical approximation is inapplicable, then in the expression (42.9) a factor will appear before the exponential function. However, the basic exponential factor does not change. Formula (42.9) is widely used for the calculation of the probabilities of transmission of particles through potential barriers.

As an example let us consider the theory of α-decay. It is well known that all heavy nuclei with mass numbers of the order of magnitude of 200 turn out to be unstable with respect to α-decay. The probability of decay strongly depends on the energy of the emitted α-particles and varies over a very wide range. Thus if the decay probability is characterized by a half-life τ, then τ is equal to 1.6×10^{-4} sec in the case of ^{234}Po which emits α-particles with an energy of 7.8 MeV, whereas τ is equal to 1.4×10^{10} years in the case of ^{232}Th which emits α-particles with an energy of 4 MeV. Such a strong dependence of the probability of α-decay on the energy is accounted for by the fact that the particle, in order to get out of the nucleus, must pass through a potential barrier*. Indeed, simplifying the treatment, we can assume that the initial

---

* R.Gurney and E.Condon, Nature 122 (1928) 439; Phys. Rev. 33 (1929) 127.

nucleus already contains an α-particle. Then the problem reduces to the calculation of the probability that the α-particle will leave the initial, parent nucleus.

We denote by $U(r)$ the energy of interaction of the α-particle with the remaining daughter nucleus. At small distances $U(r)$ amounts to the potential of nuclear forces which we shall consider constant and equal to U_0, while at large distances it is just the Coulomb interaction between the α-particle and the remaining nucleus (see fig. V.16).

$$U(r) = \begin{cases} 2Ze^2/r & (r > r_0)\,, \\ U_0 & (r \leqslant r_0)\,, \end{cases}$$

where r_0 is a distance of the order of magnitude of the nuclear size.

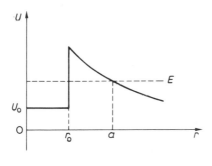

Fig. V.16

Making use of formula (42.9), we can find the probability of the α-particle passing through the potential barrier. The fact that formula (42.9) is derived for one-dimensional motion is of no importance here, because we have established in §35 that a radial motion is equivalent to a one-dimensional motion with a certain effective potential energy.

For simplicity we consider the case where $l = 0$, so that the centrifugal energy is not involved in the calculations (see also the next section). We then have

$$D = \exp\left[-2\hbar^{-1} \int_{r_0}^{a} \left\{ 2\mu\left(\frac{2Ze^2}{r} - E \right) \right\}^{\frac{1}{2}} dr \right] = e^{-2L}\,, \qquad (42.10)$$

where the turning point a is determined by the condition $a = 2Ze^2/E$, and μ is the reduced mass of the α-particle and the daughter nucleus. We calculate the integral L:

$$L = \left(\frac{4\mu Ze^2}{\hbar^2}\right)^{\frac{1}{2}} \int_{r_0}^{a} \left(1 - \frac{E}{U_{\max}} \frac{r}{r_0}\right)^{\frac{1}{2}} r^{-\frac{1}{2}} dr ,$$

where $U_{\max} = 2Ze^2/r_0$. On substituting $(Er/U_{\max} r_0)^{\frac{1}{2}} = \sin \alpha$, we easily obtain

$$L = \frac{2Ze^2}{\hbar v} (\pi - 2\alpha_0 - \sin 2\alpha_0) , \qquad (42.11)$$

where $\sin \alpha_0 = (E/U_{\max})^{\frac{1}{2}}$ and $v = (2E/\mu)^{\frac{1}{2}}$ is the velocity of the α-particle, if the difference between the reduced mass μ and the mass of the α-particle is neglected. Thus the probability for the α-particle to pass through the barrier is given by the expression

$$D = \exp \left[-\frac{4Ze^2}{\hbar v} (\pi - 2\alpha_0 - \sin 2\alpha_0) \right] . \qquad (42.12)$$

We obtain the probability λ for α-decay if we multiply the probability D of passing through the barrier by the probability for α-decay in the absence of the barrier, $\lambda = \nu D$. The quantity ν cannot be calculated with any accuracy. It is significant, however, that the very strong dependence of the decay probability on the energy of the α-particle is involved in the factor D. Qualitatively this dependence is well confirmed by experiment.

Finally we note that similar reasoning is also applicable in the case of spontaneous fission of heavy nuclei.

§43. Quasi-classical motion in a centrally symmetric field

Let us find an approximate expression for the radial component of the wave function $R(r)$ or the function $\chi(r) = rR$ if the potential energy $U(r)$ satisfies the condition for the quasi-classical approximation. We can make use of the relations already derived, since the function $\chi(r)$, as we know (see §35), is described by the one-dimensional Schrödinger wave equation with the effective potential energy

$$U_{\mathrm{eff}}(r) = U(r) + \frac{\hbar^2 l(l+1)}{2mr^2} .$$

However, in this case, of course, the fact that the coordinate r, as distinct from x, varies from 0 to ∞ must be taken into account. For $l = 0$ we have $U_{\mathrm{eff}}(r) = U(r)$. If the quasi-classical conditions are fulfilled up to the point

$r = 0$, then the corresponding wave function can easily be obtained. Indeed, the condition for the finiteness of the wave function at zero gives $\chi(0) = 0$ and, making use of (40.15), we obtain

$$\chi(r) = C[2m(E - U(r))]^{-\frac{1}{2}} \sin\left(\hbar^{-1} \int\limits_0^r [2m(E - U(r))]^{\frac{1}{2}} dr\right). \tag{43.1}$$

In the more general case $l \neq 0$ the effective potential energy $U_{eff}(r)$ must satisfy the quasi-classical condition. If it is assumed that at small distances the centrifugal energy $\hbar^2 l(l + 1)/2mr^2$ plays the basic role and that $p \sim \hbar l/r$, then it follows from the condition (39.20) that $l \gg 1$. The corresponding wave function for $r \gg a$, where a is the turning point, can be written in the form (40.8) with the condition, however, that in the centrifugal energy the quantity $l(l + 1)$ is replaced by* $(l + \frac{1}{2})^2$:

$$\chi(r) = Ap_r^{-\frac{1}{2}} \cos\left(\hbar^{-1} \int\limits_a^r p_r dr - \tfrac{1}{4}\pi\right) \tag{43.2}$$

where

$$p_r = \left[2m\left(E - U(r) - \frac{\hbar^2(l + \frac{1}{2})^2}{2mr^2}\right)\right]^{\frac{1}{2}}. \tag{43.3}$$

Thus the term $\hbar^2/8mr^2$ is added to the centrifugal energy. This addition leads to a more correct value of the phase of the wave function. Thus for free motion, $U = 0$, formula (43.2) gives for the phase of the wave function at large distances the value which we have obtained in §35**.

* H.A. Kramers, Z. Phys. 39 (1926) 828.
** See, for example, L.D.Landau and E.M.Lifshitz, *Quantum mechanics* (Pergamon Press, Oxford, 1965).

6

The Matrix Form of Quantum Mechanics

§44. Operators and matrices

The mathematical apparatus of quantum mechanics developed in the preceding chapters (the method of linear Hermitian operators) is not the only mathematical apparatus used in quantum mechanics. It turns out that all mechanical quantities in quantum mechanics can be related to so-called Hermitian matrices as well as to operators. A matrix R is understood to be the whole set of quantities forming the table

$$R = \begin{bmatrix} R_{11} & R_{12} \dots R_{1n} \\ R_{21} & R_{22} \dots R_{2n} \\ . & . & . \\ . & . & . \\ . & . & . \\ R_{n1} & R_{n2} \dots R_{nn} \end{bmatrix} \qquad (44.1)$$

The number of rows and columns in the table need not, in the general case, be the same. Each of the quantities (generally speaking complex) appearing in the table is called a matrix element. A matrix element has two indices: the first denotes the ordinal number of the row, and the second denotes the ordinal number of the column. The concept of matrices is usually introduced in connection with the linear transformation of vectors in an n-dimensional space*.

We shall see below that in quantum mechanics it is possible to make a geometrical interpretation of the wave function as a vector in a certain imaginary space. In the meantime we shall convince ourselves by means of very general reasoning of the fact that any linear operator $\hat{F}$ can be related to a matrix F with definite values of the matrix elements.

The definition of an operator means that the result of its action on the function $\psi(x)$ is given by

$$\hat{R}\psi(x) = \varphi(x) .\tag{44.2}$$

We pass over from the x-representation to the F-representation. For this we expand the functions $\psi(x)$ and $\varphi(x)$ in terms of the eigenfunctions $\psi_m(x)$ of the operator $\hat{F}$. We assume that the operator $\hat{F}$ has a discrete spectrum. For example, such a representation can be the energy representation (E-representation)

$$\psi(x) = \sum_m c_m \psi_m(x) , \qquad \varphi(x) = \sum_n b_n \psi_n(x) .\tag{44.3}$$

The whole set of amplitudes c_m (or b_n) determines the wave function ψ (or φ) in the F-representation. Sometimes this set is conveniently denoted in the form of a column

$$c = \begin{pmatrix} c_1 \\ c_2 \\ . \\ . \\ . \end{pmatrix} ; \qquad b = \begin{pmatrix} b_1 \\ b_2 \\ . \\ . \\ . \end{pmatrix}\tag{44.4}$$

We substitute the expansion (44.3) into (44.2), and obtain

$$\sum_n b_n \psi_n(x) = \sum_m c_m \hat{R} \psi_m(x) .$$

Multiplying the left-hand and right-hand sides of this equality by $\psi_l^*(x)$ and

---

* For more details see for example V.I.Smirnov, *A course of higher mathematics* (Pergamon Press, Oxford, 1964).

integrating over the entire region of variation of the independent variables, we find

$$b_l = \sum_m R_{lm} c_m \, ,$$ (44.5)

where

$$R_{lm} = \int \psi_l^*(x) \hat{R} \psi_m(x) \, \mathrm{d}V \, .$$ (44.6)

The relation (44.5) determines directly the transformation of the function ψ into the function φ in the F-representation under the action of the operator $\hat{R}$. The operator $\hat{R}$ in this representation is given by formula (44.6), i.e. in the form of a matrix. Thus the definition of the matrix R is equivalent to the definition of the operator $\hat{R}$ itself.

The matrix element R_{ik} is sometimes called the matrix element corresponding to the transition from the kth state into the ith state. Such a terminology is based on the following reasoning. We assume that the initial state of the system is the kth state, $\psi(x) = \psi_k(x)$. Under the action of the operator $\hat{R}$ the transformation (44.2) takes place. Making use of (44.3) and (44.5) and taking into account that in the given case $c_m = \delta_{mk}, b_n = R_{nk}$, we obtain

$$\varphi(x) = \hat{R}\psi_k = \sum_n b_n \psi_n = \sum_n R_{nk} \psi_n(x) \, ,$$ (44.7)

and, consequently, the square of the modulus of the matrix element R_{ik} determines the probability of finding the system in the ith state.

Knowing the matrix corresponding to the quantity R, one can also easily find the mean value of this quantity in a certain state ψ. According to the general formula (22.4) we have

$$\bar{R} = \int \psi^* \hat{R} \psi \, \mathrm{d}V \, .$$

Substituting here the expansion (44.3) instead of ψ, we obtain

$$\bar{R} = \sum_m \sum_n c_m^* c_n \int \psi_m^* \hat{R} \psi_n \, \mathrm{d}V = \sum_m \sum_n c_m^* R_{mn} c_n \, .$$ (44.8)

We note that if we determine the matrix elements (44.6) by means of the wave functions ψ_m which are the eigenfunctions of the operator $\hat{R}$, then we also determine the matrix R of the operator in its own representation

$$R_{ml} = \int \psi_m^* \hat{R} \psi_l \, \mathrm{d}V = R_l \int \psi_m^* \psi_l \, \mathrm{d}V = R_l \delta_{ml} \, .$$ (44.9)

We see that in this case only the matrix elements with $m = l$ are different from zero. Matrices of such a form are called diagonal

$$R = \begin{vmatrix} R_{11} & 0 & 0 & ... \\ 0 & R_{22} & 0 & ... \\ 0 & 0 & R_{33} & ... \\ \cdots\cdots\cdots\cdots\cdots \\ \cdots\cdots\cdots\cdots\cdots \end{vmatrix}. \tag{44.10}$$

Thus in its own representation any operator is described by a diagonal matrix, and the diagonal elements are equal to the eigenvalues of this operator.

The definition of the matrix elements R_{ml} is completely equivalent to the definition of the operator $\hat{R}$. As we shall see, it enables one to determine the eigenvalues and eigenfunctions of this operator. On the other hand, if the operator $\hat{R}$ is known, then also the matrix elements R_{ml} can be determined.

The Hermitian character of the operator $\hat{R}$ imposes a certain restriction upon the form of the matrix elements R_{ml}. Namely, for the matrix element R_{ml}^*, the complex conjugate of the element R_{ml}, we have

$$R_{ml}^* = \left(\int \psi_m^* \hat{R} \psi_l dV \right)^* = \int \psi_m \hat{R}^* \psi_l^* dV .$$

By definition of the Hermitian operator

$$\int \psi_m \hat{R}^* \psi_l^* dV = \int \psi_l^* \hat{R} \psi_m dV ,$$

so that

$$R_{ml}^* = R_{lm} .$$

We see that the Hermitian property of matrix elements

$$R_{lm} = R_{ml}^* , \tag{44.11}$$

follows from the requirement of the Hermitian character of the operator.

Every physical quantity in quantum mechanics, as well as the Hermitian operator $\hat{R}$, can be related to a Hermitian matrix R whose matrix elements are determined by formula (44.6).

As we shall see below, the matrix form of quantum mechanics is, in certain cases, more convenient than the operator form. The representation of quantum mechanics in the matrix form will enable us to formulate the equations of quantum mechanics like the equations of classical physics. The wave function will no longer appear in them. The equations in this form will coincide with those of classical mechanics, with only the basic difference that in these equations classical quantities will be replaced by the corresponding matrices. However, before passing over to a systematic exposition of quantum mechanics in the matrix form it is necessary to present the basic notions of matrix calculus.

§45. The fundamentals of matrix calculus

In the preceding section we have defined an arbitrary matrix R as the whole set of quantities R_{lm} arranged in a definite order in the form of the table

$$R = \begin{vmatrix} R_{11} & R_{12} & \cdots & \\ R_{21} & R_{22} & \cdots & \\ & & & \\ \cdot & \cdot & & \cdot \\ R_{n1} & R_{n2} & \cdots & R_{nn} & \cdots \\ \cdot & \cdot & & \cdot \\ \cdot & \cdot & & \cdot \\ \cdot & \cdot & & \cdot \end{vmatrix} \qquad (45.1)$$

Matrices in which the number of columns is equal to the number of rows are called square matrices. A matrix can be finite, if the number of columns and the number of rows is finite, as well as infinite if the number is arbitrarily large. The matrix elements $R_{11}, R_{22}, ..., R_{nn}$ ... forming the diagonal of the matrix are called the diagonal elements. The matrix for which all elements are equal to zero is called the null matrix O. The matrix for which all the diagonal elements are equal to unity is called the unit matrix. We shall denote this matrix by

$$1 = \begin{vmatrix} 1 & 0 & 0 & \cdots \\ 0 & 1 & 0 & \cdots \\ 0 & 0 & 1 & \cdots \\ \cdots\cdots\cdots \\ \cdots\cdots\cdots \end{vmatrix}, \qquad (45.2)$$

i.e. $(1)_{mn} = \delta_{mn}$. A matrix can be considered as a certain hypercomplex number, just as the set of two numbers a and b can be treated as one complex number $a + ib$. As in the case of complex numbers, one can construct an algebra of hypercomplex numbers (matrices) by defining the action of addition and multiplication of matrices.

The matrix R, each element of which is equal to the sum of the corresponding matrix elements of the matrices F and D,

$$R = F + D ,$$

if (45.3)

$$R_{ml} = F_{ml} + D_{ml} ,$$

is called the sum of the two matrices F and D. One can add or subtract only matrices having the same number of rows and columns. Two matrices F and D are equal to each other if the corresponding matrix elements are equal to each other:

$$F = D ,$$

if (45.4)

$$F_{ml} = D_{ml} .$$

Further we define the product of the number k and the matrix D as a matrix F each element of which is equal to the product of the number k and the corresponding matrix element of the matrix D:

$$F = kD ,$$
$$F_{ml} = kD_{ml} .$$

(45.5)

The matrix L is called the product of the matrices F and D

$$L = FD ,$$

if the matrix element L_{mn} is equal to

$$L_{mn} = \sum_l F_{ml} D_{ln} .$$

(45.6)

This means that each element of the matrix L is equal to the sum of the products of the elements of the mth row of the matrix F and the elements of the nth column of the matrix D.

The matrix F can be multiplied by the matrix D only in the case where the number of columns of the matrix F is equal to the number of rows of the matrix D. We stress that the product of matrices, as the product of operators, is non-commuting, i.e. in the general case

$$FD \neq DF .$$

If the unit matrix (45.2) is taken as one of the factors, then we arrive at the equalities

$$F1 = 1F = F ,\qquad\qquad (45.7)$$

i.e. multiplication by the unit matrix is commutative. We note that when two matrices with matrix elements different from zero are multiplied a null matrix can be obtained. Let, for example,

$$F = \begin{pmatrix} 1 & 0 \\ 1 & 0 \end{pmatrix} ;\qquad D = \begin{pmatrix} 0 & 0 \\ 1 & 1 \end{pmatrix} .$$

Then

$$L = FD = \begin{pmatrix} 0 & 0 \\ 0 & 0 \end{pmatrix} = 0 .$$

On the other hand

$$L' = DF = \begin{pmatrix} 0 & 0 \\ 2 & 0 \end{pmatrix} \neq 0 .$$

Analogously to (45.6) one can form the product of three and more matrices. Thus if $FD = L$, then the product RFD is equal to

$$(RFD)_{mn} = \sum_l R_{ml}L_{ln} = \sum_l \sum_p R_{ml}F_{lp}D_{pn} . \qquad\qquad (45.8)$$

We arrive at the exactly the same expression if we take the product of the matrix RF and the matrix D. Thus for the matrix product the associative law

$$(RF)D = R(FD) . \qquad\qquad (45.9)$$

holds.

It is easily shown that the distributive law

$$R(F + D) = RF + RD \qquad\qquad (45.10)$$

is also valid.

The definition of the basic actions of matrices, given by formulae (45.3)–(45.6), corresponds completely to the analogous relations for linear operators. For formulae (45.3)–(45.5) this statement is obvious. One can easily verify this correspondence also for formula (45.6). Let the operator $\hat{L}$ be equal to the product of the operators $\hat{F}$ and $\hat{D}$, i.e. $\hat{L} = \hat{F}\hat{D}$. We find the matrix corresponding to this operator in an arbitrary representation. We assume that the operator $\hat{D}$ transforms the function ψ into the function φ, and the operator $\hat{F}$ correspondingly transforms φ into χ, so that

$$\varphi = \hat{D}\psi , \qquad \chi = \hat{F}\varphi . \tag{45.11}$$

We rewrite these equations, expanding the wave functions ψ, φ and χ in a series in terms of a certain system of functions ψ_m. Let the expansions

$$\psi = \sum_n c_n \psi_n ; \qquad \varphi = \sum_l b_l \psi_l ; \qquad \chi = \sum_k d_k \psi_k .$$

hold. Comparing with (44.2) and (44.5), we obtain

$$b_l = \sum_n D_{ln} c_n , \qquad d_m = \sum_l F_{ml} b_l .$$

Substituting b_l into the second of these equations, we have

$$d_m = \sum_l \sum_n F_{ml} D_{ln} c_n . \tag{45.12}$$

On the other hand, taking into account that $\chi = \hat{L}\psi$, we can write that

$$d_m = \sum_n L_{mn} c_n . \tag{45.13}$$

Comparing (45.13) with (45.12), we arrive at eq. (45.6).

If a matrix F has an unequal number of columns and rows, then, by striking off a certain number of columns or rows, one obtains a square matrix with an equal number of rows and columns. From this table one can calculate the determinant $\det F$ of the matrix F.

The highest possible order of this determinant is obtained when the minimum number of rows or columns is struck off. The highest order of the non-zero determinant obtained from the matrix is called the rank of the matrix.

In certain applications the sum of the diagonal elements of the matrix, often called the trace, plays an important role. By definition

$$\text{Tr } F = \sum_n F_{nn} .$$

The matrix F is called non-singular if one can construct the inverse matrix. The inverse matrix, which we shall denote by F^{-1}, satisfies the equations

$$FF^{-1} = 1 ; \qquad F^{-1}F = 1 . \tag{45.14}$$

To find the elements of the matrix F^{-1} it is necessary to find the solution of the system of linear homogeneous equations which is obtained from the definition (45.14):

$$\sum_k (F)_{mk}(F^{-1})_{kn} = \delta_{mn} , \qquad \sum_l (F^{-1})_{ml}F_{ln} = \delta_{mn} \qquad (45.15)$$

for all possible values m and n.

The system of equations (45.15) can be solved only in the case where the determinant, $\det F$, of the matrix F differs from zero. (It is assumed that the matrix F is a square matrix.)

Making use of (45.14) it is easy to find the matrix which is the inverse of the product of matrices $RFD...$ (if it exists)

$$(RFD...)^{-1} = ...D^{-1}F^{-1}R^{-1} . \qquad (45.16)$$

Further we define the conjugate matrix $F^\dagger$ (or the Hermitian conjugate) of the original matrix F

$$(F^\dagger)_{mn} = (F_{nm})^* . \qquad (45.17)$$

We denote the complex conjugate matrix element $(F_{nm})^*$ by F^*_{nm}.

It follows directly from the definition (45.17) that

$$(F + D)^\dagger = F^\dagger + D^\dagger . \qquad (45.18)$$

It is also easy to define the matrix conjugate to the product of matrices. Thus, if $L = FD$, then

$$(L^\dagger)_{mn} = L^*_{nm} = \sum_k F^*_{nk}D^*_{km} = \sum_k (D^\dagger)_{mk}(F^\dagger)_{kn} \qquad (45.19)$$

and, consequently,

$$(FD)^\dagger = D^\dagger F^\dagger . \qquad (45.20)$$

In particular, if $L = kD$, where k is a number, then

$$L^\dagger = k^* D^\dagger . \qquad (45.21)$$

Of course, the relation (45.20) is immediately generalized to the product of any number of matrices

$$(FDR...)^\dagger = ...R^\dagger D^\dagger F^\dagger . \qquad (45.22)$$

If the matrix F is the same as its conjugate matrix $F^\dagger$, $F = F^\dagger$, then it is Hermitian or self-adjoint. This definition is analogous to the definition of the hermitician property of an operator (see (44.11)). For the matrix elements of a Hermitian matrix we have

$$F_{nm} = (F^\dagger)_{nm} = F^*_{mn} . \qquad (45.23)$$

The matrix F is called unitary if $F^\dagger = F^{-1}$. This condition can also be rewritten as follows:

$$F^\dagger F = F F^\dagger = 1 . \qquad (45.24)$$

If the matrices F and D are unitary, then their product is also unitary. Indeed,

$$(FD)^\dagger = D^\dagger F^\dagger = D^{-1} F^{-1} = (FD)^{-1} . \qquad (45.25)$$

We shall also encounter the simplest functions of matrices, where the definition of a function of the matrix means the definition of the law according to which one matrix is compared to another. Since the rules for the multiplication and addition of matrices are defined, one can easily introduce the notion of an integer rational function $f(D)$ of the matrix D. Furthermore, we shall also deal with more complex functions, for example of the form e^D, where D is a matrix. The function e^D is understood to be the following series:

$$e^D = 1 + D + \tfrac{1}{2} D^2 + \dots + \frac{1}{n!} D^n + \dots . \qquad (45.26)$$

We shall show, for example, that the matrix R, defined by the function e^{iF}, $R = e^{iF}$, where F is an arbitrary Hermitian matrix, is unitary. By a direct calculation it is easily checked that the matrix R^{-1}, the inverse of the matrix R, is the matrix e^{-iF}. On the other hand, for the matrix conjugate to R we have

$$R^\dagger = \left(1 + iF - \tfrac{1}{2} F^2 - \frac{i}{3!} F^3 + \dots \right)^\dagger =$$

$$= \left(1 - iF - \tfrac{1}{2} F^2 + \frac{i}{3!} F^3 + \dots \right) = e^{-iF}$$

(because $F^\dagger = F$). Thus $R^\dagger = R^{-1}$ and, consequently, the matrix R is unitary.

The rules of action formulated in this section also remain valid for matrices with an infinitely large number of columns and rows, provided all sums (45.6), are convergent.

As we have already mentioned, the introduction of matrices is closely associated with the concept of the linear transformation of an n-dimensional vector. The concept of an n-dimensional vector is a natural generalization of the ordinary concept of a vector. The vector $\mathbf{x}$ in n-dimensional space is defined by the whole set of n, in general complex, numbers which are called the components of this vector $x_1, x_2, ..., x_n$. Each of the components can be represented by a segment on one of the n mutually perpendicular axes in n-dimensional space. It is hardly necessary to mention that an n-dimensional space is not associated with physical reality, and that a vector in n-dimensional space is a mathematical generalization. As in the case of ordinary vectors, one can introduce the concept of the scalar product of two vectors $\mathbf{x}$ and $\mathbf{y}$ in n-dimensional space. Namely, the scalar product of the vectors $\mathbf{y}$ and $\mathbf{x}$ is defined as

$$(\mathbf{x} \cdot \mathbf{y}) = \sum_{i=1}^{n} x_i^* y_i. \tag{45.27}$$

If the vector $\mathbf{x}$ has real components, then the definition (45.27) is the same as the ordinary definition of the scalar product. On the contrary, if the components of one or both of the vectors $\mathbf{x}$ and $\mathbf{y}$ are complex, then the definition (45.27) leads to new important consequences.

We form the scalar product of the vector $\mathbf{x}$ and the same vector $\mathbf{x}$

$$(\mathbf{x} \cdot \mathbf{x}) = \sum_{i=1}^{n} x_i^* x_i = \sum_{i=1}^{n} |x_i|^2. \tag{45.28}$$

It represents a generalization of the concept of the square of a vector to the case of complex values of the components. The quantity

$$\left(\sum_{i=1}^{n} |x_i|^2 \right)^{\frac{1}{2}} \tag{45.29}$$

is said to be the length or norm of the vector. The scalar product of two vectors in n-dimensional space is, obviously, non-commuting.

$$(\mathbf{x} \cdot \mathbf{y}) = \sum_{i=1}^{n} x_i^* y_i \neq (\mathbf{y} \cdot \mathbf{x}) = \sum_{i=1}^{n} y_i^* x_i.$$

If the scalar product of two vectors is equal to zero $(\mathbf{x} \cdot \mathbf{y}) = 0$, then such vectors are called mutually orthogonal.

Let us consider two systems of coordinates k and k' each with mutually orthogonal axes. Let the components of a certain vector in the system k be x_i, and in the system k' be x_i'. As in three-dimensional Euclidean geometry, there is a linear relation between the components expressed by

$$x'_i = \sum_k a_{ik} x_k .$$

(45.30)

The whole set of numbers a_{ik} forms the matrix

$$\|a\| = \begin{vmatrix} a_{11} & a_{12} & \cdots & a_{1n} \\ a_{21} & a_{22} & \cdots & a_{2n} \\ \cdot & \cdot & & \cdot \\ \cdot & \cdot & & \cdot \\ \cdot & \cdot & & \cdot \\ a_{n1} & a_{n2} & \cdots & a_{nn} \end{vmatrix} ,$$

which is called the linear transformation matrix.

For the linear orthogonal transformation (45.30) the following condition holds:

$$(\mathbf{x} \cdot \mathbf{x}) = \sum_{i=1}^{n} |x_i|^2 = (\mathbf{x}' \cdot \mathbf{x}') = \sum_{i=1}^{n} |x'_i|^2 .$$

(45.31)

The transformation (45.30), satisfying the requirement (45.31), i.e. leaving the square of the length of the vector unchanged is also called a **unitary transformation**.

The concept of a vector with complex values of the components in an n-dimensional space can be generalized directly to the case of a space with an infinite number of dimensions, $n \to \infty$. A space with an infinite number of dimensions, for which the definition (45.28) of the square of the length of a vector is valid, is called a **Hilbert space**. A vector in a Hilbert space has an infinite number of components each of which can be real as well as complex.

Vectors in an n-dimensional space (with a finite as well as an infinite number of dimensions) are often written in the form of a matrix

$$\mathbf{x} = \begin{pmatrix} x_1 \\ x_2 \\ x_3 \\ \cdot \\ \cdot \\ \cdot \end{pmatrix} .$$

Then the transformation (45.30) can be written in the form $\mathbf{x}' = a\mathbf{x}$. Indeed, for the components we have $x'_i = a_{ik} x_k$, which is the same as (45.30).

§46. Geometric interpretation of the wave function.
Canonical transformations

The mathematical apparatus given briefly above, in particular the vector calculus in Hilbert space, in spite of its unusual and abstract character turns out to correspond exactly to the quantum-mechanical description of the properties of microsystems. We shall consider the wave function ψ characterizing the state of the system as a vector ψ in a Hilbert space with an infinite number of dimensions. To each quantum-mechanical quantity F characterizing a property of the system there corresponds a definite system of coordinate axes or, what is the same, a system of unit basis vectors $\psi_1(x)$, $\psi_2(x)$, ..., $\psi_n(x)$, ... This system of basis vectors (basis functions) is none other than the system of eigenfunctions of the operator $\hat{F}$ which correspond to the possible eigenvalues $F_1, F_2, ...$ (we assume that the spectrum is discrete; the generalization to a continuous spectrum is given farther). The components of the vector ψ in the chosen system of coordinates will be the amplitudes $c_1, c_2, ... c_n$ defined by the relation

$$\psi(x) = \sum_k c_k \psi_k . \qquad (46.1)$$

The amplitudes c_k, as we know (see §19), are equal to

$$c_k = \int \psi_k^*(x)\, \psi(x)\, dV .$$

On the other hand, this equality can be considered as the scalar product of the vector ψ and the vector ψ_k

$$c_k = \psi_k \cdot \psi = \int \psi_k^*(x)\, \psi(x)\, dV . \qquad (46.2)$$

The definition (46.2) corresponds to (45.27). Thus if we have two vectors $\varphi(x)$ and $\psi(x)$ with components b_k and c_k respectively, then the scalar product of the vector ψ and the vector φ is equal to

$$\varphi \cdot \psi = \sum_k b_k^* c_k = \int \varphi^*(x)\, \psi(x)\, dV . \qquad (46.3)$$

As we know, the whole set of amplitudes c_k represents the wave function in the F-representation. Thus the whole set of components of the vector ψ in the coordinate system whose unit vectors are the eigenfunctions ψ_n of the operator $\hat{F}$ is the wave function in the F-representation.

The system of basis vectors ψ_1, ψ_2, ... is a system of unit vectors which are mutually orthogonal. This follows from the condition of normalization and orthogonality of the eigenfunctions of the operator $\hat{F}$ (see §18)

$$\int \psi_k^*(x)\,\psi_m(x)\,dV = \psi_k \cdot \psi_m = \delta_{km}\ . \qquad (46.4)$$

Let us now consider the transition from one representation to another. For example, the transition from the representation in which the matrix F is diagonal (F-representation) to the representation in which the matrix D is diagonal (D-representation). Geometrically this means the transition from the coordinate system formed by the basis vectors ψ_m to the coordinate system formed by the basis vectors φ_k. The functions ψ_m and φ_k are the eigenfunctions of the operators $\hat{F}$ and $\hat{D}$ respectively. We obtain the transformation formulae if we expand the functions φ_k in terms of the system of basis functions ψ_m (assuming that there is a discrete spectrum)

$$\varphi_k = \sum_l S_{lk}\psi_l\ . \qquad (46.5)$$

It is obvious that

$$S_{lk} = \int \psi_l^*(x)\varphi_k(x)\,dV = \psi_l \cdot \varphi_k\ . \qquad (46.6)$$

We shall call the matrix S the matrix of the transformation from one representation to another (or, correspondingly, from one coordinate system to another). We can obtain definite conclusions concerning the properties of the matrix S immediately if we take into account that the system of functions ψ_k as well as the system of functions ψ_m is a system of normalized orthogonal functions. Consequently,

$$\int \varphi_m^*\varphi_k\,dV = \sum_{i,l} S_{im}^* S_{lk}\delta_{il} = \sum_i S_{im}^* S_{ik} = \sum_i (S^\dagger)_{mi}S_{ik} = \delta_{mk}\ , \qquad (46.7)$$

or in matrix form

$$S^\dagger S = 1\ . \qquad (46.8)$$

Carrying out the inverse expansion of the functions ψ_m in terms of the functions φ_k it is easy to obtain

$$SS^\dagger = 1\ . \qquad (46.9)$$

It follows from eqs. (46.8) and (46.9) that the matrix S is unitary, $S^\dagger = S^{-1}$. The transformation from one representation to another, carried out by the unitary matrix S, is called a unitary or canonical transformation. Geometrically it corresponds to a 'rotation' in the Hilbert space. It is also easy to obtain the direct relation between the components of an arbitrary vector ψ in different coordinate systems. Let

$$\psi = \sum_l c_l \psi_l = \sum_k c'_k \varphi_k .$$

Making use of (46.5), we have

$$\sum_l c_l \psi_l = \sum_{l,k} c'_k S_{lk} \psi_l .$$

Equating the expressions at equal ψ_l, we obtain

$$c_l = \sum_k S_{lk} c'_k . \tag{46.10}$$

This expression can be rewritten in the form of a matrix equation, if the whole sets of amplitudes c_l and c'_k are considered as single-column matrices c and c'. Then

$$c = Sc' . \tag{46.11}$$

Multiplying from the left by $S^\dagger$, we obtain

$$c' = S^\dagger c . \tag{46.12}$$

Further let us find how an arbitrary matrix R is transformed in the transition to another representation. We assume that in the F-representation the following relation holds:

$$b_l = \sum_m R_{lm} c_m \tag{46.13}$$

or in the matrix form

$$b = Rc . \tag{46.14}$$

In the transition to another representation the amplitudes c and b are transformed into the amplitudes c' and b' according to (46.11) and (46.12). We make use of the relation (46.11) and express c and b in eq. (46.14) in terms of c' and b'. We then obtain

$$Sb' = RSc' .$$

Multiplying this equation from the left by $S^\dagger$, we find

$$b' = S^\dagger R S c' = R' c' .$$

Thus the matrix R', i.e. the matrix R in the new representation, has the form

$$R' = S^\dagger R S$$

or

$$(R')_{mn} = \sum_{k,l} (S^\dagger)_{mk} R_{kl} S_{ln} .$$ (46.15)

Let us consider certain properties of the unitary transformation. We shall show, first of all, that if a matrix D is Hermitian in one representation, then it will also be Hermitian in another representation. Indeed, according to (46.15) and (45.22)

$$D' = S^\dagger D S ,$$

$$(D')^\dagger = S^\dagger D^\dagger (S^\dagger)^\dagger = S^\dagger D^\dagger S .$$

Since $D^\dagger = D$, it turns out that $D' = D'^\dagger$. The unitary transformation also conserves the form of matrix equations. Let us, for example, consider the equations

$$F + D = R ; \qquad PL = T .$$

Multiplying the equations from the left by $S^\dagger$ and from the right by S and making use of (45.9) and (46.9), we obtain

$$S^\dagger F S + S^\dagger D S = S^\dagger R S ,$$

$$S^\dagger P S S^\dagger L S = S^\dagger T S .$$

Using (46.15), we rewrite these equations in the new representation

$$F' + D' = R' ,$$

$$P'L' = T' .$$

We see that the form of the equations has not changed. We shall show also that in a unitary transformation the trace of the matrix does not change

$$\mathrm{Tr}\, F' = \sum_n F'_{nn} = \sum_{n,l,k} (S^\dagger)_{nl} F_{lk} S_{kn} =$$

$$= \sum_{l,k} F_{lk} \sum_n S_{kn} (S^\dagger)_{nl} = \sum_{l,k} F_{lk} \delta_{kl} = \mathrm{Tr}\, F .$$ (46.16)

Here we have made use of (46.9).

The unitary transformation also conserves the determinant of the matrix. Indeed, since the determinant of a product of matrices is equal to the product of the determinants $\det FDR = \det F \det D \det R$, we have

$$\det F' = \det S^\dagger \det F \det S = \det F \det (S^\dagger S) = \det F .$$ (46.17)

The modulus of the square of the determinant of a finite unitary matrix is equal to unity. We show this

$$|\det S|^2 = (\det S)(\det S)^*,$$

but $(\det S)^* = \det S^\dagger$, because the determinant does not change when the matrix is transposed. Consequently, we obtain

$$|\det S|^2 = \det S \det S^\dagger = \det(SS^\dagger) = 1 . \tag{46.18}$$

§47. The eigenfunctions and eigenvalues of an operator given in matrix form

Let us assume that the operator $\hat{D}$ in the F-representation is given in the form of a matrix D. We shall see how one can find the eigenfunctions and eigenvalues of this operator. The equation for the eigenfunctions and eigenvalues in the F-representation has the form

$$\sum_k D_{mk} c_k^{(n)} = D_n c_m^{(n)} . \tag{47.1}$$

Here D_n is the nth eigenvalue of the matrix D, and the whole set of amplitudes $c_1^{(n)}, c_2^{(n)}, \ldots$ is the eigenfunction of the operator $\hat{D}$ in the F-representation corresponding to the nth eigenvalue. If the eigenfunction is written in the form of a matrix with one column, $c^{(n)}$, then eq. (47.1) can be rewritten in the form

$$Dc^{(n)} = D_n c'^{(n)} , \tag{47.2}$$

It is easily seen that the magnitudes of the eigenvalues do not depend on the choice of the representation. Indeed, eq. (47.1) in another representation is written, according to (46.15) and (46.11), as follows:

$$D'c'^{(n)} = D'_n c'^{(n)}, \tag{47.3}$$

where $D' = S^\dagger DS$, $c' = S^\dagger c$. Substituting these values into (47.3), we obtain

$$S^\dagger DSS^\dagger c^{(n)} = D'_n S^\dagger c^{(n)} .$$

Multiplying from the left by S, we again arrive at eq. (47.2), whence it is seen that $D'_n = D_n$.

Thus the problem of finding the eigenvalues of a matrix D reduces to finding a unitary transformation which will bring the matrix D into diagonal form. The diagonal elements of such a matrix are, as we know (see §44), its eigenvalues. Thus if S is the required unitary transformation, then

$$S^{\dagger} DS = D' \tag{47.4}$$

or, multiplying on the left by S,

$$DS = SD' .$$

Taking into account that $(D')_{mn} = D_n \delta_{mn}$, we obtain

$$\sum_k D_{mk} S_{kn} = D_n S_{mn} \tag{47.5}$$

or

$$\sum_k (D_{mk} - D_n \delta_{km}) S_{kn} = 0 . \tag{47.6}$$

In these equations the matrix elements S_{kn} as well as the eigenvalues D_n are unknown. If D is a square matrix having N columns and as many rows as columns, then for each D_n we have a system of N equations (for $m = 1, 2, ..., N$). Since the system consists of homogeneous linear equations, it has a non-trivial solution on condition that its determinant reduces to zero

$$\det \begin{vmatrix} D_{11} - D_n & D_{12} & \cdots & D_{1N} \\ D_{21} & D_{22} - D_n & \cdots & D_{2N} \\ \cdot & & & \cdot \\ \cdots\cdots\cdots\cdots\cdots\cdots\cdots \\ \cdot & & & \cdot \\ D_{N1} & D_{N2} & \cdots & D_{NN} - D_n \end{vmatrix} = 0 . \tag{47.7}$$

This is an equation of the Nth power with respect to the unknown D_n. On solving it we find N roots which will be the eigenvalues of the matrix D. In particular, certain values can be equal to each other; then degeneracy occurs. All eigenvalues of a Hermitian matrix D will be real. Indeed, the matrix D' is also Hermitian (see §46) and, consequently, $(D')_{nn} = (D')_{nn}^*$ or $D_n = D_n^*$. Substituting the values $D_1, D_2, ..., D_N$ into the system (47.6), we determine the whole set of matrix elements $S_{kn} (S_{1n}, S_{2n}, ...)$ for each D_n, i.e. in the end we determine the unitary transformation matrix S. Comparing (47.5) with (47.1), we see that each column

$$\begin{pmatrix} S_{1n} \\ S_{2n} \\ \cdot \\ \cdot \\ \cdot \\ S_{Nn} \end{pmatrix}$$

of the matrix S is an eigenfunction of the operator $\hat{D}$ in the F-representation corresponding to a given eigenvalue D_n. Knowing the matrix S, we can find the whole set of eigenfunctions of the operator $\hat{D}$ in the x-representation. Indeed, if in the initial F-representation the whole set $\psi_1(x), \psi_2(x), \ldots$ of eigenfunctions of the operator $\hat{F}$ were the basis functions, then in the new representation in which the matrix D is diagonal (D-representation) the eigenfunctions $\varphi_1(x), \varphi_2(x), \ldots$ of the operator $\hat{D}$ will be the basis functions. The relation between them is given by formula (46.5) which determines the functions $\varphi_k(x)$

$$\varphi_k(x) = \sum_l S_{lk} \psi_l(x) . \tag{47.8}$$

If we have two matrices F and D, then by means of one and the same unitary transformation S they can be brought into diagonal form simultaneously only in the case where they commute, i.e. $FD = DF$. Indeed, suppose that F and D are brought into diagonal form, F' and D' respectively. We form the matrix $F'D'$:

$$(F'D')_{mn} = \sum_k F'_{mk} D'_{kn} = F'_{mm} D'_{mn} \delta_{mn} = (D'F')_{mn} . \tag{47.9}$$

Consequently,

$$F'D' = D'F' .$$

Since in the unitary transformation the form of the matrix equation does not change (see §46), then in the initial representation we have $FD = DF$.

Thus we have proved that commutation of matrices is necessary in order that they may be brought simultaneously into diagonal form. It is easy to show that this condition is also sufficient.

In this section we have assumed everywhere that we have been dealing with finite matrices. However, if the number N of columns and rows tends to infinity, then the mathematical problem becomes substantially more complicated. The system (47.6) will now be a system of an infinitely large number of equations. Equation (47.7) will also be of an infinitely high power. It can be shown, however, that in this case also any Hermitian matrix may be brought into a diagonal form with real eigenvalues by means of a unitary transformation. We shall not give the proof of this statement.

§48. Continuous matrices. The Dirac notation

Up to now we have assumed that variables run over a discrete sequence of values. It is clear, however, that the preceding results must be generalized to the case of continuous variables.

It turns out that this generalization can be carried out directly. All the results obtained above remain valid if all the sums in them are replaced by the corresponding integrals. For example, formula (45.6) for the matrix element of the product of two matrices will now have the form

$$L = FD ,$$

$$L_{\alpha\beta} = \int F_{\alpha\gamma} D_{\gamma\beta} d\gamma .$$

$$(48.1)$$

The integration is carried out over the entire region of variation of the corresponding variable. The unit matrix 1 is now defined by the equality

$$(1)_{\alpha\beta} = \delta(\alpha - \beta) ,$$

$$(48.2)$$

i.e. is replaced by the δ-function. As is easily seen the following relation holds: for an arbitrary matrix F

$$F1 = 1F = F .$$

The formulae of §3, which express the transformation of the wave function from the coordinate representation to the momentum representation and vice versa, can also be written in matrix form.

We note, first of all, that in its own representation the coordinate q must be expressed by a diagonal matrix (see §44). Confining ourselves for simplicity to the one-dimensional case, we have in correspondence with (48.2)

$$q_{xx'} = x\delta(x - x') .$$

$$(48.3)$$

Let us find the matrix representing the momentum of a particle in this representation. As in §26, we shall proceed from the relation

$$pq - qp = \frac{\hbar}{i} .$$

$$(48.4)$$

We shall show that this relation is satisfied if the matrix p is chosen in the form

$$p_{xx'} = \frac{\hbar}{i} \frac{\partial}{\partial x} \delta(x - x') .$$

$$(48.5)$$

First of all, equating the matrix elements of the left-hand and right-hand sides of relation (48.4), we have

$$\int (p_{xx''}q_{x''x'} - q_{xx''}p_{x''x'})\, dx'' = \frac{\hbar}{i}\delta(x - x')\,.$$

Substituting (48.3) and (48.5), we obtain

$$\int \left[x''\delta(x'' - x')\frac{\partial}{\partial x}\delta(x - x'') - x\delta(x - x'')\frac{\partial}{\partial x''}\delta(x'' - x') \right] dx'' = \delta(x - x')\,.$$

We take the integrals in accordance with the rules of action on δ-functions (see Appendix III, Vol. 1). Since $y\delta'(y) = -\delta(y)$ (see eq. (III.8)), we find that

$$-(x - x')\frac{\partial}{\partial x}\delta(x - x') = \delta(x - x')\,.$$

Consequently, we have proved that the matrices (48.3) and (48.5) satisfy relation (48.4). If we act, according to rules (44.5), with the matrix $p_{xx'}$ on a certain function $\psi(x)$, then we obtain the function $\varphi(x)$ equal to

$$\varphi(x) = \int p_{xx'}\psi(x')\, dx' = \frac{\hbar}{i}\int \frac{\partial}{\partial x}\delta(x - x')\psi(x')\, dx' =$$

$$= -\frac{\hbar}{i}\int \psi(x')\frac{\partial}{\partial x'}\delta(x - x')\, dx'\,.$$

Integrating by parts, we obtain

$$\varphi(x) = \frac{\hbar}{i}\int \delta(x - x')\frac{\partial \psi}{\partial x'}\, dx' = \frac{\hbar}{i}\frac{\partial \psi}{\partial x}\,.$$

We see that the action of the matrix $p_{xx'}$ is equivalent to the action of the operator $\hat{p} = (\hbar/i)\,\partial/\partial x$. The formula of transformation of the wave function from the coordinate representation to the momentum representation is of the form

$$c(p) = (2\pi\hbar)^{-\frac{1}{2}}\int \psi(x)\, e^{-(i/\hbar)px}\, dx\,. \tag{48.6}$$

In matrix form this relation, according to (46.12), can be rewritten as

$$c(p) = \int S_{px}^{\dagger}\psi(x)\, dx\,, \tag{48.7}$$

where $S_{px}^{\dagger} = (2\pi\hbar)^{-\frac{1}{2}}e^{-(i/\hbar)px}$ is the unitary matrix of the transformation from the x-representation to the p-representation. It is natural that the inverse transformation is accomplished by the matrix S_{xp}

$$\psi(x) = \int S_{xp}c(p)\, dp\,, \tag{48.8}$$

where

$$S_{xp} = (2\pi\hbar)^{-\frac{1}{2}} e^{(i/\hbar)px} .$$

The form of the matrix of the coordinate q in the p-representation is easily determined by means of the matrix S_{xp}. According to (46.15), we have

$$q_p = S^\dagger q_x S$$

or

$$(q_p)_{p'p''} = \int S^\dagger_{p'\tau}(q_x)_{\tau\tau'} S_{\tau'p''} \, d\tau \, d\tau' .$$

Substituting the value of the matrices S and q_x (48.3) into the integral, we obtain

$$(q_p)_{p'p''} = \frac{1}{2\pi\hbar} \int e^{-(i/\hbar)p'\tau} \tau \delta(\tau - \tau') e^{(i/\hbar)\tau'p''} \, d\tau \, d\tau' =$$

$$= \frac{1}{2\pi\hbar} \int e^{-(i/\hbar)p'\tau} \tau e^{(i/\hbar)p''\tau} \, d\tau =$$

$$= \frac{1}{2\pi i} \frac{\partial}{\partial p''} \int_{-\infty}^{\infty} e^{(i/\hbar)(p''-p')\tau} \, d\tau = \frac{\hbar}{i} \frac{\partial}{\partial p''} \delta(p'' - p') ,$$

so that we have obtained the matrix of the coordinate in the p-representation

$$(q_p)_{p'p''} = \frac{\hbar}{i} \frac{\partial}{\partial p''} \delta(p'' - p') . \qquad (48.9)$$

Of course, this result could also be obtained directly from relation (48.4), since the matrix p is diagonal in its own representation.

Knowing the expressions for the matrices q and p, we can find the matrix of an arbitrary function of q and p. Thus, if $H(p, q)$ is a certain function of p and q, then the matrix H can be obtained if, instead of p and q, we substitute the corresponding matrices and carry out the necessary operations according to the rules of matrix addition and multiplication. Here the matrix H, as well as the corresponding operator, is understood in the sense of an expansion in a power series in terms of q and p.

Suppose that $H(p, q)$ is a known function — the Hamiltonian of a system. In the coordinate representation the matrices q and p are given by expressions (48.3) and (48.5). Consequently, the matrix H is also known in this representation. By means of a certain unitary transformation S this matrix can be transformed to the diagonal form $H' = E_n \delta_{nm}$. For definiteness we shall assume that the matrix H' has a discrete spectrum. Otherwise $\delta(n - m)$ not

δ_{nm} must be written

$$H' = S^{\dagger} H(p, q) S .$$

We define the matrix S. Since $HS = SH'$, we have

$$\int H_{xx'} S_{x'n} \, dx' = \sum_m S_{xm} H'_{mn} = E_n S_{xn} \dots . \qquad (48.10)$$

We consider the integrals standing on the left for different functions $H(p, q)$. First of all,

$$\int q_{xx'} S_{x'n} \, dx' = x S_{xn} ,$$

$$\int p_{xx'} S_{x'n} \, dx' = \frac{\hbar}{i} \frac{\partial}{\partial x} S_{xn} .$$

Further, if $U(q)$ is a certain function of q, then its matrix, as is easily seen, has the form $U_{xx'} = U(x) \delta(x - x')$, and the integral is equal to

$$\int U_{xx'} S_{x'n} \, dx' = U(x) S_{xn} .$$

We also obtain analogous results for functions of p. For example, the matrix of the quantity p^2 is equal, according to the rules of matrix multiplication, to

$$\int (p^2)_{xx'} = \left(\frac{\hbar}{i} \right)^2 \frac{\partial^2}{\partial x^2} \delta(x - x') .$$

Correspondingly, the integral

$$\int (p^2)_{xx'} S_{x'n} \, dx' = \left(\frac{\hbar}{i} \right)^2 \int \frac{\partial^2}{\partial x^2} \delta(x - x') S_{x'n} \, dx' = \left(\frac{\hbar}{i} \right)^2 \frac{\partial^2}{\partial x^2} S_{xn} .$$

Consequently, for an arbitrary function $H(p, q)$ we obtain

$$\int H_{xx'} S_{x'n} \, dx' = H\left(x, \frac{\hbar}{i} \frac{\partial}{\partial x} \right) S_{xn} = \hat{H} S_{xn} . \qquad (48.11)$$

Relation (48.10) is evidently none other than the Schrödinger equation written in matrix form in the x-representation:

$$\int H_{xx'} \psi_n(x') \, dx' = E_n \psi_n(x) .$$

This equation is also easily rewritten in the p-representation in operator form as well as in matrix form. Setting

$$\hat{H} = \frac{\hat{p}^2}{2m} + U(x)$$

and denoting the function $\psi_n(x)$ in the p-representation by $c_n(p)$ (see (48.6)), we obtain

$$\left(\frac{p^2}{2m} + \hat{U}(p)\right) c_n(p) = E_n c_n(p) , \qquad (48.12)$$

where $\hat{U}(p)$ is the operator of the potential energy in the p-representation.
   In matrix form eq. (48.12) is

$$\frac{p^2}{2m} c_n(p) + \int U_{pp'} c_n(p') \, dp' = E_n c_n(p) . \qquad (48.12')$$

Here $U_{pp'}$ – the matrix of the operator $\hat{U}$ – is constructed by means of (48.9) or, what is the same, is defined by

$$U_{pp'} = \int U(x) \, e^{-(i/\hbar)(p-p')x} \, dx .$$

Making use of (48.11), we rewrite eq. (48.10) in the form

$$\hat{H} S_{xn} = E_n S_{xn} .$$

We see that the matrix S is constructed from the eigenfunctions $\psi_n(x)$ of the operator $\hat{H}$

$$S_{xn} = \psi_n(x) . \qquad (48.13)$$

As is known, if we have a certain operator $\hat{F}$, then the matrix of this operator in the energy representation is given by the relation

$$F_{nm} = \int \psi_n^* \hat{F} \psi_m \, dx .$$

On the other hand, the same relation can be considered as the unitary transformation from the coordinate representation, in which the quantity F is defined by the matrix $F_{xx'}$, to the energy representation. The matrix of the unitary transformation is given by formula (48.13). Indeed,

$$F_{nm} = \int S_{nx}^\dagger F_{xx'} S_{x'm} \, dx \, dx'$$

and

$$\int F_{xx'} S_{x'm} \, dx' = \hat{F} S_{xm}$$

by analogy with (48.11). Consequently,

$$F_{nm} = \int S_{nx}^\dagger \hat{F} S_{xm} \, dx = \int \psi_n^*(x) \hat{F} \psi_m(x) \, dx \; .$$

and we have again arrived at the preceding relation. Thus all the relations obtained in this chapter for matrices are generalized directly to the case of operators given in differential form. Keeping this fact in mind, we shall henceforth in using the word 'operator' understand that the operator can be given in differential form as well as in matrix form.

Finally, we shall dwell briefly on a certain notation proposed by Dirac, since it is frequently encountered in the literature.

The wave function ψ or, more precisely, the set of its components in a certain coordinate system (in a certain representation) is called by Dirac a ket-vector and denoted by $|\psi\rangle$. For example, the wave function ψ_{nlm}, describing the state with given quantum numbers n, l, m, is denoted by $|nlm\rangle$. On the other hand, the complex-conjugate function is said to be a bra-vector and is denoted by $\langle\psi|$ (ψ_{nlm}^* is correspondingly denoted by $\langle nlm|$). The terms bra and ket come from the word 'bracket' $\langle\rangle$. In matrix notation the ket-vector corresponds to a column, while the bra-vector corresponds to a row. The scalar product of the bra-vector $\psi_b^* = \langle b|$ and the ket-vector $\psi_a = |a\rangle$ is denoted by $\langle b|a\rangle$, i.e.

$$\int \psi_b^*(x) \psi_a(x) \, dx = \langle b|a\rangle \; . \tag{48.14}$$

On the other hand, this scalar product can evidently be treated as the wave function ψ_a in the b-representation. Indeed, if we write the expansion

$$\psi_a(x) = \int c_a(b) \psi_b(x) \, db \tag{48.15}$$

(for a discrete spectrum the integral is replaced by the sum), then $c_a(b)$ represents the wave function of the state a in the b-representation

$$c_a(b) = \int \psi_b^*(x) \psi_a(x) \, dx = \langle b|a\rangle \; . \tag{48.16}$$

Correspondingly, the wave function of the state a in the x-representation $\psi_a(x)$ in the Dirac notation is of the form

$$\psi_a(x) = \langle x|a\rangle \; . \tag{48.17}$$

In this notation the expression (48.15) can be rewritten as

$$\langle x|a\rangle = \int_b \langle x|b\rangle \langle b|a\rangle \, db \; . \tag{48.18}$$

From (48.16) there follows the relation

$$\langle b|a \rangle = \langle a|b \rangle^* , \tag{48.19}$$

connecting the wave function of the state a in the b-representation with the wave function of the state b in the a-representation. The wave function describing a state with given momentum in the coordinate representation $\psi_p(\mathbf{r})$, in the Dirac notation has the form

$$\psi_p(\mathbf{r}) = (2\pi\hbar)^{-\frac{3}{2}} e^{(i/\hbar)\mathbf{p}\cdot\mathbf{r}} = \langle \mathbf{r}|\mathbf{p} \rangle . \tag{48.20}$$

Correspondingly we write the expansion of an arbitrary function $\psi(\mathbf{r})$ in terms of plane waves as

$$\langle \mathbf{r}|\psi \rangle = \int_p \langle \mathbf{r}|\mathbf{p} \rangle \langle \mathbf{p}|\psi \rangle \, d\mathbf{p} \tag{48.21}$$

or

$$\langle \mathbf{r}| = \int_p \langle \mathbf{r}|\mathbf{p} \rangle \langle \mathbf{p}| \, d\mathbf{p} . \tag{48.22}$$

The eigenfunction of the angular momentum operator $\hat{L}^2$ in the coordinate representation in the Dirac notation has the form

$$Y_{lm}(\vartheta, \varphi) = \langle \vartheta, \varphi|l, m \rangle = \langle r^{-1}\mathbf{r}|l, m \rangle . \tag{48.23}$$

The function $\langle r^{-1}\mathbf{r}|l, m \rangle$ carries out the transition from the representation lm to the coordinate representation. On the contrary, the function $Y_{lm}^*(\vartheta, \varphi) = \langle r^{-1}\mathbf{r}|l, m \rangle^* = \langle l, m|r^{-1}\mathbf{r} \rangle$ (see (48.19)) carries out the transition from the coordinate representation to the angular representation.

In the case where the angles θ, φ define the direction of the momentum vector, the function

$$Y_{lm}(\theta, \varphi) = \langle \theta, \varphi|l, m \rangle = \langle p^{-1}\mathbf{p}|l, m \rangle \tag{48.24}$$

carries out the transition from the momentum representation to the representation lm. It is the eigenfunction of the operator $\hat{L}^2$ in the momentum representation.

The matrix element F_{ba} in the Dirac notation is of the form

$$F_{ba} = \int \psi_b^* \hat{F} \psi_a \, dV = \langle b|F|a \rangle . \tag{48.25}$$

The quantities a and b, characterizing the state of the system, can run over a discrete as well as a continuous set of values. If each of the states a and b is characterized by a set of quantum numbers, for example n', l', m' and n, l, m, then the matrix element, usually denoted by $F_{n'l'm';nlm}$ or by $F_{nlm}^{n'l'm'}$, has, in the Dirac notation the form $\langle n'l'm'|F|nlm \rangle$.

§49. The Schrödinger representation, the Heisenberg representation and the interaction representation

In this section we shall discuss certain problems connected with the further development and generalization of the mathematical apparatus of quantum mechanics. We shall consider methods for the description of the development of a process in time.

Up to now we have based our considerations exclusively on the Schrödinger equation

$$i\hbar \frac{\partial \psi}{\partial t} = \hat{H}\psi ,$$

according to which the wave function $\psi(x, t)$ of the system could be found at an arbitrary instant of time t if its initial value $\psi(x, 0)$ was known. In this approach to the development of the process in time there corresponds a change in the wave function of the system $\psi(x, t)$.

The development of a process in time can be described by means of the operator $\hat{V}(t)$ acting on the wave function defined at a certain initial instant of time

$$\psi(x, t) = \hat{V}(t) \psi(x, 0) . \tag{49.1}$$

Here we have taken as the initial time the instant $t = 0$. Of course, we could as well take as the initial time an arbitrary instant $t = t_0$. Substituting expression (49.1) into the Schrödinger equation we obtain the equation for the operator $\hat{V}(t)$

$$i\hbar \frac{\partial \hat{V}(t)}{\partial t} = \hat{H}\hat{V}(t) \tag{49.2}$$

under the condition (see (49.1)) that $\hat{V}(0) = 1$. If the operator $\hat{H}$ does not depend explicitly on time, then the solution of eq. (49.2) can be written formally as

$$\hat{V}(t) = e^{-(i/\hbar)\hat{H}t} , \tag{49.3}$$

where the exponent is understood in the sense of an expansion in a power series.

The operator $\hat{V}(t)$ is evidently unitary

$$\hat{V}^\dagger(t) \hat{V}(t) = 1 .$$

The unitary property of the operator $\hat{V}(t)$ has a simple meaning: it corresponds to the conservation in time of the normalization condition of the wave function

$$\int \psi^*(x, 0)\, \psi(x, 0)\, \mathrm{d}V = \int \psi^*(x, t)\, \psi(x, t)\, \mathrm{d}V =$$
$$= \int \hat{V}^* \psi^*(x, 0)\, \hat{V}\psi(x, 0)\, \mathrm{d}V = \int \psi^*(x, 0)\, \hat{V}^\dagger \hat{V}\psi(x, 0)\, \mathrm{d}V.$$

Thus the description of the development of a system in time amounts to the fact that the wave function or the state vector $\psi(x, t)$ changes in time. This change can be characterized by means of the unitary operator $\hat{V}(t)$ acting on the initial wave function $\psi(x, 0)$ and transforming it at every given instant to the function $\psi(x, t)$. Here the operators characterizing the system, for example the operators $\hat{x}, \hat{p}$ or any operators $\hat{F}(\hat{x}, \hat{p})$, do not explicitly change in time.

If the state of the system is characterized by means of a Hilbert space, then the trend of development of the system can be described in the following way. Let a system of unit vectors in Hilbert space be given. This system is defined by a system of eigenfunctions of the operators forming a complete set for the given system. At the initial instant of time the state of the system is defined by the state vector $\psi(x, 0)$. The development of the system in time corresponds to a rotation of the state vector ψ in Hilbert space. Its length $(\psi \cdot \psi)$ has a constant value. Such a description of the system, in which the wave function changes in time whereas the operators are time independent, is called the Schrödinger representation. We note that the word 'representation' has in this case a more general meaning than that which it has had up to now, and characterizes a method of describing the change of a state in time. In particular, one can define the state of a system in the Schrödinger coordinate representation, in the Schrödinger momentum representation, in the Schrödinger energy representation and so on. Up to now, when speaking of the definition of the wave function in one or another representation, we have borne in mind the corresponding Schrödinger representation. The operators $\hat{x}, \hat{p}$ and in general $\hat{F}$, as well as the operators of the corresponding derivatives with respect to time, $\dot{\hat{x}}, \dot{\hat{p}}$ and $\dot{\hat{F}}$, do not change in time in the Schrödinger representation (we assume that there are no non-stationary external fields). Indeed, according to (31.2) the operator $\dot{\hat{F}}$, for example, has the form

$$\dot{\hat{F}} = [\hat{H}, \hat{F}]$$

and does not change in time, since the operators $\hat{F}$ and $\hat{H}$ do not change. The matrix elements of the operator $\dot{\hat{F}}$ can also easily be determined:

$$(\dot{F})_{mn} = [\hat{H}, \hat{F}]_{mn}. \tag{49.4}$$

In the energy representation (the Schrödinger energy representation), i.e. in a representation such that the matrix $\hat{H}$ is diagonal, relation (49.4) has the form

$$(\dot{F})_{mn} = \frac{i}{\hbar} (H_{mm}F_{mn} - F_{mn}H_{nn}) = i\omega_{mn}F_{mn} , \qquad (49.5)$$

where

$$\omega_{mn} = \frac{1}{\hbar} (E_m - E_n) ,$$

$$F_{mn} = \int \psi_m^*(x) \hat{F} \psi_n(x) \, dV .$$

The matrix $(\dot{F})_{mn}$, as well as the matrix (F_{mn}), does not depend explicitly on the time.

In addition to the Schrödinger representation, use is often made in quantum mechanics of another representation, called the Heisenberg representation.

In the Heisenberg representation the development of a system in time is described by means of time-dependent operators. In this case the wave function $\Phi(x)$ itself is assumed to be dependent only on coordinates, but to be time-independent. The development in the Heisenberg representation can be pictured as a rotation of the system of basis vectors in the Hilbert space with respect to the motionless state vector $\Phi(x)$.

In the general case the transition to the Heisenberg representation is carried out by means of the unitary transformation

$$\Phi(x) = \hat{V}^{-1}(t) \psi(x, t) = \psi(x, 0) , \qquad (49.6)$$

where $\Phi(x)$ is the wave function (the state vector) in the Heisenberg representation.

Making use of expression (49.6) and taking into account that

$$\hat{V}^{-1}(t) = \hat{V}^\dagger(t) = e^{(i/\hbar)\hat{H}t} ,$$

we obtain

$$\Phi(x) = \psi(x, 0) = e^{(i/\hbar)\hat{H}t} \psi(x, t) . \qquad (49.7)$$

In accord with the general rules (46.15) an arbitrary operator $\hat{F}$ given in the Schrödinger representation will, in the Heisenberg representation (we denote it by $\hat{F}_{II}$), have the following form

$$\hat{F}_H = \hat{V}^\dagger(t) \hat{F} \hat{V}(t)$$

or

$$\hat{F}_H = e^{(i/\hbar)\hat{H}t} \hat{F} e^{-(i/\hbar)\hat{H}t} . \qquad (49.8)$$

At the initial instant of time the expressions for the wave functions as well as for the operators are the same in the two representations. We note that the operator $\hat{H}$ in the Heisenberg representation will be the same as in the Schrödinger representation. $\hat{H}_H = \hat{H}$. This immediately follows from formula (49.8) if it is taken into account that the operator $\hat{H}$ commutes with all terms of the expansion of the function $e^{(i/\hbar)\hat{H}t}$. We define the matrix elements of the operator $\hat{F}_H$ by means of the eigenfunctions of the operator $\hat{H}$ (the Heisenberg energy representation)

$$(F_H)_{mn} = \sum_{k,l} (e^{(i/\hbar)\hat{H}t})_{mk} F_{kl} (e^{-(i/\hbar)\hat{H}t})_{ln} =$$

$$= e^{(i/\hbar)E_m t} F_{mn} e^{-(i/\hbar)E_n t} = e^{i\omega mn^t} F_{mn} . \qquad (49.9)$$

In the energy representation only diagonal matrix elements differ from zero.

If the operator $\hat{F}$ is a function of $\hat{x}$ and $\hat{p}$, then, from formula (49.8), we shall obtain the operator $\hat{F}$ in the Heisenberg representation by using the operators $\hat{x}$ and $\hat{p}$ in this representation

$$\hat{F}_H = F(\hat{x}_H, \hat{p}_H) . \qquad (49.10)$$

Indeed, if, for example, $\hat{F} = \hat{p}^2$, then

$$\hat{F}_H = e^{(i/\hbar)\hat{H}t} \hat{p}^2 e^{-(i/\hbar)\hat{H}t} = e^{(i/\hbar)\hat{H}t} \hat{p} e^{-(i/\hbar)\hat{H}t} e^{(i/\hbar)\hat{H}t} \hat{p} e^{-(i/\hbar)\hat{H}t} = \hat{p}_H^2 .$$

In an analogous way it is easily verified that

$$[\hat{p}_{xH}, \hat{x}_H] = [\hat{p}_x, \hat{x}] = 1 . \qquad (49.11)$$

We shall obtain the equation of motion in the Heisenberg representation by differentiating (49.8)

$$\frac{\partial \hat{F}_H}{\partial t} = \frac{i}{\hbar}(\hat{H}\hat{F}_H - \hat{F}_H\hat{H}) = [\hat{H}, \hat{F}_H] . \qquad (49.12)$$

If the matrix elements of the left-hand and right-hand sides of this equation are taken with the functions $\psi_n(x)$, we shall obtain, analogously to (49.5),

$$\left(\frac{\partial F_H}{\partial t}\right)_{mn} = i\omega_{mn}(F_H)_{mn} . \qquad (49.13)$$

Of course, we shall arrive at exactly the same expression if we proceed from expression (49.9) and if we define the derivative of the matrix F_H with respect to time, t, as a matrix each element of which is equal to the derivative with respect to time of the corresponding matrix element of the matrix F_H, i.e.

$$\left(\frac{\partial F_H}{\partial t}\right)_{mn} = \frac{\partial (F_H)_{mn}}{\partial t} = i\omega_{mn}(F_H)_{mn} \,. \tag{49.14}$$

Thus if the operator $\hat{F}_H$ describes a certain physical quantity, then the operator $\partial \hat{F}_H/\partial t$ correspondingly describes its derivative with respect to time.

We note that eqs. (31.7) and (31.8) can also be expressed in the Heisenberg representation. Making use of (49.11) and (49.12), we have

$$\frac{\partial \hat{x}_H}{\partial t} = \frac{1}{m}\hat{p}_H \,; \qquad \frac{\partial \hat{p}_H}{\partial t} = -\frac{\partial \hat{U}_H}{\partial x} \tag{49.15}$$

or, in the Heisenberg energy representation,

$$\left(\frac{\partial x_H}{\partial t}\right)_{mn} = \frac{\partial}{\partial t}(x_H)_{mn} = i\omega_{mn}(x_H)_{mn} = \frac{1}{m}(p_H)_{mn} \,, \tag{49.16}$$

$$\left(\frac{\partial p_H}{\partial t}\right)_{mn} = \frac{\partial}{\partial t}(p_H)_{mn} = i\omega_{mn}(p_H)_{mn} = \left(-\frac{\partial U}{\partial x}\right)_{mn} \,. \tag{49.17}$$

The matrix relations (49.16) and (49.17) correspond in their external appearance to the classical laws of Newton.

Quantum mechanics was initially formulated by Heisenberg only as matrix mechanics. Heisenberg compared each mechanical variable to a certain matrix with elements which depend harmonically on time. The relations between matrices were taken in the form of classical relations, for example (49.17).

The Schrödinger and Heisenberg representations do not exhaust all the methods of describing quantum systems which are used in practice.

One very often has to deal in quantum mechanics with systems whose Hamiltonian can be divided into two parts: one of these, $\hat{H}^{(0)}$, represents the Hamiltonian of the system, while the other, $\hat{H}'$, describes the interaction of the given system with external fields or with other systems.

In this case it frequently turns out to be convenient to make use of the so-called interaction representation, introduced by Dirac.

The interaction representation is in a certain sense intermediate between the Schrödinger and Heisenberg representations. Namely, we define the wave function in the interaction representation by the relation

$$\varphi(x, t) = e^{(i/\hbar)\hat{H}_0 t}\,\psi(x, t)\,e^{-(i/\hbar)\hat{H}_0 t} \,. \tag{49.18}$$

Analogously, we define an arbitrary operator $\hat{F}$ in the interaction representation as

$$\hat{F}_{int} = e^{(i/\hbar)\hat{H}_0 t}\,\hat{F}\,e^{-(i/\hbar)\hat{H}_0 t} \,. \tag{49.19}$$

In contrast to (49.7) and (49.8), transformation formulae (49.18) and (49.19) do not involve the total Hamiltonian but only the Hamiltonian of the system without interaction, $\hat{H}_0$. The equation satisfied by the function $\varphi(x, t)$ is easily obtained. For this we differentiate relation (49.18) with respect to time and make use of the Schrödinger equation

$$i\hbar \frac{\partial \varphi(x, t)}{\partial t} = -\hat{H}_0 \varphi(x, t) + e^{(i/\hbar)\hat{H}_0 t}(\hat{H}_0 + \hat{H}') \psi(x, t) =$$

$$= e^{(i/\hbar)\hat{H}_0 t}\hat{H}'\psi(x, t) = e^{(i/\hbar)\hat{H}_0 t}\hat{H}'e^{-(i/\hbar)\hat{H}_0 t}\varphi(x, t) \qquad (49.20)$$

or, taking into account (49.19), we obtain

$$i\hbar \frac{\partial \varphi(x, t)}{\partial t} = \hat{H}'_{int}\varphi(x, t) \qquad (49.21)$$

i.e. we have obtained the Schrödinger equation with the Hamiltonian $\hat{H}'_{int}$.

From relation (49.19) we find the law of change in time of an operator given in the interaction representation

$$\frac{\partial \hat{F}_{int}}{\partial t} = \frac{i}{\hbar}(\hat{H}_0 \hat{F}_{int} - \hat{F}_{int}\hat{H}_0) = [\hat{H}_0, \hat{F}_{int}] \; . \qquad (49.22)$$

We note that the operator $\hat{H}_0$ has one and the same form in both the Schrödinger representation and the interaction representation.

If the operator $\hat{F}$ depends on $\hat{x}$ and $\hat{p}$, then, analogously to (49.10), it is easily shown that

$$\hat{F}_{int} = F(\hat{x}_{int}, \hat{p}_{int}) \; . \qquad (49.23)$$

Thus we see that in the interaction representation the dependence on time of the wave function is defined by the interaction operator $\hat{H}'_{int}$, whereas the time-dependence associated with the operator $\hat{H}_0$ is directly transferred to the operators.

§50. The linear harmonic oscillator

In §10 we considered the linear harmonic oscillator by means of the Schrödinger wave equation. In fact, this problem was initially studied by the matrix method*. We shall give this solution here. On the one hand, it is a good

* M.Born, W.Heisenberg and P.Jordan, Z. Phys. 35 (1925) 557.

illustration of the use of matrix methods, and on the other hand we shall need a number of the expressions obtained in what follows.

We proceed from the known expression for the Hamiltonian of the system

$$\hat{H} = \frac{\hat{p}^2}{2m} + \frac{m\omega^2 \hat{x}^2}{2}. \tag{50.1}$$

Here ω is the 'classical frequency of the oscillator'. The operators $\hat{p}$ and $\hat{x}$ in (50.1) are understood to be certain matrices whose mutual relation is given by eqs. (49.15). We solve the problem in the Heisenberg energy representation. Then corresponding to (49.9) we have

$$(x_H)_{nm} = x_{nm} e^{i\omega_{nm}t}. \tag{50.2}$$

The indices m, n here denote the energy levels of the system. From (49.15) and (50.1) it follows that the operator $\hat{x}_H$ satisfies the following equation of motion:

$$\frac{\partial^2 \hat{x}_H}{\partial t^2} + \omega^2 \hat{x}_H = 0. \tag{50.3}$$

We see that the equations of motion in the Heisenberg representation have the same form as the ordinary equations of motion of classical mechanics. However, in the former the classical coordinate x is replaced by the quantum-mechanical operator $\hat{x}_H$. Correspondingly, for the matrix elements x_{nm} we obtain, taking account of (50.2),

$$(\omega^2 - \omega_{nm}^2) x_{nm} = 0. \tag{50.4}$$

It follows from (50.4) that only matrix elements x_{nm} for which the condition $\omega_{nm} = \pm \omega$ or $(E_n - E_m)/\hbar = \pm \omega$ is fulfilled are different from zero. We number the states in such a way that $\omega_{n,n-1} = +\omega$, and $\omega_{n,n+1} = -\omega$. Consequently,

$$x_{nm} = 0 \quad \text{for} \quad m \neq n \pm 1 \quad \text{and} \quad x_{nm} \neq 0 \quad \text{for} \quad m = n \pm 1. \tag{50.5}$$

The matrix elements $x_{n,n\pm1}$ can be determined by proceeding from the commutation relations

$$\hat{p}\hat{x} - \hat{x}\hat{p} = -i\hbar.$$

We evaluate the following:

$$\sum_k (p_{nk} x_{km} - x_{nk} p_{km}) = -i\hbar \delta_{nm}.$$

According to (49.16) and (49.9), $p_{nk} = im\omega_{nk} x_{nk}$. Substituting p_{nk} into the above expression and taking into account (50.5), we have for $m = n$

$$x_{n,n+1}x_{n+1,n} - x_{n,n-1}x_{n-1,n} = \frac{\hbar}{2m\omega} .$$

Here we have made use of the fact that $\omega_{nm} = -\omega_{mn}$, and instead of $\omega_{n,n-1}$ and $\omega_{n,n+1}$ we have substituted respectively $+\omega$ and $-\omega$. Since the matrix of the coordinate is Hermitian, then $x_{nm} = x^*_{mn}$, and we rewrite the relation obtained in the form

$$|x_{n+1,n}|^2 - |x_{n,n-1}|^2 = \frac{\hbar}{2m\omega} . \tag{50.6}$$

It is clear from (50.6) that the squares of the moduli of the matrix elements form an arithmetical progression with the difference $\hbar/2m\omega$. Since all terms of the progression are positive, it must begin with a certain positive term to which we can assign the index $n = 0$. It is obvious that then we have $x_{1,0} \neq 0$, $x_{0,-1} \equiv 0$. Consequently, from (50.6) the equation

$$|x_{1,0}|^2 = \frac{\hbar}{2m\omega} .$$

follows. Correspondingly, for an arbitrary positive integer n we find

$$|x_{n,n-1}|^2 = \frac{n\hbar}{2m\omega} . \tag{50.7}$$

From (50.7) we obtain directly

$$x_{n,n-1} = \left(\frac{n\hbar}{2m\omega}\right)^{\frac{1}{2}} e^{i\beta} ; \qquad x_{n-1,n} = \left(\frac{n\hbar}{2m\omega}\right)^{\frac{1}{2}} e^{-i\beta} , \tag{50.8}$$

where β is an arbitrary phase factor. Making use of the arbitrariness in the choice of β, one can set it equal to zero. Taking into account (50.2), we correspondingly obtain for the time-dependent matrix elements

$$(x_H)_{n,n-1} = \left(\frac{n\hbar}{2m\omega}\right)^{\frac{1}{2}} e^{i\omega t} ; \qquad (x_H)_{n-1,n} = \left(\frac{n\hbar}{2m\omega}\right)^{\frac{1}{2}} e^{-i\omega t} . \tag{50.9}$$

Let us determine the energy levels of the system, E_n, by means of the matrices (50.8). The E_n are defined as the diagonal matrix elements of the operator $\hat{H}$. It follows from (50.1) that

$$H_{nn} = \frac{1}{2m} \sum_k p_{nk}p_{kn} + \frac{m\omega^2}{2} \sum_k x_{nk}x_{kn} .$$

Substituting $p_{mn} = im\omega_{mn}x_{mn}$, we find

$$H_{nn} = \frac{m}{2} \left[- \sum_k \omega_{nk}\omega_{kn}x_{nk}x_{kn} + \omega^2 \sum_k x_{nk}x_{kn} \right].$$

Making use of the obvious equalities $\omega_{nk} = -\omega_{kn}$, $x_{nk} = x_{kn}$ and taking into account that the matrix elements x_{nm} differ from zero only when $m = n \pm 1$, we obtain

$$H_{nn} = \frac{1}{2}m \left(\sum_k \omega_{nk}^2 x_{nk}^2 + \omega^2 \sum_k x_{nk}^2 \right) = \frac{1}{2}m \sum_k (\omega^2 + \omega_{nk}^2)x_{nk}^2 =$$

$$= \frac{1}{2}m \left[(\omega^2 + \omega_{n,n-1}^2)x_{n,n-1}^2 + (\omega^2 + \omega_{n,n+1}^2)x_{n,n+1}^2 \right].$$

Making use of (50.8) and $\omega_{n,n\pm1}^2 = \omega^2$, we have finally

$$E_n = H_{nn} = m\omega^2 \left(\frac{n\hbar}{2m\omega} + \frac{(n+1)\hbar}{2m\omega} \right) = \hbar\omega(n + \tfrac{1}{2}), \tag{50.10}$$

which, naturally, is the same as (10.13).

Instead of the operators $\hat{p}$ and $\hat{x}$, it often turns out to be convenient to introduce the operator $\hat{a}$ and the conjugate operator $\hat{a}^\dagger$ defined by the relations

$$\hat{a} = \tfrac{1}{2}\sqrt{2}[(m\omega\hbar^{-1})^{\frac{1}{2}}\hat{x} + i(m\omega h)^{-\frac{1}{2}}\hat{p}],$$
$$\hat{a}^\dagger = \tfrac{1}{2}\sqrt{2}[(m\omega\hbar^{-1})^{\frac{1}{2}}\hat{x} - i(m\omega\hbar)^{-\frac{1}{2}}\hat{p}]. \tag{50.11}$$

It follows from (50.8) that the operators $\hat{a}$ and $\hat{a}^\dagger$ have the following matrix elements different from zero:

$$(a^\dagger)_{n,n-1} = (a)_{n-1,n} = n^{\frac{1}{2}}, \tag{50.12}$$

while all the remaining matrix elements are equal to zero. We see, consequently, that for the operator $\hat{a}^\dagger$ only the matrix elements corresponding to the transition $n - 1 \rightarrow n$, i.e. to a transition with an increase of the quantum number n by one, are different from zero. For the operator $\hat{a}$ the matrix elements corresponding to the transition $n \rightarrow n - 1$ are different from zero. In this connection the operators $\hat{a}$ and $\hat{a}^\dagger$ are called the excitation annihilation operator and the excitation creation operator. From (50.12), the following commutation relation holds for the operators $\hat{a}^\dagger$ and $\hat{a}$:

$$\hat{a}\hat{a}^\dagger - \hat{a}^\dagger\hat{a} = 1. \tag{50.13}$$

The operator $\hat{H}$ expressed in terms of the operators $\hat{a}$ and $\hat{a}^\dagger$ has the form

$$\hat{H} = \tfrac{1}{2}\hbar\omega(\hat{a}^\dagger\hat{a} + \hat{a}\hat{a}^\dagger) \tag{50.14}$$

and, taking into account (50.12), we again arrive at (50.10). Making use of the matrix method, one can also obtain the expressions for the wave functions of the oscillator*.

§51. The matrix elements of the angular momentum operator**

In studying the properties of angular momentum in §30 of ch. 3, we proceeded directly from expressions (30.1) and (30.2) for the angular momentum operators. In the present section we shall base our exposition only on the commutation relations (30.3), (30.3′). It turns out that this statement of the problem is of a more general character. In particular, the concrete expressions for the operators, (30.1) and (30.2), cannot be used for the study of the properties of intrinsic angular momentum (spin) which will be considered in ch. 8. However, the commutation relations of the form of (30.3) also remain valid for the intrinsic angular momentum (see §60). The study of the properties of the angular momentum based on the corresponding commutation relations is conveniently carried out in matrix form. We shall denote the matrices corresponding to the projections of the angular momentum onto the x-, y- and z-axes by $\hat{J}_x, \hat{J}_y, \hat{J}_z$. The change in notation is connected with the fact that the results obtained in this section will be valid not only for the angular momentum associated with spatial motion, the orbital angular momentum $\hat{l} - (\hbar/i)(\mathbf{r} \times \nabla)$, but also for the angular momentum which is not associated with spatial motion, the spin, as well as for the total angular momentum (see §62). We also introduce the matrix $\hat{J}^2$ corresponding to the square of the angular momentum $\hat{J}^2 = \hat{J}_x^2 + \hat{J}_y^2 + \hat{J}_z^2$. Thus we take as our basis the following commutation relations:

$$\hat{J}_x\hat{J}_y - \hat{J}_y\hat{J}_x = i\hbar\hat{J}_z \,,$$
$$\hat{J}_y\hat{J}_z - \hat{J}_z\hat{J}_y = i\hbar\hat{J}_x \,, \qquad (51.1)$$
$$\hat{J}_z\hat{J}_x - \hat{J}_x\hat{J}_z = i\hbar J_y \,.$$

First of all, we obtain the following additional commutation rules from these

---

* L.D.Landau and E.M.Lifshitz, *Quantum mechanics* (Pergamon Press, Oxford, 1965).

** The problems touched upon in this and next sections of this chapter are considered in more detail in the books: L.D.Landau and E.M.Lifshitz, *Quantum mechanics* (Pergamon Press, Oxford, 1965), and E.Condon and H.Shortley, *The theory of atomic spectra* (University Press, Cambridge, 1951).

relations (the proof is analogous to that presented in §30):

$$\hat{J}_x \hat{J}^2 - \hat{J}^2 \hat{J}_x = 0 ,$$
$$\hat{J}_y \hat{J}^2 - \hat{J}^2 \hat{J}_y = 0 , \qquad\qquad (51.2)$$
$$\hat{J}_z \hat{J}^2 - \hat{J}^2 \hat{J}_z = 0 .$$

We choose the representation in which the matrices $\hat{J}^2, \hat{J}_z$ and $\hat{H}$ are diagonal. Indeed, in §47 we have proved that mutually commuting matrices can simultaneously be brought into diagonal form. The commutation of a given matrix with the matrix $\hat{H}$ expresses the law of conservation of the corresponding quantity (see §32). Hence the assumption of the commutation of the matrices $\hat{J}^2$ and $\hat{J}_z$ with $\hat{H}$ only means the fulfillment of certain conservation laws.

We number the columns and rows of the matrices considered by the indices m, j, n. The real number m defines the projection of the angular momentum onto the z-axis, $J_z = m\hbar$. The number j characterizes the value of the total angular momentum, and the number n is associated with the energy level of the system. Since all the matrices considered commute with the matrices $\hat{J}^2$ and $\hat{H}$, they are diagonal in the indices j and n. Consequently, making use of the Dirac notation, we can write the matrix elements of the matrices in which we are interested in the form

$$\langle m'j'n'|\hat{H}|mjn\rangle = E_{jn}\delta_{jj'}\delta_{mm'}\delta_{nn'} ,$$
$$\langle m'j'n'|\hat{J}^2|mjn\rangle = J_j^2\delta_{jj'}\delta_{mm'}\delta_{nn'} ,$$
$$\langle m'j'n'|\hat{J}_z|mjn\rangle = m\hbar\delta_{jj'}\delta_{mm'}\delta_{nn'} , \qquad (51.3)$$
$$\langle m'j'n'|\hat{J}_x|mjn\rangle = (J_x)_{m'm}\delta_{jj'}\delta_{nn'} ,$$
$$\langle m'j'n'|\hat{J}_y|mjn\rangle = (J_y)_{m'm}\delta_{jj'}\delta_{nn'} .$$

Here we have denoted the eigenvalue of the square of the angular momentum by J_j^2. For what follows it will be convenient for us to introduce also the matrices $\hat{J}_+ = \hat{J}_x + i\hat{J}_y$ and $\hat{J}_- = \hat{J}_x - i\hat{J}_y$. It is obvious that these matrices, as well as the initial matrices $\hat{J}_x$ and $\hat{J}_y$, are diagonal in the indices j and n. Taking this into account, we shall in what follows drop the indices j and n.

Our problem is the determination of the spectrum of possible values of the projection of the angular momentum onto an arbitrarily oriented axis, the establishment of the relation of these quantities to the absolute value of the angular momentum $(J_j^2)^{\frac{1}{2}}$, and finding the matrices $(J_x)_{m'm}$ and $(J_y)_{m'm}$. First of all we shall show that the spectrum of possible values of the projection of the angular momentum for a given total angular momentum is bounded both above and below. For this we make use of the matrix relation

$$\hat{\mathbf{J}}^2 - \hat{J}_z^2 = \hat{J}_x^2 + \hat{J}_y^2 \, .$$

Equating the diagonal matrix elements of the left-hand and right-hand sides, we obtain

$$\mathbf{J}_j^2 - m^2\hbar^2 = \sum_k \left[(J_x)_{mk}(J_x)_{km} + (J_y)_{mk}(J_y)_{km} \right] = \sum_k |(J_x)_{mk}|^2 + |(J_y)_{mk}|^2 \, . \tag{51.4}$$

Here we have made use of the Hermitian property of the matrices J_x and J_y. Thus the right-hand side of eq. (51.4) is undoubtedly not negative. Whence the inequality

$$m^2\hbar^2 \leqslant \mathbf{J}_j^2 \tag{51.5}$$

or $-(\mathbf{J}_j^2)^{\frac{1}{2}} \leqslant m\hbar \leqslant (\mathbf{J}_j^2)^{\frac{1}{2}}$ follows. We denote the values of the quantum number m corresponding respectively to the largest and smallest possible values of the component of the angular momentum along the z-axis as m_1 and m_2. We find the spectrum of possible values of the number m by means of the matrices $\hat{J}_+$ and $\hat{J}_-$. For this we find the commutator of these matrices with the matrix $\hat{J}_z$. Making use of (51.1), we obtain

$$\begin{aligned} \hat{J}_z\hat{J}_+ - \hat{J}_+\hat{J}_z &= \hbar\hat{J}_+ \, , \\ \hat{J}_z\hat{J}_- - \hat{J}_-\hat{J}_z &= -\hbar\hat{J}_- \, . \end{aligned} \tag{51.6}$$

We evaluate the first of these relations

$$(J_zJ_+)_{m'm''} - (J_+J_z)_{m'm''} = \hbar(J_+)_{m'm''} \, .$$

Calculating the matrix element of the derivative according to rule (45.6) and taking into account that the matrix J_z is diagonal, we find

$$\hbar(m' - m'') (J_+)_{m'm''} = \hbar(J_+)_{m'm''} \, . \tag{51.7}$$

It follows from eq. (51.7) that the matrix $\hat{J}_+$ has non-zero matrix elements $(J_+)_{m'm''}$ only under the condition that $m' - m'' = 1$, i.e. for transitions corresponding to an increase of the quantum number m by unity, $m \to m + 1$. In an analogous way it is easily shown from the second equation, (51.6), that the matrix $\hat{J}_-$ has non-zero matrix elements only for transitions with a decrease of the quantum number m by unity, i.e. $m \to m - 1$. Thus we arrive at the conclusion that, if for a given $\mathbf{J}_j^2$ a certain value $m\hbar$ of the z-component of the angular momentum is possible, then the values $(m + 1)\hbar$, $(m - 1)\hbar$, $(m + 2)\hbar$, $(m - 2)\hbar$ and so on are also possible. However, we have explained before that the spectrum of possible values of the number m must be bounded:

$m_2 \leqslant m \leqslant m_1$. Setting $m'' = m_1$ in equality (51.7) and taking into account that in it m' cannot assume the value $m_1 + 1$, we see that it is fulfilled only when the matrix element $(J_+)_{m_1+1,m_1}$ reduces to zero. Consequently,

$$\langle m_1 + 1 | J_+ | m_1 \rangle = 0 \ . \tag{51.8}$$

We have an analogous situation for the minimum possible value of the number m. The corresponding equality is fulfilled here when the matrix element $(J_-)_{m_2-1,m_2}$ reduces to zero,

$$\langle m_2 - 1 | J_- | m_2 \rangle = 0 \ . \tag{51.9}$$

Thus the possible values of the angular momentum component are equal to $m_2\hbar$, $(m_2 + 1)\hbar$, $(m_2 + 2)\hbar$, ..., $(m_1 - 1)\hbar$, $m_1\hbar$. Here the difference $m_1 - m_2$ can only be equal to a positive integer (including zero). We shall show that the values of the numbers m_1 and m_2 determine the quantity J_j^2. Indeed, the matrix $\hat{J}^2$ can be written in the form

$$\hat{J}^2 = \hat{J}_- \hat{J}_+ + J_z^2 + \hbar \hat{J}_z \ . \tag{51.10}$$

Taking the diagonal matrix elements of the left-hand and right-hand sides, corresponding to the transition $m_1 \to m_1$, we have

$$J_j^2 = \sum_k (J_-)_{m_1 k}(J_+)_{k m_1} + \hbar^2 m_1^2 + \hbar^2 m_1 \ .$$

Here only $k = m_1 + 1$ is possible, but then the matrix element of J_+ (51.8) reduces to zero. Consequently,

$$J_j^2 = \hbar^2 m_1 (m_1 + 1) \ .$$

On the other hand, eq. (51.10) can also be rewritten in the form:

$$\hat{J}^2 = \hat{J}_+ \hat{J}_- + \hat{J}_z^2 - \hbar \hat{J}_z \ . \tag{51.11}$$

If in the above expression one equates the diagonal matrix elements $m_2 \to m_2$, then

$$J_j^2 = \hbar^2 m_2 (m_2 - 1)$$

and, consequently,

$$m_1(m_1 + 1) = m_2(m_2 - 1) \ .$$

This equation is satisfied under the condition that $m_2 = m_1 + 1$ and $m_2 = -m_1$. Since, however, $m_2 \leqslant m_1$ always, we must retain only the second root, $m_2 = -m_1$. Consequently, the maximum (equal to $m_1\hbar$) and minimum (equal to $m_2\hbar$) possible values of the projection of the angular momentum onto the z-axis are equal in absolute value. As we have shown, the square of the total angular momentum is equal to $\hbar^2 m_1(m_1 + 1)$. On the other hand, we decided to characterize this quantity by the quantum number j. Hence it is natural to set $m_1 = j$. In this case we have

$$\mathbf{J}_j^2 = \hbar^2 j(j + 1) . \tag{51.12}$$

The possible values of the angular momentum component J_z are correspondingly equal to

$$J_z = j\hbar , \quad (j-1)\hbar , \quad (j-2)\hbar, \dots , (-j+1)\hbar , \quad -j\hbar . \tag{51.13}$$

On the whole the angular momentum component assumes $2j + 1$ values. We note that, since $2j + 1$ is a positive integer, the quantum number j can take on only integer or half-integer values, $j = 0, \frac{1}{2}, 1, \frac{3}{2}$ and so on. For the orbital angular momentum this number, as we have explained in §30, takes on only integer values $j = l$. We shall see however, in ch. 8, that for intrinsic angular momentum j can also take on half-integer values.

Since the z-axis has in no way been singled out beforehand, the angular momentum projection onto any other axis is also given by formula (51.13). We note that if the number j is integer, then the angular momentum projections onto any axis are also integer (in units of $\hbar$); but if j is half-integer, then the angular momentum projections take on half-integer values.

Let us now find the matrices $\hat{J}_x$ and $\hat{J}_y$. For this we can make use, for example, of the relation (51.10), taking the diagonal matrix elements of the left-hand and right-hand sides. Also taking into account (51.12) we have

$$\hbar^2 j(j + 1) = \sum_k (J_-)_{mk}(J_+)_{km} + \hbar^2 m^2 + \hbar^2 m ,$$

where only the term of the sum with $k = m + 1$ differs from zero.

From the Hermitian property of the matrices J_x and J_y it follows that

$$(J_+)_{km} = (J_-)_{mk}^* .$$

Consequently, the preceding equation gives

$$|(J_+)_{m+1,m}|^2 = \hbar^2[j(j + 1) - m(m + 1)] = \hbar^2(j - m)(j + m + 1) .$$

For the matrix element $(J_+)_{m+1,m}$ we have

$$\langle m + 1|J_+|m\rangle = \hbar[(j - m)(j + m + 1)]^{\frac{1}{2}} e^{i\beta} .$$

Without restricting the generality, the phase β can be put equal to zero. We finally obtain

$$\langle m + 1|J_+|m\rangle = (J_x + iJ_y)_{m+1,m} = \hbar\,[(j - m)(j + m + 1)]^{\frac{1}{2}}\,,$$

$$\langle m|J_-|m + 1\rangle = (J_x - iJ_y)_{m,m+1} = \hbar\,[(j - m)(j + m + 1)]^{\frac{1}{2}}\,,$$

(51.14)

i.e.

$$(J_+)_{m+1,m} = (J_-)_{m,m+1}\,.$$

(51.15)

From the definition of the matrices J_+ and J_- it follows that

$$\hat{J}_x = \tfrac{1}{2}(\hat{J}_+ + \hat{J}_-)\,,\qquad \hat{J}_y = \tfrac{1}{2}i^{-1}(\hat{J}_+ - \hat{J}_-)\,.$$

Making use of (51.14), we get

$$\langle m + 1|J_x|m\rangle = \langle m|J_x|m + 1\rangle = \tfrac{1}{2}\hbar\,[(j - m)(j + m + 1)]^{\frac{1}{2}}\,.$$

$$\langle m + 1|J_y|m\rangle = -\langle m|J_y|m + 1\rangle = -\tfrac{1}{2}i\hbar\,[(j - m)(j + m + 1)]^{\frac{1}{2}}\,.$$

(51.16)

As an example we write the matrices which are obtained for $j = 1$:

$$J_x = \begin{pmatrix} (J_x)_{11} & (J_x)_{10} & (J_x)_{1,-1} \\ (J_x)_{01} & (J_x)_{00} & (J_x)_{0,-1} \\ (J_x)_{-1,1} & (J_x)_{-1,0} & (J_x)_{-1,-1} \end{pmatrix} = \frac{\hbar}{\sqrt{2}}\begin{pmatrix} 0 & 1 & 0 \\ 1 & 0 & 1 \\ 0 & 1 & 0 \end{pmatrix}.$$

$$J_y = \frac{\hbar}{\sqrt{2}}\begin{pmatrix} 0 & -i & 0 \\ i & 0 & -i \\ 0 & i & 0 \end{pmatrix}; \qquad J_z = \hbar\begin{pmatrix} 1 & 0 & 0 \\ 0 & 0 & 0 \\ 0 & 0 & -1 \end{pmatrix},\qquad (51.17)$$

$$\mathbf{J}^2 = \hbar^2 \cdot 2\begin{pmatrix} 1 & 0 & 0 \\ 0 & 1 & 0 \\ 0 & 0 & 1 \end{pmatrix}.$$

§52. The addition of angular momenta

We now determine the possible values of an angular momentum $\mathbf{J}$ which is equal to the sum of two angular momenta, $\mathbf{J} = \mathbf{J}_1 + \mathbf{J}_2$. Let $\mathbf{J}_1$ and $\mathbf{J}_2$ be the angular momenta referring to two sub-systems whose mutual interaction can be disregarded. This means that the operators $\hat{\mathbf{J}}_1$ and $\hat{\mathbf{J}}_2$ act on variables referring to different sub-systems and, consequently, commute with each other,

$\hat{\mathbf{J}}_1\hat{\mathbf{J}}_2 = \hat{\mathbf{J}}_2\hat{\mathbf{J}}_1$. Since each of the operators $\hat{\mathbf{J}}_1$ and $\hat{\mathbf{J}}_2$ satisfies the commutation relations (51.1), the operator $\hat{\mathbf{J}}$ also satisfies the same commutation relations. The state of the system will be defined if the quantum numbers j_1, j_2 and m_1, m_2 characterizing the total angular momenta $\mathbf{J}_1^2$ and $\mathbf{J}_2^2$ and their projections onto an arbitrarily oriented z-axis are defined (we digress from other quantities contained in the total set, since they are not essential for what follows). For given j_1 and j_2 each of the numbers m_1 and m_2 runs respectively over $(2j_1 + 1)$ and $(2j_2 + 1)$ values. Consequently there correspond $(2j_1 + 1)(2j_2 + 1)$ states to given numbers j_1 and j_2. However, the state of the system may be characterized by the numbers j_1, j_2, j and m, where j and m are the quantum numbers corresponding to the total angular momentum $\hat{\mathbf{J}}$ and its projection on the z-axis instead of by the four numbers j_1, j_2, m_1, m_2. This means, in essence, the transition from the representation j_1, j_2, m_1, m_2 to the representation j_1, j_2, j, m. Indeed, the operators corresponding to the four latter quantities can enter into the total set just as well as the operators corresponding to the quantities j_1, j_2, m_1, m_2. Since the obvious equality $\hat{J}_z = \hat{J}_{1z} + \hat{J}_{2z}$ holds, then the quantum number m is equal to the sum $m = m_1 + m_2$. Of course, such a simple relation does not exist for the squares of the angular momenta and we have to determine the possible values of the number j for given j_1 and j_2. First of all we note that the maximum value of the number j is obtained if we take the largest m_1, equal to j_1, and the largest m_2, equal to j_2. Consequently, in this case $j = j_1 + j_2$. Further, we consider the following possible value of the number m: $m = j_1 + j_2 - 1$. Such a value of m can be realized either for $m_1 = j_1$, $m_2 = j_2 - 1$, or for $m_1 = j_1 - 1$, $m_2 = j_2$. Thus two independent states correspond to the given value of $m = j_1 + j_2 - 1$. Consequently, two possible values of the number j must correspond to the given m. But because the largest possible value of j is equal to $j_1 + j_2$ and because the number m cannot be larger than the number j, it is clear that to the chosen m there can correspond only $j = j_1 + j_2$ and $j = j_1 + j_2 - 1$. Choosing m one unit smaller, we obtain three states corresponding to given m:

(1) $m_1 = j_1$, $m_2 = j_2 - 2$;

(2) $m_1 = j_1 - 1$, $m_2 = j_2 - 1$;

(3) $m_1 = j_1 - 2$, $m_2 = j_2$.

By analogy with the foregoing we arrive at the fact that the number j can assume the values $j = j_1 + j_2, j = j_1 + j_2 - 1$ and $j = j_1 + j_2 - 2$. Going on with this reasoning we find that for given j_1 and j_2 the number j can assume the values

$$j = j_1 + j_2; \quad j_1 + j_2 - 1; \quad j_1 + j_2 - 2, ..., j_1 - j_2. \tag{52.1}$$

On the whole the number j assumes $2j_2 + 1$ values (under the condition that

$j_2 \leqslant j_1$; otherwise the indices 1 and 2 must be exchanged). The total number of states corresponding to given j_1 and j_2 is equal to

$$\sum_{j=j_1-j_2}^{j=j_1+j_2} (2j+1) = (2j_1+1)(2j_2+1) ,$$

as it should be. This result was obtained earlier in the so-called 'vector model' which was introduced before the appearance of quantum mechanics. In the vector model it is assumed that the length of the vector $\mathbf{j}$ formed by adding two angular momentum vectors $\mathbf{j}_1$ and $\mathbf{j}_2$ can change by unity in a discontinuous manner. The length is maximum when these vectors are 'parallel': $\mathbf{j} = \mathbf{j}_1 + \mathbf{j}_2$, and minimum when they are 'antiparallel': $\mathbf{j} = \mathbf{j}_1 - \mathbf{j}_2$.

In the case where three or more angular momenta are to be added, we apply the rule which has been derived, adding them in pairs.

As well as the possible values of j one can find the probability that the total angular momentum of the system is equal to one or other possible value of j for given j_1 and j_2. For this, according to general rules (§21), it is necessary to expand the wave function of the system describing the state with given values of j_1, j_2, m_1, m_2 in terms of the wave functions ψ_{jm} of the states with given j_1, j_2, j, m. Since the initial system is divided into two non-interacting sub-systems, its wave function with given j_1, m_1, j_2, m_2 can be written in the form of the product of two functions which refer respectively to each of the sub-systems $\psi_{j_1 j_2 m_1 m_2} = \psi_{j_1 m_1}(1) \psi_{j_2 m_2}(2)$. The expansion has the form

$$\psi_{j_1 m_1} \psi_{j_2 m_2} = \sum_j C^j_{m_1 m_2} \psi_{j, m_1 + m_2} . \tag{52.2}$$

The squares of the moduli of the coefficients $C^j_{m_1 m_2}$ determine the probabilities sought. We note that, since the transformation of wave functions from one representation to another is carried out by unitary matrices, we can write the expansion which is the converse of (52.2) in the form

$$\psi_{jm} = \sum_{m_2} (C^j_{m-m_2, m_2})^* \psi_{j_1, m-m_2}(1) \psi_{j_2 m_2}(2) . \tag{53.3}$$

The coefficients $C^j_{m_1 m_2}$ were calculated by Wigner by a group theory method. The reader will find a sufficiently extensive table of these coefficients in, for example, the book of Condon and Shortley†. We shall confine ourselves to the consideration of the simplest case where one of the angular

† E.Condon and H.Shortley, *The theory of atomic spectra* (University Press, Cambridge, 1951).

momenta is equal to one half, and the other is arbitrary. Thus we shall assume that $j_2 = \frac{1}{2}$. For given j_2 the number m_2 runs over only two values, namely $m_2 = \frac{1}{2}$ and $m_2 = -\frac{1}{2}$. We can correspondingly rewrite the expansion (52.3), dropping all superfluous indices, in the form

$$\psi_{jm} = C_{\frac{1}{2}} \psi_{j_1, m-\frac{1}{2}}(1) \psi_{\frac{1}{2}\frac{1}{2}}(2) + C_{-\frac{1}{2}} \psi_{j_1, m+\frac{1}{2}}(1) \psi_{\frac{1}{2}, -\frac{1}{2}}(2). \quad (52.4)$$

We act on the left-hand and right-hand sides of this expansion with the operator $\hat{J}^2 = (\hat{J}_1 + \hat{J}_2)^2 = \hat{J}_1^2 + \hat{J}_2^2 + 2\hat{J}_1\hat{J}_2$. The scalar product on the right is conveniently transformed so that

$$\hat{J}^2 = \hat{J}_1^2 + \hat{J}_2^2 + (\hat{J}_{1x} + i\hat{J}_{1y})(\hat{J}_{2x} - i\hat{J}_{2y}) + (\hat{J}_{1x} - i\hat{J}_{1y})(\hat{J}_{2x} + i\hat{J}_{2y}) + 2\hat{J}_{1z}\hat{J}_{2z} =$$
$$= \hat{J}_1^2 + \hat{J}_2^2 + \hat{J}_{1+}\hat{J}_{2-} + \hat{J}_{1-}\hat{J}_{2+} + 2\hat{J}_{1z}\hat{J}_{2z}. \quad (52.5)$$

Acting on the left-hand side of (52.4) with this operator we obtain $\hbar^2 j(j+1)$. When acting on the right-hand side it is convenient to write the operator $\hat{J}^2$ in the form (52.5). In this case it should be recalled that each of the operators $\hat{J}_1, \hat{J}_2$ acts only on the wave function of the corresponding sub-system. It follows from the matrix relations (51.14) that

$$\hat{J}_{2-} \psi_{\frac{1}{2}, -\frac{1}{2}}(2) = \hat{J}_{2+} \psi_{\frac{1}{2}\frac{1}{2}}(2) = 0,$$
$$\hat{J}_{2-} \psi_{\frac{1}{2}\frac{1}{2}}(2) = \hbar \psi_{\frac{1}{2}, -\frac{1}{2}}(2),$$
$$\hat{J}_{2+} \psi_{\frac{1}{2}, -\frac{1}{2}}(2) = \hbar \psi_{\frac{1}{2}\frac{1}{2}}(2),$$
$$\hat{J}_{1+} \psi_{j_1, m-\frac{1}{2}}(1) = \hbar [(j_1 - m + \frac{1}{2})(j_1 + m + \frac{1}{2})]^{\frac{1}{2}} \psi_{j_1, m+\frac{1}{2}}(1),$$
$$\hat{J}_{1-} \psi_{j_1, m+\frac{1}{2}}(1) = \hbar [(j_1 - m + \frac{1}{2})(j_1 + m + \frac{1}{2})]^{\frac{1}{2}} \psi_{j_1, m-\frac{1}{2}}(1).$$

Making use of these relations, we obtain the following equation:

$$j(j+1) \psi_{jm} = [C_{\frac{1}{2}}(j_1(j_1+1) + \tfrac{1}{4} + m) +$$
$$+ C_{-\frac{1}{2}}\{(j_1 + m + \tfrac{1}{2})(j_1 - m + \tfrac{1}{2})\}^{\frac{1}{2}}] \, \psi_{j_1, m+\frac{1}{2}}(1) \psi_{\frac{1}{2}\frac{1}{2}}(2) +$$
$$+ [C_{-\frac{1}{2}}(j_1(j_1+1) + \tfrac{1}{4} - m) +$$
$$+ C_{\frac{1}{2}}\{(j_1 - m + \tfrac{1}{2})(j_1 + m + \tfrac{1}{2})\}^{\frac{1}{2}}] \, \psi_{j_1, m+\frac{1}{2}}(1) \psi_{\frac{1}{2}, -\frac{1}{2}}(2).$$

Again substituting the expansion (52.4) for ψ_{jm} into the left-hand side and equating the coefficients of the same functions $\psi(1)\psi(2)$, we obtain two equations with respect to $C_{\frac{1}{2}}$ and $C_{-\frac{1}{2}}$. However, of these two equations only one will be independent. It gives

$$C_{-\frac{1}{2}} = C_{\frac{1}{2}} \frac{j(j+1) - j_1(j_1+1) - \frac{1}{4} - m}{[(j_1 + m + \frac{1}{2})(j_1 - m + \frac{1}{2})]^{\frac{1}{2}}}. \quad (52.6)$$

We shall obtain the second relation if we take into account that the squares of the moduli of these coefficients are equal to the corresponding probabilities

$$|C_{-\frac{1}{2}}|^2 + |C_{\frac{1}{2}}|^2 = 1 . \tag{52.7}$$

The relations (52.6) and (52.7) determine the coefficients $C_{\frac{1}{2}}$ and $C_{-\frac{1}{2}}$ to within the immaterial phase factor $e^{i\alpha}$ (we choose the phase in correspondence with that taken in tables of the coefficients $C^j_{m_1 m_2}$; see the book of Condon and Shortley). Since for $j_2 = \frac{1}{2}$ the total angular momentum j can take on only two values $j_1 + \frac{1}{2}$ and $j_1 - \frac{1}{2}$, we obtain the following values of the coefficients $C^j_{m_1 m_2}$ (see table 1).

Table 1
The coefficients $C^j_{m_1 m_2}$

j	$m_2 = \frac{1}{2}$	$m_2 = -\frac{1}{2}$
$j_1 + \frac{1}{2}$	$\left(\dfrac{j_1 + m + \frac{1}{2}}{2j_1 + 1}\right)^{\frac{1}{2}}$	$\left(\dfrac{j_1 - m + \frac{1}{2}}{2j_1 + 1}\right)^{\frac{1}{2}}$
$j_1 - \frac{1}{2}$	$-\left(\dfrac{j_1 - m + \frac{1}{2}}{2j_1 + 1}\right)^{\frac{1}{2}}$	$\left(\dfrac{j_1 + m + \frac{1}{2}}{2j_1 + 1}\right)^{\frac{1}{2}}$

7

Perturbation Theory

§53. The theory of time-independent perturbations

The Schrödinger equation is a linear differential equation in partial derivatives with variable coefficients. Its exact solution can only be found for simple problems, some of which were considered in preceding sections.

In general the exact solution of the Schrödinger equation is associated with considerable mathematical difficulties. Hence a number of approximate methods of solving it have been devised. One such method is that of the quasi-classical approximation already considered. Another very important approximate method is the so-called perturbation theory. The term 'perturbation' and the idea of this method, which consists of a particular variant of the method of expansion in terms of a small parameter familiar in mathematics, were introduced into quantum mechanics by analogy with the perturbation method of classical mechanics. The latter played a particularly important role in solving problems of celestial mechanics.

We shall discuss perturbation theory in a general form. Its applications to the solution of actual problems will be illustrated in what follows by numerous examples.

Let us first of all consider the simplest case of a quantum-mechanical system in which the Hamiltonian operator $\hat{H}$ does not depend explicitly on the time.

We assume that the operator $\hat{H}$ can be written in the form

$$\hat{H} = \hat{H}_0 + \hat{H}' , \tag{53.1}$$

where the operator $\hat{H}'$ can be considered to be small in comparison with the operator $\hat{H}_0$ (the meaning of the word 'small' will be explained below). Then the Schrödinger equation takes the form

$$(\hat{H}_0 + \hat{H}') \psi = E\psi . \tag{53.2}$$

We further assume that the solution of the equation

$$\hat{H}_0 \psi^{(0)} = E^{(0)} \psi^{(0)} \tag{53.3}$$

is known. Then for the solution of eq. (53.2) use can be made of what is, in essence, a method of successive approximation. In what follows the Hamiltonian $\hat{H}_0$ and the wave function $\psi^{(0)}$ will be called unperturbed, while the operator $\hat{H}'$ will be called the perturbation operator. The 'smallness' of the operator $\hat{H}'$ means that under the action of a perturbation the state of the system changes relatively little. Our problem is to find the solution of the Schrödinger equation assuming that the wave function $\psi^{(0)}$ of the unperturbed system is known. We shall consider the perturbations of states belonging to the discrete spectrum of the operator $\hat{H}_0$. However, the operator $\hat{H}_0$ can have eigenvalues corresponding to a continuous spectrum as well as ones belonging to the discrete spectrum.

We seek a solution of eq. (53.2) in the form of a series in terms of the eigenfunctions of the operator $\hat{H}_0$

$$\psi(x) = \sum_k c_k \psi_k^{(0)} . \tag{53.4}$$

If the operator $\hat{H}_0$ also possesses a continuous spectrum, then we have to add the corresponding integral taken over the continuous spectrum to the sum (53.4). Substituting the sum (53.4) into eq. (53.2) and taking into account (53.3), we obtain

$$\sum_k \hat{H}' c_k \psi_k^{(0)} = \sum_k c_k (E - E_k^{(0)}) \psi_k^{(0)} .$$

We multiply the left- and right-hand sides of the equation by $\psi_m^{(0)*}$ and integrate it over the entire region of variation of the independent variables. Making use of the orthogonality of the functions $\psi_k^{(0)}$, we find

$$c_m(E - E_m^{(0)}) = \sum_k H'_{mk} c_k , \qquad m = 1, 2, 3, \dots \tag{53.5}$$

where

$$H'_{mk} = \int \psi_m^{(0)*} \hat{H}' \psi_k^{(0)} \, dV \tag{53.6}$$

is the matrix element of the perturbation operator calculated using the unperturbed wave functions. The system of equations (53.5) is exactly equivalent to eq. (53.2). It represents the Schrödinger equation in the energy representation. We now make use of our assumption of the smallness of the perturbation operator. The energy levels and wave functions in our problem will be close to those for the unperturbed system. Hence we shall look for them in the form of the following series:

$$E = E^{(0)} + E^{(1)} + E^{(2)} + \dots ,$$
$$c_m = c_m^{(0)} + c_m^{(1)} + c_m^{(2)} + \dots . \tag{53.7}$$

Here $E^{(0)}$ and $c_m^{(0)}$ are unperturbed values. The corrections $E^{(1)}$ and $c_m^{(1)}$ are of the same order of small quantities as the perturbation; $E^{(2)}$ and $c_m^{(2)}$ are quadratic in the perturbation and so on.

We find the correction to the nth energy level and correspondingly to the nth eigenfunction of the unperturbed system, confining ourselves to terms up to the second order of small quantities:

$$E = E_n^{(0)} + E_n^{(1)} + E_n^{(2)} ,$$
$$c_m = c_m^{(0)} + c_m^{(1)} + c_m^{(2)} . \tag{53.8}$$

In this section we shall assume that the nth energy level of the unperturbed system is not degenerate. For the other levels this assumption is unnecessary. In the zero order approximation the wave function is the same as the function $\psi_n^{(0)}$. This gives

$$\psi = \sum_k c_k^{(0)} \psi_k^{(0)} = \psi_n^{(0)} , \quad \text{i.e.} \quad c_k^{(0)} = \delta_{kn} . \tag{53.9}$$

Substituting (53.8) into eq. (53.5), we get

$$(\delta_{mn} + c_m^{(1)} + c_m^{(2)})(E_n^{(0)} - E_m^{(0)} + E_n^{(1)} + E_n^{(2)}) =$$
$$= \sum_k H'_{mk}(\delta_{kn} + c_k^{(1)} + c_k^{(2)}) . \tag{53.10}$$

In eq. (53.10) one has to equate terms of the same order of small quantities. For the terms of the first order we obtain the relation

$$(E_n^{(0)} - E_m^{(0)}) c_m^{(1)} + E_n^{(1)} \delta_{mn} = \sum_k H'_{mk} \delta_{kn} = H'_{mn} . \tag{53.11}$$

Setting $m = n$, we find

$$E_n^{(1)} = H'_{nn} = \int \psi_n^{(0)*} \hat{H}' \psi_n^{(0)} \, dV . \tag{53.12}$$

We see that the first order correction to the energy level is equal to the mean value of the perturbation energy in the unperturbed state $\psi_n^{(0)}$. From eq. (53.11) for $m \neq n$ we find the correction of first order to the wave function

$$c_m^{(1)} = \frac{H'_{mn}}{E_n^{(0)} - E_m^{(0)}} . \tag{53.13}$$

We now write the equation for terms to the second order of small quantities:

$$(E_n^{(0)} - E_m^{(0)}) c_m^{(2)} + c_m^{(1)} E_n^{(1)} + E_n^{(2)} \delta_{mn} = \sum_k H'_{mk} c_k^{(1)} . \tag{53.14}$$

Setting $m \neq n$, we find from eq. (53.14) the correction of second order of smallness to the unperturbed wave function

$$c_m^{(2)} = \frac{1}{E_n^{(0)} - E_m^{(0)}} \left(\sum_k H'_{mk} c_k^{(1)} - E_n^{(1)} c_m^{(1)} \right) . \tag{53.15}$$

The value of the amplitudes $c_n^{(1)}$ and $c_n^{(2)}$ can be obtained from the normalization condition which, taking into account (53.4), can be written in the form

$$\sum_k |c_k|^2 = 1 . \tag{53.16}$$

Substituting expansion (53.8) into (53.6), we obtain

$$\sum_k' |\delta_{kn} + c_k^{(1)} + c_k^{(2)}|^2 = 1 . \tag{53.17}$$

We equate quantities of the same order on the left and on the right. Then we have

$$c_n^{(1)} + c_n^{(1)*} = 0 , \qquad c_n^{(2)} + c_n^{(2)*} + \sum_k |c_k^{(1)}|^2 = 0 . \tag{53.18}$$

It follows from the relations (53.18) that the imaginary parts of the amplitude $c_n^{(1)}$ and $c_n^{(2)}$ are arbitrary quantities. The appearance of this arbitrariness is associated with the fact that the wave function is determined to within the phase factor $e^{i\alpha}$, where α can also be written in the form of a series. In correspondence with this, without restricting the generality we can assume that

$$c_n^{(1)} = 0, \qquad c_n^{(2)} = -\frac{1}{2} \sideset{}{'}\sum_k \frac{|H'_{kn}|^2}{(E_n^{(0)} - E_k^{(0)})^2}. \tag{53.19}$$

Here the prime on the sum indicates that the term with $k = n$ is excluded in the summation.

From (53.15) we find $c_m^{(2)}$

$$c_m^{(2)} = \sideset{}{'}\sum_k \frac{H'_{mk}H'_{kn}}{(E_n^{(0)} - E_k^{(0)})(E_n^{(0)} - E_m^{(0)})} - \frac{H'_{mn}H'_{nn}}{(E_n^{(0)} - E_m^{(0)})^2}, \qquad m \neq n. \tag{53.20}$$

Setting $m = n$ in eq. (53.14), we find the second order correction to the energy level of the system:

$$E_n^{(2)} = \sideset{}{'}\sum_k \frac{H'_{nk}H'_{kn}}{E_n^{(0)} - E_k^{(0)}}. \tag{53.21}$$

The second order correction to the basic energy level turns out to be negative irrespective of the character of the perturbation. As follows from (53.8), (53.12) and (53.21), the energy of the system, to an accuracy within terms of the second order of small quantities, is equal to

$$E = E_n^{(0)} + H'_{nn} + \sideset{}{'}\sum_k \frac{|H'_{nk}|^2}{E_n^{(0)} - E_k^{(0)}}. \tag{53.22}$$

In an analogous way we obtain the expression for the perturbed wave function of the system

$$\psi = \psi_n^{(0)} + \sideset{}{'}\sum_k \frac{H'_{kn}}{E_n^{(0)} - E_k^{(0)}} \psi_k^{(0)} + \dots . \tag{53.23}$$

(We have written this formula only to within an accuracy of terms of the first order of small quantities.)

It follows from expression (53.23) that the first order correction will indeed be small if the following inequality is satisfied:

$$|H'_{kn}| \ll |E_n^{(0)} - E_k^{(0)}|. \tag{53.24}$$

Thus the perturbation theory method developed above is applicable if the matrix elements of the perturbation operator are small in comparison with the spacing between the corresponding energy levels of the unperturbed system.

§54. **Perturbation theory in the presence of degeneracy**

We now assume that the eigenvalues of the unperturbed operator $\hat{H}_0$ are degenerate and that the multiplicity of the degeneracy of the nth level (with energy $E_n^{(0)}$) is equal to s.

This means that the state of the unperturbed system with energy E_n is described by mutually orthogonal wave functions $\psi_{n1}^{(0)}, \ldots, \psi_{ns}^{(0)}$ or by arbitrary linear combinations of them which can be chosen in such a way that the wave functions are, as before, orthogonal. When a perturbation is imposed, the eigenvalues of the operator $\hat{H}_0$ as a rule turn out to be non-degenerate or in any case the multiplicity of the degeneracy decreases. This fact is closely associated with the very nature of degeneracy. We have already pointed out in §35 that degeneracy is always associated with a symmetry of the Hamiltonian with respect to a definite class of transformations of the coordinates of the system. The perturbation, as a rule, does not possess the same symmetry. Hence the resulting Hamiltonian of the perturbed system will not have the previous symmetry and its energy levels will not be degenerate. Thus the perturbation removes the degeneracy. For example, in considering motion in a centrally symmetric field we have seen that the $(2l+1)$-fold degeneracy of the energy levels is associated with the symmetry (invariance) of the Hamiltonian with respect to rotation of the system about the centre of force. If the system is now placed in an external field, then the total Hamiltonian will no longer possess spherical symmetry. The perturbation (in the given case, the external field) removes the $(2l+1)$-fold degeneracy corresponding to the components of the angular momentum.

On imposing the perturbation the degenerate energy level splits into s close levels. To each of these energy levels there corresponds a wave function which is a linear combination of the functions $\psi_{nr}^{(0)}$

$$\psi = \sum_{m,r} c_{mr}\psi_{mr}^{(0)}. \tag{54.1}$$

As before, we consider the perturbation to be small and, in the first approximation of perturbation theory, we seek the nearby energy levels (they are often called sub-levels) into which the degenerate level splits. At the same time we seek the corresponding set of wave functions in the zero order approximation. That is, we have to find, in the zero order approximation, correct expressions for the amplitudes c_{mr} in the sum (54.1) such that the linear combination (54.1) will correspond to one of the sub-levels into which the initial energy level splits and that it will undergo only a small change when the perturbation is taken into account in the next approximation.

Let us first consider the case of two close levels. In this case formula (54.1) gives $\psi = c_1\psi_1 + c_2\psi_2$. Substituting this value into the Schrödinger equation (53.2), we find

$$-c_1(E - E^{(0)}) + H_{12}c_2 + H_{11}c_1 = 0 ,$$

$$c_1 H_{21} - c_2(E - E^{(0)}) + c_2 H_{22} = 0 .$$

Setting $E = E^{(0)} + E^{(1)}$, we obtain the system of homogeneous equations

$$(H_{11} - E^{(1)})c_1 + H_{12}c_2 = 0 ,$$

$$H_{21}c_1 + (H_{22} - E^{(1)})c_2 = 0 .$$

The condition for this system to have a solution is that the determinant of the coefficients be equal to zero,

$$\begin{vmatrix} H_{11} - E^{(1)} & H_{12} \\ H_{21} & H_{22} - E^{(1)} \end{vmatrix} = 0 .$$

Hence

$$(H_{11} - E^{(1)})(H_{22} - E^{(1)}) = |H_{12}|^2 ,$$

or

$$E^{(1)}_{1,2} = \tfrac{1}{2}(H_{11} + H_{22}) \pm \tfrac{1}{2}[(H_{11} - H_{22})^2 + 4|H_{12}|^2]^{\frac{1}{2}} .$$

We see that the degenerate level splits into two levels corresponding to the two different signs in front of the square root.

If the perturbation is small, so that $|H_{12}|^2 \ll (H_{11} - H_{22})^2$, then we come back to the case of two independent levels whose energies are equal to

$$E_1 = E^{(0)} + E^{(1)} = E^{(0)} + H_{11} + \frac{|H_{12}|^2}{H_{11} - H_{22}}$$

$$E_2 = E^{(0)} + H_{22} - \frac{|H_{12}|^2}{H_{11} - H_{22}} .$$

But if the levels are lying so close to each other that $|H_{12}|^2 \gg (H_{11} - H_{22})^2$, then we obtain

$$E_1 = E^{(0)} + \tfrac{1}{2}(H_{11} + H_{22}) + |H_{12}|^2 + \frac{(H_{11} - H_{22})^2}{8|H_{12}|} \, ,$$

$$E_2 = E^{(0)} + \tfrac{1}{2}(H_{11} + H_{22}) - |H_{12}|^2 - \frac{(H_{11} - H_{22})^2}{8|H_{12}|} \, .$$

Analogous results are also obtained in the general case of s-fold degeneracy.

Substituting (54.1) into the Schrödinger equation (53.2) we obtain analogously to (53.5)

$$c_{mp}(E - E_m^{(0)}) = \sum_{k,r} H'_{mp;kr} c_{kr} \, , \tag{54.2}$$

where

$$H'_{mp;kr} = \int \psi_{mp}^{(0)*} \hat{H}' \psi_{kr}^{(0)} \, dV \, .$$

If we are interested in perturbation of the energy level E_n, then we have to put $m = n$ and to equate terms of the first order of magnitude. But in the zero order approximation the wave function ψ is the superposition of the functions $\psi_{nr}^{(0)}$, i.e. the $c_{kr}^{(0)}$ are different from zero only for $k = n$. Writing the energy E in eq. (54.2) in the form $E = E_n^{(0)} + E^{(1)}$, we get

$$c_p^{(0)} E^{(1)} = \sum_{r=1}^{s} H'_{pr} c_r^{(0)} \tag{54.3}$$

(we have dropped the fixed index n in the notation).

The system of homogeneous equations (54.3) has a non-trivial solution only in the case where the determinant of the coefficients of the unknown quantities is equal to zero, i.e. under the condition

$$\begin{vmatrix} H'_{11} - E^{(1)} & H'_{12} & \cdots & H'_{1s} \\ H'_{21} & H'_{22} - E^{(1)} & \cdots & H'_{2s} \\ H'_{s1} & \cdots\cdots\cdots\cdots\cdots & & H'_{ss} - E^{(1)} \end{vmatrix} = 0 \, . \tag{54.4}$$

This equation is called the secular equation. The secular equation is an equation of the sth order with respect to $E^{(1)}$ and has, consequently, s roots. Solving it with respect to $E^{(1)}$, we find s values for this quantity. This means that the nth energy level splits into s sublevels $E_n^{(0)} + E_1^{(1)}$, $E_n^{(0)} + E_2^{(1)}$, ..., $E_n^{(0)} + E_s^{(1)}$. In particular cases certain roots of the secular equation turn out to be equal to one another. In this case the perturbation only partially removes the degeneracy in the system.

Substituting the values $E^{(1)}$ found into eq. (54.3), we can determine the amplitudes $c_{nr}^{(0)}$ corresponding to a given correction to the energy $E^{(1)}$. By this means we find, in the zeroth approximation, the correct wave functions corresponding to the energy sub-levels into which the level $E_n^{(0)}$ splits. These wave functions are slightly distorted under the action of the perturbation.

The method discussed is also applicable in the case where the eigenvalues of the operator $\hat{H}_0$ are not degenerate but are so closely spaced that the inequality (53.24) is not satisfied*.

As an example of the application of the methods discussed in this and the preceding sections we shall consider the displacement of the lowest energy level of the hydrogen-like atom, and the splitting of the first excited level, caused by the finite size of the nucleus.

In considering hydrogen-like atoms we assumed that the electron was in the Coulomb field of the nucleus. However, the difference between the correct field and a Coulomb field in the region of the nucleus itself was not taken into account. We now assume the nucleus to be a uniformly charged sphere of radius r_0. Then the potential energy of the electron for $r \leqslant r_0$ has the form

$$U(r) = -\frac{Ze^2}{r_0}\left(\frac{3}{2} - \frac{r^2}{2r_0^2}\right). \tag{54.5}$$

The difference between the potential energy of the electron and its value for a pure Coulomb field is the perturbation Hamiltonian

$$\hat{H}' = \begin{cases} -\dfrac{Ze^2}{r_0}\left(\dfrac{3}{2} - \dfrac{r^2}{2r_0^2}\right) + \dfrac{Ze^2}{r}, & r \leqslant r_0 \\ 0 & r > r_0. \end{cases} \tag{54.6}$$

We define the correction to the ground energy level in the first approximation:

$$E^{(1)} = H'_{00} = \int \psi_0^* \hat{H}' \psi_0 \, dV. \tag{54.7}$$

The wave function of the ground state (38.22) is

$$\psi_0 = 2(Z/a)^{\frac{3}{2}} e^{-Zr/a}(4\pi)^{-\frac{1}{2}}. \tag{54.8}$$

* For more details see V.A.Fok, *Nachala kvantovoi mekhaniki* (*Principles of quantum mechanics*) (KUBUCH, 1932) p. 92.

Substituting (54.8) into (54.7), we obtain

$$E^{(1)} = \frac{Z^3}{a^3} 4 \int_0^{r_0} e^{-2Zr/a} \left[\frac{Ze^2}{r} - \frac{Ze^2}{r_0} \left(\frac{3}{2} - \frac{r^2}{2r_0^2} \right) \right] r^2 \, dr . \quad (54.9)$$

Since the radius of the first Bohr orbit is $a \sim 10^{-8}$ cm and $r_0 = 10^{-12}$ cm, the exponent of the exponential in (54.9) is very small, and the exponential can be replaced by unity. Integrating the integral in (54.9), we find

$$E^{(1)} = \frac{2Z^4 e^2}{5a} \left(\frac{r_0}{a} \right)^2 = -\frac{4}{5} E_1^{(0)} \left(\frac{r_0}{a} \right)^2 Z^2 . \quad (54.10)$$

Even for the heaviest atoms, $Z \sim 100$, and the ratio $E^{(1)}/E_1^{(0)} \sim 10^{-4}$.

Let us now consider the first excited level $n = 2$. As we have shown in §38 this level will be 4-fold degenerate (the states $\psi_{200}, \psi_{211}, \psi_{210}, \psi_{21-1}$). We shall number these wave functions by the index $s = 1, 2, 3, 4$ respectively. It is clear already from general considerations that the perturbation will partially remove the degeneracy. Indeed, in the Coulomb field we have degeneracy in the two quantum numbers l and m. The degeneracy in the quantum number l is specific for the Coulomb field. However, the degeneracy in the magnetic quantum number m occurs in an arbitrary centrally symmetric field. In view of the fact that when the perturbation is taken into account the field is no longer strictly a Coulomb field, although it will remain a central field, the degeneracy in the quantum number l is removed.

Thus we can expect the level with $n = 2$ to split into 2 levels, with $n = 2$, $l = 0$ and $n = 2, l = 1$. We shall show that a calculation does indeed lead to this splitting.

The secular equation in this case will have the form

$$\begin{vmatrix} H'_{11} - E^{(1)} & H'_{12} & H'_{13} & H'_{14} \\ H'_{21} & H'_{22} - E^{(1)} & H'_{23} & H'_{24} \\ H'_{31} & H'_{32} & H'_{33} - E^{(1)} & H'_{34} \\ H'_{41} & H'_{42} & H'_{43} & H'_{44} - E^{(1)} \end{vmatrix} = 0 . \quad (54.11)$$

The matrix elements are taken with respect to the functions ψ_{nlm}: $\psi_1 = \psi_{200}$, $\psi_2 = \psi_{211}, \psi_3 = \psi_{210}$ and $\psi_4 = \psi_{21-1}$.

In view of the fact that the perturbation operator $\hat{H}'$ (54.6) depends only on the coordinate r, all non-diagonal matrix elements in (54.11) reduce to zero because of the orthogonality of the spherical functions (30.18). Indeed, integrating with respect to the angular variables, we obtain

$$\int Y_{l'm'}^* Y_{lm} \sin \vartheta \, d\vartheta \, d\varphi = \delta_{ll'} \delta_{mm'} \, .$$

Making use of (38.22)–(38.24), we obtain for the diagonal matrix elements (since the integral with respect to angular variables is equal to unity):

$$H_{11}' = \frac{Z^3}{2a^3} \int_r^{r_0} \left(1 - \frac{Zr}{2a}\right)^2 e^{-Zr/a} \left(\frac{Ze^2}{r} - \frac{3Ze^2}{2r_0} + \frac{Ze^2 r^2}{2r_0^3}\right) r^2 \, dr \, , \tag{54.12}$$

$$H_{22}' = H_{33}' = H_{44}' = \frac{Z^3}{24a^3} \int_0^{r_0} e^{-Zr/a} \frac{Z^2 r^2}{a^2} \left(\frac{Ze^2}{r} - \frac{3Ze^2}{2r_0} + \frac{Ze^2 r^2}{2r_0^3}\right) r^2 \, dr \, . \tag{54.13}$$

Neglecting terms of the order r_0/a in comparison with unity, we get

$$E_1^{(1)} = \frac{1}{20} \frac{Z^4 e^2}{a} \left(\frac{r_0}{a}\right)^2 \, , \tag{54.14}$$

$$E_2^{(1)} = \frac{1}{1120} \frac{Z^2 e^2}{a} \left(\frac{Zr_0}{a}\right)^4 \, . \tag{54.15}$$

We see that the original level is split into two sub-levels. The displacement of each of them with respect to the position of the original level is given by formulae (54.14) and (54.15). The value of the shift of the level $n = 2, l = 0$ is smaller by about an order of magnitude than the shift of the level $n = 1, l = 0$. The shift of the level $n = 2, l = 1$ is even smaller owing to the small factor $10^{-3}(Zr_0/a)$. This is due to the fact that the electron in the state $n = 2, l = 1$ is, in the main, outside the region of the nucleus and that the distortion of the Coulomb field in this region affects its state very little.

Finally, we note that the corrections considered here turn out to be considerably more substantial for mesic atoms. This is associated with the fact that mesons are much heavier than electrons and hence are in the main much more near the nucleus (see §38). Thus for the μ-mesic atom the relative shift of the level with $l = 0$ is larger by a factor of about 4×10^4 than for the ordinary atom, and becomes an appreciable quantity.

§55. The theory of time-dependent perturbations

Perturbations acting on a quantum-mechanical system very often have a nonstationary character (i.e. depend on time). This means that the perturbation operator $\hat{H}'$ is an explicit function of time $\hat{H}'(t)$. Numerous examples of such perturbations will be given below. We assume that the stationary states of the unperturbed system are known, i.e. that the wave functions $\psi_n^{(0)}(x, t) = \psi_n^{(0)}(x) \exp[-(i/\hbar) E_n t]$ the unperturbed equation

$$i\hbar \frac{\partial \psi_n^{(0)}(x, t)}{\partial t} = \hat{H}_0 \psi_n^{(0)}(x, t) . \tag{55.1}$$

are known. We restrict ourselves first to the simple case where the states of the unperturbed system belong to a discrete spectrum.

If the system is acted upon by a small perturbation described by the operator $\hat{H}'(t)$, then the wave function of the perturbed system ψ satisfies the equation

$$i\hbar \frac{\partial \psi}{\partial t} = (\hat{H}_0 + \hat{H}') \psi . \tag{55.2}$$

The method of approximate solution of this equation was worked out by Dirac and is often called the Dirac perturbation theory or the method of variation of constants. The state of a perturbed system depends on time and its energy is not a constant of the motion. Our problem now is not to find the stationary states of the perturbed system, because they do not exist, but the calculation of the time-dependent wave function of the system. Hence the perturbation theory method must be modified. The solution of eq. (55.2) in the method of variation of constants is written in the form of an expansion in terms of the eigenfunctions of the unperturbed problem

$$\psi(x, t) = \sum_k c_k(t) \psi_k^{(0)}(x, t) . \tag{55.3}$$

Since the wave functions $\psi_k^{(0)}(x, t)$ form a complete system of functions, such an expansion is always possible. The coefficients $c_k(t)$ of the expansion are functions of time only and not of the coordinates. Substituting the expansion (55.3) into eq. (55.2), we obtain

$$i\hbar \sum_k \left(\frac{dc_k}{dt} \psi_k^{(0)}(x, t) + c_k \frac{d\psi_k^{(0)}(x, t)}{\partial t} \right) = \sum_k c_k (\hat{H}_0 + \hat{H}') \psi_k^{(0)}(x, t) . \tag{55.4}$$

We multiply eq. (55.4) from the left by $\psi_m^{(0)*}(x, t)$ and integrate over all space. Then, taking into account eq. (55.1) and the orthogonality of the wave functions of the unperturbed system $\psi_k^{(0)}(x, t)$, we have

$$i\hbar \frac{dc_m}{dt} = \sum_k H'_{mk} \exp(i\omega_{mk} t) c_k , \tag{55.5}$$

where H'_{mk} is the matrix element of the perturbation operator

$$H'_{mk} = \int \psi_m^{(0)*}(x)\hat{H}'\psi_k^{(0)}(x)\,dV\,,$$

and

(55.6)

$$\omega_{mk} = \hbar^{-1}(E_m - E_k)\,.$$

The system of equations (55.5) is exact. It is equivalent to the initial equation (55.2), since the whole set of coefficients c_k completely determines the wave function ψ. However, it is clear that the solution of the infinite system of equations (55.5) is no simpler a problem than the solution of the initial equation (55.2). Hence for a simplification of the system of equations (55.5) we have to make use of the fact that the perturbation acting on the system is small. We assume initially that for $t \leqslant 0$ the system was in a state with the wave function $\psi_n^{(0)}$. Then for $t \leqslant 0$ all coefficients in the expansion (55.3), with the exception of the coefficients with the index n, are equal to zero, i.e.

$$c_k(0) = \delta_{kn}\,.$$

(55.7)

Beginning with time $t = 0$ the system undergoes the action of a small perturbation. We assume that owing to the weakness of the perturbation the wave function of the initial state, $\psi_n(0)$, changes little with time. Correspondingly, we seek the coefficients $c_k(t)$ at an instant of time $t > 0$ in the form

$$c_k(t) = c_k^{(0)}(t) + c_k^{(1)}(t) + c_k^{(2)}(t) + \ldots$$

(55.8)

where

$$c_k^{(0)}(t) = c_k(0) = \delta_{nk}\,.$$

The correction $c_k^{(1)}(t)$ is of the same order of small quantities as the perturbation, $c_k^{(2)}(t)$ is quadratic in the perturbation and so on. Substituting the expansion (55.8) into eq. (55.5), we find

$$i\hbar\frac{dc_m^{(1)}}{dt} = \sum_k H'_{mk}\exp(i\omega_{mk}t)c_k^{(0)} = H'_{mn}\exp(i\omega_{mn}t)\,.$$

(55.9)

Here all terms of the second and higher order of small quantities in the perturbation have been dropped. Integrating (55.9), we get

$$c_m^{(1)}(t) = \frac{1}{i\hbar}\int_0^t H'_{mn}\exp(i\omega_{mn}t)\,dt\,.$$

(55.10)

In an analogous way one can find the corrections to $c_m^{(0)}$ of the second and higher orders of small quantities. For example, for the correction of the second order $c_m^{(2)}$ one easily obtains the expression

$$c_m^{(2)} - \frac{1}{i\hbar} \sum_k \int_0^t H'_{mk} \exp(i\omega_{mk}t) c_k^{(1)} \, dt \, . \tag{55.11}$$

If the perturbation is sufficiently small, then one can restrict oneself to a small number of terms in the expansion. Thus, in principle, the wave function at any instant of time $t > 0$ can be found with the desired degree of accuracy.

§56. The transition of a system into new states under the action of perturbations

We have found that if a system in a definite energy state for $t \leqslant 0$ described by the wave function $\psi_n^{(0)}$ is acted upon by a perturbation $\hat{H}'(t)$, then for $t > 0$ the system turns out to be in a new state with the wave function (55.3). This means that for $t > 0$ the system can be found in any of its possible stationary quantum states; the probability of finding the system in a certain quantum state m is defined according to the general rules of quantum mechanics by the value of the quantity $|c_m|^2$. Since at the initial instant $t = 0$ the system was in the nth stationary state, then, consequently, $|c_m(t)|^2$ defines the probability of the transition of the system from the nth state into the mth state in time t, $W_{mn}(t) = |c_m(t)|^2 \equiv |c_{mn}(t)|^2$. Here we have denoted the initial state of the system by the second subscript.

Thus the perturbation turns out to give rise to the transition of the system from one quantum state into another. A characteristic property of this process, which has no analogy in classical physics, is the fact that a given perturbation gives rise to the transition of the system from a stationary state with definite energy into a new state in which the energy has no clearly defined value. This is often understood in such a way that under the action of the perturbation the system goes over by a discontinuous process into one of the possible energy states. The state into which the system goes will be a matter of chance. However, such an idea is incorrect and contradicts the physical basis of quantum mechanics. As a matter of fact, the final state is described by a wave function ψ and is hence a definite state (in the quantum-mechanical sense).

The transition from the initial to the final state is not carried out discontinuously, but proceeds in time. Indeed, as we shall see below, the transition probability is determined by the character of the perturbation and by its dependence on time.

Transitions from a discrete into a continuous spectrum are of the greatest interest, and we shall consider such transitions in what follows.

To determine the transition probability it is evidently necessary to know the dependence of the matrix element of the perturbation operator $\hat{H}'_{vn}$ on time. Here we characterize a state in the continuous spectrum by the index v.

Let us consider, first of all, the important case where the perturbation operator is a harmonic function of time. Then the matrix element of the perturbation operator (taken with respect to time-independent unperturbed wave functions) is also a harmonic function of time, i.e.

$$H'_{vn}(t) = H'_{vn}(0) \cos \omega t . \tag{56.1}$$

We shall assume that the frequency ω satisfies the relation

$$h\omega > E_0 - E_n^{(0)} ,$$

where E_0 denotes that energy value of the system with which the continuous spectrum begins. Substituting (56.1) into (55.10), we find

$$c_{vn}^{(1)} = -\frac{1}{2h} H'_{vn}(0) \left[\frac{\exp\left[i(\omega_{vn} + \omega)t\right] - 1}{\omega_{vn} + \omega} + \frac{\exp\left[i(\omega_{vn} - \omega)t\right] - 1}{\omega_{vn} - \omega} \right] . \tag{56.2}$$

Here we denote the initial state of the system by the second index in $c_{vn}^{(1)}$. Since the continuous spectrum lies in a range of energies higher than the discrete spectrum, then $\omega_{vn} > 0$. From the structure of expression (56.2) it follows that for $\omega_{vn} \approx \omega$ the denominator of one of the terms is close to zero. Transitions into states for which the condition $\omega_{vn} \approx \omega$ is fulfilled occur with a low probability. From what follows it will be seen that the probability of the transition into such states increases linearly with time. Hence, dropping the first term in formula (56.2), we have

$$c_{vn}^{(1)} = -\frac{1}{2h} H'_{vn}(0) \frac{\exp\left[i(\omega_{vn} - \omega)t\right] - 1}{\omega_{vn} - \omega} . \tag{56.3}$$

Correspondingly for the square of the modulus $|c_{vn}^{(1)}|^2$ we obtain

$$|c_{vn}^{(1)}|^2 = \frac{1}{4h^2} |H'_{vn}(0)|^2 \frac{\sin^2 \frac{1}{2}(\omega_{vn} - \omega)t}{\frac{1}{4}(\omega_{vn} - \omega)^2} = \frac{\pi|H'_{vn}(0)|^2}{4\hbar^2} tf(\alpha, t) , \tag{56.4}$$

where

$$\alpha = \tfrac{1}{2}(\omega_{vn} - \omega) \quad \text{and} \quad f(\alpha, t) = \frac{\sin^2 \alpha t}{\pi \alpha^2 t} .$$

Usually in practice it is of interest to know the magnitude of $|c_{vn}^{(1)}|^2$ for large values of time t (we recall that the instant of switching on the perturbation is taken as the zero of time $t = 0$). Therefore it is necessary to consider the behaviour of the function $f(\alpha, t)$ when $t \to \infty$. It is easily seen that for $\alpha \neq 0$ and $t \to \infty$, $f(\alpha, t) \to 0$. For $\alpha = 0$, $f(0, t) = t/\pi$ and increases indefinitely with increasing time t. Finally, integrating $f(\alpha, t)$ over all values of α, we find

$$\int_{-\infty}^{\infty} \frac{\sin^2 \alpha t}{\pi \alpha^2 t} \, d\alpha = \frac{1}{\pi} \int_{-\infty}^{\infty} \frac{\sin^2 x}{x^2} \, dx = 1 . \tag{56.5}$$

Comparing the above properties of the function $f(\alpha, t)$ with the properties of the δ-function, we see that they are identical (see Appendix III). Thus $\lim_{t \to \infty} f(\alpha, t)$ is one of the possible concrete forms of the δ-function, and we can write

$$\lim_{t \to \infty} \frac{\sin^2 \alpha t}{\pi \alpha t} = \delta(\alpha) = \delta\left(\frac{\omega_{vn} - \omega}{2}\right) .$$

Substituting this expression into formula (56.4) and making use of the known properties of the δ-function (see Appendix III, Vol. 1), we obtain

$$|c_{vn}^{(1)}|^2 = \frac{\pi}{4\hbar^2} |H'_{vn}(0)|^2 \, t \delta\left(\frac{\omega_{vn} - \omega}{2}\right) =$$

$$= \frac{\pi}{2\hbar} |H'_{vn}(0)|^2 \, t \delta \, (E_v - E_n^{(0)} - \hbar\omega) . \tag{56.6}$$

Formulae (56.4) and (56.6) will be valid only under the condition that the probability of the transition from the given nth state into any vth state is low, i.e.

$$\int |c_{vn}^{(1)}|^2 \, dv \ll 1 .$$

Only in this case is the initial assumption of the smallness of change of the wave function of the initial state fulfilled. Since the transition probability increases linearly with time, it is necessary, for perturbation theory to be applicable, that the time of action of the perturbation t be not too large. Therefore we shall find out what conclusions can be drawn concerning the probability of a transition in a finite time interval t. For this we study formula (56.4) without passing to the limit $t \to \infty$, i.e. we consider the behaviour of the function $f(\alpha, t)$. The plot of this function with respect to time is shown in fig. V.17.

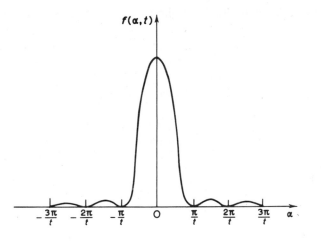

Fig. V.17

From the form of the function $f(\alpha, t)$ it follows that, in the main, transitions are realized into those states for which the quantity α lies within the limits of the principal maximum, i.e. $\Delta\alpha \sim t^{-1}$. The straggling of the values of the parameter α determines the straggling of the energy values of the final state of the system

$$\Delta E_v \sim \hbar \Delta\alpha \sim \hbar/t .\tag{56.7}$$

Thus we arrive at the conclusion that in time t the system can, under the action of perturbation (56.1), make transitions into states with energy $E_v = E_n^{(0)} + \hbar\omega + \Delta E_v$, where $\Delta E_v \sim \hbar/t$.

The uncertainty in the energy of the final state is $\Delta E_v \to 0$ as $t \to \infty$. We note that, proceeding from the uncertainty relation for time and energy (see §34), just such a value of the uncertainty in the energy of the final state $\Delta E_v \sim \hbar/t$ was to be expected.

From the requirement that the uncertainty in the energy of the final state ΔE_v be small in comparison with the energy $\hbar\omega$ the following inequality arises: $t \gg \omega^{-1}$. Consequently, $\Delta E_v \ll \hbar\omega$ if the time of action of the perturbation is large as compared to the period of perturbation.

The transition to the δ-function in formula (56.6) means that the time t must be sufficiently large for the uncertainty in the energy of the final state to be disregarded, but that nevertheless the condition of applicability of perturbation theory is still fulfilled.

Formula (56.6), containing the δ-function, has, of course, a meaning only because integration with respect to the argument of the δ-function is subsequently implied.

We note that the conditions of applicability of perturbation theory are violated in considering transitions in a discrete spectrum in the so-called resonance case, i.e. when $|\omega_{kn}| \approx \omega$. Under these conditions the corrections to the wave function $\psi_n^{(0)}$ become large and the problem must be solved precisely*.

The probability of the transition per unit time from a quantum state with energy $E_n^{(0)}$ into a state of the continuous spectrum in the interval dv is defined by the formula

$$dW_{vn} = \frac{1}{t} |c_{vn}^{(1)}|^2 \, dv = \frac{\pi}{2\hbar} |H'_{vn}(0)|^2 \delta(E_v - E_n^{(0)} - \hbar\omega) \, dv . \qquad (56.8)$$

In this case the wave functions of the continuous spectrum must be normalized to the δ-function in the v-space. Formula (56.8) shows that under the action of a perturbation harmonically dependent on time the system may carry out transitions only into states with energy $E_v = E_n^{(0)} + \hbar\omega$.

The transition probability is defined by the square of the matrix element of the perturbation operator and depends, of course, on the choice of quantities characterizing the state of the continuous spectrum. The energy of the particle is often chosen as one of the parameters characterizing a state in the continuous spectrum. Then, integrating with respect to other parameters, we have

$$dW_{En} = \frac{\pi}{2\hbar} |H'_{En}(0)|^2 \rho(E) \delta(E - E_n^{(0)} - \hbar\omega) \, dE , \qquad (56.9)$$

where $\rho(E) \, dE$ is the number of states with an energy in the interval from E to $E+dE$, and the following notation is introduced:

$$dE \int |H'_{vn}|^2 \frac{dv}{dE} = |H'_{En}|^2 \rho(E) \, dE .$$

Integrating over energy we find the total probability of the transition per unit time from a state with energy $E_n^{(0)}$ to a state of the continuous spectrum under the action of a harmonic perturbation:

$$W = \frac{\pi}{2\hbar} |H'_{En}(0)|^2 \rho(E) , \qquad (56.10)$$

---

* See L.D.Landau and E.M.Lifshitz, *Quantum mechanics* (Pergamon Press, Oxford, 1965).

where $E = E_n^{(0)} + \hbar\omega$. We note that if, as distinct from (56.1), we denote the matrix element of the perturbation, introducing exponential functions, by

$$H'_{vn}(t) = H'_{vn}(0)(e^{i\omega t} + e^{-i\omega t}),$$

then the numerical coefficient in formulae (56.8)–(56.10) would evidently change by a factor of 4. For example, formula (56.8) would then be rewritten in the form

$$\mathrm{d}W_{vn} = \frac{2\pi}{\hbar} |H'_{vn}(0)|^2 \delta(E_v - E_n^{(0)} - \hbar\omega)\,\mathrm{d}v \qquad (56.8')$$

and formulae (56.9) and (56.10) change analogously.

Another particularly important case is the transition caused by a time-independent perturbation. The expression for the transition probability can be obtained from formula (56.8) by setting the frequency in it to $\omega = 0$ and doubling the matrix element of the perturbation. This is associated with the fact that a term, which for $\omega = 0$ is the same as the term retained, was dropped in the transition from (56.2) to (56.3). For the transition probability we have

$$\mathrm{d}W_{vv_0} = \frac{2\pi}{\hbar} |H'_{vv_0}|^2 \delta(E_v - E_{v_0})\,\mathrm{d}v . \qquad (56.11)$$

A time-independent perturbation can give rise to transitions only to states with the same energy. In other words, it can cause transitions only between degenerate states. We have here denoted the initial state by the index v_0, since transitions in the continuous spectrum are of the greatest interest in the case of the action of a constant perturbation. Of course, all the above reasoning associated with the transition to the δ-function is also valid in this case.

Integrating (56.11) over final state energies, we can write the transition probability in another form:

$$\mathrm{d}W_{vv_0} = \frac{2\pi}{\hbar} \int |H'_{vv_0}|^2 \frac{\mathrm{d}v}{\mathrm{d}E} \delta(E - E_{v_0})\,\mathrm{d}E = \frac{2\pi}{\hbar} |H'_{vv_0}|^2 \frac{\mathrm{d}v}{\mathrm{d}E} . \qquad (56.12)$$

We write the total transition probability, analogously to (56.10), in the form

$$W = \frac{2\pi}{\hbar} |H'_{Ev_0}|^2 \rho(E) , \qquad (56.13)$$

where $E = E_{v_0}$. Let, for example, the final state be characterized by defining the momentum of the particle, so that

$$\mathrm{d}v = \mathrm{d}p_x\,\mathrm{d}p_y\,\mathrm{d}p_z = p^2\mathrm{d}p\,\mathrm{d}\Omega = pm\,\mathrm{d}E\,\mathrm{d}\Omega ,$$

where $E = \mathbf{p}^2/2m$ is the energy of the final state of the particle, and $d\Omega$ is an element of solid angle. Formula (56.12) is, in this case, rewritten in the form

$$dW_{\mathbf{p}v_0} = \frac{2\pi}{\hbar} |H'_{\mathbf{p}v_0}|^2 mp \, d\Omega , \quad \text{where} \quad p = (2mE_{v_0})^{\frac{1}{2}} . \quad (56.14)$$

Here the wave functions of the final state must be normalized to the δ-function in momentum space.

In another method for the normalization of these functions, for example normalization in a "box" (see (26.16) and (26.17)), the interval of final states dv' will have the form

$$dv' = dn_x \, dn_y \, dn_z = \frac{dp_x \, dp_y \, dp_z \, V}{(2\pi\hbar)^3} . \quad (56.15)$$

Of course the expression for the transition probability (56.14) will not change in this case. Finally, we note that expressions (56.11)–(56.14) depend on the method of normalizing the wave function of the initial state, which also belongs to the continuous spectrum.

The matrix element of the perturbation operator very often turns out to be equal to zero. In this case the transition probability reduces to zero. This means that the corresponding transition is impossible in the first approximation of perturbation theory. In the next higher approximation the probability of the corresponding transition may turn out to be different from zero.

Let us find the probability of transition caused by a time-independent perturbation in the second order approximation for such a case.

Formula (55.11) gives

$$c_v^{(2)} = \frac{1}{i\hbar} \sum_k \int_0^t H'_{vk} c_k^{(1)}(t) \exp(i\omega_{vk}t) \, dt . \quad (56.16)$$

The sum (or over the continuous spectrum, the integral) involved here is taken over intermediate or, as they are often called, virtual states, so that the transition itself can be treated as a transition via intermediate states. It should be stressed that the transition of the system via intermediate states is not a real physical process, but serves only as a way of dealing with the formulae. Hence, for example, in transitions into virtual states the energy of the system does not need to be conserved. Substituting the expression for $c_k^{(1)}$ from (56.2) ($\omega = 0$) into (56.16) and integrating, we obtain

$$c_v^{(2)} = \frac{1}{\hbar^2} \sum_k H'_{vk} H'_{kv_0} \left[\frac{\exp(i\omega_{vv_0}t) - 1}{\omega_{kv_0} \, \omega_{vv_0}} - \frac{\exp(i\omega_{vk}t) - 1}{\omega_{kv_0} \, \omega_{vk}} \right] .$$

$$(56.17)$$

Since, by assumption, transitions are absent in the first approximation of perturbation theory, the matrix element of the perturbation operator $H'_{kv_0} = 0$ for transitions proceeding with energy conservation $\omega_{kv_0} = 0$. In correspondence with this those intermediate states for which $\omega_{kv_0} = 0$ give no contribution to the amplitude (56.17). For transitions proceeding with energy conservation* $(\omega_{vv_0} = 0)$ the second term in the bracket of formula (56.17) is not large. Indeed, it might give an appreciable contribution only for $\omega_{vk} = 0$. But $\omega_{vv_0} = \omega_{vk} + \omega_{kv_0}$, and for $\omega_{vv_0} = 0$ and $\omega_{vk} = 0$ it turns out that also ω_{kv_0} reduces to zero. For such transitions $H'_{kv_0} = 0$ and, consequently, they can be disregarded. Proceeding from this, we can rewrite (56.17) as

$$c_v^{(2)} = -\frac{1}{\hbar} \Lambda_{vv_0} \frac{\exp(i\omega_{vv_0}t) - 1}{\omega_{vv_0}}, \tag{56.18}$$

where

$$\Lambda_{vv_0} = \sum_k \frac{H'_{vk}H'_{kv_0}}{E_{v_0} - E_k} \tag{56.19}$$

(integration is implied over the continuous spectrum).

We see that in the notation in the form of (56.18) the expression for the amplitude $c_v^{(2)}$ is the same as (56.2) $(\omega = 0)$. Therefore the results obtained, in particular the formula for the transition probability (56.8), are conserved under the condition that the matrix element H'_{vv_0} be replaced by the matrix element Λ_{vv_0}.

§57. The adiabatic theory of perturbations

In certain cases the perturbation acting on a quantum system is associated with a slow, adiabatic change of the parameters on which the state of the system depends.

In the case of an adiabatic change of certain of the parameters which characterize a system it turns out that it is possible to develop a special approximate method of calculation called the adiabatic theory of perturbations. We shall encounter this method below in studying the properties of molecules and solid bodies. In such systems there are particles of two kinds:

---

* The possibility of transitions with non-conservation of energy is associated with the assumption of suddenness of switching on the perturbation (see (56.1)). For a more detailed discussion of this problem see, for example, L.Schiff, *Quantum mechanics* (McGraw-Hill, New York, 1949).

light electrons moving with large velocities, and heavy nuclei performing relatively slow movements. We shall call the electrons of the system the fast sub-system, and the heavy nuclei the slow sub-system. Roughly speaking, the characteristic time needed for a change of state of the fast sub-system is very small in comparison with the corresponding time for the slow sub-system. The essence of the adiabatic theory of perturbations amounts to the fact that the motion of the fast sub-system is considered in the first order approximation for given coordinates of the slow sub-system.

In other words, the motions of the fast and slow sub-systems are to a certain degree independent.

Let us consider the motion of a system consisting of electrons and nuclei. We write the Schrödinger equation in the form

$$\left[-\frac{\hbar^2}{2M} \sum_i \mathbf{V}_i^2 - \frac{\hbar^2}{2m} \sum_k \mathbf{V}_k^2 + U(\mathbf{r}_k, \mathbf{R}_i) \right] \psi(\mathbf{r}_k, \mathbf{R}_i) = E\psi(\mathbf{r}_k, \mathbf{R}_i) . \qquad (57.1)$$

Here m and M are the masses of the electrons and nuclei respectively. The sum over k is carried out with respect to the coordinates of the electrons, while the sum over i corresponds to the coordinates of the nuclei. U is the operator corresponding to the mutual interaction energy of the particles.

Further, we assume that it is possible to find the solution of the following Schrödinger equation:

$$\left[-\frac{\hbar^2}{2m} \sum_k \mathbf{V}_k^2 + U(\mathbf{r}_k, \mathbf{R}_i) \right] \varphi_n(\mathbf{r}_k, \mathbf{R}_i) = E_n(\mathbf{R}_i)\varphi_n(\mathbf{r}_k, \mathbf{R}_i) . \qquad (57.2)$$

Equation (57.2) has the following physical meaning. The nuclei are assumed to be fixed at points $\mathbf{R}_i$. Finding the solution of eq. (57.2) comes down to the determination of the electron wave function φ_n and the energy levels of the electron system. As is seen from eq. (57.2), the energy levels $E_n(\mathbf{R}_i)$ of the electron sub-system depend on the coordinates of the nuclei (the heavy sub-system) as parameters.

Geometrically the electron energy $E_n(\mathbf{R}_i)$ forms a certain surface in space R_i. This surface is called the electron term.

We write the solution of eq. (57.1) in the form of an expansion in a series in terms of the complete system of wave functions φ_n,

$$\psi(\mathbf{r}_k, \mathbf{R}_i) = \sum_n \alpha_n(\mathbf{R}_i)\varphi_n(\mathbf{r}_k, \mathbf{R}_i) . \qquad (57.3)$$

We substitute (57.3) into (57.1), and then multiply eq. (57.1) by φ_m^* and integrate with respect to the coordinates of the electrons $dV = dV_1 dV_2 \dots$.

Taking into account the formula

$$\mathbf{V}_i^2 \alpha_n \varphi_n = \varphi_n \mathbf{V}_i^2 \alpha_n + \alpha_n \mathbf{V}_i^2 \varphi_n + 2\mathbf{V}_i \varphi_n \cdot \mathbf{V}_i \alpha_n$$

we find the following equation:

$$-\frac{\hbar^2}{2M} \sum_i \mathbf{V}_i^2 \alpha_m + E_m(\mathbf{R}_i) \alpha_m =$$

$$= E\alpha_m + \sum_n \sum_i \left[\frac{\hbar^2}{2M} \alpha_n \int \varphi_m^* \mathbf{V}_i^2 \varphi_n \, dV + \right.$$

$$\left. + \frac{\hbar^2}{M} \int \varphi_m^* \mathbf{V}_i \varphi_n \cdot \mathbf{V}_i \alpha_n \, dV \right]. \tag{57.4}$$

Here $\mathbf{V}_i$ is calculated with respect to the coordinates $\mathbf{R}_i$ of the nuclei. We rewrite eq. (57.4) in the form

$$\left[-\frac{\hbar^2}{2M} \sum_i \mathbf{V}_i^2 + E_m(\mathbf{R}_i) \right] \alpha_m(\mathbf{R}_i) = E\alpha_m(\mathbf{R}_i) + \hat{C}\alpha_m , \tag{57.5}$$

where the operator $\hat{C}$ is defined in the following way:

$$\hat{C}\alpha_m = \sum_i \sum_n \left(\frac{\hbar^2}{M} \mathbf{V}_i \alpha_n \int \varphi_m^* \mathbf{V}_i \varphi_n \, dV + \frac{\hbar^2}{2M} \alpha_n \int \varphi_m^* \mathbf{V}_i^2 \varphi_n \, dV \right). \tag{57.6}$$

The operator $\hat{C}$ is called the non-adiabatic operator.

If one assumes that the operator $\hat{C}$ is small and neglects it in eq. (57.5) then the equations for the functions φ_m and α_m assume the form

$$\left[-\frac{\hbar^2}{2m} \sum_k \mathbf{V}_k^2 + U(\mathbf{r}_k, \mathbf{R}_i) \right] \varphi_m = E_m(\mathbf{R}_i) \varphi_m , \tag{57.7}$$

$$\left[-\frac{\hbar^2}{2M} \sum_i \mathbf{V}_i^2 + E_m(\mathbf{R}_i) \right] \alpha_m = E\alpha_m . \tag{57.8}$$

Thus we obtain an important result in the zero order approximation with respect to the operator $\hat{C}$. Equation (57.7) represents a Schrödinger equation. The coordinates of the nuclei are involved in this equation as parameters. The function $\varphi_m(\mathbf{r}_k, \mathbf{R}_i)$ describes the motion of the electrons for motionless nuclei. Equation (57.8) contains only operators acting on the coordinates of the nuclei. It can be considered as the Schrödinger equation for the heavy sub-system, the nuclei. Then the energy $E_m(\mathbf{R}_i)$ of the electron sub-system plays the role of the potential energy of the nuclei.

The total wave function of the system in the zero order approximation

$\hat{C} = 0$ can be written in the form of a product of the wave functions α_m and φ_m, i.e. it has the same form as if the two sub-systems were quite independent:

$$\psi = \alpha_m(\mathbf{R}_i)\varphi_m(\mathbf{r}_k, \mathbf{R}_i) .$$

In the approximation described it can be said that the electron sub-system follows the motion of the nuclei adiabatically in the sense that the electron sub-system remains in the same quantum state E_m when the position $\mathbf{R}_i$ of the nuclei is changed. However, its energy level E_m changes in correspondence with the motion of the nuclei.

The condition of the smallness of the operator $\hat{C}$ cannot be formulated in general form. In every actual problem this condition must be considered separately. Examples of such a consideration can be found in the books of Pauli, and Born and Huan Kun*.

§58. Perturbation theory in integral form

Perturbation theory can easily be developed within the framework of the Feynman formalism**. For this it is convenient to use as a basis the integral equation (29.5) for the Green's function $K(\mathbf{r}_2 t_2; \mathbf{r}_1 t_1)$ which we shall denote by $K(2, 1)$

$$K(2, 1) = K_0(2, 1) - \frac{i}{\hbar} \int_{-\infty}^{\infty} K_0(2, 3)\hat{H}'(3)K(3, 1)\,\mathrm{d}^4 x_3 . \quad (58.1)$$

Here we have denoted the Green's function of the unperturbed problem $\hat{H} = \hat{H}_0, \hat{H}' = 0$ by $K_0(2, 1)$.

Making use of the smallness of perturbation, we solve eq. (58.1) by a method of successive approximations. In the zeroth approximation, i.e. assuming $\hat{H}' = 0$, we have

$$K(2, 1) = K_0(2, 1) . \tag{58.2}$$

---

* W.Pauli, *Die allgemeinen Prinzipien der Wellenmechanik (General principles of wave mechanics)*, Handbuch der Physik V/1, 1958; M.Born and Huang Kun, *Dynamical theory of crystal lattices* (University Press, Oxford, 1954).

** R.P.Feynman, Phys. Rev. 76 (1949) 740. See also S.Schweber, H.Bethe and F. Hofman, *Mesons and fields* (Row, Peterson and Co., Evanston, Ill. and White Plains, N.Y., 1956).

We shall obtain the next approximation if we substitute into the integral (58.1) the zeroth order approximation of the Green's function $K(3, 1)$, i.e.

$$K^{(1)}(2, 1) = -\frac{i}{\hbar} \int K_0(2, 3) \hat{H}'(3) K_0(3, 1) \, d^4 x_3 \,. \tag{58.3}$$

To obtain the correction to the Green's function in the second order approximation, we have to substitute into integral (58.1) the function $K(3, 1)$ with an accuracy to within terms of the first order of small quantities:

$$K^{(2)}(2, 1) = \left(\frac{-i}{\hbar}\right)^2 \int K_0(2, 3) \hat{H}'(3) K_0(3, 4) \hat{H}'(4) K_0(4, 1) \, d^4 x_3 \, d^4 x_4 \,. \tag{58.4}$$

The correction to any order can be obtained in an analogous way. Finally, we have

$$K(2, 1) = K_0(2, 1) - \frac{i}{\hbar} \int K_0(2, 3) \hat{H}'(3) K_0(3, 1) \, d^4 x_3 +$$

$$+ \left(-\frac{i}{\hbar}\right)^2 \int K_0(2, 3) \hat{H}'(3) K_0(3, 4) \hat{H}'(4) K_0(4, 1) \, d^4 x_3 \, d^4 x_4 + \dots \,. \tag{58.5}$$

Formula (58.5) can be interpreted in the following way. The zero order term describes the motion of the unperturbed particle from point 1 to point 2. The next term describes the motion of the free particle from point 1 to point 3. At point 3 a perturbation acts. Thereupon the particle, again as a free particle, moves from point 3 to point 2. The integration means that we sum the contribution of all possible points 3. The term of second order smallness takes into account the action of the perturbation at two points, 3 and 4, and so on. By calculating the Green's function K from eq. (58.5) to a given approximation, we also know the wave function in this approximation. The convenience of the integral equation (58.1) lies in the fact that it makes it possible to obtain very simple a perturbation theory series. Examples of the use of the integral from the perturbation theory will be considered in ch. 14.

8

Spin and Identity of Particles

§59. The spin of elementary particles

Up to now we have assumed that the state of an individual microparticle is defined if its three space coordinates, or three momentum components, or in general three quantities forming a complete set are known. A number of experimental results indicated that many microparticles, for example electrons, protons, neutrons, have a specific intrinsic degree of freedom. This intrinsic degree of freedom is associated with an intrinsic angular momentum of the particle which does not depend on its orbital motion. This angular momentum of the particle is called the spin. The fact that the electron has a spin was established before the development of quantum mechanics. Attempts were made to interpret the spin as a manifestation of the rotation of a particle about its own axis (whence its name arose). However, this classical interpretation turned out to be untenable. All attempts to obtain the correct value of the ratio of the angular momentum to the magnetic moment for a system of a distributed rotating charge failed. As for the model of a rigid rotating sphere, (for which any value of this ratio can be obtained), such a model, as was explained in §13 of Part II, contradicts the general propositions of the theory of relativity. This contradiction was resolved in quantum mechanics. As we shall see below, this intrinsic degree of freedom and the spin associated with it have a specific quantum character. In the transition $\hbar \to 0$ to classical mechanics the spin reduces to zero. Hence the spin has no classical analogue and does not allow interpretations of a classical character. The hypothesis of the existence of spin was initially put forward in connection with the interpretation of the spectra of alkali metals. Subsequently a number of facts were established enabling one to state unambiguously that this hypothesis is correct.

In the experiments of Stern and Gerlach the magnetic moment which is not associated with the orbital motion of electrons was observed directly. Namely, in these experiments it was established that when a beam of hydrogen atoms in an S-state was passed through a non-uniform magnetic field, then this beam split into two beams. However, in the S-state there is no orbital angular momentum, and consequently no orbital magnetic moment, so that the beam should pass through the magnetic field without undergoing any deflection.

The two-fold splitting is indicative of two possible orientations of the magnetic moment of the electron. The value of the spin magnetic moment can be determined from the magnitude of the splitting.

Direct experiments carried out by Einstein and de Haas made it possible to determine the ratio of the intrinsic angular momentum to the magnetic moment.

The spin of the electron (the intrinsic angular momentum) possesses the general properties of a quantum-mechanical moment which were discussed in §51. This was proved rigorously by the mathematical technique of group theory. In particular, the eigenvalue of the operator of the square of the spin moment $\hat{s}^2 = \hat{s}_x^2 + \hat{s}_y^2 + \hat{s}_z^2$ is expressed by the formula

$$s(s+1)\hbar^2 , \tag{59.1}$$

where s denotes the corresponding quantum number, the intrinsic or spin quantum number of the particles. This quantum number is often called briefly the spin of the particle.

The number of possible spin projections onto an arbitrarily oriented z-axis is equal to $2s + 1$. The value of the intrinsic number s for each elementary particle must be determined experimentally. For the electron the existence and the value of the spin follows strictly from relativistic quantum mechanics, (Dirac's theory) to which ch. 13 is devoted.

The spins of the elementary particles which are most often encountered are the following: for the electron $s = \frac{1}{2}$, for the proton and neutron $s = \frac{1}{2}$, for the π-meson $s = 0$, for the muon $s = \frac{1}{2}$. This means that the possible values of the projections of the intrinsic angular momentum onto an axis arbitrarily oriented in space, for example for the electron and other particles with spin $\frac{1}{2}$, are

$$s_z = \pm\tfrac{1}{2}\hbar .$$
(59.2)

From the experiment of Stern and Gerlach, the corresponding projections of the intrinsic magnetic moment of the electron are equal in absolute value to the Bohr magneton μ_0

$$\mu_z = \mp \frac{|e|\hbar}{2mc} = \mp\mu_0 .$$
(59.3)

It is of great importance that the ratio of the intrinsic magnetic moment to the intrinsic angular momentum is equal to e/mc

$$\boldsymbol{\mu} = \frac{e}{mc}\,\mathbf{s} ,$$
(59.4)

whereas for the orbital motion this ratio is smaller by a factor of 2 (see §63).

In §118 we shall show that this value of the intrinsic magnetic moment can be derived theoretically from Dirac's relativistic wave equation.

The spin properties of elementary particles play a very important role in the realm of microphenomena as well as in the behaviour of macroscopic bodies. This is associated with the fact that the spin determines directly the statistical properties of systems of quantum particles.

§60. Spin operators

Although, as we shall see below, the existence of spin for the electron and all the properties associated with it can be established theoretically from the propositions of relativistic quantum mechanics, a number of the properties of particles having spin can also be obtained without referring to relativistic theory, on the basis of general quantum-mechanical considerations and a relatively small number of experimental data. Since such a semi-empirical theory of particles with spin has a rather simple character but still makes it possible to obtain important results, we shall dwell on it below.

The wave function of a particle with spin depends not only on its three spatial coordinates but also on a fourth coordinate characterizing the intrinsic state of the particle. The value of the spin projection onto an arbitrarily oriented z-axis in space can be chosen as the fourth coordinate. Then the wave function can be written in the form

$$\psi = \psi(x, s_z, t) .$$ (60.1)

As distinct from the spatial coordinates x, the 'spin coordinate' s_z takes on only a discrete sequence of values. The number of possible values of s_z is determined by the properties of the given elementary particle. As was mentioned above, the spin of most elementary particles is equal to one half. Since in this case the spin projection can take on only two values, the wave function (60.1) is conveniently written in the form of a column with two rows:

$$\psi = \begin{pmatrix} \psi(x, & \tfrac{1}{2}\hbar, & t) \\ \psi(x, & -\tfrac{1}{2}\hbar, & t) \end{pmatrix} = \begin{pmatrix} \psi_1 \\ \psi_2 \end{pmatrix} .$$ (60.2)

We can then interpret $|\psi_1|^2 dV$ as the probability that the electron at the instant of time t will be found in the volume element dV and that it will have a spin component along the z-axis equal to $\tfrac{1}{2}\hbar$. Correspondingly, $|\psi_2|^2 dV$ is the probability that for the electron found in the volume element dV the spin component along the z-axis is equal to $-\tfrac{1}{2}\hbar$. The wave function $\psi(x, s_z, t)$ is assumed to be normalized in such a way that

$$\sum_{s_z} \int |\psi(x, s_z, t)|^2 dV = 1 ,$$

where the sum is taken over all possible values of the spin projection s_z. If the probability of one or another spin projection does not depend on the coordinates of the particle, then the wave function (60.2) can be rewritten in the form

$$\psi = \psi(x, t)\varphi ,$$ (60.3)

where $\psi(x, t)$ is the ordinary (coordinate) wave function, $\varphi = \begin{pmatrix} c_1 \\ c_2 \end{pmatrix}$ is the spin wave function, and c_1 and c_2 are numbers. $|c_1|^2$ and $|c_2|^2$ give the probabilities that the spin projection s_z is equal respectively to $+\tfrac{1}{2}\hbar$ and $-\tfrac{1}{2}\hbar$.

By virtue of the normalization condition for the wave function, we have

$$|c_1|^2 + |c_2|^2 = 1 .$$ (60.4)

Having defined the concept of the spin wave function, we have to introduce the spin operators acting on it. In general the action of an operator on the spin function $\varphi = \begin{pmatrix} c_1 \\ c_2 \end{pmatrix}$ amounts to replacement of the components c_1 and c_2 by some linear combination of them

$$\begin{pmatrix} c_1 \\ c_2 \end{pmatrix} \rightarrow \begin{pmatrix} a_{11}c_1 + a_{12}c_2 \\ a_{21}c_1 + a_{22}c_2 \end{pmatrix} \tag{60.5}$$

Corresponding with this the spin operator can be written in the form of a matrix

$$\hat{a} = \begin{pmatrix} a_{11} & a_{12} \\ a_{21} & a_{22} \end{pmatrix}. \tag{60.6}$$

The action of such an operator on the wave function is given by formula (60.5), i.e.

$$\hat{a}\varphi = \begin{pmatrix} a_{11} & a_{12} \\ a_{21} & a_{22} \end{pmatrix}\begin{pmatrix} c_1 \\ c_2 \end{pmatrix} = \begin{pmatrix} a_{11}c_1 + a_{12}c_2 \\ a_{21}c_1 + a_{22}c_2 \end{pmatrix}. \tag{60.7}$$

If the division of the wave function into a coordinate component and a spin component is not allowed, then formula (60.7) is rewritten in the form

$$\hat{a}\psi = \begin{pmatrix} a_{11} & a_{12} \\ a_{21} & a_{22} \end{pmatrix}\begin{pmatrix} \psi_1 \\ \psi_2 \end{pmatrix} = \begin{pmatrix} a_{11}\psi_1 + a_{12}\psi_2 \\ a_{21}\psi_1 + a_{22}\psi_2 \end{pmatrix}. \tag{60.8}$$

The mean value of the operator $\hat{a}$ taken in the state ψ is determined according to the general formula (44.8)

$$\bar{a}(x,t) = \psi_1^* a_{11}\psi_1 + \psi_1^* a_{12}\psi_2 + \psi_2^* a_{21}\psi_1 + \psi_2^* a_{22}\psi_2 . \tag{60.9}$$

This equation can be rewritten in matrix form

$$\bar{a}(x,t) = \psi^\dagger \hat{a}\psi , \tag{60.10}$$

where $\psi^\dagger$ is the matrix consisting of one row with the elements ψ_1^* and ψ_2^*:

$$\psi^\dagger = (\psi_1^* \psi_2^*) . \tag{60.11}$$

Relation (60.10) determines the mean value of the quantity a at the instant of time t at the given point of space x. If this expression is averaged over all positions of the particle, then we obtain

$$\bar{a}(t) = \int \psi^\dagger \hat{a}\psi \mathrm{d}V . \tag{60.12}$$

We now introduce the operators corresponding to the spin components $\hat{s}_x$, $\hat{s}_y$, $\hat{s}_z$. In §51 it was shown that the form of these operators and all the properties of the spin can be obtained if the commutation relations

$$\hat{s}_x\hat{s}_y - \hat{s}_y\hat{s}_x = i\hbar\hat{s}_z \ ,$$
$$\hat{s}_y\hat{s}_z - \hat{s}_z\hat{s}_y = i\hbar\hat{s}_x \ , \qquad\qquad (60.13)$$
$$\hat{s}_z\hat{s}_x - \hat{s}_x\hat{s}_z = i\hbar\hat{s}_y \ ,$$

are taken as the basis. The fact that the spin component operators must satisfy the same commutation relations as the operators of the components of the orbital angular momentum is, of course, not accidental. In §30 it was shown that the operator corresponding to the component of orbital angular momentum along any axis is associated with the operator corresponding to an infinitesimal rotation about this axis. The commutation relations (30.3) and (30.3′) are a consequence of this fact, i.e. a consequence of the commutation relations between the infinitesimal rotation operators. In the next section we shall show that the spin component operators are also associated with rotation operators which act, however, not on the coordinate function but on the spin function. The commutation relations (60.13) are a consequence of the commutation relations between the infinitesimal rotation operators about the x-axis, y-axis and z-axis. The above considerations are rigorously substantiated by the methods of group theory*.

By analogy with (30.4), it follows from the relations (60.13) that

$$\hat{s}_x\hat{s}^2 - \hat{s}^2\hat{s}_x = 0 \ ,$$
$$\hat{s}_y\hat{s}^2 - \hat{s}^2\hat{s}_y = 0 \ , \qquad\qquad (60.14)$$
$$\hat{s}_z\hat{s}^2 - \hat{s}^2\hat{s}_z = 0 \ ,$$

Thus the square of the total spin and one of its projections onto an arbitrary axis can be measured simultaneously. The other two projections have no simultaneously sharp values.

For $s = \frac{1}{2}$ the matrices corresponding to the total spin, $\hat{s}^2 = \hat{s}_x^2 + \hat{s}_y^2 + \hat{s}_z^2$, and its projection on the z-axis have, in their own representation, the form

$$\hat{s}^2 = \tfrac{3}{4}\hbar^2 \begin{pmatrix} 1 & 0 \\ 0 & 1 \end{pmatrix}; \qquad \hat{s}_z = \tfrac{1}{2}\hbar \begin{pmatrix} 1 & 0 \\ 0 & -1 \end{pmatrix} = \tfrac{1}{2}\hbar\sigma_z \ , \qquad (60.15)$$

(the diagonal matrix elements are equal to the eigenvalues of the corresponding operators). According to the general formula (51.16), the matrices $\hat{s}_x$, $\hat{s}_y$ in this representation are written as

---

* W.Pauli, *Die allgemeinen Prinzipien der Wellenmechanik (General principles of wave mechanics)*, Handbuch der Physik V/1, (Springer, Berlin, 1958).

$$\hat{s}_x = \tfrac{1}{2}\hbar \begin{pmatrix} 0 & 1 \\ 1 & 0 \end{pmatrix} = \tfrac{1}{2}\hbar\sigma_x \; ; \qquad \hat{s}_y = \tfrac{1}{2}\hbar \begin{pmatrix} 0 & -i \\ i & 0 \end{pmatrix} = \tfrac{1}{2}\hbar\sigma_y \; . \tag{60.16}$$

The matrices $\sigma_x, \sigma_y, \sigma_z$, which differ from the matrices $\hat{s}_x, \hat{s}_y, \hat{s}_z$ by a constant factor $\tfrac{1}{2}\hbar$, are called the Pauli matrices. They satisfy the following commutation relations:

$$\begin{aligned} \sigma_x\sigma_y &= -\sigma_y\sigma_x = i\sigma_z \, , \\ \sigma_y\sigma_z &= -\sigma_z\sigma_y = i\sigma_x \, , \\ \sigma_z\sigma_x &= -\sigma_x\sigma_z = i\sigma_y \, , \\ \sigma_x^2 = \sigma_y^2 = \sigma_z^2 &= \begin{pmatrix} 1 & 0 \\ 0 & 1 \end{pmatrix} = 1 \, . \end{aligned} \tag{60.17}$$

An arbitrary matrix of the second rank can be expressed in terms of the matrices $\sigma_x, \sigma_y, \sigma_z$ and the unit matrix.

As well as the similarity between the orbital and intrinsic angular momenta there is also a basic difference between them. The orbital angular momentum is characterized by the quantum number l which can take on any integer values irrespective of the nature of the particle, whereas the spin quantum number s takes on a limited number of values, for example $s = \tfrac{1}{2}$ for most elementary particles. Every kind of elementary particles has its own characteristic value of the spin. If the transition to classical mechanics is made by assuming $\hbar \to 0$, then, as was explained in §41, one must pass simultaneously to the limit of large quantum numbers. Hence, although according to formula (30.15) $l^2 = \hbar^2 l(l+1)$, it still does not follow that as $\hbar \to 0$ $k \to 0$, because simultaneously with $\hbar \to 0$ one has to assume that $l \to \infty$. In the case of the intrinsic angular momentum the situation is different. Since s takes on only a limited number of values, the transition to classical mechanics always leads to the spin value $s = 0$. We see that in classical mechanics there is no quantity which could serve as the classical analogue of the spin. The spin is a purely quantum concept characterizing the specific properties of microparticles.

§61. The eigenfunctions of the operators of the components of the spin of a particle. The rotation matrix

Let us find the eigenfunctions and eigenvalues of the operators $\hat{s}_x, \hat{s}_y, \hat{s}_z$. The equation for the eigenfunctions $\begin{pmatrix} c_1 \\ c_2 \end{pmatrix}$ and the eigenvalues s_x of the operator $\hat{s}_x$ has the form

$$\hat{s}_x \begin{pmatrix} c_1 \\ c_2 \end{pmatrix} = s_x \begin{pmatrix} c_1 \\ c_2 \end{pmatrix}.$$

Taking into account (60.16) and carrying out the multiplication, we obtain

$$\begin{pmatrix} \frac{1}{2}\hbar c_2 \\ \frac{1}{2}\hbar c_1 \end{pmatrix} = \begin{pmatrix} s_x c_1 \\ s_x c_2 \end{pmatrix}.$$

We evaluate this equality:

$$\tfrac{1}{2}\hbar c_2 = s_x c_1, \qquad \tfrac{1}{2}\hbar c_1 = s_x c_2. \tag{61.1}$$

Hence, upon multiplying, we find $s_x = \pm\tfrac{1}{2}\hbar$.

The eigenvalues of the spin component operator, as was to be expected, turn out to be equal to $\pm\tfrac{1}{2}\hbar$. We determine the form of the eigenfunctions corresponding to these eigenvalues.

For $s_x = +\tfrac{1}{2}\hbar$ we have from (61.1) $c_1 = c_2$. Taking into account the normalization condition (60.4), we finally get

$$\varphi_{s_x = +\hbar/2} = \frac{1}{\sqrt{2}}\, e^{i\alpha_1} \begin{pmatrix} 1 \\ 1 \end{pmatrix}, \tag{61.2}$$

where α_1 is an arbitrary phase.

Correspondingly, for $s_x = -\tfrac{1}{2}\hbar$, $c_1 = -c_2$, and the spin wave function is written in the form

$$\varphi_{s_x = -\hbar/2} = \frac{1}{\sqrt{2}}\, e^{i\alpha_2} \begin{pmatrix} 1 \\ -1 \end{pmatrix}. \tag{61.3}$$

Naturally, the eigenvalues of the operators $\hat{s}_y$ and $\hat{s}_z$ are also equal to $\pm\tfrac{1}{2}\hbar$. We also find their eigenfunctions in an analogous way

$$\varphi_{s_y = \hbar/2} = \frac{1}{\sqrt{2}}\, e^{i\alpha_3} \begin{pmatrix} 1 \\ i \end{pmatrix}; \qquad \varphi_{s_y = -\hbar/2} = \frac{1}{\sqrt{2}}\, e^{i\alpha_4} \begin{pmatrix} 1 \\ -i \end{pmatrix}, \tag{61.4}$$

$$\varphi_{s_z = \hbar/2} = \varphi_{\frac{1}{2}} = e^{i\alpha_5} \begin{pmatrix} 1 \\ 0 \end{pmatrix}; \qquad \varphi_{s_z = -\hbar/2} \equiv \varphi_{-\frac{1}{2}} = e^{i\alpha_6} \begin{pmatrix} 0 \\ 1 \end{pmatrix}. \tag{61.5}$$

The arbitrary phase factors α_i can, in particular, be set equal to zero.

We now consider a certain rotation of the system of coordinates $x,y,z \to x',y',z'$. Then the spin wave functions also change, $\varphi \to \varphi'$. Indeed, the transition from the coordinate system x,y,z to the system x',y',z' means a corresponding transition from one representation to another. Such a transition, as we know (see §46), is carried out by means of a certain unitary matrix $\hat{T}$, so that $\varphi' = \hat{T}\varphi$. In the given case it is natural to call the unitary matrix $\hat{T}$ the rotation matrix. Let us define this matrix. We first consider a rotation about the z-axis through an angle η. According to (46.15), the operators of the spin components in the new representation have the form

$$
\begin{aligned}
\hat{s}'_x &= \hat{T}_z \hat{s}_x \hat{T}_z^{-1} , \\
\hat{s}'_y &= \hat{T}_z \hat{s}_y \hat{T}_z^{-1} , \\
\hat{s}'_z &= \hat{T}_z \hat{s}_z \hat{T}_z^{-1} .
\end{aligned}
\tag{61.6}
$$

Here $\hat{s}'_x$, $\hat{s}'_y$, $\hat{s}'_z$ are the operators of the spin projections onto the old coordinate axes x, y, z but taken in the new representation connected with the system x', y', z'. Since we consider a rotation about the z-axis, the z-axis and the z'-axis coincide and, consequently, the operators of the spin projections onto these axes are the same in the two representations (expressed by the diagonal matrix (60.15)). However, the conditions for the operator of the spin projection onto the z-axis to be chosen in the form of the diagonal matrix (60.15) are still insufficient for the definition of the operators of spin projections along other directions. The operator of the spin projection along an arbitrary direction will be known if we also give the operators $\hat{s}_x$ and $\hat{s}_y$ in the form of matrices, i.e. if we give the system of x,y,z coordinates with which we connect the representation. We can, in particular, choose the representation connected with the system x', y', z'. In this representation (which we denote by primes) the operators of spin projections onto the x'-axis, y'-axis and z'-axis $\hat{s}'_{x'}$, $\hat{s}'_{y'}$, $\hat{s}'_{z'}$ have the form (60.16), (60.15).

Because the operators $\hat{s}_x$, $\hat{s}_y$, $\hat{s}_z$ correspond to the projections of the spin moment, when the coordinate system rotates they must transform as the projections of an angular momentum, i.e. as the components of an axial vector. Since we consider the rotation through an angle η about the z-axis, the operators of spin projections onto primed and non-primed coordinate axes are connected with each other by the relations

$$
\begin{aligned}
\hat{s}_x &= \hat{s}_{x'} \cos\eta - \hat{s}_{y'} \sin\eta , \\
\hat{s}_y &= \hat{s}_{x'} \sin\eta + \hat{s}_{y'} \cos\eta , \\
\hat{s}_z &= \hat{s}_{z'} .
\end{aligned}
\tag{61.7}
$$

The equalities (61.7) hold in any representation. In particular, in the representation connected with the rotated system of coordinates we have

$$\hat{s}'_x = \hat{s}'_{x'} \cos\eta - \hat{s}'_{y'} \sin\eta = \hat{s}_x \cos\eta - \hat{s}_y \sin\eta \ ,$$

$$\hat{s}'_y = \hat{s}'_{x'} \sin\eta + \hat{s}'_{y'} \cos\eta = \hat{s}_x \sin\eta + \hat{s}_y \cos\eta \ , \qquad (61.8)$$

$$\hat{s}'_z = \hat{s}'_{z'} = \hat{s}_z \ .$$

Indeed, the two systems of coordinates, the primed and the non-primed, are completely equivalent. As we have already noted, the operators of the spin projections onto the axes of the primed coordinate system, taken in the representation which is just connected with this rotated system of coordinates, have the ordinary form (60.15) and (60.16), i.e. are the same as the operators of the spin projections onto the axes of the non-primed system of coordinates, taken in the representation connected with the non-primed system of coordinates:

$$\hat{s}'_{x'} = \hat{s}_x \ ; \qquad \hat{s}'_{y'} = \hat{s}_y \ ; \qquad \hat{s}'_{z'} = \hat{s}_z \ .$$

Considering the equalities (61.6) and (61.8) together, we find the matrix $\hat{T}_z$. First of all, it follows from (61.6) and (61.8) that $\hat{s}_z = \hat{T}_z \hat{s}_z \hat{T}_z^{-1}$, i.e. the matrix $\hat{T}_z$ commutes with $\hat{s}_z$. Since the matrix $\hat{s}_z$ is diagonal, the matrix $\hat{T}_z$ is also diagonal (see §47). Consequently, the matrix $\hat{T}_z$ has the form

$$\hat{T}_z = \begin{pmatrix} a_1 & 0 \\ 0 & a_2 \end{pmatrix}.$$

The condition for the unitarity of the matrix $\hat{T}_z$, $\hat{T}_z \hat{T}_z^\dagger = \hat{T}_z^\dagger \hat{T}_z = 1$, leads to the equality

$$\begin{pmatrix} |a_1|^2 & 0 \\ 0 & |a_2|^2 \end{pmatrix} = \begin{pmatrix} 1 & 0 \\ 0 & 1 \end{pmatrix}$$

or $|a_1|^2 = 1$ and $|a_2|^2 = 1$. Consequently, the matrix $\hat{T}_z$ has the form

$$\hat{T}_z = \begin{pmatrix} e^{i\alpha_1} & 0 \\ 0 & e^{i\alpha_2} \end{pmatrix}, \qquad (61.9)$$

where α_1 and α_2 are real. We rewrite eq. (61.6) in the form

$$\hat{s}'_x \hat{T}_z = \hat{T}_z \hat{s}_x \ ,$$

$$\hat{s}'_y \hat{T}_z = \hat{T}_z \hat{s}_y \ .$$

Substituting the values of $\hat{s}'_x$ and $\hat{s}'_y$ from (61.8) into these expressions and equating the corresponding matrix elements, we find that the equalities are satisfied under the condition $\alpha_1 - \alpha_2 = \eta$.

Thus for the two phases α_1 and α_2 we have obtained one condition connecting them. This fact is not accidental. The point is that the matrix $\hat{T}_z$ can contain an arbitrary phase factor which will not affect the results because the wave functions themselves are defined to within an arbitrary phase factor. In accordance with this we write the matrix $\hat{T}_z$ in the form

$$\hat{T}_z(\eta) = \begin{pmatrix} e^{\frac{1}{2}i\eta} & 0 \\ 0 & -e^{-\frac{1}{2}i\eta} \end{pmatrix}. \tag{61.10}$$

We can express the matrix $\hat{T}_z$ also in terms of the matrix σ_z

$$\hat{T}_z(\eta) = \cos\tfrac{1}{2}\eta + i\sigma_z \sin\tfrac{1}{2}\eta , \tag{61.11}$$

or in a somewhat different form

$$\hat{T}_z = \exp\left(\tfrac{1}{2}i\eta\hat{\sigma}_z\right). \tag{61.12}$$

The above expressions should be understood as

$$\hat{T}_z = 1 + \left(\frac{i}{2}\eta\sigma_z\right) + \frac{1}{2}\left(\frac{i}{2}\eta\sigma_z\right)^2 + \dots + \frac{1}{n!}\left(\frac{i}{2}\eta\sigma_z\right)^n + \dots .$$

Since $\sigma_z^2 = 1$, $\sigma_z^3 = \sigma_z$ and so on, the series is easily evaluated, and we again arrive at formula (61.11). If the angle of rotation is small, then the rotation matrix (61.12) has the form

$$\hat{T}_z = 1 + i\frac{\eta}{\hbar}\hat{s}_z , \tag{61.13}$$

where we have introduced the matrix $\hat{s}_z = \tfrac{1}{2}\hbar\sigma_z$ instead of σ_z. We have obtained an expression which has the same structure as the rotation operator obtained in §30 which acts on a function depending on spatial coordinates.

Of course, a relation of the type (61.11) is also valid for a rotation about any other axis, since all directions are equivalent. Thus, for example, we write the matrix $\hat{T}_x$ of a rotation about the x-axis through a certain angle ϑ:

$$\hat{T}_x(\vartheta) = \cos\tfrac{1}{2}\vartheta + i\sigma_x \sin\tfrac{1}{2}\vartheta . \tag{61.13'}$$

The arbitrary rotation of one coordinate system with respect to another can be characterized by three Euler angles ϑ, φ, ψ (fig. V.18). The angle ψ is the angle between the axis Ox and the straight line ON forming the intersection of the planes xOy and $x'Oy'$, ϑ is the angle between the axes Oz and Oz' and, finally, φ is the angle between ON and the axis Ox'. The total rotation can be resolved into three consecutive rotations: I about the z-axis through the angle ψ; II about the new position of the axis $x(ON)$ through the angle ϑ, and III

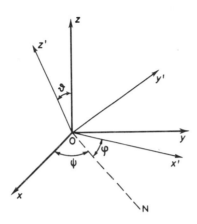

Fig. V.18

about the new position of the axis Oz through the angle φ. The rotation matrix $\hat{T}$ will be equal to the product of three matrices $\hat{T} = \hat{T}_z(\varphi)\hat{T}_x(\vartheta)\hat{T}_z(\psi)$. Making use of (61.10) and (61.13') and multiplying the matrices, we find

$$\hat{T} = \begin{pmatrix} e^{\frac{1}{2}i(\psi+\varphi)}\cos\frac{1}{2}\vartheta & ie^{-\frac{1}{2}i(\psi-\varphi)}\sin\frac{1}{2}\vartheta \\ ie^{\frac{1}{2}i(\psi-\varphi)}\sin\frac{1}{2}\vartheta & e^{-\frac{1}{2}i(\psi+\varphi)}\cos\frac{1}{2}\vartheta \end{pmatrix}. \tag{61.14}$$

This matrix was first obtained by Pauli. The two-component wave function $\binom{c_1}{c_2}$ which transforms, when the system of coordinates rotates, according to the law

$$\begin{pmatrix} c_1' \\ c_2' \end{pmatrix} = \hat{T}\begin{pmatrix} c_1 \\ c_2 \end{pmatrix}$$

is called a spinor.

The spinor components which we have introduced are usually denoted as

$$\varphi = \begin{pmatrix} \varphi^1 \\ \varphi^2 \end{pmatrix}.$$

We note that for any given spinor $\varphi = \binom{\varphi^1}{\varphi^2}$ one can always find a matrix $\hat{T}(\vartheta,\varphi,\psi)$ such that $\varphi^{1'} = 1$, $\varphi^{2'} = 0$, i.e. it is always possible to determine the direction, characterized by the angles (φ,ϑ,ψ), along which the spin of the particle is oriented.

We rewrite the matrix $\hat{T}$ in the form

$$\hat{T} = \begin{pmatrix} \alpha & \beta \\ \gamma & \delta \end{pmatrix},$$

so that

$$\varphi^{1'} = \alpha\varphi^1 + \beta\varphi^2 ,$$
$$\varphi^{2'} = \gamma\varphi^1 + \delta\varphi^2 ,$$

(61.15)

where, as follows from (61.14),

$$\alpha = \delta^* , . \qquad \beta = -\gamma^* , \qquad \alpha\delta - \beta\gamma = 1 .$$

(61.16)

The transformation (61.15) is usually called a bilinear transformation.

The bilinear transformation leaves certain bilinear forms invariant. Indeed, making use of (61.15) and (61.16) we easily obtain for two arbitrary spinors φ and η

$$\varphi^{1'}\eta^{2'} - \varphi^{2'}\eta^{1'} = (\alpha\delta - \beta\gamma)(\varphi^1\eta^2 - \varphi^2\eta^1) = \varphi^1\eta^2 - \varphi^2\eta^1 ,$$
$$\varphi^{1*}\varphi^1 + \varphi^{2*}\varphi^2 = \text{const} .$$

(61.17)

The above relation expresses the conservation of normalization under the rotation of a coordinate system.

By means of the spinor components η^1, η^2 and ζ^1, ζ^2 one can construct quantities which transform under a rotation of the coordinate system as the components of a vector, i.e. according to the law

$$a_i' = \sum_{k=1}^{3} \alpha_{ik} a_k ,$$

where α_{ik} are the cosines of the angles between the old and new coordinate axes. By a direct check, making use of (61.14) and (61.15), one can verify that the vector components are determined by the relations

$$a_z = (\eta^1\zeta^2 + \eta^2\zeta^1) ,$$
$$a_x = (\eta^2\zeta^2 - \eta^1\zeta^1) ,$$

(61.18)

$$a_y = -i(\eta^1\zeta^1 + \eta^2\zeta^2) .$$

Correspondingly for the square of the vector we have

$$a^2 = a_x^2 + a_y^2 + a_z^2 = (\eta^1 \zeta^2 - \eta^2 \zeta^1)^2 \tag{61.19}$$

We have, as was to be expected, obtained a scalar quantity.

The components of a tensor of arbitrary rank $B_{ikl\dots}$ can be defined as the product $a_i b_k c_l \dots$ of the corresponding components of vectors. By means of formulae (61.18) we can identify the components $B_{ikl\dots}$ with the products of the components of spinors.

Finally, we shall define the law of transformation of a spinor under the inversion of the coordinate system, i.e. under the transformation $x \to -x$, $y \to -y$, $z \to -z$.

We denote the inversion operator by $\hat{I}$, so that

$$\varphi' = \hat{I}\varphi . \tag{61.20}$$

The transformation (61.20) can now be considered as a transition to another representation. The corresponding transformation of the operators $\hat{s}_x$, $\hat{s}_y$, $\hat{s}_z$ is analogous to (61.6). On the other hand, we can consider the operators $\hat{s}_x$, $\hat{s}_y$, $\hat{s}_z$ as the components of an axial vector (as the components of an orbital angular momentum). Consequently, these operators must not change sign under reflection. Based on this, we obtain, in analogy with formula (61.6)

$$\hat{s}_x = \hat{I}\hat{s}_x\hat{I}^{-1} ,$$
$$\hat{s}_y = \hat{I}\hat{s}_y\hat{I}^{-1} , \tag{61.21}$$
$$\hat{s}_z = \hat{I}\hat{s}_z\hat{I}^{-1} .$$

When the matrix $\hat{I}$ is applied twice we come back to the initial state and, consequently, we must obtain the spinor φ. Furthermore, we can obtain the spinor $-\varphi$ if we consider a double reflection as a rotation through the angle 2π. As is seen from formulae (61.10) and (61.13'), under such a rotation the spinor changes sign. Correspondingly we have

$$\hat{I}^2 = \pm \begin{pmatrix} 1 & 0 \\ 0 & 1 \end{pmatrix}. \tag{61.22}$$

We see, consequently, that the matrix $\hat{I}$ must commute with the matrices $\hat{s}_x$, $\hat{s}_y$, $\hat{s}_z$, and that its square must give the unit matrix multiplied by ± 1. These requirements will be fulfilled under the condition that

$$\hat{I} = \pm \begin{pmatrix} 1 & 0 \\ 0 & 1 \end{pmatrix} \quad \text{or} \quad \hat{I} = \pm i \begin{pmatrix} 1 & 0 \\ 0 & 1 \end{pmatrix}. \tag{61.23}$$

Consequently, under reflection the spinor can transform in the following way:

$$\varphi' = \pm\varphi \tag{61.24}$$

or

$$\varphi' = \pm i\varphi . \tag{61.25}$$

If the law of transformation is determined by the upper sign in formulae (61.24) and (61.25), then φ is sometimes called a polar spinor; if the law of transformation is determined by the lower sign, then φ is called a pseudo-spinor.

§62. The total angular momentum

The total angular momentum of a particle is made up of the orbital angular momentum and the spin angular momentum. According to the rules for the addition of vector operators we have for the total angular momentum operator $\hat{\mathbf{j}}$

$$\hat{\mathbf{j}} = \hat{\mathbf{l}} + \hat{\mathbf{s}} . \tag{62.1}$$

The orbital and intrinsic angular momentum operators act on different variables. The first acts on space variables, while the second acts only on spin variables. Hence the two operators commute with each other. It follows directly from this that the components of the total angular momentum operator satisfy the same commutation rules as the components of the orbital and intrinsic angular momenta. These commutation relations also follow from the connected between the total angular momentum operator and the rotation operator (see below)

$$\hat{j}_x\hat{j}_y - \hat{j}_y\hat{j}_x = i\hbar\hat{j}_z ,$$

$$\hat{j}_y\hat{j}_z - \hat{j}_z\hat{j}_y = i\hbar\hat{j}_x , \tag{62.2}$$

$$\hat{j}_z\hat{j}_x - \hat{j}_x\hat{j}_z = i\hbar\hat{j}_y ,$$

and also

$$\hat{j}_x\hat{\mathbf{j}}^2 - \hat{\mathbf{j}}^2\hat{j}_x = 0 ,$$

$$\hat{j}_y\hat{\mathbf{j}}^2 - \hat{\mathbf{j}}^2\hat{j}_y = 0 , \tag{62.3}$$

$$\hat{j}_z\hat{\mathbf{j}}^2 - \hat{\mathbf{j}}^2\hat{j}_z = 0 .$$

It follows from the relations (62.2) (see §51) that the eigenvalues of the operator $\hat{j}^2$ have the form

$$j^2 = \hbar^2 j(j+1) . \tag{62.4}$$

The value of the quantum number j determines the value of the total angular momentum. According to the rules of addition of angular momenta in quantum mechanics (52.1) it follows that for a given l and s the number j runs over the sequence of values

$$j = |l-s| , \quad |l-s| + 1 , ..., \quad l+s-1 , \quad l+s .$$

The quantum number j takes on integer values if the spin has integer values, and half-integer values if the spin is half-integer.

We shall show that for a particle moving in free space or in a centrally symmetric field it is the total angular momentum which is the constant of the motion. For proof we introduce the rotation operator $\hat{R}$ taking into account the change in the spin coordinates as well the spatial coordinates of the wave function. We consider a rotation of the coordinate system about the z-axis through a small angle $\delta\varphi$. Proceeding from the results obtained in §30 and §61, it is easy to determine the change in the total wave function under such a rotation

$$\psi' = \hat{R}_z \psi = \hat{T}_z \hat{W}_z \psi = \left[1 + \frac{i}{\hbar} \delta\varphi (\hat{s}_z + \hat{l}_z) \right] \psi = \left(1 + \frac{i}{\hbar} \delta\varphi \hat{j}_z \right) \psi .$$

Of course, an analogous relation also holds for a rotation about any other axis. Consequently we see that it is simply the total angular momentum operator which is connected with the rotation operator. But the operation of rotation, by virtue of the isotropy of space, must not change the Hamiltonian of a closed system (or a system in a centrally symmetric field). Mathematically this is shown by the fact that the operator $\hat{R}_z$ and, consequently, the operator $\hat{j}_z$ will commute with the Hamiltonian $\hat{H}$ of the particle. Thus the total angular momentum conservation law is a consequence of the isotropy of space. For the intrinsic and orbital angular momenta separately the conservation laws hold only approximately, to the extent that we neglect the spin–orbit interaction.

If we have a system of non-interacting particles, then the total angular momentum of the entire system $\hat{J}$ is made up of the angular momenta $\hat{j}_k$ of the individual particles according to the rules of addition of angular momenta in quantum mechanics

$$\hat{J} = \sum \hat{j}_k . \tag{62.5}$$

One can also introduce the total orbital angular momentum operator $\hat{\mathbf{L}}$

$$\hat{\mathbf{L}} = \sum_k \hat{\mathbf{l}}_k , \tag{62.6}$$

and the total spin operator of the system $\hat{\mathbf{S}}$

$$\hat{\mathbf{S}} = \sum_k \hat{\mathbf{s}}_k . \tag{62.7}$$

Since $\hat{\mathbf{j}}_k = \hat{\mathbf{l}}_k + \hat{\mathbf{s}}_k$, then, obviously, we have

$$\hat{\mathbf{J}} = \hat{\mathbf{L}} + \hat{\mathbf{S}} . \tag{62.8}$$

The operators referring to different particles commute with each other because they act on different variables. Hence for the operators of the components of the total angular momenta $\hat{\mathbf{J}}$, $\hat{\mathbf{L}}$, $\hat{\mathbf{S}}$ there are the same commutation relations as for the operators referring to individual particles. For example, for the operators $\hat{J}_x$, $\hat{J}_y$ of the components of the total angular momentum we have

$$\hat{J}_x \hat{J}_y - \hat{J}_y \hat{J}_x = \sum_{k,i} (\hat{j}_{kx} \hat{j}_{iy} - \hat{j}_{iy} \hat{j}_{kx}) =$$

$$= \sum_k (\hat{j}_{kx} \hat{j}_{ky} - \hat{j}_{ky} \hat{j}_{kx}) = i\hbar \sum_k \hat{j}_{kz} = i\hbar \hat{J}_z .$$

An analogous result is also obtained for other projections of the operator $\hat{\mathbf{J}}$, as well as for the operators $\hat{\mathbf{L}}$ and $\hat{\mathbf{S}}$.

For given eigenvalues of the operators referring to individual particles the eigenvalues of the operators $\hat{J}^2$, $\hat{L}^2$, $\hat{S}^2$ and of the operators of the components of the angular momenta are determined according to the rules of addition of angular momenta in quantum mechanics. The total angular momentum $\mathbf{J}$ is conserved quantity for a closed system of particles.

§63. The Pauli equation. The probability current density vector

In ch. 13, devoted to relativistic quantum mechanics, we shall establish the exact relativistic equation of quantum mechanics and show that the Schrödinger equation is obtained from it in the limit when $v/c \to 0$.

In this case it turns out that if terms of higher orders of small quantities are taken into account, then additional terms will arise in the Hamiltonian, describing a number of important properties of quantum systems.

In particular, the existence of spin, as well as the fact that the electron has a magnetic moment, follows from the relativistic wave equation by expanding in powers of v/c and preserving terms of the first order of small quantities.

Deferring the proof of this statement to §118, we introduce the intrinsic magnetic moment operator in correspondence with (59.4) by the relation

$$\hat{\boldsymbol{\mu}} = \alpha \hat{\mathbf{s}} = \frac{e}{mc} \hat{\mathbf{s}} . \tag{63.1}$$

Then the Hamiltonian for an electron in an electromagnetic field assumes the form

$$\hat{H} = \frac{1}{2m} \left(\hat{\mathbf{p}} - \frac{e}{c} \mathbf{A} \right)^2 + U(\mathbf{r}) - (\hat{\boldsymbol{\mu}} \cdot \boldsymbol{\mathcal{H}}) =$$

$$= \frac{1}{2m} \left(\hat{\mathbf{p}} - \frac{e}{c} \mathbf{A} \right)^2 + U(\mathbf{r}) - \frac{e}{mc} (\hat{\mathbf{s}} \cdot \boldsymbol{\mathcal{H}}) , \tag{63.2}$$

where $\boldsymbol{\mathcal{H}}$ is the magnetic field strength. Since the Hamiltonian depends on the spin, the wave function of the electron also depends on the spin variable, i.e. $\psi = \psi(x,y,z,t,s_z)$. The equation for the wave function $\psi(x,y,z,t,s_z)$ in a magnetic field, first introduced by Pauli and called the Pauli equation, is of the form

$$i\hbar \frac{\partial \psi}{\partial t} = \frac{1}{2m} \left(\hat{\mathbf{p}} - \frac{e}{c} \mathbf{A} \right)^2 \psi - \frac{e}{mc} (\hat{\mathbf{s}} \cdot \boldsymbol{\mathcal{H}}) \psi + U\psi . \tag{63.3}$$

Let us find the probability current density vector. For this we write the equation for the function $\psi^\dagger$

$$-i\hbar \frac{\partial \psi^\dagger}{\partial t} = \frac{1}{2m} \left(\hat{\mathbf{p}} + \frac{e}{c} \mathbf{A} \right)^2 \psi^\dagger - \frac{e}{mc} [(\hat{\mathbf{s}} \cdot \boldsymbol{\mathcal{H}}) \psi]^\dagger + U\psi^\dagger . \tag{63.3'}$$

We multiply (63.3) from the left by $\psi^\dagger$ and (63.3') from the right by ψ and subtract one from the other.

Taking into account that $\hat{\mathbf{p}} \cdot \mathbf{A} - \mathbf{A} \cdot \hat{\mathbf{p}} = (\hbar/i) \nabla \mathbf{A}$, we obtain after simple calculation

$$\frac{\partial (\psi^\dagger \psi)}{\partial t} = -\frac{i\hbar}{2m} \nabla [(\nabla \psi^\dagger)\psi - \psi^\dagger \nabla \psi] +$$

$$+ \frac{e}{mc} \nabla (\mathbf{A}\psi^\dagger \psi) + \frac{ie}{mc\hbar} [\psi^\dagger (\hat{\mathbf{s}} \cdot \boldsymbol{\mathcal{H}})\psi - ((\hat{\mathbf{s}} \cdot \boldsymbol{\mathcal{H}})\psi)^\dagger \psi] . \tag{63.4}$$

It follows from rule (45.20) that

$$((\hat{s} \cdot \mathcal{H})\psi)^\dagger = \psi^\dagger (\hat{s}^\dagger \cdot \mathcal{H}) \tag{63.5}$$

and, furthermore, $\hat{s}^\dagger = \hat{s}$ by virtue of the hermitian property of the spin operator. Thus the term in square brackets reduces to zero and we obtain an expression for the probability current density vector. (It should be borne in mind that it contains two-component wave functions). Multiplying it by the charge e, we obtain the electric current density vector

$$\mathbf{j} = \frac{i\hbar e}{2m}((\nabla \psi^\dagger)\psi - \psi^\dagger \nabla \psi) - \frac{e^2}{mc} \mathbf{A}\psi^\dagger \psi \ . \tag{63.6}$$

The expression (63.4) determines the probability current density vector $\mathbf{j}$ with an accuracy to within $\nabla \times \mathbf{B}$, where $\mathbf{B}$ is an arbitrary vector. It can be shown that $\mathbf{B} = (e/m)\psi^\dagger \hat{s}\psi$, so that the total expression for the electric current density will have the form*

$$\mathbf{j}_n = -\frac{i\hbar e}{2m}(\psi^\dagger \nabla \psi - (\nabla \psi^\dagger)\psi) - \frac{e^2}{mc} \mathbf{A}(\psi^\dagger \psi) + \frac{e}{m} \nabla \times (\psi^\dagger \hat{s}\psi) \ . \tag{63.7}$$

Let us consider the case where the particle moves in a constant magnetic field $\mathcal{H}$ and the electric field is absent. Then the vector potential $\mathbf{A}$ can be chosen in the form

$$\mathbf{A} = \tfrac{1}{2}(\mathcal{H} \times \mathbf{r}) \ . \tag{63.8}$$

We transform the Pauli eq. (63.3), taking into account that the following relation holds for the vector potential (63.8)

$$\hat{\mathbf{p}} \cdot \mathbf{A} = \mathbf{A} \cdot \hat{\mathbf{p}} \ .$$

Furthermore we assume that the magnetic field $\mathcal{H}$ is relatively weak, and in correspondence with this we drop terms in the Pauli equation which are quadratic in $\mathcal{H}$. We then have

$$i\hbar \frac{\partial \psi}{\partial t} = \frac{\hat{\mathbf{p}}^2}{2m} \psi - \frac{e}{2mc}(\mathcal{H} \times \mathbf{r}) \cdot \hat{\mathbf{p}}\psi - \frac{e}{mc}(\hat{s} \cdot \mathcal{H})\psi \ .$$

Since

$$(\mathcal{H} \times \mathbf{r}) \cdot \hat{\mathbf{p}} = \mathcal{H} \cdot (\mathbf{r} \times \hat{\mathbf{p}}) = \mathcal{H} \cdot \hat{\mathbf{l}} \ ,$$

where $\hat{\mathbf{l}}$ is the orbital angular momentum, then

---

* See L.D.Landau and E.M.Lifshitz, *Quantum mechanics* (Pergamon Press, Oxford, 1965).

$$i\hbar \frac{\partial \psi}{\partial t} = \frac{\hat{p}^2}{2m}\psi - (\hat{\mu}_l \cdot \mathcal{H})\psi - (\hat{\mu}_s \cdot \mathcal{H})\psi \ . \tag{63.9}$$

It is natural to call the operator $\hat{\mu}_l$,

$$\hat{\mu}_l = \frac{e}{2mc}\hat{l} \ ,$$

the orbital magnetic moment operator (as distinct from $\hat{\mu}_s$, the spin magnetic moment operator).

We see that the ratio of the orbital magnetic moment $\hat{\mu}_l$ to the orbital angular momentum $\hat{l}$ is, as in classical physics, equal to $e/2mc$. For spin moments this ratio is larger by a factor of two.

Eqs. (63.3) and (63.9) are naturally generalized to the case of a system of particles. Thus for a system of particles (of charge e and mass m) placed in a weak magnetic field $\mathcal{H}$ the Pauli equation (63.9) has the form

$$i\hbar \frac{\partial \psi}{\partial t} = \frac{1}{2m}\sum_k \hat{p}_k^2 \psi - \frac{e}{2mc}(\hat{L}\cdot\mathcal{H})\psi - \frac{e}{mc}(\hat{S}\cdot\mathcal{H}) + U_{\text{int}}\psi \ , \tag{63.10}$$

where $\hat{L} = \Sigma_k \hat{l}_k$ is the operator of the total orbital angular momentum of the system, $\hat{S} = \Sigma_k \hat{s}_k$ is the operator of the total spin angular momentum of the system (the summation is carried out over all particles of the system), and U_{int} is the potential energy of interaction of the particles with each other.

The term which is quadratic in the magnetic field is also easily taken into account. In this case the Pauli equation has the form

$$i\hbar \frac{\partial \psi}{\partial t} = \left\{ \frac{1}{2m}\sum_k \hat{p}_k^2 - \frac{e}{2mc}(\hat{L}+2\hat{S})\cdot\mathcal{H} + \frac{e^2}{8mc^2}\sum_k [\mathcal{H}\times r_k]^2 + U_{\text{int}} \right\}\psi \ . \tag{63.11}$$

We see that in certain cases (see §75) the quadratic term plays an important role.

If the magnetic field is absent and the potential energy of interaction U_{int} does not depend on the spin of the particles, for example in the case of the Coulomb interaction of charged particles, then the Hamiltonian of the system does not contain any spin variables. In this case the wave function can be written in the form of a product of the coordinate wave function and the spin wave function. The particles can be in an arbitrary spin state, while the coordinate function satisfies the ordinary Schrödinger equation.

§64. The identity of particles. The principle of identity of particles. Symmetric and antisymmetric states

We now pass on to the construction of the wave function of a system of particles of one kind, for example a system of electrons, or protons, or photons.

In such systems new important specific properties having no analogue in classical mechanics appear. These properties will become clear from the comparison of the process of collision of two particles — macroscopic and microscopic.

In classical mechanics the properties of every particle are characterized by one quantity — its mass. If the masses of two particles are the same, then the particles can be assumed to be the same. The state of each of the particles at the instant of time $t = 0$ is defined by initial conditions.

Moving in definite trajectories the particles collide elastically at a certain point of space and diverge along separate trajectories.

If the initial conditions are given, then the trajectories of each particle are completely defined and one can follow the motion of each particle. Hence particles in classical mechanics, even if they are the same, conserve their individuality. One can always establish which one of the colliding particles one has at a given point of space.

In the case of two microparticles the situation in a collision is quite different. At the instant of collision let the two particles be at definite points of space. According to the uncertainty relation, their momenta do not have sharp values. After the collision we can fix 'the trajectories' of the particles, for example two tracks in a cloud chamber. However, it is clear that if the two colliding particles are of the same kind, for example two electrons or two protons, then it is in principle impossible to establish which one of these particles is associated with a given track.

As a second example let us consider a system consisting of two hydrogen atoms.

If the atoms are at sufficiently large distances from each other, so that the electron clouds do not overlap, then each of the electrons is in effect localized by its nucleus. As the atoms approach, an overlapping of the electron clouds occurs. This means that in the region of overlap there is a certain probability of finding both electrons.

As a result of a measurement let an electron be observed in this region. It is clear that there is no way which would make it possible to establish which one of the electrons, that which earlier belonged to nucleus no. 1 or that which earlier belonged to nucleus no. 2, is observed.

The examples given above show that the 'sameness' of quantum particles has a much more profound nature than the 'sameness' of classical particles. Quantum particles are not simply the same, but are completely identical.

If we were able to change the initial state of the system by replacing each electron by the other, no physical changes would occur in the system, and this replacement could not be observed by any physical experiment.

It should also be stressed that in the examples given we have somewhat idealized the situation. If, for example, the two colliding particles have definite values of their momenta, they have no definite values of their coordinates. Hence one cannot indicate the region of collision.

Thus we arrive at the principle of identity of particles, which can be formulated in the following way: in a system of identical particles only states which do not change under the exchange of any two identical particles can be realized.

The identity of microparticles leads to very important and profound consequences. We recall that we have already encountered this property of microparticles in statistical physics. We shall now discuss the effect of the identity of particles on their collective properties in a more consistent way.

We consider a system consisting of N identical particles. The wave function of this system ψ will have the form

$$\psi(\xi_1,\xi_2,...,\xi_i,...,\xi_k,...,\xi_N,t) \ .$$

Here ξ_i is understood to be the whole set of the coordinates and spin variables characterizing the ith particle.

If two particles are exchanged, i.e. if the coordinates and the spin of the ith particle are replaced by the corresponding values for the kth particle and vice versa, then by virtue of the principle of identity the state of the system cannot change. Consequently, the wave function of the system can change only by an immaterial phase factor.

After the exchange of two particles the wave function can be expressed in terms of the initial wave function by the relation

$$\psi(\xi_1,\xi_2,...,\xi_k,...,\xi_i,...,\xi_N,t) = e^{i\alpha}\psi(\xi_1,\xi_2,...,\xi_i,...,\xi_k,...,\xi_N,t) \ , \qquad (64.1)$$

where α is a real quantity. If the operation of exchange of the ith and kth particles is carried out once more, then the system will come back to its initial state.

On the other hand, repeating operation (64.1), we can write that

$$\psi(\xi_1,\xi_2,...,\xi_k,...,\xi_i,...,\xi_N,t) = e^{i\alpha}\psi(\xi_1,\xi_2,...,\xi_i,...,\xi_k,...,\xi_N,t) =$$
$$= e^{2i\alpha}\psi(\xi_1,...,\xi_k,...,\xi_i,...,\xi_N,t) \ .$$

Hence it follows that $e^{2i\alpha} = 1$ and

$$e^{i\alpha} = \pm 1 .$$

Thus when two identical particles are exchanged the wave function of the system can either remain unchanged, or change its sign. Wave functions of the first type are called symmetric, while those of the second type are called antisymmetric. We introduce the particle exchange operator $\hat{P}_{ik}$, which is important in what follows. By definition, when the operator $\hat{P}_{ik}$ acts on the wave function of a system of particles $\psi(\xi_1, \xi_2, ..., \xi_i, ..., \xi_k, ..., \xi_N, t)$ it transforms it into the new function

$$\hat{P}_{ik} \psi(\xi_1, ..., \xi_i, ..., \xi_k, ..., \xi_N, t) = \psi(\xi_1, ..., \xi_k, ..., \xi_i, ..., \xi_N, t) . \qquad (64.2)$$

There corresponds to this transformation a transition of the ith particle into the state which had previously been occupied by the kth particle, and a transition of the kth particle into the state which had previously been occupied by the ith particle.

Comparing (64.2) with (64.1) we see that the eigenvalues of the operator $\hat{P}_{ik}$ are equal to $e^{i\alpha} = \pm 1$. The symmetric and antisymmetric functions are the eigenfunctions of the operator $\hat{P}_{ik}$ corresponding to the eigenvalues $+1$ and -1 respectively.

We shall show by means of the exchange operator that the properties of symmetry are conserved in time. This means that if at the initial instant of time the system was in either the symmetric or antisymmetric state, then no subsequent action whatever will change the character of its symmetry. In other words, the system will always remain either in the symmetric state or in the antisymmetric state. For the proof of this statement it is necessary to show that the operator $\hat{P}_{ik}$ commutes with the Hamiltonian operator. The exchange of two identical particles corresponds only to an exchange of terms in the sum forming the Hamiltonian of the system. This is easily seen in the example of a system consisting of two identical particles. In this case the Hamiltonian can be written in the form

$$\hat{H} = -\frac{\hbar^2}{2m} \nabla_1^2 - \frac{\hbar^2}{2m} \nabla_2^2 + U(\xi_1, t) + U(\xi_2, t) + U_{12}(\xi_1, \xi_2, t) . \qquad (64.3)$$

Here $U_{12}(\xi_1, \xi_2, t)$ is the energy of interaction of the particles and U corresponds to the interaction with the external field and has, obviously, the same form for the two identical particles. When the particles are exchanged, we have for the new Hamiltonian

$$\hat{H} = -\frac{\hbar^2}{2m} \nabla_2^2 - \frac{\hbar^2}{2m} \nabla_1^2 + U(\xi_2, t) + U(\xi_1, t) + U_{12}(\xi_2, \xi_1, t) . \qquad (64.4)$$

It is clear that this is the same Hamiltonian as before the exchange. The result obtained is easily applied to the case of a system of N particles. We see that the exchange of particles does not change the Hamiltonian. Hence we obtain

$$\hat{H}\hat{P}_{ik} - \hat{P}_{ik}\hat{H} = 0 . \tag{64.5}$$

Consequently, the symmetry properties of the system are a constant of motion and are conserved in time.

Thus it is natural to think that the symmetry is determined by the properties of the elementary particles which make up the system. Pauli managed to show that particles having an integer spin are described by symmetric functions, whereas particles having a half-integer spin are described by antisymmetric functions. The first particles are called Bose–Einstein particles, or bosons, while the latter are called Fermi–Dirac particles, or fermions. Examples of the first group of particles are light quanta (see ch. 12) and π-mesons. The second group of particles includes neutrons, protons, positrons, electrons, neutrinos and muons (all having spin $\frac{1}{2}$).

To elucidate the problem of the symmetry properties of a system consisting of identical complex particles, it is necessary to determine the total spin of the complex particle. As in the case of elementary particles, when the complex particle has an integer spin the wave function is symmetric under the exchange of the complex particles, and is antisymmetric when the complex particle has a half-integer spin.

As an example let us consider a system consisting of α-particles. For the determination of the symmetry properties of the wave function of the system it is necessary to calculate the total spin of the α-particle. The α-particle consists of two neutrons and two protons. Since the spins of the particles constituting it are equal to $\frac{1}{2}\hbar$ and the number of the particles is even, then the total spin of the α-particle is equal to an integer multiple of $\hbar$. Thus the wave function of a system of α-particles is a symmetric wave function.

§65. Wave functions for systems of fermions and bosons. The Pauli principle

Let us consider a system consisting of N non-interacting identical particles. The Schrödinger equation for the stationary states of such a system has the form

$$\sum_{i=1}^{N} \left[-\frac{\hbar^2}{2m}\nabla_i^2 + U(\xi_i) \right] \psi(\xi_1,\xi_2,...,\xi_N) = E\psi(\xi_1,\xi_2,...,\xi_N) . \tag{65.1}$$

In §14 it was shown that the solution of this equation is the function

$$\psi = \psi_{k_1}(\xi_1)\psi_{k_2}(\xi_2) \dots \psi_{k_N}(\xi_N) \,. \tag{65.2}$$

Here k_1, k_2, ... are the quantum numbers of the states available to the particles. Each k_i represents the complete set of quantum numbers which characterize the state of an individual particle. The function ψ_{k_i} is the solution of the Schrödinger equation for one particle

$$-\frac{\hbar^2}{2m} \nabla_i^2 \psi_{k_i}(\xi_i) + U(\xi_i)\psi_{k_i}(\xi_i) = E_{k_i}\psi_{k_i}(\xi_i) \,.$$

However, function (65.2) does not satisfy the requirements of symmetry. In general it is neither a symmetric nor an antisymmetric function. Since eq. (65.1) is linear, then a superposition of solutions of the type (65.2) will also be a solution. To obtain a wave function possessing the required symmetry one has to take the corresponding superposition of wave functions.

For simplicity we consider a system consisting of only two non-interacting particles. It is obvious that the functions

$$\psi_1(\xi_1,\xi_2) = \psi_1(\xi_1)\psi_2(\xi_2) \,, \qquad \psi_2(\xi_1,\xi_2) = \psi_2(\xi_1)\psi_1(\xi_2) \,,$$

where the subscripts 1 and 2 of the wave functions $\psi_1(\xi_1)$, $\psi_1(\xi_2)$ and $\psi_2(\xi_1)$, $\psi_2(\xi_2)$ denote two different states of a particle. The wave functions $\psi_1(\xi_1,\xi_2)$, $\psi_2(\xi_1,\xi_2)$ correspond to one and the same energy of the system. From these functions one can make up two symmetrized combinations corresponding to the same energy:

$$\psi_s = C_1[\psi_1(\xi_1)\psi_2(\xi_2) + \psi_2(\xi_1)\psi_1(\xi_2)] \,,$$

$$\psi_a = C_2[\psi_1(\xi_1)\psi_2(\xi_2) - \psi_2(\xi_1)\psi_1(\xi_2)] \,.$$

The first wave function is symmetric with respect to exchange of the particles, while the second is antisymmetric. The constants C_1 and C_2 can be determined from the normalization condition. If the functions $\psi_1(\xi_1)$ and $\psi_2(\xi_2)$ are normalized to unity, while ψ_s (and ψ_a) is normalized by the condition $\int |\psi_s|^2 dV_1 dV_2 = 1$, a simple calculation gives in both cases

$$C_1 = C_2 = \frac{1}{\sqrt{2}} \,.$$

Hence the normalized and symmetrized function can be written in the form

$$\psi_s = \frac{1}{\sqrt{2}}[\psi_1(\xi_1)\psi_2(\xi_2) + \psi_2(\xi_1)\psi_1(\xi_2)] \,, \tag{65.3}$$

$$\psi_a = \frac{1}{\sqrt{2}}[\psi_1(\xi_1)\psi_2(\xi_2) - \psi_2(\xi_1)\psi_1(\xi_2)] \,. \tag{65.4}$$

It is now easy to generalize formulae (65.3) and (65.4) to the case of an arbitrary number of non-interacting particles. Namely, for a system described by symmetric functions one can write that

$$\psi_s = \left(\frac{N!}{n_1! n_2! ...}\right)^{-\frac{1}{2}} \sum_p \psi_{k_1}(\xi_1)\psi_{k_2}(\xi_2) ... \psi_{k_N}(\xi_N) . \qquad (65.5)$$

Here n_k is the number of indices which have the same i-value. Thus the n_k show the number of particles in a given ψ_k-state. Evidently $\Sigma n_k = N$.

The wave function $\psi_k(\xi_k)$ is normalized to unity, all functions $\psi_k(\xi_k)$ are orthogonal to each other. Therefore in the normalization condition $\int |\psi_s|^2 \, dV$ only terms of the type $\int |\psi_k(\xi)|^2 \, d\xi$ contribute. Thus

$$\int |\psi_s|^2 \, dV = \sum \int |\psi_k(\xi)|^2 \, d\xi = \frac{N!}{n_1! n_2! ...} .$$

Here $N!/(n_1! n_2! ...)$ is equal to the number of rearrangements of different indices n_i. Similarly for particles with antisymmetric functions

$$\psi_a = (N!)^{-\frac{1}{2}} \begin{vmatrix} \psi_{k_1}(\xi_1) & \cdots & \psi_{k_1}(\xi_N) \\ \psi_{k_2}(\xi_1) & \cdots & \psi_{k_2}(\xi_N) \\ \cdots \cdots \cdots \cdots \cdots \cdots \\ \psi_{k_N}(\xi_1) & \cdots & \psi_{k_N}(\xi_N) \end{vmatrix} \qquad (65.6)$$

The symmetrized normalized wave functions ψ_s and ψ_a describe the state of systems of N non-interacting bosons and N non-interacting fermions respectively.

Let us now consider the change in the wave function of the system when there is an interaction between the identical particles. We assume that the interaction depends on time. The exact wave function can be written in the form of one of the superpositions

$$\psi = \sum_i c_i(t)(\psi_s)_i , \qquad \psi = \sum_k c_k(t)(\psi_a)_k .$$

The coefficients c_i and c_k represent the time-dependent amplitudes of the probability of the corresponding ith and kth symmetric and antisymmetric states.

The interaction gives rise to transitions in the system. As follows directly from the symmetry conservation law described in the preceding section, the system will go over into states with the same symmetry under any external action.

Thus the wave function describing a system of interacting particles is expressed in terms of the wave functions of a system of non-interacting particles with definite symmetry.

The wave functions (65.5) and (65.6) found above make it possible to obtain a number of very important results.

Let us first of all consider a system of fermions. We assume that two particles in the system are in one and the same quantum state, i.e. that $k_1 = k_2$. This means that the two particles have the same complete set of quantum numbers, for example one and the same value of the quantum numbers n, l, m, s_z for motion in a centrally symmetric field, or p_x, p_y, p_z, s_z for free motion with a definite momentum.

Then in the determinant of (65.6) two rows turn out to be the same, and the wave function reduces identically to zero. This proves the following statement: In a system of identical fermions one cannot simultaneously have two or more particles in one and the same quantum state. This is the well-known Pauli principle, established by Pauli before the appearance of quantum mechanics on the basis of an analysis of experimental data.

The Pauli principle is often conveniently formulated in terms of the quasi-classical approximation: 'No more than one particle with given spin orientation can be found in each cell of phase space of volume $(2\pi\hbar)^3$.'

As we have seen in statistical physics, the Pauli principle determines the statistical behaviour of systems made up of identical particles with half-integer spin. The Pauli principle is of no less importance in understanding the regularities of the structure of many-electron atoms and of complex nuclei to which the next chapter is devoted.

For what follows we consider the following problem. Let a system consist of N identical particles (bosons). At a given instant of time each of the bosons is in one and the same state with the wave function $\psi(\xi)$ which is normalized in the following way:

$$\int \psi^*(\xi)\psi(\xi)\mathrm{d}V = N .$$

Let us determine the mean energy of the system in this state. We write the Hamiltonian of such a system of particles in the form

$$\hat{H} = \sum_{i=1}^{N} \hat{H}_i(\xi_i) + \tfrac{1}{2} \sum_i \sum_j \hat{W}_{ij}(\xi_i,\xi_j) , \tag{65.7}$$

where $\hat{H}_i$ is the energy operator of the ith boson, and $\hat{W}_{ij}$ is the operator corresponding to the energy of interaction of the ith and jth bosons. The wave function of the system of bosons, normalized to unity, at this instant of time has the form

$$\psi(\xi_1,\xi_2,...,\xi_N) = N^{-\frac{1}{2}N}\ \psi(\xi_1)\psi(\xi_2) \ ... \ \psi(\xi_N)\ .$$

The mean energy of the system in this state is equal to

$$\bar{H} = \int \psi^*(\xi_1,\xi_2,...,\xi_N)\hat{H}\psi(\xi_1,\xi_2,...,\xi_N)\mathrm{d}V_1\mathrm{d}V_2 \ ... \ \mathrm{d}V_N\ .$$

Taking into account the identity of bosons and assuming $N \gg 1$, we obtain

$$\bar{H} = \int \psi^*(\xi_i)\hat{H}_i\psi(\xi_i)\mathrm{d}V_i + \tfrac{1}{2}\int \psi^*(\xi_i)\psi^*(\xi_j)\hat{W}_{ij}(\xi_i,\xi_j)\psi(\xi_i)\psi(\xi_j)\mathrm{d}V_i\mathrm{d}V_j\ . \quad (65.8)$$

If the particles do not interact, then $\hat{W} \equiv 0$, and the mean value of the energy has the form

$$\bar{H} = \int \psi^*(\xi_i)\hat{H}_i\psi(\xi_i)\mathrm{d}V_i\ . \tag{65.9}$$

We shall need (65.8) and (65.9) for what follows.

§66. The wave function of a system of two identical particles with spin $\frac{1}{2}$

Bearing in mind further applications, let us consider the wave function of a system consisting of two particles with spin $\frac{1}{2}$, for example two electrons or two protons, in more detail.

The total wave function $\psi_n(\mathbf{r}_1,s_{1z},\mathbf{r}_2,s_{2z})$ depends on the spatial and spin coordinates of the two particles and is antisymmetric in these variables. Assuming that there is no external magnetic field and that the interaction between the particles does not depend on their spins, we write the total wave function in the form of a product of wave functions which depend only on spatial variables and spin variables

$$\psi_n(\mathbf{r}_1,s_{1z},\mathbf{r}_2,s_{2z}) = \Phi(\mathbf{r}_1,\mathbf{r}_2)\varphi(s_{1z},s_{2z})\ . \tag{66.1}$$

We write the total spin function of the system φ in the form of a product of the eigenfunctions of the operators corresponding to the square of the spin and its z-component for each of the particles, i.e. the functions $\varphi_{\frac{1}{2}}(1)$, $\varphi_{-\frac{1}{2}}(1)$, $\varphi_{\frac{1}{2}}(2)$, $\varphi_{-\frac{1}{2}}(2)$, where the subscript denotes the spin projection of the z-axis, and the number in parentheses corresponds to the particle.

The function φ can be written in the most general form as follows:

$$\varphi(1,2) = c_1 \varphi_{\frac{1}{2}}(1)\varphi_{\frac{1}{2}}(2) + c_2 \varphi_{-\frac{1}{2}}(1)\varphi_{-\frac{1}{2}}(2) +$$
$$+ c_3 \varphi_{\frac{1}{2}}(1)\varphi_{-\frac{1}{2}}(2) + c_4 \varphi_{-\frac{1}{2}}(1)\varphi_{\frac{1}{2}}(2) , \tag{66.2}$$

where c_1, c_2, c_3 and c_4 are arbitrary amplitudes.

We determine the spin wave functions describing the states of the system with given total spin and given z-component of spin.

Since spins are added according to the general rules of addition of angular momenta (see §52), the total spin of a system of two particles has two possible values, $S = 1$ and $S = 0$. Its z-component correspondingly has the values $1, 0$ and -1 for $S = 1$, and $S_z = 0$ for $S = 0$ (in units of $\hbar$).

The functions φ describing a state with given S and S_z satisfy the equations

$$\hat{S}^2\varphi = \hbar^2 S(S+1)\varphi , \qquad \hat{S}_z\varphi = \hbar S_z\varphi , \tag{66.3}$$

where $\hat{S} = \hat{s}_1 + \hat{s}_2$ is the total spin operator of the system. The coefficients c_1, c_2, c_3 and c_4 in the spin function of eq. (66.2) must be chosen in a way such that both equations (66.3) for φ are automatically satisfied.

One can easily convince oneself by a direct check that the spin function of a system corresponding to all the conditions mentioned above can be written in the form

$$\varphi_1^1 = \varphi_{\frac{1}{2}}(1)\varphi_{\frac{1}{2}}(2) , \qquad S = 1, S_z = 1 ,$$
$$\varphi_0^1 = \frac{1}{\sqrt{2}} [\varphi_{\frac{1}{2}}(1)\varphi_{-\frac{1}{2}}(2) + \varphi_{-\frac{1}{2}}(1)\varphi_{\frac{1}{2}}(2)] , \qquad S = 1, S_z = 0 , \tag{66.4}$$
$$\varphi_{-1}^1 = \varphi_{-\frac{1}{2}}(1)\varphi_{-\frac{1}{2}}(2) , \qquad S = 1, S_z = -1 ,$$

$$\varphi_0^0 = \frac{1}{\sqrt{2}} [\varphi_{\frac{1}{2}}(1)\varphi_{-\frac{1}{2}}(2) - \varphi_{-\frac{1}{2}}(1)\varphi_{\frac{1}{2}}(2)] , \qquad S = 0, S_z = 0 , \tag{66.5}$$

where the superscript indicates the total spin of the two particles, and the subscript indicates the z-component of the spin. The expressions (66.4) and (66.5) follow from formula (52.3) and table 1 at the end of ch. 6 of coefficients C for $j_1 = \frac{1}{2}$.

This result can also be obtained from the relations (61.18). We note that the spin functions (66.4) do not change when the two particles are exchanged, i.e. under the exchange $1 \to 2, 2 \to 1$. Consequently, these functions are symmetric in the spins of the particles. The spin function (66.5) changes sign under such an exchange and is antisymmetric in the spins.

The spin functions (66.4) form a spin triplet. The whole set of the three components of a triplet is equivalent to the three-component spin function of a particle with spin one. The spin function (66.5) describing the state with spin zero forms a spin singlet.

Let us determine the eigenvalues of the scalar product $\hat{s}_1 \cdot \hat{s}_2$ in the singlet and triplet states. We shall have to deal with this product in what follows. Since

$$\hat{S}^2 = (\hat{s}_1 + \hat{s}_2)^2 = \hat{s}_1^2 + \hat{s}_2^2 + 2\hat{s}_1 \cdot \hat{s}_2 ,$$

then

$$\hat{s}_1 \cdot \hat{s}_2 = \tfrac{1}{2}(\hat{S}^2 - \hat{s}_1^2 - \hat{s}_2^2) . \tag{66.6}$$

Substituting the eigenvalues of the operators $\hat{S}^2$, $\hat{s}_1^2$ and $\hat{s}_2^2$ into the right-hand side, we have

$$(\hat{s}_1 \cdot \hat{s}_2) \varphi^s = \tfrac{1}{2}\hbar^2 (S(S+1) - \tfrac{3}{2}) \varphi^s . \tag{66.7}$$

For the triplet state $S = 1$, and thus

$$(\hat{s}_1 \cdot \hat{s}_2) \varphi^1 = \tfrac{1}{4}\hbar^2 \varphi^1 . \tag{66.8}$$

Correspondingly for the singlet state $S = 0$ and

$$(\hat{s}_1 \cdot \hat{s}_2) \varphi^0 = -\tfrac{3}{4}\hbar^2 \varphi^0 . \tag{66.9}$$

We now consider the function of spatial variables $\Phi(\mathbf{r}_1, \mathbf{r}_2)$. Since the total wave function (66.1) is antisymmetric, then the coordinate wave function will be antisymmetric in the state $S = 1$ and symmetric in the state $S = 0$. If the particles do not interact and are in states ψ_n and ψ_m, then the coordinate function has the form

$$\Phi_a(\mathbf{r}_1, \mathbf{r}_2) = \frac{1}{\sqrt{2}} [\psi_n(\mathbf{r}_1)\psi_m(\mathbf{r}_2) - \psi_m(\mathbf{r}_1)\psi_n(\mathbf{r}_2)] , \qquad S = 1 , \tag{66.10,}$$

$$\Phi_s(\mathbf{r}_1, \mathbf{r}_2) = \frac{1}{\sqrt{2}} [\psi_n(\mathbf{r}_1)\psi_m(\mathbf{r}_2) + \psi_m(\mathbf{r}_1)\psi_n(\mathbf{r}_2)] , \qquad S = 0 . \tag{66.11}$$

As was shown in §14, it is convenient in the general case to use the coordinates $\mathbf{R} = \tfrac{1}{2}(\mathbf{r}_1 + \mathbf{r}_2)$ and $\mathbf{r} = \mathbf{r}_1 - \mathbf{r}_2$, describing respectively the motion of the centre of mass of the system and the relative motion of the particles

$$\Phi(\mathbf{r}_1, \mathbf{r}_2) = \psi_0(\mathbf{R})\psi(\mathbf{r}) . \tag{66.12}$$

Let us find the consequences of the requirement of symmetry or antisymmetry of the wave function (66.12). We note first of all that if the potential of interaction of the particles depends only on the distance between them, $U(|\mathbf{r}_1 - \mathbf{r}_2|)$, then the orbital angular momentum associated with the relative motion of the particles is conserved in such a system (see §35). We now exchange the coordinates of the particles; $\mathbf{r}_1 \rightarrow \mathbf{r}_2$, $\mathbf{r}_2 \rightarrow \mathbf{r}_1$. The radius vector of the centre of mass $\mathbf{R}$ does not change under this exchange. Consequently, the wave function $\psi_0(\mathbf{R})$ does not change. The radius vector of the relative motion $\mathbf{r}$ changes sign, $\mathbf{r} \rightarrow -\mathbf{r}$. As we have explained in §33, if the orbital angular momentum associated with the relative motion of the particles is given and is determined by the quantum number l, then the law of transformation of the function $\psi(\mathbf{r})$ under the replacement of $\mathbf{r}$ by $-\mathbf{r}$ will be

$$\psi(-\mathbf{r}) = (-1)^l \psi(\mathbf{r}) . \tag{66.13}$$

We see that in this case the coordinate wave function (66.12) transforms under the exchange of the particles $\mathbf{r}_1 \rightarrow \mathbf{r}_2$, $\mathbf{r}_2 \rightarrow \mathbf{r}_1$ according to the law

$$\Phi(\mathbf{r}_1, \mathbf{r}_2) \rightarrow (-1)^l \Phi(\mathbf{r}_1, \mathbf{r}_2) . \tag{66.14}$$

It follows immediately from (66.14) that if the particle is in the triplet state $S = 1$, then the quantum number l can take on only odd values. On the contrary, the number l can take on only even values if the particle is in the single state $S = 0$.

§67. Exchange interaction and the concept of the chemical and strong nuclear interactions

The identity of quantum particles leads to a fundamental change in the concept of the interaction between particles.

We dwell first of all on a simple example which allows us to understand the essence of these changes. We assume that two identical particles with half-integer spin do not interact with each other in the classical sense of the word. This means that there are no terms in the Hamiltonian of the system describing an interaction between the particles.

Let one of the particles be in a given cell of phase space with the linear dimension of the cell $\sim d$. The obvious relation

$$(\Delta q_1 \Delta p_1)^3 \sim d^3 (\Delta p_1)^3 \sim \hbar^3 ,$$

holds, where $\Delta q_1 \sim d$ is the uncertainty in the coordinate of the particle, and $\Delta p_1 \sim \hbar/d$ is the uncertainty in its momentum.

According to the Pauli principle, the second particle cannot be in the same cell of phase space. Hence it must either be at a distance larger than d from the first particle, or have a momentum p_2 such that $|\mathbf{p}_2 - \mathbf{p}_1|$ exceeds Δp_1, i.e. $|\mathbf{p}_2 - \mathbf{p}_1| > \hbar/d$. Only under this condition can it approach the first particle to within a distance smaller than d and be in a different cell of phase space.

We see that particles with parallel spins cannot approach each other unless they possess a sufficiently large difference of momentum. Such behaviour of the particles is equivalent to the appearance of a repulsive force between them. If the spins of the particles are antiparallel this reasoning is not valid, since the Pauli principle does not forbid such particles to be in the same cell of phase space.

Thus from the Pauli principle, which imposes restrictions upon the states of particles, it follows that there exists an interaction between the particles which depends on the orientation of their spins.

The interaction between bosons cannot be illustrated by such an obvious example. Nevertheless, it is clear that the requirement of the symmetrization of the wave function corresponds to a definite dependence of the energy of a system of particles on its total spin, i.e. it leads to an interaction between the particles.

We now assume that there is a certain weak interaction between two particles with spin $\frac{1}{2}$, described by the operator $\hat{H}'(r_{12})$, where r_{12} is the distance between the particles. For clarity, and keeping in mind further applications, we assume that $\hat{H}'$ represents the Coulomb repulsion of two charges $\hat{H}'(r_{12}) = e^2/r_{12}$. Then the mean energy of interaction in the first approximation is equal to

$$E^{(1)} = \sum \int \psi_0^* \hat{H}' \psi_0 \, dV_1 dV_2 \ . \tag{67.1}$$

Here ψ_0 is the normalized function of the unperturbed state, and the summation is carried out over all values of the spin variables.

Since in the zero order approximation particles are assumed to be non-interacting, then the spin wave functions and the coordinate wave functions are separable. The latter can be written in terms of the symmetrized or anti-symmetrized products (66.10) and (66.11).

Substituting the value of the operator $\hat{H}'$ and the wave functions into (67.1), we have

$$E^{(1)} = \frac{1}{2} \int \frac{e^2}{r_{12}} |\psi_{n_1}(1)\psi_{n_2}(2) \pm \psi_{n_1}(2)\psi_{n_2}(1)|^2 dV_1 dV_2 =$$

$$= e^2 \int \frac{|\psi_{n_1}(1)|^2 |\psi_{n_2}(2)|^2}{r_{12}} \, dV_1 dV_2 \pm$$

$$\pm e^2 \int \frac{\psi_{n_1}^*(1)\psi_{n_2}^*(2)\psi_{n_1}(2)\psi_{n_2}(1)}{r_{12}} \, dV_1 dV_2 \ ,$$

where the numerals 1 and 2 denote respectively the coordinates of the first and sec. ¹ electrons, and r_{12} is the distance between them. The signs + and − refer to the states of the particles which are respectively symmetric and antisymmetric with respect to exchange of the particles. In this formula the summation has been carried out over the spin variables, which give unity. Furthermore, we have made use of the obvious equality

$$\int \psi^*_{n_1}(1)\psi^*_{n_2}(2) \frac{e^2}{r_{12}} \psi_{n_1}(2)\psi_{n_2}(1)dV_1 dV_2 =$$

$$= \int \psi_{n_1}(1)\psi_{n_2}(2) \frac{e^2}{r_{12}} \psi^*_{n_1}(2)\psi^*_{n_2}(1)dV_1 dV_2 \ .$$

In this equality one integral goes over into the other when the integration indices 1 and 2 are interchanged.

Introducing the notation

$$C = \int |\psi_{n_1}(1)|^2 \frac{e^2}{r_{12}} |\psi_{n_2}(2)|^2 \ dV_1 dV_2 =$$

$$= \int |\psi_{n_1}(1)|^2 \ \hat{H}'(r_{12})|\psi_{n_2}(2)|^2 \ dV_1 dV_2 \ , \tag{67.2}$$

$$A = \int \psi^*_{n_1}(1)\psi^*_{n_2}(2) \frac{e^2}{r_{12}} \psi_{n_1}(2)\psi_{n_2}(1) \ dV_1 dV_2 =$$

$$= \int \{\psi^*_{n_1}(1)\psi^*_{n_2}(2) \ \hat{H}'(r_{12})\psi_{n_1}(2)\psi_{n_2}(1)\}dV_1 dV_2 \ , \tag{67.3}$$

we write the energy of interaction (67.1) in the form

$$E^{(1)}_{\uparrow\downarrow} = C + A \ , \tag{67.4}$$

$$E^{(1)}_{\uparrow\uparrow} = C - A \ . \tag{67.5}$$

The sign ↑↓ denotes antiparallel spins (spin singlet), while the sign ↑↑ denotes parallel spins (spin triplet).

It is clear from the derivation that the general form of formulae (67.4) and (67.5) obtained above is not specific, that is, it does not refer only to the case of the Coulomb interaction, but could be obtained for any interaction which depends on the coordinates of the particles.

It is interesting to compare this result with an analogous calculation for two different kinds of particles. We would then write for the non-symmetrized wave function $\Psi = \psi_{n_1}(1)\psi_{n_2}(2)$ and would correspondingly obtain

$$E' = \int |\psi_{n_1}|^2 |\psi_{n_2}|^2 \frac{e^2}{r_{12}} \, dV_1 dV_2 . \tag{67.6}$$

Formula (67.6) has a simple meaning. It represents the mean value of the energy of the Coulomb repulsion of two particles. The position of one of the particles in the state n_1 is characterized by the probability density $|\psi_{n_1}(1)|^2$, while the position of the other particles is characterized by $|\psi_{n_2}(2)|^2$.

In formulae (67.4) and (67.5) the integral C has a structure analogous to (67.6) and is often called the Coulomb integral. However, strictly speaking, it does not allow such an interpretation, since it cannot be indicated which one of the identical particles is in the state n_1 and which one is in the state n_2.

The integral A, usually called the exchange integral (in German Austausch, which means exchange) has no classical analogues. Calculations carried out for actual systems show that the integrals C and A are always positive. It follows directly from formulae (67.4) and (67.5) that the correction to the mean energy, determined by the interaction of the particles, depends on the orientation of their spins.

First of all we stress that it would be incorrect to assume that the interaction is made up of two parts, a classical part and an exchange part, as is often done for the sake of clarity. The contribution to the energy determined by the Coulomb integral C is called the classical part of the interaction, while the corresponding contribution of the exchange integral A is called the exchange part. In reality it is impossible to divide the interaction into the two parts, since the quantity A itself does not allow a classical interpretation.

The most characteristic part of the exchange interaction is expressed by the integral A (67.3). This integral can be treated as the matrix element corresponding to the transition of the first particle from the state n_2 into the state n_1, and to the transition of the second particle from the state n_1 to the state n_2. Indeed, we introduce the operator $\hat{P}_{12}$ defined by formula (64.2), which carries out the exchange of particles, so that

$$\hat{P}_{12}\psi_{n_1}(1)\psi_{n_2}(2) = \psi_{n_1}(2)\psi_{n_2}(1) ,$$
$$\hat{P}_{12}\psi_{n_2}(1)\psi_{n_1}(2) = \psi_{n_1}(1)\psi_{n_2}(2) .$$

Thus the operator $\hat{P}_{12}$ is the exchange operator of the first and second particles. By means of this operator the integral A can be written as

$$A = \int \psi_{n_1}^*(1)\psi_{n_2}^*(2)\hat{H}_{12}\psi_{n_1}(2)\psi_{n_2}(1) \, dV_1 dV_2 =$$
$$= \int \psi_{n_1}^*(1)\psi_{n_2}^*(2)\hat{H}_{12}\hat{P}_{12}\psi_{n_1}(1)\psi_{n_2}(2) \, dV_1 dV_2 . \tag{67.7}$$

Thus the exchange interaction corresponds to the replacement of the operator $\hat{H}_{12}$ by the operator $\hat{H}_{12}\hat{P}_{12}$. The total energy of interaction can be written by means of the exchange operator in the form

$$E' = C \pm A = \int \psi_{n_1}^*(1)\psi_{n_2}^*(2)\,(\hat{H}_{12}\pm\hat{H}_{12}\hat{P}_{12})\psi_{n_1}(1)\psi_{n_2}(2)\,\mathrm{d}V_1\mathrm{d}V_2 \ . \quad (67.7')$$

We see that the identity of quantum-mechanical particles essentially changes their interaction. If particles of different kinds possess an arbitrary interaction characterized by the operator $\hat{H}_{12}$, then in the case of identical particles the operator of this interaction must be replaced by $\hat{H}_{12} \pm \hat{H}_{12}\hat{P}_{12}$. This inference does not depend on the nature of the interaction, i.e. on the character of the operator $\hat{H}_{12}$. Thus the electrical interaction of two identical particles (for example, two positrons) differs from the electrical interaction of different particles (for example, a positron and a proton).

Thus the fact that one has identical particles whose state is characterized by symmetrized wave functions leads to a very important general consequence: the state of the system turns out to be dependent on the total spin of the system.

This fact is a quantitative expression of the qualitative considerations presented in the beginning of this section.

The dependence of the energy of a system of particles on the total spin is equivalent to the statement of the existence of an interaction between the particles. This interaction is called the exchange interaction.

The exchange interaction has a specific quantum character. This follows formally from the fact that in the classical limit the spin of the system reduces to zero (see §60). Hence in the transition to the classical limit any difference between states with different spin, in particular the difference in their energies, vanishes.

It should be stressed that although up to now we have dealt with particles with half-integer spin, the quantitative conclusion is equally applicable to particles with integer spin, i.e. bosons. In a system of two bosons having spin zero not all states obtained as a result of the formal solution of the corresponding Schrödinger equation are realized. Only states for which the wave function is symmetric in the particles correspond to physical states of a system and to definite values of its energy. In the case of two bosons with spin 1 the energy of the system also turns out to be dependent on the total spin.

The results obtained for a system consisting of two particles, fermions or bosons, are directly applicable to the general case of a system with an arbitrary number of identical particles.

Coming back to the example of the interaction of two electrons, we shall show that exchange forces allow the following obvious, although not rigorous, interpretation. We assume that at the instant of time $t = 0$ the first electron was in the state n_1 and the second in the state n_2. It should be stressed once more that in reality such a formulation refers to the instant of time $t = 0$ and that the reasoning to come serves only to clarify the effect of the exchange interaction. Then the initial wave function has the form

$$\Phi(0) \equiv \frac{1}{\sqrt{2}} \left[\psi_a(t{=}0) + \psi_s(t{=}0) \right] = \psi_{n_1}(1)\psi_{n_2}(2) .$$

The states described by the symmetric ψ_s and antisymmetric ψ_a wave functions are stationary states with energies respectively

$$E_s = E + C + A ,$$
$$E_a = E + C - A .$$

Hence the dependence of the wave functions ψ_s and ψ_a on time is given by the formulae

$$\psi_s = \psi_s(0)\, e^{(-i/\hbar)(E+C+A)t} ,$$
$$\psi_a = \psi_a(0)\, e^{(-i/\hbar)(E+C-A)t} .$$

The total wave function $\Phi(t)$ for $t > 0$ is their superposition and, consequently, does not describe a stationary state.

$$\Phi(t) = \frac{1}{\sqrt{2}} (\psi_s(t) + \psi_a(t)) =$$

$$= \tfrac{1}{2}\{[\psi_{n_1}(1)\psi_{n_2}(2) + \psi_{n_2}(1)\psi_{n_1}(2)]\, e^{(-i/\hbar)(E+C+A)t} +$$

$$+ [\psi_{n_1}(1)\psi_{n_2}(2) - \psi_{n_2}(1)\psi_{n_1}(2)]\, e^{(-i/\hbar)(E+C-A)t}\} = \qquad (67.8)$$

$$= \{\psi_{n_1}(1)\psi_{n_2}(2)\cosh^{-1} At - i\psi_{n_2}(1)\psi_{n_1}(2)\sin\hbar^{-1}At\}\, e^{(-i/\hbar)(E+C)t}.$$

Formula (67.8) shows that if at the instant of time $t = 0$ electron 1 was in the state n_1 and electron 2 in the state n_2, then after the lapse of a time interval

$$\tau = \frac{\pi\hbar}{2A} \qquad (67.9)$$

the electrons exchange states. The wave function

$$-i\psi_{n_2}(1)\psi_{n_1}(2)\, e^{(-i/\hbar)(E+C)t}$$

corresponds to the first electron being in the state n_2 and the second electron in the state n_1. After the lapse of a time inverval 2τ they come back into the initial states and so on. Thus the electrons exchange states with a period τ.

Such an exchange of states is often presented in a concrete way as follows. One of the electrons of a system, for example an atom, is emitted and is then absorbed by another atom. The latter, in its turn, emits an electron which is absorbed by the first atom. In the process of 'emission' and 'capture' of the electrons a change in the momentum of the corresponding atoms occurs. The change in the momentum of the atoms means that a certain interaction between them exists. This schematic and obvious consideration of the exchange interaction justifies the term 'exchange'. However, it should not be taken literally.

This is seen particularly clearly from the following reasoning. Let the states n_1 and n_2 correspond to the bound states of the electrons in two atoms. If we tried to understand the process described above literally, in the classical sense, then a contradiction would arise. Indeed, the electrons would not manage to exchange states or be 'emitted' and 'captured' by the atoms, because for this they would need to obtain from outside a certain energy in excess of their binding energy in the atoms. In reality each of the two atoms between which there is an exchange interaction is not in a state with definite energy. The uncertainty in the energy of the system ΔE is of the order of magnitude of $\Delta E \sim A$. It makes no sense to speak of the constancy of the energy during the time interval τ, the exchange time which is in order of magnitude equal to $\Delta E \Delta t \sim A t \sim \hbar$. During the time τ the system is not in a state with definite energy and momentum. The two electrons are in a state with the wave function $\Phi(t)$.

In this connection it is clear that it would be inadmissible to indicate the direction of the momentum of recoil of the atoms in the 'transfer' of the electrons and to try, proceeding from this, to determine the sign of the energy of interaction. Thus when speaking of the exchange of particles it should be kept in mind that this exchange has a virtual and not a real character. The word virtual means that only the initial and final states of the system have a direct meaning.

In order that the exchange integral A may have values different from zero the wave functions ψ_{n_1} and ψ_{n_2} must overlap sufficiently, i.e. must both be different from zero in one and the same region of space. If, on the contrary, the wave functions ψ_{n_1} and ψ_{n_2} are different from zero only in different regions of space, then the exchange integral reduces to zero. If, in particular, ψ_{n_1} and ψ_{n_2} are the wave functions describing the bound states of electrons in different atoms, then the exchange interaction is possible only when the atoms are in direct contact. Further, let the wave function correspond to two bound states in an atom. For example, n_1 is the normal state, and n_2 is one of the excited states. Then the value of the exchange integral decreases extremely rapidly in the transition to higher excited states, when the states n_1 and n_2 possess substantially different energies. Finally, when one speaks of the interaction of free particles which are described by plane waves, then the exchange integral differs from zero only for particles having similar values of the momenta. If, for example, the momenta of the particles differ appreciably, and the energy of interaction changes relatively slowly with coordinates, then A under the integral contains the product of a smoothly varying function and a rapidly oscillating function. The entire integral is then small. Thus an appreciable exchange interaction can take place only for identical particles which are in similar states, i.e. which are localized in a small region of space or have similar values of the energy and momentum.

The following important consequence results from this property of the exchange interaction; the exchange interaction possesses the property of saturation, so that in a system of a large number N of identical particles the total energy of the exchange interaction is proportional to the number of particles N. Indeed, two particles connected by the exchange interaction, for example two electrons with antiparallel spins, cannot themselves interact with a third particle.

If the ordinary energy of a pair interaction is proportional to the number of pairs, i.e. to $\frac{1}{2}N(N-1)$, then not all pairs are involved in the exchange interaction but only those which contain particles in 'similar' states (in the sense specified above). Hence the total number of particles connected by the exchange interaction is equal to the number of pairs made up of particles in similar states. It is obvious that this number of pairs is equal to $\frac{1}{2}N$.

In conclusion we point out the following fact. The derivation of the formulae for the energy of interaction was carried out with the assumption that the interaction operator does not contain any quantities which depend on the spin of the particles. However, one can also arrive at the same results in the case where the interaction operator contains spin operators.

The exchange interaction between identical particles plays a very important role in nature.

It is sufficient to point out that an exchange character is possessed by the forces to which homopolar chemical binding is due, the interaction which is responsible for the formation of crystals, the phenomenon of ferromagnetism and, finally, the interaction between particles in atomic nuclei, i.e. nuclear forces. We shall come back to the problem of chemical binding in §79, and in the meantime we shall dwell briefly on the problem of nuclear forces.

Up to now it has not been possible to construct a consistent theory of nuclear forces. The development of this theory is one of the major tasks of contemporary theoretical physics. At present the theory of nuclear forces has a semi-empirical character and is based on a number of experimental data. The totality of the available data has made it possible to establish the following properties of the nuclear interaction:

1. Experiments on the scattering of neutrons on protons show that very strong attractive forces exist between nuclear particles at distances from 1×10^{-13} cm to 2×10^{-13} cm. These forces decrease very rapidly with increasing distance and are not appreciable at distances larger than 2×10^{-13} cm. At very small distances, smaller than 1×10^{-13} cm, the attraction is replaced by a repulsion.

2. Nuclear forces turn out to be independent of the charge of the particles, i.e. the nuclear forces acting between two protons, a neutron and a proton and two neutrons are the same. The charge independence of nuclear forces follows from direct experiments on the scattering of fast neutrons and protons on protons, as well as from an analysis of the properties of the so-called mirror nuclei. Mirror nuclei are nuclei differing from each other by the interchange of neutrons and protons (nuclei with atomic numbers Z and $A - Z$, where A is the mass number, are mirror nuclei).

The identity of neutrons and protons in nuclear interactions points to a profound symmetry existing between these particles. The inequality of the masses and the presence of an electric charge on the proton are facts of relatively minor importance. Hence, according to the modern point of view, the proton and neutron should be considered as different charge states of one particle — the nucleon.

The nucleon has spin $\frac{1}{2}$ and in a given charge state obeys the Pauli exclusion principle. The nuclear interaction between nucleons is called the strong interaction (see §112 and §130).

3. The nucleon can be in two different charge states, the proton state and the neutron state, between which transitions are possible.

In free motion the proton state, which has a smaller mass and energy, is more stable. Hence the free neutron decays according to the scheme

$$n \to p + e + \bar{\nu},$$

where $\bar{\nu}$ is the antineutrino.

In atomic nuclei, where there is a nuclear interaction between the particles, the transformation of neutrons and protons into each other occurs (see below).

4. The presence of a charge on the proton entails two consequences: (1)

the proton state and the neutron state are different states of the nucleon; (2) besides the nuclear interaction there are the forces of Coulomb repulsion between two protons. These forces become important in the case of heavy nuclei, determining their instability.

5. The nuclear interaction depends not only on the distance but also on the mutual orientation of the spins of the interacting particles, as well as on the orientation of the spins with respect to the axis joining the two nucleons.

The dependence of the nuclear interaction on the orientation of the spins follows directly from experiments on the scattering of very slow neutrons on orthohydrogen and parahydrogen.

The existence of the dependence on the orientation of the spins with respect to the axis follows from an analysis of the properties of the deuteron, in particular from its possession of a quadrupole moment.

6. The nuclear interaction has exchange character. This fundamental conclusion follows, first of all, from the very fact of the stability of nuclei.

If the nuclear (strong) interaction depended only on the distance between the particles, then the potential energy of a system with mass number A would be proportional to A^2 – to the number of pairs which attract each other. However, the kinetic energy of a gas of Fermi particles confined in a given volume increases with the number of particles, according to (79.4) of Part III, as $A^{5/3}$.

Thus for a sufficiently large mass number the potential energy would turn out to be larger than the kinetic energy, the nucleus would have to contract and the particles would have to merge with each other. The volume of the nucleus would be a constant quantity which does not depend on A, and its binding energy would be proportional to A^2. In reality the data on scattering show that the volume of the nucleus increases in proportion to A, and that the binding energy is also proportional to A. This means that nuclear forces possess the property of saturation. Saturation, as we have seen above, is a characteristic property of exchange forces.

We shall dwell in somewhat more detail on the description of the modern concepts of the nature of nuclear forces.

The following simple picture of nuclear forces results from the assumption that the proton and neutron are different states of one particle and from the exchange character of the nuclear interaction: between two nucleons at very small distances there is a virtual exchange of a certain particle which is called the 'carrier' of the interaction. This exchange is in principle similar to the virtual exchange of the electron which was considered in detail above in the example of the exchange interaction.

It turned out (see below) that the particle responsible for the nucleon—nucleon interaction is the π-meson.

Three types of exchange are possible:

$$p \rightleftharpoons n + \pi^+ ,$$
$$n \rightleftharpoons p + \pi^- ,$$
$$\left. \begin{array}{l} p \rightleftharpoons p + \pi^0 , \\ n \rightleftharpoons n + \pi^0 . \end{array} \right\}$$

In the first two virtual processes the nucleon goes over from the proton state into the neutron state and vice versa; in the last virtual process the charge state of the nucleon does not change. The process of exchange of the charged meson can clearly be interpreted in the same way as the exchange of electrons: each of the nucleons spends a part of its time in the charged state and a part of its time in the neutral state. The exchange of π-mesons gives rise to the attraction between nucleons. We have emphasized the virtual character of the exchange, since an energy not less than $m_\pi c^2$, where m_π is the mass of the π-meson, would be necessary for the production of real π-mesons.

All π-mesons, positive, negative and neutral, should be considered as different charge states of one particle.

Further, it turns out that the masses of π-mesons cannot be arbitrary. They can be connected with the range of nuclear forces. Since nuclear forces do not depend on the electric charge and have a purely quantum nature, the range of the forces can depend only on the mass of the carrier particles m_π and the universal constants $\hbar$ and c.

From the above three quantities one can construct only one constant with the dimensionality of length; the Compton wavelength of the meson

$$R \sim \frac{\hbar}{m_\pi c}. \tag{67.10}$$

On the basis of the following reasoning one can ascribe an obvious meaning to expression (67.10). In the virtual exchange of the meson the energy of each of the nucleons must have an uncertainty $\Delta E \sim m_\pi c^2$. The exchange time τ must have an order of magnitude of $\tau \sim \hbar/m_\pi c^2$. If it is assumed that the meson moves with a velocity $\sim c$, then during the time τ it traverses the path

$$R \sim c\tau \sim \frac{\hbar}{m_\pi c}.$$

This distance is just the range of nuclear forces. Giving the range of nuclear forces, one can find the mass of the carrier particles

$$m_\pi \sim \frac{\hbar}{Rc} \approx 300 m_{\text{el}} \, ,$$

where m_{el} is the mass of the electron. This is of the same order of magnitude as the mass of the π-meson. We note that π-mesons were discovered experimentally after having been introduced in the theory as the hypothetical particles responsible for the strong nuclear interaction. The most convincing proof that π-mesons are the carriers of nuclear forces is the experimentally established extremely strong interaction of π-mesons with nucleons.

If the energy of a system of nucleons exceeds $m_\pi c^2$, then π-mesons can really be produced. The appearance of π-mesons was observed in collisions of fast nucleons (see §136) as well as in the action of γ-rays on nucleons. The reaction

$$p + \gamma \rightarrow n + \pi^+ \, ,$$

representing the elementary act of the nuclear photoeffect was observed. Later, in particular in §112, we shall come back to the problem of nuclear forces.

9

Applications of Quantum Mechanics to Atomic and Nuclear Systems

§68. The helium atom

The hydrogen atom, which was considered in detail in ch. 4, is the simplest one-electron system. Passing to the study of many-electron systems it is natural to investigate, first of all, the properties of the helium atom, in which two electrons revolve around the nucleus. We assume that the nucleus has an infinitely large mass. Hence, assuming the nucleus to be at rest, we write the Hamiltonian of the system of two electrons in the form

$$\left(-\frac{\hbar^2}{2m} \nabla_1^2 - \frac{\hbar^2}{2m} \nabla_2^2 - \frac{2e^2}{r_1} - \frac{2e^2}{r_2} + \frac{e^2}{r_{12}} \right) \psi = E\psi . \tag{68.1}$$

Here $\mathbf{r}_1$ and $\mathbf{r}_2$ are the radius vectors of the first and second electrons, and $\mathbf{r}_{12}$ is the distance between them. The third and fourth terms of (68.1) express the potential energy of the electrons in the field of the nucleus, and the last term expresses the energy of the Coulomb interaction between the electrons.

It should be noted that a number of approximations are associated with the representation of the Hamiltonian in such a form. Electrons possess a magnetic moment whose interaction is of a more complex character than the Coulomb interaction. Further, magnetic moments (the spin magnetic moment and the orbital magnetic moment) also interact with each other. However, we shall not study these effects, which have the character of small corrections, in detail.

Since the Hamiltonian of the system does not contain any spin operators, the solution of eq. (68.1) must be sought in the form of a product of two functions, one of which depends only on the coordinates and the other only on the spin

$$\psi = \Phi(\mathbf{r}_1,\mathbf{r}_2)\varphi(s_{z_1},s_{z_2}) . \tag{68.2}$$

In §66 it was shown that the spin function of two electrons is symmetric with respect to the exchange of two particles if the total spin of the system is equal to one, and that it is antisymmetric if the total spin is equal to zero. Thus it is seen that the states of the helium atom are divided into two groups. States with zero spin are called parastates, and those with unity spin are called orthostates. If the Hamiltonian (68.1) described the system exactly, then the three orthostates differing in the z-component of the spin would have the same energy. However, the weak interaction between the spin magnetic moment and the orbital magnetic moment removes the degeneracy and three close sub-levels arise. Thus the energy spectrum of helium consists of a set of singlet and triplet levels.

From general considerations it is easy to establish the group of states to which the ground state of helium belongs. As is known, the ground state of helium is described by a wave function having no nodes (see §10). It is obvious that this function cannot be an antisymmetric coordinate function, since the latter reduces to zero for $\mathbf{r}_1 = \mathbf{r}_2$.

Indeed, if $\Phi(\mathbf{r}_1,\mathbf{r}_2)$ is an antisymmetric function of the two variables $\mathbf{r}_1$ and $\mathbf{r}_2$, then it satisfies the relation

$$\Phi(\mathbf{r}_1,\mathbf{r}_2) = -\Phi(\mathbf{r}_2,\mathbf{r}_1) . \tag{68.3}$$

For $\mathbf{r}_1 = \mathbf{r}_2 = \mathbf{r}$ we have $\Phi(\mathbf{r},\mathbf{r}) = 0$.

Thus we see that in the normal state the wave function is symmetric in the coordinates and, consequently, is antisymmetric in the spins.

The normal state of helium is a parastate.

In eq. (68.1) the variables are not separable and it is impossible to obtain the exact solution. Therefore a number of approximate methods have been devised for its solution. The application of perturbation theory makes it possible to obtain the wave functions and, in a rather rough approximation, the energy of the ground state of helium.

We shall assume that the interaction between the electrons is the perturbation in eq. (68.1).

Then in the zero order approximation (68.1) can be written in the form

$$\left(-\frac{\hbar^2}{2m}\nabla_1^2 - \frac{\hbar^2}{2m}\nabla_2^2 - \frac{2e^2}{r_1} - \frac{2e^2}{r_2}\right)\Phi_0 = E_0\Phi_0 .\tag{68.4}$$

This equation can be solved by the method of separation of variables. For the normal state of the helium atom we have

$$\Phi_0 = \psi_1(\mathbf{r}_1)\psi_1(\mathbf{r}_2), \qquad E_0 = 2E_1 ,$$

where E_1 and ψ_1 denote respectively the energy and the wave function of the normal state of a hydrogen-like atom with charge $Z = 2$. As we have pointed out before, the function Φ_0 is symmetric in the coordinates of the electrons.

The energy level of the ground state in the first approximation of perturbation theory is given by the formula

$$E = E_0 + E^{(1)} ,$$

where the quantity $E^{(1)}$ in accordance with formula (53.12) is determined by the matrix element

$$E^{(1)} = \int |\psi_1(\mathbf{r}_1)|^2\, |\psi_1(\mathbf{r}_2)|^2 \frac{e^2}{r_{12}}\, dV_1 dV_2 .\tag{68.5}$$

In calculating higher energy levels of the helium atom the solution of the unperturbed wave equation can be written in the form

$$\Phi_0 = \psi_n(\mathbf{r}_1)\psi_m(\mathbf{r}_2), \qquad E_0 = E_n + E_m .\tag{68.6}$$

Here ψ_n and ψ_m are the wave functions of the hydrogen-like atom in the nth and mth states respectively. (We shall not take the degeneracies in the orbital and magnetic quantum numbers further into account in obtaining a qualitative picture.)

It is easily understood that the function

$$\Phi_0 = \psi_n(\mathbf{r}_2)\psi_m(\mathbf{r}_1), \qquad E_0 = E_n + E_m \tag{68.7}$$

will also be a solution of the unperturbed equation.

Thus a two-fold degeneracy arises. The two solutions (68.6) and (68.7) differ from each other by the exchange of electrons. For further calculations we shall need to have recourse to perturbation theory in the presence of degeneracy (see §54).

The correction to the energy E_0 in the first approximation is determined in this case from the condition that the determinant below be zero (see §54)

$$\begin{vmatrix} H'_{11} - E^{(1)} & H'_{12} \\ H'_{21} & H'_{22} - E^{(1)} \end{vmatrix} = 0 .$$

The quantities $H'_{11}, H'_{12}, H'_{21}$ and H'_{22} represent the matrix elements

$$H'_{11} = \int \psi_n^*(\mathbf{r}_1)\psi_m^*(\mathbf{r}_2)\frac{e^2}{r_{12}}\psi_n(\mathbf{r}_1)\psi_m(\mathbf{r}_2)\, dV_1 dV_2 \, ,$$

$$H'_{22} = \int \psi_m^*(\mathbf{r}_1)\psi_n^*(\mathbf{r}_2)\frac{e^2}{r_{12}}\psi_m(\mathbf{r}_1)\psi_n(\mathbf{r}_2)\, dV_1 dV_2 \, ,$$

$$H'_{12} = \int \psi_n^*(\mathbf{r}_1)\psi_m^*(\mathbf{r}_2)\frac{e^2}{r_{12}}\psi_n(\mathbf{r}_2)\psi_m(\mathbf{r}_1)\, dV_1 dV_2 \, ,$$

$$H'_{21} = \int \psi_m^*(\mathbf{r}_1)\psi_n^*(\mathbf{r}_2)\frac{e^2}{r_{12}}\psi_n(\mathbf{r}_1)\psi_m(\mathbf{r}_2)\, dV_1 dV_2 \, .$$

(68.8)

It is easily seen that $H'_{11} = H'_{22}$, $H'_{12} = H'_{21}$. Indeed, if $\mathbf{r}_1$ is replaced by $\mathbf{r}_2$ and $\mathbf{r}_2$ by $\mathbf{r}_1$ in the expression for the matrix element H'_{11}, then we obtain exactly the expression H'_{22}. An analogous statement also holds for the matrix elements H'_{12} and H'_{21}. Evaluating the determinant, we get

$$(H'_{11} - E^{(1)})^2 - H'^2_{12} = 0 \, .$$

Hence we find two values for the correction to the energy:

$$E_1^{(1)} = H'_{11} + H'_{12} \, , \tag{68.9}$$

$$E_2^{(1)} = H'_{11} - H'_{12} \, . \tag{68.10}$$

The expressions (68.9) and (68.10) are the same as the general formula found in §67. To the two values of the energy (68.9) and (68.10) there correspond two wave functions of the type $\psi = a\psi_n(\mathbf{r}_1)\psi_m(\mathbf{r}_2) + b\psi_n(\mathbf{r}_2)\psi_m(\mathbf{r}_1)$. The coefficients a and b are determined from the equations

$$(H'_{11} - E^{(1)})a + H'_{12}b = 0 \, , \qquad H'_{21}a + (H'_{22} - E^{(1)})b = 0 \, .$$

For the value of $E_1^{(1)}$ we obtain $a = b$, while in the case of $E_2^{(1)}$ we have $a = -b$. Thus under the action of the perturbation the degeneracy is removed and we obtain two different states

$$\Phi_0^{(s)} = a_1 [\psi_n(\mathbf{r}_1)\psi_m(\mathbf{r}_2) + \psi_n(\mathbf{r}_2)\psi_m(\mathbf{r}_1)] \, ,$$

$$\Phi_0^{(a)} = a_2 [\psi_n(\mathbf{r}_1)\psi_m(\mathbf{r}_2) - \psi_n(\mathbf{r}_2)\psi_m(\mathbf{r}_1)] \, .$$

(68.11)

where a_1 and a_2 are the normalization constants.

As is seen from formulae (68.11), the states $\Phi_0^{(s)}$ and $\Phi_0^{(a)}$ are respectively symmetric and antisymmetric in the coordinates of the electrons. In correspondence with the aforesaid, the function $\Phi_0^{(s)}$ describes the state of the helium atom with spin zero, while the wave function $\Phi_0^{(a)}$ corresponds to the state with total spin equal to one.

The wave function $\Phi_0^{(s)}$ corresponds to a parastate with zero spin and higher energy $E_1^{(1)}$. Correspondingly, $\Phi_0^{(a)}$ corresponds to an orthostate with unit spin and lower energy $E_2^{(1)}$.

The matrix elements $H'_{11} = H'_{22}$, as is seen from their definition, represent the Coulomb integral, while $H'_{12} = H'_{21}$ represent the exchange integral. All that was said in §67 applies to the system of two electrons in helium. For example, if electron 1 is in the ground state, while electron 2 is in an excited state, then after a lapse of time $\tau = \pi\hbar/2|H'_{12}|$ they exchange states.

The relations (68.9) and (68.10) still do not give a complete picture of the levels of the atom. Indeed, in the calculation carried out above we have not taken into account the degeneracy in the quantum number l of the levels of hydrogen-like atoms. The interaction between the electrons removes this degeneracy, and the levels turn out to be dependent not only on the principal quantum numbers but also on the orbital angular momentum.

We note that the method of calculation given above does not give high accuracy. The energy of the ground level of the helium atom obtained according to the above theory differs by about 20% from the value found experimentally. This strong disagreement is due to the choice of the perturbation, which is not sufficiently small.

§69. The variational principle

We have already seen that the two-electron problem, the helium atom, cannot be solved accurately and calls for the use of approximate methods.

This applies even more to complex atoms containing many electrons. Despite the complexity of many-electron atoms, effective approximate methods of solution allow one to get a very detailed idea of their properties. The effective approximate methods are to a large degree connected with the extremal properties of the Schrödinger equation. Namely, it turns out that the Schrödinger equation can be obtained from a variational principle.

We introduce the functional

$$J = \int \varphi^* \hat{H} \varphi \, dV , \qquad (69.1)$$

where the restriction

$$\int \varphi^* \varphi \, dV = 1 \qquad (69.2)$$

is imposed upon the function φ. In other respects φ remains an arbitrary complex function having the same dimensionality as the eigenfunctions of the operator $\hat{H}$. The minimum value of the functional J under the condition (69.2) can be found by Lagrange's method. In varying the complex function it is possible to vary φ and φ^* independently. For concreteness, let us vary φ^*. The variation of φ leads to the same result.

Clearly we have

$$\delta \int \varphi^* \hat{H} \varphi \, dV + E_0 \delta \int \varphi^* \varphi \, dV = 0 ,$$

where E_0 is the Lagrange multiplier. Hence we have

$$\int \delta \varphi^* (\hat{H} - E_0) \varphi \, dV = 0 , \qquad (69.3)$$

or in view of the arbitrariness of $\delta \varphi^*$,

$$(\hat{H} - E) \varphi = 0 . \qquad (69.4)$$

Thus, if $\varphi = \psi_0$, where ψ_0 is the normalized solution of the Schrödinger equation corresponding to the eigenvalue E_0 of the operator $\hat{H}$, then the functional J is equal to

$$J(\psi_0) = \int \psi_0^* \hat{H} \psi_0 \, dV = E_0 . \qquad (69.5)$$

We shall show that E_0 is the minimum eigenvalue of $\hat{H}$, i.e. the energy of the ground state. Let $\varphi = \psi_0 + \Sigma c_n \psi_n$. Then for J we find

$$J = \int (\psi_0^* + \Sigma c_n^* \psi_n^*) \hat{H} (\psi_0 + \Sigma c_n \psi_n) dV = E_0 + \Sigma |c_n|^2 E_n \geqslant E_0 . \quad (69.6)$$

The wave functions of the excited stationary states ψ_n must satisfy not only condition (69.2) but also the condition of orthogonality

$$\int \psi_0 \psi_n \, dV = 0 . \qquad (69.7)$$

They correspond to extrema but not the minimum of $J(\psi_n)$.

The variational properties of the Schrödinger equation are widely used for obtaining approximate solutions of it. Defining the form of a trial function on the basis of physical considerations or experimental data one seeks the minimum value of the integral $J(\varphi)$.

Let us consider as an example the harmonic oscillator. Choosing as the trial function the normalized function $\varphi = (2\alpha/\pi)^{1/4} e^{-\alpha x^2}$ we have

$$J(\varphi) = \left(\frac{2\alpha}{\pi}\right)^{\frac{1}{2}} \int e^{\alpha x^2} \left(-\frac{\hbar^2}{2m} \frac{d^2}{dx^2} + \frac{m\omega^2 x^2}{2}\right) e^{-\alpha x^2} \, dx = \frac{\hbar^2 \alpha}{2m} + \frac{m\omega^2}{8\alpha} .$$

The condition of minimum gives $\alpha = m\omega/2\hbar$. Hence

$$J_{min} = E_0 = \tfrac{1}{2}\hbar\omega$$

and

$$\varphi_{min} = \psi_0 = (m\omega/\hbar\pi)^{\frac{1}{4}} \exp(-m\omega x^2/2\hbar) .$$

The appropriate choice of the trial function φ led us to an accurate value of E_0 and ψ_0. Had we chosen another trial function, then we would have obtained another, although similar, value E_0' and ψ_0'. A shortcoming of the variational method is the fact that it gives an unpredictable error.

Other examples of the use of the variational principle will be given in the next section.

§70. The self-consistent field method (Hartree–Fock method)

For the calculation of many-electron systems wide use is made of the self-consistent field method which we have already encountered (see §41 of Part IV). The idea of the method (often called the Hartree–Fock method) is as follows. In the zero order approximation all the electrons are assumed to move independently of each other in the field of the nucleus. By means of the wave functions of the zero order approximation one finds the charge density and the mean electrostatic field produced by all the electrons.

In the next approximation each of the electrons is assumed to move in the field of the nucleus and the field produced by all the other electrons. The solution of the Schrödinger equation in this field gives the wave function in the first order approximation. Introducing the correction into the charge distribution and field distribution and solving the Schrödinger equation in the new field, one can find the correction of the second order approximation and so on.

To obtain the Schrödinger equation in the self-consistent field method we shall make use of the variational principle. To abbreviate the notation we shall carry out the calculations for the example of a two-electron system (the helium atom), confining ourselves to the calculation of the ground state. Therefore we shall not take into account the requirement of symmetrization of the wave function of the system of electrons. This will be done somewhat later. In the zero order approximation both electrons are described by identical real wave functions ψ, and the wave function of the atom has the form

$$\Psi = \psi_1\psi_2 . \tag{70.1}$$

The variational principle reads

$$\delta \int \Psi^*(\hat{H} - E)\Psi \, dV = \delta \int \psi_1 \psi_2 (\hat{H} - E)\psi_1 \psi_2 \, dV =$$
$$= \int \psi_2 \delta \psi_1 (\hat{H} - E)\psi_1 \psi_2 \, dV = 0 . \tag{70.2}$$

Hence

$$\int \psi_2 (\hat{H} - E)\psi_1 \psi_2 \, dV = 0 . \tag{70.3}$$

Substituting the value of H from (68.1) into (70.3), we obtain

$$\left\{ -\frac{\hbar^2}{2m} \nabla_1^2 + E_1 - \frac{2e^2}{r_1} + \int \psi_2^2 \frac{e^2}{r_{12}} \, dV_2 \right\} \psi_1 = 0 . \tag{70.4}$$

Here the additional term in the potential energy has the simple meaning

$$C(\mathbf{r}_1) = \int \psi_2^2 \frac{e^2}{r_{12}} \, dV_2 = \int \frac{e\rho(\mathbf{r}_2)}{r_{12}} \, dV_2 , \tag{70.5}$$

where $\rho(\mathbf{r}_2)$ is the charge density produced by the second electron. The same equation is obtained for ψ_2. The total energy of the atom turns out not to be equal to twice the value of E_1 but is given by the formula

$$E = 2E_1 - \int \frac{\rho_1 \rho_2}{r_{12}} \, dV_1 dV_2 .$$

Indeed, by definition

$$E = \int \Psi^* \hat{H} \Psi \, dV = \int \Psi^* \left[\hat{H}_1 + \hat{H}_2 + \frac{e^2}{r_{12}} - C(\mathbf{r}_1) - C(\mathbf{r}_2) \right] \Psi \, dV = 2E_1 - \bar{C} . \tag{70.6}$$

Here

$$E_1 = E_2 = \int \Psi^* \hat{H}_1 \Psi \, dV ,$$

where the operator $\hat{H}_1$ is equal to

$$\hat{H}_1 = -\frac{\hbar^2}{2m} \nabla_1^2 - \frac{2e^2}{r_1} + C(\mathbf{r}_1) .$$

The quantity $\bar{C}$ is equal to

$$\bar{C} = \int \frac{\rho_1 \rho_2}{r_{12}} \, dV .$$

It is obvious that $\bar{C}$ represents the mean energy of the electrostatic interaction between the electrons. To obtain the correct value of the energy E it is necessary to subtract $\bar{C}$ from $2E_1$, since this quantity is involved in the Hamiltonian of each of the electrons. In the case of a system of N electrons an analogous derivation gives for the ith electron in the nth quantum state

$$\left\{-\frac{\hbar^2}{2m} \nabla_1^2 + U(\mathbf{r}_i) + \sum e_i e_k \int \frac{|\psi_{nk}|^2 \, dV_k}{|\mathbf{r}_i - \mathbf{r}_k|}\right\} \psi_{ni} = E_n \psi_{ni} . \tag{70.7}$$

The structure of the general equation does not differ from that of equation (70.4). The complexity of the Schrödinger equations in the self-consistent field approximation is associated with the fact that the equation for ψ_i involves the wave functions of all the other electrons. Therefore, even in the simplest case of a two-electron system, one has to solve eq. (70.4) either by numerical or by approximate methods; for example the variational method. In this latter case it is natural to choose as the trial functions the hydrogen-like functions for a certain effective nuclear charge. The value of this charge is found from the condition of minimum of the integral (70.2). These calculations, as well as a summary of the numerical solutions, can be found in the book of Bethe and Salpeter*.

So far we have not taken into account the symmetry of the wave function. It is clear, however, that from the theoretical point of view the symmetrization of the wave function must be carried out from the very beginning of the calculation. For example, if no account is taken of the symmetry of the wave function no difference appears in the energy of orthohelium and para-helium.

The self-consistent field method taking account of the requirements of symmetry of the wave function is called the Hartree—Fock method. In the simplest case of two electrons all the preceding calculations can easily be carried out for the symmetrized wave function

$$\Psi(1,2) = \frac{1}{\sqrt{2}} [\psi_1(1)\psi_2(2) - \psi_2(1)\psi_1(2)] .$$

Substituting this expression into (70.2) we have to vary the wave functions ψ_1 and ψ_2 independently of each other.

Then instead of (70.3) we obtain

$$\sum_{i=1}^{2} \int\int dV_1 \, \delta\psi_i(1) dV_2 \, \{\psi_k(2)(\hat{H} - E)[\psi_k(2)\psi_i(1) \pm \psi_i(2)\psi_k(1) = 0 .$$

$$\tag{70.8}$$

* H.A.Bethe and E.E.Salpeter, *Quantum mechanics of one and two electron systems*, Handbuch der Physik, vol. 35 (Springer, Berlin, 1957).

In this case in (70.8) $i = 1$ for $k = 2$ and $i = 2$ for $k = 1$. In view of the arbitrariness of $\delta\psi_1$ and $\delta\psi_2$ we arrive at two equations. On substituting the total operator $\hat{H}$ from (68.1), these equations assume the form

$$\left\{-\frac{\hbar^2}{2m}\nabla_1^2 - E - \frac{2e^2}{r_{12}} + H_{22} + C_{22}\right\}\psi_1(\mathbf{r}) - [H_{12} + C_{12}]\psi_2(\mathbf{r}) = 0,$$

$$\tag{70.9}$$

$$\left\{-\frac{\hbar^2}{2m}\nabla_2^2 - E - \frac{2e^2}{r_{12}} + H_{11} + C_{11}\right\}\psi_2(\mathbf{r}) - [H_{12} + C_{12}]\psi_1(\mathbf{r}) = 0,$$

where

$$C_{12} = \int \psi_1(2)\psi_2(1)\frac{e^2}{r_{12}}\,dV_2, \qquad H_{ik} = \int \psi_i\left(-\frac{\hbar^2}{2m}\nabla_2^2 - \frac{2e^2}{r_{12}}\right)\psi_k\,dV.$$

By taking into account the symmetry of the wave function the number of unknown wave functions is doubled, and a system of simultaneous equations is obtained. The main difference between the Hartree–Fock equations and the Hartree equations consists in the appearance of exchange integrals, i.e. terms of the form C_{12}.

In the general case of many-electron atoms the wave function of the system of electrons which is to be substituted into the equation of the variational principle must be written in the form (65.6). We shall not give the cumbersome equations which are then obtained. Although in solving the Hartree–Fock equations numerically one has to carry out very laborious calculations, it is possible to find with a high degree of accuracy the energy of the ground and excited states, and the distribution of the charge and of the field for helium, as well as for a number of other atoms and ions. Naturally the number of numerical calculations necessary in integrating the Hartree–Fock equations increases rapidly with increasing number of electrons.

§71. The statistical model of the atom

For heavy atoms, when the calculation of the many-electron system according to the Hartree–Fock method becomes very time-consuming, a statistical method is widely adopted.

Let a system of a large number of electrons move in a spherically symmetric field $\varphi(r)$. By virtue of the Pauli principle a large fraction of these electrons will be in states with large quantum numbers. If the potential $\varphi(r)$

changes sufficiently slowly in space, then the electrons can be considered in the quasi-classical approximation. If, furthermore, the interaction between the electrons is sufficiently weak, then the whole set of electrons can be considered to be an ideal Fermi gas at absolute zero.

In a degenerate Fermi gas (see §79 of Part III) the electrons occupy quantum states in pairs, so that there is a phase-space cell of volume $(2\pi\hbar)^3$ per pair. In this case, all cells in momentum space with a momentum lying in the interval $0 \leqslant p \leqslant p_{max}$ are filled. The value p_{max} is easily expressed in terms of the electron gas density n (i.e. in terms of the mean number of electrons per unit volume). The number of electrons per unit volume with a given value of the momentum is evidently equal to

$$dn = 2 \frac{4\pi}{(2\pi\hbar)^3} p^2 \, dp \ .$$

Integrating from $p = 0$ to $p = p_{max}$ we have

$$p_{max}^3 = \frac{3}{8\pi} (2\pi\hbar)^3 n \ . \tag{71.1}$$

This formula allows the charge density to be expressed in terms of the momentum

$$\rho = \frac{8\pi e}{3(2\pi\hbar)^3} \, p_{max}^3 \ . \tag{71.2}$$

On the other hand, p_{max} can be related to the potential by means of the following simple reasoning. The energy of an electron bound in an atom, E, is always negative, i.e.

$$E = \frac{p^2}{2m} + e\varphi(r) \leqslant 0 \ .$$

We assume that the potential $\varphi(r)$ reduces to zero outside the atom. Hence for the maximum momentum compatible with the requirement $E = 0$ we find

$$p_{max} = [-2me\varphi(r)]^{\frac{1}{2}} \ .$$

Hence the electron charge density is connected with the potential by the relation

$$\rho = \frac{8\pi e(-2me)^{\frac{3}{2}} \varphi^{\frac{3}{2}}}{3(2\pi\hbar)^3} \ . \tag{71.4}$$

In the self-consistent field approximation one can write for the potential of the electrostatic field $\varphi(r)$ the Poisson equation

$$\nabla^2\varphi = -4\pi\rho \;,$$

or, taking into account the spherical symmetry of the atom,

$$\frac{1}{r}\frac{d^2(r\varphi)}{dr^2} = -\frac{32\pi^2 e(-2me)^{\frac{3}{2}}\varphi^{\frac{3}{2}}}{3(2\pi\hbar)^3} = \\ = -\frac{4e(-2me)^{\frac{3}{2}}\varphi^{\frac{3}{2}}}{3\pi\hbar^3} \;. \tag{71.5}$$

The equation obtained is called the Thomas–Fermi equation. To obtain the distribution of the potential $\varphi(r)$ it is necessary to supplement this equation with boundary conditions. Let us first consider the case of neutral atoms. Then one of the boundary conditions is $\varphi \to 0$ as $r \to \infty$. The second condition follows from the requirement that near the nucleus, when its charge is not screened by electrons, the field be a purely Coulomb field, i.e. that

$$\varphi(r) \to \frac{Ze}{r} \qquad \text{as} \qquad r \to 0 \;. \tag{71.6}$$

To obtain the solution of eq. (71.5) with boundary conditions (71.3) and (71.6) it is convenient to pass to dimensionless quantities, defining them by the relations

$$\chi = \frac{r\varphi}{Z|e|} \;, \qquad x = \frac{r}{d} \;,$$

where d is a constant quantity with the dimensionality of length. For χ we find the equation

$$\frac{d^2\chi}{dx^2} = \frac{|e|^3(2m)^{\frac{3}{2}}Z^{\frac{1}{2}}d^{\frac{3}{2}}}{3\pi\hbar^3}\frac{\chi^{\frac{3}{2}}}{x^{\frac{1}{2}}} \;.$$

Setting d equal to

$$d = \frac{1}{2}\left(\frac{9\pi^2}{16}\right)^{\frac{1}{3}}\frac{\hbar^2}{me^2}\frac{1}{Z^{\frac{1}{3}}} = 0.88\frac{a}{Z^{\frac{1}{3}}} \;, \tag{71.7}$$

where a is the radius of the Bohr orbit, we arrive at the equation

$$\frac{d^2\chi}{dx^2} = \frac{\chi^{\frac{3}{2}}}{x^{\frac{1}{2}}} \;. \tag{71.8}$$

In this case it is obvious that

$$\chi \to 1 \quad \text{as} \quad x \to 0,$$

$$\chi \to 0 \quad \text{as} \quad x \to \infty.$$

(71.9)

The integration of eq. (71.8) with boundary conditions (71.9) has been carried out numerically. Since the boundary value problem does not depend on the atomic number, the integration of this system allows one to find the universal distribution of the dimensionless potential in an atom.

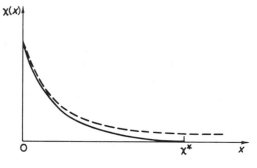

Fig. V.19

The behaviour of the function $\chi(x)$ for an atom is shown in fig. V.19 by the dotted line*. Since the function $\chi(x)$ for $x \to \infty$ only reduces to zero asymptotically, the potential, and also the electron density, nowhere reduces to zero. This means that in the approximation considered a finite value of the atomic radius cannot be found.

In fig. V.20 the curve of the radial electron density $D = 4\pi r^2 \rho(r)$ for the argon atom according to Thomas–Fermi (solid line) is compared with the result of the Hartree–Fock method (dotted line).

Fig. V.20 illustrates in an obvious way the merits and shortcomings of the Thomas–Fermi method. It does not give all the details of the behaviour of the electron density inside the atom, but it makes it possible to establish sufficiently accurately its general trend.

---

* The tabulated values of the function $\chi(x)$ can be found in the following books: L.D.Landau and E.M.Lifshitz, *Quantum mechanics* (Pergamon Press, Oxford, 1965) and P.Gombas, *Die statistische Theorie des Atoms und Ihre Anwendungen* (Springer Verlag, Wien, 1949).

In the outer parts of the atom, at a large distance from the nucleus, the electron density as calculated by the Thomas–Fermi method is overestimated.

The fact that the Thomas–Fermi method gives poor results for the peripheral regions of the atom follows from the conditions of its applicability (see below). The numerical calculation of the behaviour of the electron density with distance from the nucleus shows that one half of the total electron charge is contained in a sphere of radius $R \approx 1.33aZ^{-1/3}$.

Therefore, qualitatively, the quantity R can be considered to be the effective radius of the atom. It decreases with increasing Z.

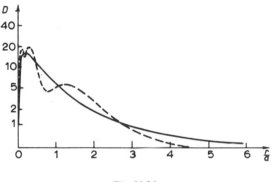

Fig. V.20

The total energy of all electrons in the atom is of the order of magnitude of the mean electrostatic energy of one electron $Ze^2/R \sim Z^{4/3}e^2/a$ multiplied by their total number Z, i.e. is of the order of $e^2 Z^{7/3} a$. These mean values, as well as all quantities referring to the properties of the inner regions of atoms, for example the structure of X-ray levels, are in good agreement with experimental data.

On the contrary, quantities which depend on the properties of the peripheral electrons, for example the ionization potentials of the atoms, cannot be determined satisfactorily by the Thomas–Fermi method. At the periphery of the atom the electron density is insufficiently large for the electrons to be considered as a degenerate electron gas.

The main merit of the Thomas–Fermi method is its simplicity. As an example we present an important result which also follows from calculations by the Hartree–Fock method, but which in that case requires very cumbersome calculations. The question is that of finding those values of the atomic number Z for which states with a given value of the orbital angular momentum begin to be occupied. If the electron moves with angular momentum l

in the self-consistent field $\varphi(r)$, then its effective potential energy can be given by formula (35.18). In the quasi-classical approximation $l(l+1)$ can be replaced by $(l+\frac{1}{2})^2$. We then have

$$U_{\text{eff}} = -|e|\varphi(r) + \frac{\hbar^2}{2m} \frac{(l+\frac{1}{2})^2}{r^2} ,$$

where $\varphi(r)$ is the potential found from the Thomas–Fermi equation. Since the total energy E is always negative, the total potential energy must be essentially negative, $U_{\text{eff}} < 0$, or

$$|e|\varphi(r)r^2 > \frac{\hbar^2}{2m}(l+\tfrac{1}{2})^2 . \tag{71.10}$$

Passing to the dimensionless quantities χ and x, we have instead of (71.10)

$$Z^{\frac{2}{3}} \frac{x\chi(x)}{(\frac{3}{4}\pi)^{\frac{2}{3}}} > (l+\tfrac{1}{2})^2 \tag{71.11}$$

From fig. V.19 it is seen that the quantity $x\chi(x)$ is limited and has a gently sloping maximum. For large x the potential $\chi(x)$ decreases more rapidly than x^{-1}; for $x \to 0$, $x\chi(x)$ is also equal to zero. Hence inequality (71.11) for given l is fulfilled only for a sufficiently large value of Z. This means that the curve U_{eff} lies entirely above the abscissa for a sufficiently small Z and goes below the axis for a sufficiently large Z. There cannot be states with $U_{\text{eff}} > 0$. Hence the limit of realizable states is determined by the condition of the curve U_{eff} being tangent to the abscissa, i.e. by the fulfillment of the conditions

$$U_{\text{eff}} = 0 , \qquad \frac{dU_{\text{eff}}}{dr} = 0 , \tag{71.12}$$

or

$$Z^{\frac{2}{3}}_{\text{crit}} x\chi(x) = \left(\frac{4}{3\pi}\right)^{\frac{2}{3}} (l+\tfrac{1}{2})^2 ,$$

$$Z^{\frac{2}{3}}_{\text{crit}} [x^2\chi'(x) - x\chi(x)] = -2 \left(\frac{4}{3\pi}\right)^{\frac{2}{3}} (l+\tfrac{1}{2})^2 .$$

To each value of l there corresponds a certain critical value of nuclear charge Z_{crit} for which the conditions (71.12) are fulfilled.

It is easy to eliminate χ' and χ from these equations, after which we find the relation between l and Z_{crit}

$$Z_{\text{crit}} = 0.155 (2l+1)^3 . \tag{71.13}$$

Setting $l = 1, 2, 3, \ldots$ in this formula and rounding off the result to the closest integer, we find the values Z_{crit} for which the states with the above angular momenta begin to be occupied. These values are respectively

$$Z_{crit} = 5, 21, 58, 124 .\tag{71.14}$$

In §73 it will be shown that this result is of great importance for understanding the properties of complex atoms.

Another important application of the Thomas–Fermi method is the study of the properties of positive ions. In this case it is to be expected that because of the predominance of the nuclear charge the electron shell will be compressed and the electron density will decrease so rapidly that one can introduce a finite radius of the electron shell, R^*. Outside the ion, for $r > R^*$, an electric field with the potential

$$\varphi = \frac{|e|Z(1-\sigma)}{r}, \qquad r > R^* ,$$

must exist, where the quantity $\sigma = |$charge of the shell$|/$charge of the nucleus is called the degree of ionization.

For $r = R^*$ the potential is equal to

$$\varphi_0 = \frac{Z|e|(1-\sigma)}{R^*} .$$

Correspondingly, the energy of an electron at the surface of the ion is equal to $e\varphi_0$.

The condition that the electron be bound in the ion assumes the form

$$E = \frac{p^2}{2m} + e\varphi \leqslant e\varphi_0$$

instead of $E \leqslant 0$ for the neutral atom. Correspondingly, the maximum momentum p_{max} is equal to

$$p_{max} = [2me(\varphi_0 - \varphi)]^{\frac{1}{2}}$$

and eq. (71.5) for the ion assumes the form

$$\frac{1}{r}\frac{d^2(r\varphi)}{dr^2} = -\frac{4e(2me)^{\frac{3}{2}} (\varphi_0 - \varphi)^{\frac{3}{2}}}{3\pi\hbar^3}$$

Its integration taking into account the boundary condition at the surface of the ion $\varphi = \varphi_0$ and condition (71.6) may be carried out numerically, as was done for the atom. The curve $\chi(x)$ for the rubidium ion is shown in fig. V.19 by the solid line. The curve $\chi(x)$ intersects the abscissa at the point $x^* = R^*/d$, where R^* is determined by the condition

$$4\pi \int_0^{R^*} \rho r^2 \, dr = -Ze\sigma .$$

Fig. V.21 shows the radial electron density distribution for the ion Rb^+ calculated by the Thomas–Fermi method and according to Hartree (dotted line). We see that in the peripheral part of the ion the agreement between the curves is better than for the atom.

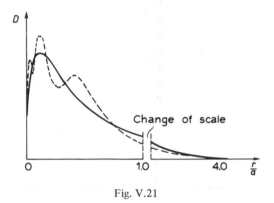

Fig. V.21

The limits of applicability of the Thomas–Fermi method are closely related to those of the quasi-classical approximation.

Upon substituting the expressions

$$U = e\varphi = \frac{Ze^2}{r} , \qquad p \approx p_{max} = (2me\varphi)^{\frac{1}{2}}$$

into formula (39.23) the criterion of applicability of the Thomas–Fermi method is given by the condition

$$r \gg \frac{\hbar^2}{Ze^2 m} \sim \frac{a}{Z} .$$

At large distances $r \sim a$ the quasiclassical approximation is invalid. Thus the Thomas–Fermi method is useful for r in the interval

$$a/Z \ll r \ll a . \tag{71.15}$$

§72. The quantum numbers characterizing the states of electrons in atoms

We now turn to a discussion of the properties of many-electron atoms. It is obvious that in such an atom (a system consisting of a nucleus and several electrons) the laws of conservation of total energy, total angular momentum, and the component of the angular momentum along an arbitrary axis must be fulfilled. By analogy with the theory of the hydrogen atom one can introduce quantum numbers defining the values of the conserved quantities. At first sight it seems that the quantum numbers must characterize the system as a whole, since, generally speaking, neither the energy nor the angular momentum of an individual electron is conserved. However, the self-consistent field method allows one to consider electrons as independent particles (the wave function of the system is the product of the wave functions of individual particles) in an external field. Each of the electrons moves in the self-consistent spherically symmetric field of the nucleus and of other electrons. Since for motion in a spherically symmetric field the energy, the angular momentum and a component of it along an arbitrary axis are conserved, then not only the atom as a whole but also an individual electron can be characterized by quantum numbers n, l, m. The self-consistent field of the atom is not a Coulomb field, hence the energy levels will depend on l as well as on n. The energy of the electron does not depend on the orientation of the angular momentum in space and, consequently, cannot depend on the quantum number m.

Thus we see that in order to characterize the state of an atom it is necessary to indicate the state of each atomic electron. States with angular momentum $l = 0, 1, 2, 3, ...$ are denoted respectively by s, p, d, f and so on. The principal quantum number is indicated in the form of a number standing in front of the letter. For example, the notation 5f indicates that in the given state the electron is characterized by the quantum number $n = 5$ and has orbital angular momentum $l = 3$. If several electrons are in a state with the same numbers n and l, then for simplicity their number is indicated in the form of superscript. For example, the normal state of nitrogen is characterized by $1s^2 2s^2 2p^3$. This means that two electrons have the quantum numbers $n = 1, l = 0$; two other electrons are in the state $n = 2, l = 0$ and, finally, three electrons are in the 2p-state. However, this information is insufficient for the complete description of the state of the atom, because it does not tell us how the orbital and spin angular momenta of the individual electrons are combined and what the total angular momentum of the atom is.

We have already mentioned that the total angular momentum of the atom is conserved in time and that because of this it can characterize stationary states. Furthermore, it can be assumed (neglecting the weak interaction) that the total spin and the total orbital angular momentum of the system are separately conserved. It is just these three quantities which are chosen to characterize the system as a whole. The symbols for the states of atoms with different total angular momenta L are introduced by analogy with the symbols for individual electrons. Namely, for $L = 0, L = 1, L = 2, L = 3$ and so on the states are called respectively S-state, P-state, D-state, F-state and so on. The value of the total angular momentum J is indicated in the form of a subscript on the right of the symbol of the orbital angular momentum. For example, $P_{3/2}$ means that the atom is in a state with orbital angular momentum $L = 1$ and total angular momentum $J = 3/2$. Usually a quantity equal to $2S+1$, where S is the total spin of the atom, is also indicated. The value of $2S+1$ is indicated in the form of a superscript on the left of L. The quantity $2S+1$ for $L \geqslant S$ shows the number of close levels of the atom constituting its fine structure. Indeed, from the rule of addition of angular momenta it follows that if $L \geqslant S$, then only $2S+1$ different states may arise when the orbital angular momentum and the spin angular momentum are combined to obtain the total angular momentum of the system.

It turns out that these $2S+1$ states have close but different energies. In other words, $2S+1$ levels form a multiplet. The difference in the energies of the components of the multiplet is associated with the so-called spin–orbit interaction. This is the interaction dependent on the mutual orientation of the orbital and spin angular momentum vectors.

In §118 it will be shown that the relativistic equation for the electron (the Dirac equation) enables one to calculate the spin–orbit interaction.

If, however, one allows for the very fact of the existence of this interaction, i.e. if one assumes that the orientations of the orbital and spin angular momentum vectors are not independent, then the law of interaction can be established from very general considerations. The spin–orbit interaction operator must be a scalar made up of the vectors $\mathbf{L}$ and $\mathbf{S}$. The only scalar combination is the quantity $\mathbf{L \cdot S}$. For the mean energy of the spin–orbit interaction we then obtain

$$E_{\mathbf{S-L}} = A\overline{\mathbf{L} \cdot \mathbf{S}}$$

where the coefficient A can be either positive or negative. The mean value of $\mathbf{L \cdot S}$ will be calculated in §74. Formula (74.4) gives

$$E_{\mathbf{S-L}} = A'[J(J+1) - L(L+1) - S(S+1)]$$

For different levels belonging to a given multiplet the quantities L and S do not change. Hence for the multiplet splitting we obtain

$$\Delta E_{S-L} = A'J(J+1) .$$

The spacing between the neighbouring components of a multiplet is

$$\Delta E = E_{S-L}^{(J)} - E_{S-L}^{(J-1)} = A'J .$$

If $A' > 0$, then the lowest level in the multiplet is the level with the lowest possible value of J (i.e. $J = |L - S|$). These are the so-called normal multiplets. This case is realized for those atoms whose open shell is more than half filled. Otherwise it turns out that $A' < 0$. These are the so-called inverted multiplets, in which the lowest level has the largest total angular momentum J (i.e. $J = L + S$).

The absolute value of multiplet splitting is proportional to Z^2 and rapidly increases in going to heavy atoms.

It is often important to know the total number of possible states of an tom when the quantum numbers n and l of each electron are given. For this purpose it is convenient to use the concept of equivalent electrons which was irst introduced by Pauli.

Equivalent electrons are those electrons which have the same quantum numbers n and l.

Where the electrons are not equivalent the calculation of the possible terms is extremely simple.

Let us consider as an example two electrons in a state with $n = 3$, $l = 2$ and $n = 2$, $l = 1$. On the basis of the rule of addition of angular momenta the orbital angular momentum of this system can take on the values $L = 1, 2, 3$, while the total spin of the system can assume two values $S = 0, 1$. Thus we have the terms ^1P, ^3P, ^1D, ^3D, ^1F, ^3F. However for equivalent electrons one has to take into account the Pauli principle in calculating possible terms, and this makes the calculations somewhat more complicated. We consider first the following simple example. Let two electrons be in a state $n_1 = n_2$, and $l_1 = 0$, $l_2 = 0$. In this case the components of the angular momenta in an arbitrary direction are also equal to zero, i.e. $m_1 = 0$, $m_2 = 0$. To satisfy the Pauli principle, s_{z_1} and s_{z_2} must have opposite signs. Consequently, we may have, for example, $s_{z_1} = \frac{1}{2}$, $s_{z_2} = -\frac{1}{2}$. But in correspondence with the principle of identity $s_{z_1} = -\frac{1}{2}$, $s_{z_2} = \frac{1}{2}$ also represents the same state.

The states $s_{z_1} = \frac{1}{2}$ and $s_{z_2} = \frac{1}{2}$ are forbidden by the Pauli principle. Hence only the term ^1S can be realized. The term ^3S is forbidden. This calculation shows that He, as well as Be, Mg and Ca and analogous elements cannot have a triplet ground level. We note here that historically Pauli arrived at the exclusion principle by investigating atomic spectra. The exclusion principle

was discovered as a result of the necessity to account for the absence of
certain terms.

For what follows we shall need one more example. Let a system of two
electrons have the quantum numbers $n_1 = n_2$; $l_1 = l_2 = 1$. In this case each
electron can be in the following states:

1) $m = 1, s_z = \frac{1}{2}$;   2) $m = 0, s_z = \frac{1}{2}$;   3) $m = -1, s_z = \frac{1}{2}$;

4) $m = 1, s_z = -\frac{1}{2}$;   5) $m = 0, s_z = -\frac{1}{2}$;   6) $m = -1, s_z = -\frac{1}{2}$.

In calculating the possible states of the whole system one can combine only
different states of individual electrons. This must be done so as not to violate
the Pauli principle. According to the rule of addition of angular momenta we
obtain the following possible states with $M_z = m_1 + m_2$; $S_z = s_{z_1} + s_{z_2}$:

1) $M_z = 2, S_z = 0$;   2) $M_z = 1, S_z = 1$;   3) $M_z = 1, S_z = 0$;

4) $M_z = 0, S_z = 0$;   5) $M_z = 0, S_z = 0$;   6) $M_z = 0, S_z = 1$;

7) $M_z = 1, S_z = 0$;   8) $M_z = 0, S_z = 0$.

We have not written down analogous states having negative values of the component M_z.

In analyzing the results one has to begin with the state with the largest
component M_z. In the case given we have the state with $M_z = 2$, $S_z = 0$.
Hence we conclude that there must be a ^{1}D term (to which there correspond
also states with $M_z = 1, S_z = 0; M_z = 0, S_z = 0$).

After eliminating the states numbered 1), 3) and 4) from the table we
again choose the state with the largest component M_z. In this case $M_z = 1$,
$S_z = 1$. The term ^{3}P corresponds to this state and also to the states denoted
by 2), 6), 3) and 4). Finally, only the state $M_z = 0, S_z = 0$, corresponding to
the term ^{1}S, remains in the table.

Thus we see that a system of two equivalent electrons with $l_1 = 1, l_2 = 1$
can be in the states ^{1}D, ^{3}P and ^{1}S. In computing states in more complex
cases one has to proceed in an analogous way.

We now discuss some general regularities in the ordering of the energy
levels of an atom. If the electrons are in states with definite numbers n and l
(in such cases one says that the electron configuration is given), then to such
a distribution there may correspond several different energy levels differing
in the total orbital angular momentum as well as in the total spin of the
system.

Taking into account the multiplet splitting, the state of the atom turns out
to depend on the quantities J, L and S.

The state of the atom as a whole, represented by the atomic term, is determined by these quantities. The symbol for the term is $^{2S+1}L_J$. For example, the normal term of the nitrogen atom ($L = 0$, $S = \frac{3}{2}$, $J = \frac{3}{2}$) is written as $^4S_{3/2}$.

The ordering of terms of different multiplicity was obtained from calculations carried out by the Hartree–Fock method (although historically it was established much earlier by Hund).

It turns out that of all the terms of a given configuration the lowest energy is possessed by the term with the largest value of the total spin S.

For a given S the term with the largest value of L has the lowest energy.

In the example just given of the terms ^1D, ^1S and ^3P the order of the terms in increasing energy will be ^3P, ^1D and ^1S. As to the ordering of levels within a given multiplet, there are two cases. The first of these, called the multiplet with normal structure, is characterized by an increase in the energy of the levels with increasing total angular momentum L. In the second case, where the energy of the levels decreases with increasing L, the structure of the multiplet is said to be inverted. It turns out that if the number of equivalent electrons in the atom or ion is lower than the total number of electrons, then the multiplets have a normal structure. In atoms or ions in which the number of equivalent electrons is larger than or equal to one half of the total number of electrons the multiplets are inverted. For example, in the oxygen atom of the eight electrons four are in the 2p-state (structure $2p^4$) and are equivalent. Hence in the case of the oxygen atom the multiplets are inverted. In the oxygen ion O^{2-} there are 2 electrons in the 2p-state (configuration $2p^2$), and the multiplets have a normal structure.

We now introduce the concept of an atomic shell. This is the set of all the electron states with the same values of the quantum numbers n and l. If all states with the quantum numbers n and l are occupied, then the corresponding shell is closed. It is known that for given n and l there are in all $2l+1$ different states differing in the quantum number m. If the spin is also taken into account, then the total number of electrons necessary to fill the shell will be equal to $2(2l+1)$. If the shell is closed, then the total spin of the system, as well as the components of the orbital angular momentum, must be equal to zero. In this case $S = 0$, $L = 0$, $J = 0$. This can be shown by taking into account the Pauli principle and recalling that in a completely filled shell all possible states with positive as well as negative projections of the angular momentum onto the z-axis are occupied. The term $^1S_0$ corresponds to a closed shell.

We stress that our preceding considerations were based on the assumption that the orbital angular momenta of the electrons were combined into a total orbital angular momentum of the system, and that the spin angular momenta were combined into a total spin angular momentum of the system. Such an assumption corresponds to the statement that the interaction between the spin and orbital motions of electrons is much weaker than the interaction between the spins. Then one can speak of approximate conservation of the total orbital angular momentum and of the total spin of the system. This type of interaction is called normal or Russell–Saunders coupling. On the basis of the assumption of normal coupling it turns out to be possible to systematize the lowest energy levels of most atoms. Deviations from normal coupling are observed in the seventh and eighth groups of the periodic system of the elements.

In principle another limiting form of coupling, usually called jj coupling, is also possible. In jj coupling the orbital angular momentum and the spin angular momentum of each electron are added to give the total angular momentum j of that electron (in this case the orbital angular momentum of an individual electron is not conserved). In their turn the total angular momenta of the individual electrons are added together to give the total angular momentum of the atom, J. Such coupling is not encountered in atoms in its pure form.

Let us consider some examples of different modifications of the basic forms of coupling in atoms. If an electron is in a highly excited state and, consequently, is sufficiently far from the nucleus and other electrons of the atom, then the behaviour of this electron can be considered to be independent of the rest of the atom. In this case the total angular momentum of the individual electron can be considered to be conserved independently of the total angular momentum of the rest of the atom.

Consider another example. For atoms with a large charge Z the inner electrons interact strongly with the nuclear charge and relatively weakly with the outer atomic electrons. Hence one can assume approximately that the inner electrons do not interact with the other electrons (the total angular momentum of such electrons is conserved). In this case one can speak of jj coupling. We note that such electrons must be characterized not by the quantum numbers n and l but by the quantum numbers n and j.

For use in what follows we shall now show that the electric dipole moment of an atom

$$\mathbf{d} = \sum_{i=1}^{N} \int e_i \mathbf{r}_i |\psi(\mathbf{r}_1, \mathbf{r}_2, ..., \mathbf{r}_i, ..., \mathbf{r}_n)|^2 \, dV_1 dV_2 ... dV_n \ , \qquad (72.1)$$

in a stationary state with definite parity is equal to zero. Indeed, since the parity operator commutes with the Hamiltonian, the wave function ψ is an eigenfunction of the parity operator. In other words, it satisfies the relation

$$\psi(\mathbf{r}_1,\mathbf{r}_2,...,\mathbf{r}_n) = -\psi(-\mathbf{r}_1,-\mathbf{r}_2,...,-\mathbf{r}_n)$$

or

$$\psi(\mathbf{r}_1,\mathbf{r}_2,...,\mathbf{r}_n) = \psi(-\mathbf{r}_1,-\mathbf{r}_2,...,-\mathbf{r}_n) \ .$$

In both cases the function $|\psi|^2$ is an even function. It is now obvious that the integrand in (72.1) is odd and, consequently, the dipole moment of an atom is equal to zero.

§73. The periodic system of the elements

In its time the theoretical construction of the Mendeleyev periodic system, carried out by Bohr in 1922, was one of the most effective results obtained by means of the quantum theory.

The construction of the periodic system of the elements is based on three assumptions:

(1) The structure of atoms is determined by the atomic number Z (the charge of the nucleus).

(2) As the atomic number and the number of electrons in the atom increase the electrons fill the states with the lowest possible energy.

(3) The occupation of energy states is limited by the Pauli principle.

In §72 we have defined the term atomic shell. A closed shell contains $2(2l+1)$ electrons. The energy of an atom depends only on the quantum numbers n and l. Thus all $2(2l+1)$ electrons in a shell have the same energy (if we do not take into account the weak spin—orbit interaction).

The set of sub-shells with fixed principal quantum number n is called a shell. The number of electrons filling a shell is equal to

$$2 \sum_{l=0}^{n-1} (2l+1) = 2n^2 \ .$$

Each shell is denoted by letters taken from the classification adopted in X-ray spectroscopy. That is, as follows:

n	1	2	3	4	5
Symbol of the shell	K	L	M	N	O
Possible number of electrons in the shell	2	8	18	32	50

In contrast to the hydrogen atom, in other atomic systems, the energy of states is defined by both the principal quantum number n and the orbital number l. The dependence of the energy on n is, generally speaking, stronger than the dependence on l. This means that for all values of l states with a given value of n lie lower than states with the quantum number $n + 1$. The sequence of energy states is

$$1s, \ 2s, \ 2p, \ 3s, \ 3p, \ \dots \ .$$

However, in going to d-states and particularly to f-states the situation changes. For large values of the angular momentum the dependence of the energy on the orbital quantum number l turns out to be most important.

The effective potential energy of the electrons arises from the Coulomb field of the nucleus screened by electrons, and the centrifugal force. The screened potential of the nucleus decreases at large distances substantially more slowly than the Coulomb potential and still more slowly than the centrifugal potential.

Comparison between the total effective energy of electrons with small angular momenta (s- and p-states) and large angular momenta (d- and f-states) shows that there is an essential difference between them. Namely, the curve U_{eff} for $l = 2$ and 3 lies higher than for states with $l = 0$ and 1.

Because of this the minimum of the energy for d- and f-states lies closer to the nucleus than for s- and p-states. This means that on the average d-electrons and particularly f-electrons move closer to the nucleus, in deeper parts of the shell, than s-electrons and p-electrons; d-electrons and f-electrons are often said to be penetrating. This general property of states with large angular momenta leads to the fact that 3d-electrons on the average move closer to the nucleus than 4s-electrons.

On the other hand, the energy in the screened field increases with increasing angular momentum. Experiment and calculations by the Hartree—Fock method show that the energy of the 4s-state lies below the energy of the 3d-state. Hence the order of filling states lying above 3p turns out to be as follows:

$$4s, \ 3d, \ 4p, \ 5s, \ 4d, \ 5p, \ 6s, \ 4f, \ 5d, \ 6p, \ 7s, \ 6d, \ 5f \ .$$

Elements in which the 3d-shell and particularly the 4f-shell and the 5f-shell are partly filled possess special properties. Since the motion in a non-Coulomb field penetrates closer to the nucleus for states with large angular momenta, the addition of electrons to the d-shell and particularly to the f-shell does not change those properties of atoms which depend on the peripheral electrons.

Let us analyze in more detail the order in which states are occupied. This will allow us to find out which properties of atoms should display a periodic trend and which a monotonic trend with increasing atomic number Z.

The first element of the periodic system is hydrogen. Its normal term is $^2S_{1/2}$.

In the next element (helium) the K-shell having two electrons is filled. It is easy to find that, in accordance with the rules discussed in the preceding section, the normal term of helium is $^1S_0$.

The building of the L-shell begins in the next element (lithium). The third electron of lithium goes into a 2s-state.

In calculating the normal term one need not take into account the electrons of the filled shell: their spin orbital and total angular momenta are equal to zero.

The normal term of lithium is defined by the single electron of the L-shell. Lithium is in the state $^2S_{1/2}$.

In beryllium the fourth electron fills the 2s-shell. The normal term of beryllium is $^1S_0$, as for helium.

The fifth electron of boron goes into the 2p-state. Thus the boron atom has the following distribution of electrons: $1s^2 2s^2 2p$. Since the 1s and 2s shells are filled, the normal term of the boron atom is easily found. It is $^2P_{1/2}$.

In the case of carbon, six electrons are distributed as follows: $1s^2 2s^2 2p^2$. In order to determine the normal term of carbon, we turn to the example discussed in §72 of the determination of terms for two equivalent p-electrons. Making use of the Hund rule, we see that the normal term of carbon is ^3P. In this case the atom contains less than one half of all possible equivalent p-electrons. Hence the multiplet structure of the lower level corresponds to a minimum J, in the given case $J = 0$. Thus, finally, for the normal term we have the symbol $^3P_0$.

The order of further filling the terms of normal states is shown in table 2. We see that in neon the L-shell is complete and that, like helium, it has no electrons in unfilled shells and sub-shells.

It is natural to identity filled shells with the periods of the Mendeleyev system of the elements. Each period begins to be filled by one electron in an s-state and is completed when a filled shell is formed.

The first period of the periodic system of the elements contains elements in which the K-shell is filled. It comprises two elements ($n=1, l=0$). The second

Table 2

The distribution of electrons in the periodic system of elements

Element		K	L		M			N		Normal term	Ionization potential (eV)
		1s	2s	2p	3s	3p	3d	4s	4p		
H	1	1	—	—	—	—	—	—	—	$^2S_{1/2}$	13.595
He	2	2		—	—	—	—	—	—	$^1S_0$	24.58
Li	3	2	1	—	—	—	—	—	—	$^2S_{1/2}$	5.39
Be	4	2	2	—	—	—	—	—	—	$^1S_0$	9.32
B	5	2	2	1	—	—	—	—	—	$^2P_{1/2}$	8.30
C	6	2	2	2	—	—	—	—	—	$^3P_0$	11.26
N	7	2	2	3	—	—	—	—	—	$^4S_{3/2}$	14.53
O	8	2	2	4	—	—	—	—	—	$^3P_2$	13.61
F	9	2	2	5	—	—	—	—	—	$^2P_{3/2}$	17.42
Ne	10	2	2	6	—	—	—	—	—	$^1S_0$	21.56
Na	11				1	—	—	—	—	$^2S_{1/2}$	5.14
Mg	12				2	—	—	—	—	$^1S_0$	7.64
Al	13				2	1	—	—	—	$^2P_{1/2}$	5.98
Si	14		Neon		2	2	—	—	—	$^3P_0$	8.15
P	15		configuration		2	3	—	—	—	$^4S_{3/2}$	10.48
S	16				2	4	—	—	—	$^3P_2$	10.36
Cl	17				2	5	—	—	—	$^2P_{3/2}$	13.01
Ar	18				2	6	—	—	—	$^1S_0$	15.76
K	19						—	1	—	$^2S_{1/2}$	4.34
Ca	20						—	2	—	$^1S_0$	6.11
Sc	21						1	2	—	$^2D_{3/2}$	6.54
Ti	22		Argon				2	2	—	$^3F_2$	6.82
V	23		configuration				3	2	—	$^4F_{3/2}$	6.74
Cr	24						5	1	—	$^7S_3$	6.76
Mn	25						5	2	—	$^6S_{5/2}$	7.43
Fe	26						6	2	—	$^5D_4$	7.90
Co	27						7	2	—	$^4F_{9/2}$	7.86
Ni	28						8	2	—	$^3F_4$	7.63
Cu	29						10	1	—	$^2S_{1/2}$	7.72
Zn	30						10	2	—	$^1S_0$	9.39
Ga	31		Argon				10	2	1	$^2P_{1/2}$	6.00
Ge	32		configuration				10	2	2	$^3P_0$	7.88
As	33						10	2	3	$^4S_{3/2}$	9.81
Se	34						10	2	4	$^3P_2$	9.75
Br	35						10	2	5	$^2P_{3/2}$	11.84
Kr	36						10	2	6	$^1S_0$	14.00

Table 2 (continued)

Element		Configuration of inner shells	N		O			P	Normal term	Ionization potential (eV)
			4d	4f	5s	5p	5d	6s		
Rb	37		–	–	1	–	–	–	$^2S_{1/2}$	4.19
Sr	38		–	–	2	–	–	–	$^1S_0$	5.69
Y	39		1	–	2	–	–	–	$^2D_{3/2}$	6.38
Zr	40		2	–	2	–	–	–	$^3F_2$	6.84
Nb	41	Krypton configuration	4	–	1	–	–	–	$^6D_{1/2}$	6.88
Mo	42		5	–	1	–	–	–	$^7S_3$	7.10
Tc	43		5	–	2	–	–	–	$^6S_{5/2}$	7.28
Ru	44		7	–	1	–	–	–	$^5F_5$	7.36
Rh	45		8	–	1	–	–	–	$^4F_{9/2}$	7.46
Pd	46		10	–	–	–	–	–	$^1S_0$	8.33
Ag	47			–	1	–	–	–	$^2S_{1/2}$	7.57
Cd	48			–	2	–	–	–	$^1S_0$	8.99
In	49			–	2	1	–	–	$^2P_{1/2}$	5.78
Sn	50	Paladium configuration		–	2	2	–	–	$^3P_0$	7.34
Sb	51			–	2	3	–	–	$^4S_{3/2}$	8.64
Te	52			–	2	4	–	–	$^3P_2$	9.01
I	53			–	2	5	–	–	$^2P_{3/2}$	10.45
Xe	54			–	2	6	–	–	$^1S_0$	12.13
Cs	55			–			–	1	$^2S_{1/2}$	3.89
Ba	56			–			–	2	$^1S_0$	5.21
La	57			–			1	2	$^2D_{3/2}$	5.61
Ce	58			2			–	2	$^3H_4$	6.91
Pr	59			3			–	2	$^4I_{9/2}$	5.76
Nd	60			4			–	2	$^5I_4$	6.31
Pm	61			5			–	2	$^6H_{5/2}$	6.30
Sm	62			6	The 5s		–	2	$^7F_0$	5.10
Eu	63	The sub-shells from 1s to 4d contain 46 electrons		7	and 5p		–	2	$^8S_{7/2}$	5.67
Gd	64			7	sub-shells		1	2	$^9D_2$	11.40
Tb	65			8	together		1	2	$^8H_{17/2}$	6.74
Dy	66			10	contain		–	2	$^5I_8$	6.82
Ho	67			11	8 electrons		–	2	$^4I_{15/2}$	6.90
Er	68			12			–	2	$^3H_6$	6.90
Tm	69			13			–	2	$^2F_{7/2}$	6.90
Yb	70			14			–	2	$^1S_0$	6.20
Lu	71			14			1	2	$^2D_{3/2}$	5.00

Table 2 (continued)

Element		Configuration of inner shells	O		P			Q	Normal term	Ionization potential (eV)
			5d	5f	6s	6p	6d	7s		
Hf	72		2	–	2	–	–	–	$^3F_2$	7.00
Ta	73		3	–	2	–	–	–	$^4F_{3/2}$	7.88
W	74	The sub-shells from 1s to 5p contain 68 electrons	4	–	2	–	–	–	$^5D_0$	7.98
Re	75		5	–	2	–	–	–	$^6S_{5/2}$	7.87
Os	76		6	–	2	–	–	–	$^5D_4$	8.70
Ir	77		7	–	2	–	–	–	$^4F_{1/2}$	9.00
Pt	78		9	–	1	–	–	–	$^3D_3$	9.00
Au	79		–	1	–	–	–	$^2S_{1/2}$	9.22
Hg	80		–	2	–	–	–	$^1S_0$	10.44
Tl	81		–	2	1	–	–	$^2P_{1/2}$	6.11
Pb	82	The sub-shells from 1s to 5d contain 78 electrons	–	2	2	–	–	$^3P_0$	7.42
Bi	83		–	2	3	–	–	$^4S_{3/2}$	7.29
Po	84		–	2	4	–	–	$^3P_2$	8.43
At	85		–	2	5	–	–	$^2P_{3/2}$	9.40
Rn	86		–	2	6	–	–	$^1S_0$	10.75
Fr	87		–	2	6	–	1	$^2S_{1/2}$	4.00
Ra	88		–	2	6	–	2	$^1S_0$	5.28
Ac	89		–	2	6	1	2	$^2D_{3/2}$	5.5
Th	90		–	2	6	2	2	$^3F_2$	5.7
Pa	91		2	2	6	1	2	$^4K_{11/2}$	5.7
U	92		3	2	6	1	2	$^5L_6$	4.0
Np	93		4	2	6	1	2	$^6L_{11/2}$	
Pu	94	The sub-shells from 1s to 5d contain 78 electrons	6	2	6	–	2	$^7F_0$	
Am	95		7	2	6	–	2	$^8S_{7/2}$	
Cm	96		7	2	6	1	2	$^9D_2$	
Bk	97		8	2	6	1	2	$^8H_{17/2}$	
Cf	98		10	2	6	–	2	$^6I_8$	
Es	99		11	2	6	–	2	$^4I_{15/2}$	
Fm	100		12	2	6	–	2	$^3H_6$	
Md	101		13	2	6	–	2	$^2F_{7/2}$	
No	102		14	2	6	–	2	$^1S_0$	
Lw	103		14	2	6	1	2	$^2D_{1/2}$	
Ku	104		14	2	6	2	2	$^3F_2$	

period contains elements with the L-shell filled. It comprises 8 elements ($n=2$, $l=0,1$) from lithium to neon. The 3s-states and 3p-states of the M-shell are filled from sodium to argon. Up to now periods ending with the noble gases He, Ne, Ar have been formed in the Mendeleyev periodic system. In the next period, beginning with potassium and ending with krypton, there is a departure from the simple rules of successive filling. Namely, as we have seen in §71, electrons with angular momentum $l = 2$, i.e. in the d-state, must begin at the element with $Z = 21$. Hence in Sc the additional twenty-first electron does not go into a 4p-state but into a 3d-state, which proceeds to fill from Sc to Ni. It is interesting to note that in Cr the tendency to fill a 3d-state is so strong that one of the 4s-electrons goes over into a 3d-state. The filling of 4s-states and 4p-states of the fourth period of the system of elements, which ends with krypton, begins again after Ni.

The further simple building of the N-shell, i.e. the fifth period containing 18 elements up to Xe, proceeds after krypton. In the sixth period, containing 32 elements, the filling of 6s-, 4f- and 6p-states proceeds. Here again there is a more complex order of filling. In Ce ($Z=58$) the electrons begin to fill the 4f-states. We note that according to a calculation in the Thomas–Fermi approximation the electrons with $l = 3$ should appear beginning with the element $Z = 55$.

The building of the seventh period for elements existing in nature remains incomplete. The filling of the deep 5f-state begins with Pa. Up to now there are artificially produced elements from Np ($Z = 93$) to kourchatovium (Ku, $Z=104$). In all these elements, called the actinides, the filling of the 5f-states proceeds.

As the atomic number Z increases all the properties of atoms determined by the inner electrons display monotonic changes. As an example we can cite the characteristic X-ray spectra. Characteristic X-rays arise when a vacancy in one of the inner shells is filled. The X-ray spectrum evidently has a character similar to that of hydrogen, but with the Rydberg constant multiplied by Z^2. The frequency of emission lines increases in proportion to Z^2 (the Moseley law).

On the other hand, all the properties of atoms determined by the peripheral electrons have a periodic behaviour with increasing Z. For example, the ionization potential of the atom (see table 2) is one such property. It has its smallest value for the first element and reaches its largest value for the last element of a period.

Another property displaying periodicity is the atomic volume. The largest volume is possessed by the alkali metal atoms, in which there is one electron outside a filled shell.

From the point of view of chemistry the most important characteristic of an atom is its valence. Chemical binding only involves unpaired electrons. Electrons which are in filled states and have a total spin equal to zero do not take part in the chemical interaction.

Hence it follows that the chemical interaction and its qualitative characteristic (the valence) are determined exclusively by the number of unpaired electrons which are in unfilled states. The numerical value of the valence of an atom in a given state is equal to $r = 2S$, where S is the spin of the atom in that state.

We particularly stress that the valence of an atom is related to its state because the atom, when making a transition from one state into another, may change its valence.

In §79 we shall discuss this problem in somewhat more detail. Here we shall confine ourselves to the following remark: if the first excited term lies close to the normal term, then the atom may be involved in chemical binding in the excited state. From the above it follows that the elements of the first group of the periodic system, with which all 7 periods begin, in the $^2S_{1/2}$ state have the valence $r = 1$.

The elements of the second group which are in the $^1S_0$ state have zero valence. They would be chemically inert if the excited term did not lie close to the normal state. In the excited state the two outer electrons have the configuration s and p, so that the atom has a total spin 1 and a valence $r = 2$.

In the atoms of the third group three electrons are outside filled shells. Their configuration s^2p corresponds to a total spin $S = \frac{1}{2}$ and to a valence one. However, those of these atoms which have a small excitation energy may make a transition into the state with the configuration sp^2 and spin $S = \frac{3}{2}$. In the excited state their valence is $r = 3$. The elements of the first three groups are regarded chemically as metals. From the chemical point of view metals are characterized by the ability to lose electrons when forming ionic chemical compounds.

The elements of the fourth group enter into a chemical bond in the normal and excited states with the configurations s^2p^2 and sp^3. The corresponding values of the spin and valence are equal respectively to $S = 1, r = 2$ and $S = 2, r = 4$.

The excited state corresponds to the configuration s^2p^2s and to the transition of the fifth electron into the s-state of the next shell (i.e. to a transition with an increase in the principal quantum number by one). The spin and valence in the excited state are equal respectively to $S = \frac{5}{2}$ and $r = 5$.

In the sixth group the atoms in the normal state have the configuration s^2p^4 with spin $S = 1$. Their valence is $r = 2$. When excited, one of the electrons makes a transition from the p-state to the s-state of the next shell. In this excited state $r = 4$.

In addition to this type of excitation, excitations with a transition of two electrons into the next shell, one from the s-state and the second from the p-state, are often brought about. In this excited state the atom has the valence $r = 6$.

In the atoms of the seventh group the normal configuration is s^2p^5, the spin is $S = \frac{1}{2}$ and the valence is $r = 1$. However, transitions of one, two and three electrons into the next layer are possible. Hence also the valences $r = 3$, $r = 5$ and $r = 7$ are realized.

The elements of the fourth, fifth, sixth and seventh groups which stand at the beginning of the groups are non-metals. In compounds of ionic type they gain electrons (are oxidizers), having the tendency to form a filled state.

The elements of the transition groups — iron, palladium and platinum — as well as the lanthanides (rare earths) and actinides, have special chemical properties.

The completion of deep d-states and f-states takes place in the atoms of the group of iron, the lanthanides and the actinides, d-electrons and f-electrons usually do not take part in valence bonds and the valence of the atoms is determined by the electrons in the outer states. However, this is not a strict law, since in certain cases of chemical compound formation, electrons from inner states make transitions into outer states and contribute to the valence. This is particularly clearly displayed in the case of some actinides. Hence the chemical properties of elements of the groups with special properties are rather complex.

Thus we see that not only is the theoretical substantiation of the distribution of atoms in the periodic system of the elements possible, but further a relatively detailed prediction of their chemical properties can be given.

§74. The Zeeman effect

We have seen in §31 of Part I that the full theory of the Zeeman effect could not be constructed on the basis of classical electrodynamics. After analyzing a vast amount of experimental data, Landé introduced a parameter which quantitatively determines the characteristics of the Zeeman effect. This parameter is called the Landé g-factor.

The quantum theory of the Zeeman effect makes it possible to find the value of the Landé g-factor and the form of the Zeeman splitting without any new assumptions. Let us consider how the positions of the energy levels of an atom change if the atom is placed in a constant external magnetic field. The wave function ψ for the stationary states of an atom can, as usual, be written in the form

$$\psi(\mathbf{r}_i, t) = \psi(\mathbf{r}_i)\, e^{-iEt/\hbar} . \tag{74.1}$$

On substituting (74.1) into the Pauli equation (63.11) the latter transforms into the form

$$\left\{ \frac{1}{2m} \sum_i \hat{\mathbf{p}}_i^2 + \frac{|e|}{2mc}(\hat{\mathbf{L}}+2\hat{\mathbf{S}}) \cdot \boldsymbol{\mathscr{H}} + \frac{e^2}{8mc^2} \sum_i [\boldsymbol{\mathscr{H}} \times \mathbf{r}_i]^2 + U \right\} \psi = E\psi , \tag{74.2}$$

where U takes into account the interaction of the electrons with each other and with the nuclei. We assume that the external magnetic field strength is sufficiently small that the term containing the square of the field can be dropped in (74.2).

We introduce the quantity H' given by

$$\hat{H}' = \frac{|e|}{2mc}(\hat{\mathbf{L}}+2\hat{\mathbf{S}}) \cdot \boldsymbol{\mathscr{H}} = \frac{-e}{2mc}(\hat{\mathbf{J}}+\hat{\mathbf{S}}) \cdot \boldsymbol{\mathscr{H}} , \tag{74.3}$$

where $\hat{\mathbf{J}}$ is the total angular momentum operator, $\hat{H}'$ is the small perturbation acting on the atom.

The Hamiltonian $\hat{H}$ can then be written in the form

$$\hat{H} = \hat{H}_0 + \hat{H}' ; \qquad \hat{H}_0 = \frac{1}{2m} \sum_i \hat{\mathbf{p}}_i^2 + U . \tag{74.4}$$

In the unperturbed state the atom is characterized by a definite total angular momentum of the system J and by a definite z-component of the total angular momentum, M_z. Clearly, one has to apply perturbation theory for degenerate states. Indeed, the energy of the unperturbed state does not depend on the value of the total angular momentum component M_z. Since the perturbation operator represents the projection of a certain vector onto the z-axis and is brought to diagonal form simultaneously with the operator of the z-component of the total angular momentum, one needs to calculate only the diagonal elements of the perturbation operator

$$\hat{H}' = \frac{|e|\mathscr{H}}{2mc}(\hat{J}_z+\hat{S}_z) . \tag{74.5}$$

The diagonal matrix element is taken with respect to the quantum numbers of the total angular momentum J and the z-component of the angular momentum $M_z \equiv M$.

The diagonal matrix element of the operator $\hat{J}_z$ is equal to

$$(\hat{J}_z)_{JM;JM} = \hbar M . \tag{74.6}$$

Consequently, we have to determine the expression

$$(\hat{S}_z)_{JM;JM} = \bar{S}_z .$$

The calculation of this matrix element by means of the commutation rules is a good example of the practical use of the matrix method. We note that the value of $\bar{S}_z$ is usually found from obvious but not quite rigorous considerations associated with the precession of the vector $\mathbf{S}$ with respect to the vector $\mathbf{J}^*$. We note at first that by definition the operator $\hat{\mathbf{J}}$ is equal to $\hat{\mathbf{J}} = \hat{\mathbf{L}} + \hat{\mathbf{S}}$. The operators $\hat{\mathbf{L}}$ and $\hat{\mathbf{S}}$ commute with each other, since they act on different variables.

Knowing the commutation rules for $\hat{L}_x, \hat{L}_y, \hat{L}_z; \hat{S}_x, \hat{S}_y, \hat{S}_z$ we easily find the following relations:

$$\{\hat{J}_x, \hat{S}_x\} = 0 , \qquad \{\hat{J}_x, \hat{S}_y\} = i\hbar\hat{S}_z , \qquad \{\hat{J}_x, \hat{S}_z\} = -i\hbar\hat{S}_y .$$

Other commutation rules can be obtained by cyclic permutation. Then we obtain

$$\begin{aligned}
\{\hat{J}_y, \hat{S}_y\} &= 0 , & \{\hat{J}_y, \hat{S}_z\} &= i\hbar\hat{S}_x , & \{\hat{J}_y, \hat{S}_x\} &= -i\hbar\hat{S}_z , \\
\{\hat{J}_z, \hat{S}_z\} &= 0 , & \{\hat{J}_z, \hat{S}_y\} &= -i\hbar\hat{S}_x , & \{\hat{J}_z, \hat{S}_x\} &= i\hbar\hat{S}_y .
\end{aligned} \tag{74.7}$$

From these rules there results the relation

$$(\hat{J}_x + i\hat{J}_y)(\hat{S}_x + i\hat{S}_y) - (\hat{S}_x + i\hat{S}_y)(\hat{J}_x + i\hat{J}_y) = 0 . \tag{74.8}$$

We calculate the following matrix element of the right-hand and left-hand sides of this relation:

$$[(\hat{J}_x + i\hat{J}_y)(\hat{S}_x + i\hat{S}_y)]_{J,M+1;J,M-1} = [(\hat{S}_x + i\hat{S}_y)(\hat{J}_x + i\hat{J}_y)]_{J,M+1;J,M-1} .$$

In correspondence with formula (51.14) the matrix element of the operator $\hat{J}_x + i\hat{J}_y$ differs from zero only in the case of the transition $J, M \to J, M - 1$. Hence

$$(\hat{J}_x + i\hat{J}_y)_{JM;JM-1} = \hbar\,[(J+M)(J-M+1)]^{\frac{1}{2}} . \tag{74.9}$$

Then making use of the rule of multiplication for matrices (45.6), we obtain

---

* See L.D.Landau and E.M.Lifshitz, *Quantum mechanics* (Pergamon Press, Oxford, 1965).

$$(\hat{J}_x+i\hat{J}_y)_{M+1;M}(\hat{S}_x+i\hat{S}_y)_{M;M-1} - (\hat{S}_x+i\hat{S}_y)_{M+1;M}(\hat{J}_x+i\hat{J}_y)_{M;M-1} = 0 \,. \tag{74.10}$$

In this formulae we have dropped the suffix J. Using relation (74.9) we find

$$\frac{(\hat{S}_x+i\hat{S}_y)_{M+1;M}}{[(J+M+1)(J-M)]^{\frac{1}{2}}} = \frac{(\hat{S}_x+i\hat{S}_y)_{M;M-1}}{[(J+M)(J-M+1)]^{\frac{1}{2}}} \equiv A \,.$$

Analogously one can obtain

$$\frac{(\hat{S}_x+i\hat{S}_y)_{M+2;M+1}}{[(J+M+2)(J-M-1)]^{\frac{1}{2}}} = A \,.$$

Hence we see that the quantity A does not depend on M and, consequently, we have

$$(\hat{S}_x+i\hat{S}_y)_{M;M-1} = A\,[(J+M)(J-M+1)]^{\frac{1}{2}} \,. \tag{74.11}$$

The matrix elements of the operator $\hat{S}_z$ which we need can be found by making use of the following formula:

$$(\hat{J}_x-i\hat{J}_y)(\hat{S}_x+i\hat{S}_y) - (\hat{S}_x+i\hat{S}_y)(\hat{J}_x-i\hat{J}_y) = -2\hbar\hat{S}_z \,. \tag{74.12}$$

Calculating the diagonal matrix element of relation (74.12), we obtain as a result

$$(\hat{J}_x-i\hat{J}_y)_{M;M+1}(\hat{S}_x+iS_y)_{M+1;M} - (\hat{S}_x+i\hat{S}_y)_{M;M-1}(\hat{J}_x-i\hat{J}_y)_{M-1;M} =$$

$$= -2\hbar(\hat{S}_z)_{M;M} \,.$$

Making use of (74.11) and knowing that

$$(\hat{J}_x-i\hat{J}_y)_{M;M+1} = \hbar\,[(J+M+1)(J-M)]^{\frac{1}{2}} \,,$$

we can easily determine the diagonal element $(\hat{S}_z)_{M;M}$ which is equal to

$$-2(\hat{S}_z)_{M;M} = (J+M+1)(J-M)\,A - (J+M)(J-M+1)\,A = -2AM \,,$$

$$(\hat{S}_z)_{M;M} = AM \,. \tag{74.13}$$

We now turn to finding the quantity A. From the relation

$$\hat{J}^2 = (\hat{L}+\hat{S})^2 = \hat{L}^2 + 2(\hat{L}\cdot\hat{S}) + \hat{S}^2 = \hat{L}^2 + 2(\hat{J}\cdot\hat{S}) - S^2$$

it immediately follows that

$$\hat{J}\cdot\hat{S} = \tfrac{1}{2}(\hat{J}^2+\hat{S}^2-\hat{L}^2) \,.$$

In the case of Russell–Saunders coupling the diagonal matrix element of the scalar product $\mathbf{J}\cdot\mathbf{S}$ is equal to

$$(\hat{\mathbf{J}} \cdot \hat{\mathbf{S}})_{JM; JM} = \tfrac{1}{2}\hbar^2 [J(J+1) - L(L+1) + S(S+1)] . \tag{74.14}$$

On the other hand, the matrix element can be found if the scalar expression $\hat{\mathbf{J}} \cdot \hat{\mathbf{S}}$ is transformed to the form

$$(\hat{\mathbf{J}} \cdot \hat{\mathbf{S}})_J = \tfrac{1}{2}(\hat{S}_x + i\hat{S}_y)(\hat{J}_x - i\hat{J}_y) + \tfrac{1}{2}(\hat{S}_x - i\hat{S}_y)(\hat{J}_x + i\hat{J}_y) + \hat{J}_z \hat{S}_z . \tag{74.15}$$

We find the diagonal matrix element of the right-hand and left-hand sides of relation (74.15). We then have

$$(\hat{\mathbf{J}} \cdot \hat{\mathbf{S}})_{M; M} = \tfrac{1}{2}(\hat{S}_x + i\hat{S}_y)_{M; M-1}\, (\hat{J}_x - i\hat{J}_y)_{M-1; M} +$$
$$+ \tfrac{1}{2}(\hat{S}_x - i\hat{S}_y)_{M; M+1}\, (\hat{J}_x + i\hat{J}_y)_{M+1; M} + \hbar M(\hat{S}_z)_{M; M} . \tag{74.16}$$

We now need to find the matrix element $(\hat{S}_x - i\hat{S}_y)_{M; M+1}$. This is easily done by means of relation (74.11). We note beforehand that the constant A is real. This is seen from formula (74.13), in which all the quantities determining the quantity A are real.

Carrying out the complex conjugation of the right-hand and left-hand sides of the relation

$$(\hat{S}_x + i\hat{S}_y)_{M+1; M} = A\,[(J+M+1)(J-M)]^{\frac{1}{2}} ,$$

we obtain

$$(\hat{S}_x)^*_{M+1; M} - i(\hat{S}_y)^*_{M+1; M} = A\,[(J+M+1)(J-M)]^{\frac{1}{2}} .$$

Making use of the hermiticity of the operators, we find

$$(\hat{S}_x)^*_{M+1; M} - i(\hat{S}_y)^*_{M+1; M} = (\hat{S}_x)_{M; M+1} - i(\hat{S}_y)_{M; M+1} =$$
$$= (\hat{S}_x - i\hat{S}_y)_{M; M+1} = A\,[(J+M+1)(J-M)]^{\frac{1}{2}} . \tag{74.17}$$

Substituting the values of $(\hat{\mathbf{J}} \cdot \hat{\mathbf{S}})_{M; M}$ and $(\hat{S}_x + i\hat{S}_y)_{M; M-1}$ into relation (74.16) in correspondence with formulae (74.14) and (74.11), and also using (74.17) and (74.13), we find the quantity A:

$$A = \hbar\, \frac{J(J+1) + S(S+1) - L(L+1)}{2J(J+1)} . \tag{74.18}$$

Using the value found for A and also formula (74.13), we obtain the diagonal matrix element $(\hat{S}_z)_{JM; JM}$ which has the form

$$(\hat{S}_z)_{JM; JM} = \hbar M\, \frac{J(J+1) + S(S+1) - L(L+1)}{2J(J+1)} . \tag{74.19}$$

The correction to the energy levels of the atom due to the magnetic field $\mathcal{H}$ is given by the expression

$$\Delta E = \frac{|e|\mathcal{H}\hbar M}{2mc} \left(1 + \frac{J(J+1) + S(S+1) - L(L+1)}{2J(J+1)}\right) \equiv \frac{|e|\mathcal{H}\hbar M}{2mc}\, g . \tag{74.20}$$

The factor g is called the Landé g-factor.

For singlet levels $J = L$, $S = 0$ we have

$$\Delta E = \frac{|e| \mathcal{H} \hbar M}{2mc} .$$ (74.21)

Before turning to the discussion of formulae (74.20) and (74.21) we shall establish the limits of applicability of the above derivation.

The unperturbed Pauli equation

$$\hat{H}_0 \psi = E\psi ; \qquad \hat{H}_0 = \frac{1}{2m} \sum_i \hat{\mathbf{p}}_i^2 + U$$

defines the energy levels of the atom including also its multiplet structure. Thus for the theory described above to be applicable it is necessary that the matrix element of the perturbation (74.3) be smaller than the spacing between the levels corresponding to the fine structure of the atom.

It should also be pointed out that in the calculations it was assumed that Russell–Saunders coupling holds in the atom, i.e. that the quantity J is conserved in time as well as L and S.

Let us now turn to the discussion of the formulae obtained, which define the Zeeman effect.

From formula (74.20) it is seen that each component of the multiplet splits into $2J+1$ levels. Indeed, for given J the component of the total angular momentum M can take on $2J+1$ different values. This is in accordance with the result of §54: perturbation removes the degeneracy. As to the distribution of the newly arising terms, the following can be stated.

If J is an integer, then in a magnetic field a level corresponding to the value $M = 0$ arises at the place of the unperturbed level.

Of the remaining $2J$ levels J levels are distributed above and J levels below at equal distances from the level with $M = 0$.

If J is a half integer, then the levels are also distributed symmetrically with respect to the old position of the unperturbed level, and the closest levels are at the distance $|e| \hbar \mathcal{H} g / 4mc$ from the initial position.

We also note that if there is jj coupling, the character of the Zeeman effect is much modified. This coupling is seldom encountered in the pure form, and we shall not carry out the corresponding calculations here.

§75. The Paschen–Back effect and the diamagnetism of atoms

In strong magnetic fields the character of the Zeeman effect changes. Namely, as the magnetic field strength increases the spacing between the multiplet levels increases. In very strong fields the splitting of a level is so large that the spacings between the components of the multiplet arising in the field turn out to be large in comparison with the natural multiplet spacing. We recall that the latter arises from the spin–orbit interaction. In this case formula (74.20) is no longer applicable, and the character of the spectrum changes. This change in the spectrum in a strong magnetic field is called the Paschen–Back effect.

We shall carry out the calculation for the case where the splitting due to a magnetic field is large in comparison with the natural multiplet spacing. This means that the energy acquired in the magnetic field is large compared to the spin–orbit interaction. Then the term taking the spin–orbit interaction into account can be left out of the unperturbed Hamiltonian H_0 in formula (74.4). Therefore the unperturbed states of the atom can be characterized by the total angular momentum J as well as by the component L_z of the orbital angular momentum and the component S_z of the spin angular momentum.

The perturbation operator has, as before, the form

$$\hat{H}' = \frac{|e|\hbar}{2mc}(\hat{J}_z + \hat{S}_z)\mathscr{H} = \frac{|e|\hbar}{2mc}(\hat{L}_z + 2\hat{S}_z)\mathscr{H} . \tag{75.1}$$

The correction to the energy is equal to the mean value of the operator $\hat{H}'$ over states with definite components of the orbital and spin angular momenta, i.e.

$$E' = \frac{|e|\hbar\mathscr{H}}{2mc}(\bar{L}_z + 2\bar{S}_z) = \frac{|e|\hbar\mathscr{H}}{2mc}(L_z + 2S_z) . \tag{75.2}$$

Formula (75.2) defines the fine structure of the spectrum in strong magnetic fields.

Let us now consider the effect of the neglected quadratic term in the magnetic field in formula (74.2). Taking this quantity into account is particularly important for terms with $L = S = 0$. In this case no splitting of levels occurs on account of the term linear in $\mathscr{H}$. This can be seen from the general formula (74.20). In this case the correction due to the quadratic term cannot be disregarded. As the perturbation operator, in correspondence with formula (74.2), one has to take

$$\hat{H}_1' = \frac{e^2}{8mc^2} \sum_i |\mathscr{H} \times \mathbf{r}_i|^2 . \tag{75.3}$$

The sum over i corresponds to summing over all the electrons of the atom.

The correction to the energy levels due to the operator $\hat{H}'_1$ is again defined by the diagonal matrix element

$$E'_2 = \frac{e^2}{8mc^2} \sum_i |\mathcal{H} \times \mathbf{r}_i|^2 = \frac{e^2}{8mc^2} \sum_i \overline{(\mathcal{H} r_i \sin\theta)^2} .$$

In calculating $\overline{[\mathcal{H} \times \mathbf{r}]^2}$ it should be recalled that the wave function of the system $L = 0, S = 0$ is spherically symmetric, hence

$$\overline{\sin^2\theta} = 1 - \overline{\cos^2\theta} = \tfrac{2}{3} .$$

Thus for the shift of the levels we obtain

$$\Delta E = \frac{e^2 \mathcal{H}^2}{12mc^2} \sum_i \overline{r_i^2} . \tag{75.4}$$

Since the magnetic moment of the atom can be calculated by means of the formula $\mathbf{M} = -\partial \Delta E/\partial \mathcal{H}$ (see (18.1) of Part IV), we obtain

$$\mathbf{M} = \chi \mathcal{H}; \qquad \chi = -\frac{e^2}{6mc^2} \sum_i \overline{r_i^2} . \tag{75.5}$$

Thus atoms possess diamagnetic susceptibility. Since the diamagnetic susceptibility is in the main determined by the mean square distance of the electrons from the nucleus, χ is particularly large for many-electron atoms. For such atoms good results are obtained by the Thomas–Fermi method. Hence the diamagnetic susceptibilities are often calculated by this method.

On the other hand, measurements of χ represent one of the best ways of finding the effective size of atoms. We stress that all atoms and ions have a diamagnetic susceptibility. However, in certain ions the paramagnetic susceptibility, associated with the magnetic moment, exceeds the diamagnetic susceptibility.

§76. Deuteron theory

The deuteron, consisting of a proton and a neutron, plays the same role in nuclear theory as the hydrogen atom in atomic theory.

The nuclear interaction between a proton and a neutron may depend on their separation r and the relative orientation of the spins, $\mathbf{s}_1$ and $\mathbf{s}_2$, of the two particles. The explicit form of the potential energy of the nuclear interaction is at present unknown. Hence one has to confine oneself to writing the most general expression for the potential energy operator depending on $\mathbf{r}$, $\mathbf{s}_1$ and $\mathbf{s}_2$. This interaction operator must not change under rotation of the coordinate system. Furthermore, as shown by experiment, the parity conservation law holds for nuclear forces (see §33). This means that the interaction operator must not change under reflection of coordinates (the interaction operator must commute with the parity operator). Thus we have to make up all possible scalars of the three vectors $\mathbf{r}$, $\mathbf{s}_1$ and $\mathbf{s}_2$. The following scalars do not change under rotation of the coordinate system: $\mathbf{s}_1 \cdot \mathbf{s}_2$, $\mathbf{s}_1 \cdot \mathbf{r}$ and $\mathbf{s}_2 \cdot \mathbf{r}$.

The products $\mathbf{s}_1 \cdot \mathbf{r}$ and $\mathbf{s}_2 \cdot \mathbf{r}$ cannot be involved separately in the potential energy, because the spin vector is an axial vector and the product $\mathbf{s} \cdot \mathbf{r}$ is a pseudoscalar which changes sign under reflection of coordinates. The product $(\mathbf{s}_1 \cdot \mathbf{r})(\mathbf{s}_2 \cdot \mathbf{r})$ does not change sign under the reflection and, consequently, can be involved in the potential energy. Spin operators in higher powers are not involved in the operator of the interaction energy U, because the higher powers of the spin operators may be reduced by means of formula (60.17) to linear combinations of $\mathbf{s}$.

Thus the expression for the potential energy has the form

$$U \doteq U_1(r) + U_2(r)(\mathbf{s}_1 \cdot \mathbf{s}_2) + U_3(r)(\mathbf{s}_1 \cdot \mathbf{r})(\mathbf{s}_2 \cdot \mathbf{r}) , \qquad (76.1)$$

where U_1, U_2, U_3 are certain functions depending on the distance between the particles. Besides the operator (76.1), representing a potential energy of the ordinary type, the interaction between a proton and a neutron may also have the character of an exchange force. According to the results of §67, U_{exch} can be written by means of the exchange operator $\hat{P}_{12}$ in the form

$$U_{\text{exch}} = \hat{P}_{12}[U_4(r) + U_5(r)(\mathbf{s}_1 \cdot \mathbf{s}_2) + U_6(r)(\mathbf{s}_1 \cdot \mathbf{r}_1)(\mathbf{s}_2 \cdot \mathbf{r}_2)] . \quad (76.2)$$

Here U_4, U_5 and U_6 are functions of the distance between the particles independent of their spins. For generality it is assumed that the form of these functions is different from the form of the functions U_1, U_2, U_3 involved in the potential energy of the ordinary interaction. The total interaction energy is equal to the sum of expressions (76.1) and (76.2). Available data on the stable states of the deuteron, the study of neutron–proton scattering etc. do not, as yet, allow one to determine the form of these functions. Moreover, there are no grounds for considering any of these functions to be small in comparison with the others. Thus even the simplest nuclear system turns out to be immeasurably more complex than atomic systems.

Experimental data already makes it possible to carry out a classification of the states of the deuteron. As is easily seen, the Hamiltonian of a system of two nucleons (a proton and a neutron) with the interaction energy written

above leads to two conservation laws: the total angular momentum conservation law and the parity conservation law.

The states of the deuteron are denoted by the same symbols as the states
of atoms. States with the orbital angular momentum $L = 0$, 1, 2, ... are
denoted respectively by S, P, D and so on. The multiplicity of the $(2S+1)$th
term is denoted by a superscript on the left (S is the total spin of the
deuteron). The subscript on the right denotes the total angular momentum J
of the deuteron. For example, in the state $^3P_0$ the total spin is equal to one,
$L = 1$, and the total angular momentum is equal to zero.

Let us discuss the possible states of the system taking into account the
fact that the spins of the neutron and proton are equal to $\frac{1}{2}$. Formal application of the rule of addition of angular momenta leads to the following
possible states of the system:

$$^1S_0, \quad ^1P_1, \quad ^1D_2 \quad \text{(singlets)},$$

$$^3S_1, \quad ^3P_0, \quad ^3P_1, \quad ^3P_2, \quad ^3D_1, \quad ^3D_2, \quad ^3D_3 \quad \text{(triplets)}.$$

The S- and D-states are even, whereas the P-state is odd. We have not
written down states with $L > 2$. The states realized in nature can be deduced
only from experimental data. Experiment shows that the ground state of the
deuteron is an even state with $J = 1$*.

Further, making use of the rules of addition of angular momenta, we shall
establish the possible states of a system with total angular momentum $J = 1$.

The total spin of a system consisting of a neutron and a proton can be
equal either to zero or to one. If the spin is equal to zero, then only one state
$L = 1$ leading to the total angular momentum $J = 1$ is possible. For a spin
equal to one the orbital angular momentum can take on three values: $L = 0$,
1, 2. Consequently, four states are in all possible: $^1P_1, ^3S_1, ^3P_1, ^3D_1$; the
states $^1P_1$ and $^3P_1$ cannot be realized, because they are odd.

Further, it is easily seen that superpositions of states such as $^3S_1 + ^3P_1$ or
$^3S_1 + ^1P_1$ are impossible, since S- and P-states have different parities and the
wave function corresponding to their superposition is not an eigenfunction of
the parity operator.

Thus the deuteron can be either in the state $^3S_1$ or in the state $^3D_1$ or in
a state which is a superposition of these two states. The S-state is spherically

---

* See L.D.Landau and Ya.Smorodinskii, *Lectures on nuclear theory* (Consultants
Bureau, New York, 1958); A.I.Akhiezer and I.Ya.Pomeranchuk, *Nekotorye voprosy
teorii yadra (Some problems of nuclear theory)* (Gostekhizdat, Moscow, 1950) § 2 and
§ 5.

symmetric. If the deuteron were in this state, then its quadrupole moment would be equal to zero. Experiment shows, however, that the quadrupole moment of the deuteron is different from zero, although it is small. This means that the normal state of the deuteron represents the superposition of the spherically symmetric $^3S_1$-state and the asymmetric $^3D_1$-state.

Knowing the experimental value of the quadrupole moment of the deuteron, one can estimate the contribution given by the $^3D_1$-state to the wave function of the deuteron. This contribution turns out to be small. Thus it can be assumed that the deuteron is a spherically symmetric system with a small admixture of asymmetry brought about by the D-state.

We shall consider further a rough model of the deuteron in which we assume that the potential energy of the interaction between the neutron and the proton depends only on the distance between them. In other words, we shall retain only the first term $U_1(r) \equiv U(r)$ in formula (76.1). We shall disregard the asymmetry of the deuteron, assuming it to be in the ground state. The equation for the relative motion of the neutron and proton can be written, in correspondence with formula (14.11), in the form

$$\left[-\frac{\hbar^2}{2\mu}\nabla^2 + U(r)\right]\psi_0 = \epsilon\psi_0 . \tag{76.3}$$

In this case the reduced mass of the system is equal to

$$\frac{1}{\mu} = \frac{1}{m_p} + \frac{1}{m_n} ,$$

where m_p is the mass of the proton, and m_n is the mass of the neutron. Since $m_p \approx m_n$, we obtain

$$\frac{1}{\mu} = \frac{2}{m_p} .$$

As to the potential energy $U(r)$ we confine ourselves only to the general assumption that it tends to 0 rapidly as $r \to r_0$, where r_0 is the range of nuclear forces. We cannot give the concrete form of $U(r)$ for $r < r_0$ since we do not know the law of interaction of nuclear forces.

If the function ψ_0 is sought in the form

$$\psi_0(r) = \frac{\chi(r)}{r} , \tag{76.4}$$

then making use of formula (35.16) with $l = 0$ we obtain the equation for the function $\chi(r)$

$$\left[-\frac{\hbar^2}{m_{\mathrm{p}}} \frac{d^2}{dr^2} + U(r) \right] \chi(r) = \epsilon \chi(r) . \qquad (76.5)$$

For $r > r_0$ eq. (76.5) is written in the form

$$-\frac{\hbar^2}{m_{\mathrm{p}}} \frac{d^2 \chi}{dr^2} = \epsilon \chi . \qquad (76.6)$$

We seek the solution decreasing at infinity in the form

$$\chi = C e^{-\alpha r} . \qquad (76.7)$$

Substituting (76.7) into (76.6) we obtain the relation for α

$$-\frac{\hbar^2}{m_{\mathrm{p}}} \alpha^2 = \epsilon = -|\epsilon| , \qquad \alpha = \left[\frac{m_{\mathrm{p}} |\epsilon|}{\hbar^2} \right]^{\frac{1}{2}} . \qquad (76.8)$$

Then for the wave function we have

$$\psi_0 = C \frac{e^{-\alpha r}}{r} . \qquad (76.9)$$

As a characteristic of the size of the deuteron one can choose the quantity $r_1 = \alpha^{-1}$, i.e. the distance at which the wave function χ decreases by a factor e. The distance r_1 is easily determined from relation (76.8), since the binding energy of the deuteron is well known from experimental data to be $|\epsilon| = 2.19$ MeV. Substituting the value of $\hbar$ and m_{p} into formula (76.8), we obtain $r_1 = 4.3 \times 10^{-13}$ cm. Consequently, the wave function ψ_0 of the deuteron differs from zero in a range considerably larger than the range of nuclear forces ($r_0 \approx 2 \times 10^{-13}$ cm). Thus we see that the neutron and proton can be observed with a high probability at distances from each other which substantially exceed the size of the sphere of action of nuclear forces.

The dependence of the wave function ψ_0 on the distance cannot be determined in the region $r < r_0$, since the potential energy in this region is unknown.

However, from the general theory of motion in a spherically symmetric field it follows that for $r \to 0$ the function χ is proportional to r^{l+1} (see §35) and, consequently, in the S-state is proportional to r. Thus at small distances the function χ tends to zero.

The constant C contained in the ψ-function can be found from the normalization condition. For the wave function ψ_0 for $r < r_0$, we take it in the form (76.9), which we assume to be valid over all space. This does not introduce a substantial error, since a large part of the normalization integral refers to the region $r > r_0$. Substituting (76.9) into the normalization condition, we find

$$C = (\alpha/2\pi)^{\frac{1}{2}} . \tag{76.10}$$

Let us now establish the general relation between the width of the well r_0 and its depth. For this we integrate eq. (76.5) in the range from zero to $r = r_0$. As a result of the integration we obtain

$$\chi'_{r=r_0} - \chi'_{r=0} = \frac{m_p}{\hbar^2} \int_0^{r_0} U(r)\chi(r)\, dr + \frac{m_p|\epsilon|}{\hbar^2} \int_0^{r_0} \chi\, dr . \tag{76.11}$$

As can be seen from fig. V.22, the value of the derivative $|\chi'|$ taken at the point $r = r_0$ is considerably smaller than that of the derivative $|\chi'_{r=0}|$. Furthermore, one can disregard the binding energy in comparison with the potential energy of interaction, i.e. one can assume that $|\epsilon| < |U(r)|$ for $r < r_0$. At small distances $\chi = Nr$, where N is a constant. Then (76.11) transforms into the form

$$-N = \frac{m_p}{\hbar^2} N \int_0^{r_0} U(r)r\, dr \tag{76.12}$$

or

$$\int_0^{r_0} U(r)r\, dr = -\frac{\hbar^2}{m_p} . \tag{76.13}$$

Replacing the integral in (76.13) by $U_0 r_0^2$, where U_0 is a mean energy of interaction, i.e. the mean depth of the well, we obtain in order of magnitude

$$U_0 \approx -\frac{\hbar^2}{m_p r_0^2} \sim -40\,\text{MeV} .$$

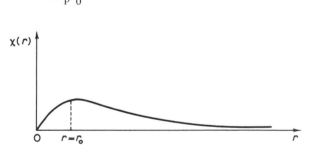

Fig. V.22

§77. Nuclear-shell theory

In contrast to atoms, in which the interaction between electrons is of a secondary character and takes place against the background of the principal interaction (the attraction towards the nucleus) there is no single centre of interaction in atomic nuclei.

On the contrary, all the nuclear particles (nucleons) interact intensively with each other through strong nuclear forces. Hence for a long time it seemed that it made no sense to distinguish between the states of the individual particles in the nucleus, and that one could speak only of the state of the system as a whole.

It turned out, however, that a number of the observed properties of atomic nuclei pointed to the conservation of the individuality of nucleons in nuclei. Apparently, the conservation of individuality of particles in nuclei is associated with the fact that the nuclear forces decrease very rapidly with increasing distance, and that the kinetic energy of the nucleons in nuclei is very large.

Proceeding from the assumption that each of the nucleons in a nucleus moves in the self-consistent field produced by all the other nucleons, it was possible to account for a number of important properties of nuclei.

The self-consistent field of most nuclei is spherically symmetric. Of course, the precise law of the potential distribution inside the nucleus is unknown. It turns out, however, that the character of the distribution of levels depends relatively little on the model of the potential field adopted, provided it gives correctly the basic feature of the field of the nucleus as a whole: the sharp increase of the potential at its surface $r = R$. The simplest nuclear model is a spherical potential well of infinite depth. In this model the self-consistent potential field U in which an individual nucleon moves has the form

$$U = \begin{cases} 0 & \text{for} \quad r < R \\ \infty & \text{for} \quad r \geqslant R. \end{cases}$$

Already this very simplified model allows one to get a general idea of the properties and distribution of levels. The wave function satisfies the equation

$$\frac{d^2\psi}{dr^2} + \frac{2}{r}\frac{d\psi}{dr} + \frac{l(l+1)}{r^2}\psi - k^2\psi = 0, \tag{77.1}$$

where $k^2 = 2mE/h^2$. The solution of this equation is expression (36.9). The boundary condition at the surface of the nucleus

$$\psi = 0 \quad \text{for} \quad r = R \tag{77.2}$$

leads to the condition

$$J_{l+\frac{1}{2}}(kR) = 0 . \tag{77.3}$$

For a given value of the orbital angular momentum the energy levels, which are the roots of the transcendental equation (77.3), are classified by means of the principal quantum number n. To the smallest root of this transcendental equation there corresponds a wave function having no nodes for $r < R$. This level is classified as a level with $n = 1$. The next root of (77.3) corresponds to a wave function having one node for $r < R$. The corresponding state is denoted by $n = 2$ and so on. For a given n, l can take on any value. The energy of the nucleon increases with increasing orbital angular momentum.

The ordering of levels is given by the sequence

$$1s, 1p, 1d, 2s, 1f, 2p, 1g, 2d \dots .$$

It turned out that in atomic nuclei an important role is played by the spin–orbit interaction, which has so far not been taken into account at all. The special role of the spin–orbit interaction is associated with the fact that, owing to the rapid decrease of nuclear forces with increasing distance between the particles, the energy of pair interaction is on the average small. The considerable magnitude of the spin–orbit interaction leads to the establishment of jj coupling in nuclei. The spin and orbital angular momenta of each nucleon are added up into the total angular momentum j. The energy of a nucleon turns out to depend on its spin. In nuclei there exists the inverted structure of levels, in which levels with large j lie below those with smaller j. The states of nucleons are denoted by the symbol nl_j; for example $1s_{1/2}$ or $2p_{3/2}$. Since nucleons obey the Pauli principle, $2j+1$ neutrons and $2j+1$ protons can be in each state with given values of n, l and j. Because of this there is a situation in nuclei very similar to that in atoms: the states of the nucleons can be divided into groups or shells. When each shell is filled a closed configuration arises, possessing the greatest stability − in the given case the largest energy of binding of the nucleon in the nucleus. The energy distribution of the states and the number of nucleons in these states is given by the table

$1s_{1/2}$						2
$1p_{3/2}$,	$1p_{1/2}$					6
$1d_{3/2}$,	$1d_{1/2}$,	$2s_{1/2}$,	$1f_{1/2}$			20
$2p_{3/2}$,	$1f_{5/2}$,	$2p_{1/2}$,	$1g_{3/2}$			22
$2d_{5/2}$,	$1g_{7/2}$,	$1h_{11/2}$,	$2d_{3/2}$,	$3s_{1/2}$		32
$2f_{7/2}$,	$1h_{9/2}$,	$2i_{13/2}$,	$2f_{3/2}$,	$3p_{3/2}$,	$3p_{1/2}$	44

From this table it is seen that in the closed shells there are consecutively 2, 8, 28, 50, 82 and 126 nucleons.

Correspondingly, nuclei with the total number of nucleons given by these numbers, which are called magic numbers, possess particular stability.

Nuclei in which both the number of protons and the number of neutrons are magic numbers are particularly stable. Such nuclei are, for example, $^4_2$He and $^{16}_8$O. They are often said to be doubly magic. This very simple scheme makes it possible to account not only for a particular stability and natural abundance of certain isotopes but also a number of other properties of atomic nuclei, for example their magnetic moments. However, in a number of cases it turns out to be inadequate. Thus, for example, nuclei with open shells display deviations from spherical form. This is exhibited in the presence of rotation of the nucleus as a whole. The experimental proof of rotation of the nucleus is the presence in nuclear spectra of a structure similar to that of the spectra of diatomic molecules.

We cannot dwell on details here, but the reader may find these in the specialized literature*.

---

* See, for example, P.E.Nemirovskii, *Contemporary models of the atomic nucleus* (Pergamon Press, Oxford, 1963).

10

The Theory of Diatomic Molecules

§78. The adiabatic approximation and the classification of electron terms

We now turn to the study of the properties of a more complex system, i.e. a molecule. We shall confine ourselves to the consideration of the simplest diatomic molecule. The basic properties of molecular systems will be illustrated in this simple example.

We have seen in the preceding chapter that the calculation for atoms is carried out by approximate methods. It is natural that approximate methods of calculation should also be widely used in the theory of molecules. The Hamiltonian of the diatomic molecule is of the form

$$\left[-\frac{\hbar^2}{2M_1} \nabla_1^2 - \frac{\hbar^2}{2M_2} \nabla_2^2 - \sum_{k=1}^{n} \frac{\hbar^2}{2m} \nabla_k^2 + U(\mathbf{r}_k, \mathbf{R}_i) \right] \psi_{\text{mol}}(\mathbf{r}_k, \mathbf{R}_i) =$$

$$= E\psi_{\text{mol}}(\mathbf{r}_k, \mathbf{R}_i) . \tag{78.1}$$

Here M_1 and M_2 are the masses of the nuclei, m is the mass of an electron, $\mathbf{r}_k$ are the coordinates of the electrons, and $\mathbf{R}_i$ are the coordinates of the nuclei. The potential energy $U(\mathbf{r}_k, \mathbf{R}_i)$ involves the interaction of the electrons with the nuclei, the interaction of the electrons with each other and the interaction of the nuclei with each other. The summation over k is carried out over all the electrons of the molecule. The velocities of the nuclei, which have masses larger than the electron mass by a factor of several thousand, are substantially lower than the velocities of the electrons. Correspondingly, in the theory of molecules use is made of the adiabatic approximation (cf. §57). The wave function of the system is written in the form

$$\psi_{mol} = \alpha_n(\mathbf{R}_i)\psi_n(\mathbf{r}_k, \mathbf{R}_i) .$$

Using formulae (57.7) and (57.8), we can write the equations for the functions α and ψ:

$$\left[-\frac{\hbar^2}{2m} \sum \nabla_k^2 + U(\mathbf{r}_k, \mathbf{R}_i) \right] \psi_n = E_n(\mathbf{R}_i)\psi_n , \quad \hat{H}_{el}\psi_n = E_n\psi_n , \tag{78.2}$$

$$\left[-\frac{\hbar^2}{2M_1} \nabla_1^2 - \frac{\hbar^2}{2M_2} \nabla_2^2 + E_n(\mathbf{R}_i) \right] \alpha_n(\mathbf{R}_i) = E\alpha_n(\mathbf{R}_i) . \tag{78.3}$$

Eq. (78.2) describes the motion of electrons for nuclei at rest. The quantity $E_n(\mathbf{R}_i)$ defines the energy levels of the system for motionless nuclei which are at a fixed distance from each other. The energy E_n for a fixed distance between the nuclei is called the electron term.

The Schrödinger equation (78.3) describes the motion of the nuclei. The quantity $E_n(\mathbf{R}_i)$ in it is the potential energy of the nuclei. From eq. (78.3) we see that the total energy of the nuclei E depends on the state of the electron part of the system, i.e. on the electron term.

The number of electrons in a molecule is always greater than one. Hence even the solution of the approximate equation (78.2) is associated with great mathematical difficulties which are insuperable in the case of many-electron molecules. (An exception to this is the ion H_2^+, for which a precise solution of the Schrödinger equation for the electron part of the wave function has been obtained.) We are forced, without trying to solve eq. (78.2), to find the most general properties of a system of electrons moving in the field of two nuclei.

For this purpose we find, as usual, quantities which commute with the Hamiltonian $\hat{H}_{el}$, in other words we find quantities which simultaneously have definite values in stationary states of the system. In contrast to the atomic field, which possesses spherical symmetry, the field of a diatomic molecule has cylindrical symmetry. The symmetry axis is the straight line joining the two nuclei, which we shall choose to be the z-axis in what follows. The potential energy of the interaction of the electrons with the nuclei, as well as that of the interaction of the electrons with each other, does not change under a rotation through the angle φ with respect to the z-axis. Hence the Hamiltonian of the system of electrons

$$\hat{H}_{el} = -\frac{\hbar^2}{2m} \sum_k \nabla_k^2 + U$$

does not depend on the angle φ.

Thus we arrive at the conclusion that the component of the total angular momentum of the electrons along the axis of the molecule is conserved. Disregarding the weak spin—orbit interaction, it can be assumed that the component of the orbital angular momentum of the electrons in the z-direction is also conserved. The states of the electrons are classified according to the eigenvalues of the operator L_z. The eigenvalues of the z-component of the orbital angular momentum of the electrons are denoted by the letter Λ. States with $\Lambda = 0, 1, 2$ are called Σ-, Π- and Δ-states (in analogy with the S-, P- and D-states of atoms). The total spin of the electrons S is also conserved for a system of electrons in a molecule.

All the reasoning about the total spin which was presented in the theory of the atom applies also to the molecule.

As in the case of an atom, the multiplicity $2S+1$ of an electron term of a molecule is indicated in the form of a superscript on the left of the quantum number Λ, i.e. in the form $^{2S+1}\Lambda$.

Further, we shall show that the Hamiltonian $\hat{H}_{el}$ does not change under the reflection of the coordinates of the electrons in any plane passing through the nuclei of the molecule (i.e. through the z-axis). In other words, we shall show that the Hamiltonian commutes with the reflection operator. This can easily be seen if, for example, the plane in which the reflection takes place is chosen in such a way that it passes through the z-axis and y-axis. In this case the reflection corresponds to the replacement of all coordinates $x_i \rightarrow -x_i$. But, since the interaction depends only on the distance between the particles $(x_1 - x_2)^2$, it becomes evident that the Hamiltonian commutes with the reflection operator. Hence it follows that the operator $\hat{H}_{el}$ and the reflection operator in a plane passing through the axis of the molecule have common eigenfunctions. Therefore stationary states can be characterized, in addition to the eigenvalues Λ, by the eigenvalues P_i of the reflection operator. The latter, as is easily seen, takes on two values $P_i = \pm 1$. However, things are complicated by the fact that the angular momentum component operator $\hat{L}_z$ does not commute with the reflection operator. Indeed, the operator of the z-component of the angular momentum has the form

$$\hat{L}_z = -i\hbar y \frac{\partial}{\partial x} + i\hbar x \frac{\partial}{\partial y}.$$

The x-coordinate will change sign under reflection in the zy-plane, whereas the y-coordinate will not. Hence it follows directly that the operators $\hat{P}_i$ and $\hat{L}_z$ do not commute. Therefore molecular terms cannot be simultaneously characterized by means of the quantities Λ and P_i, except for terms for which $L_z = 0$. In this last case, states with parity $P_i = 1$ and $P_i = -1$, which are denoted by Σ^+ and Σ^-, are possible.

The wave functions corresponding to these states change and do not change sign respectively under the action of the operator $\hat{P}_i$, corresponding to reflection in a plane passing through the nuclei of the molecule.

Let us now consider the particular case of a molecule with identical nuclei. If the origin is chosen to be at the point which lies on the z-axis halfway between the nuclei, then it is easily seen that the operator of inversion of the electron coordinates (corresponding to the replacement of all coordinates of the electrons by the inverse coordinates $\mathbf{r}_i \to -\mathbf{r}_i$) commutes with the Hamiltonian $\hat{H}_{\mathrm{el}}$. Since at the same time the operator $\hat{P}_i$ of the reflection of electron coordinates in a plane passing through the nuclei of the molecule commutes with $\hat{H}_{\mathrm{el}}$, the state with $L = 0$ can be characterized by three eigenvalues $\Lambda = 0$, $P_i = \pm 1$ and the eigenvalues of the inversion operator (see §33). The latter has two values ± 1 denoted by the letters g (even state) and u (odd state). These indices are written as subscripts on the right. For example, $^1\Sigma_g^+$ corresponds to a term whose wave function is even and does not change sign under the action of the operator of reflection in a plane passing through the z-axis; the z-component of the angular momentum is equal to zero; the term is singlet. Further, we know that the inversion operator commutes with the operator L_z. Hence the states Π and Δ also can be both even and odd. In other words, the states Π_u, Π_g, Δ_u, Δ_g and so on are possible.

Let us dwell on the problem of degeneracy of the electron terms. If Λ is defined, then this means that the absolute value of the z-component of the angular momentum is defined. Since the energy of the system cannot depend on the orientation of the angular momentum component with respect to the z-axis, i.e. is the same for $L_z = +\Lambda$ and $L_z = -\Lambda$, we arrive at the conclusion that each term with $L_z \neq 0$ is two-fold degenerate. Finally, we point out that the energy of the electron terms of the molecule is of the same order as the energy of atomic terms.

§79. The hydrogen molecule. Ideas of the theory of chemical binding

The only molecule for which one can obtain a reasonably accurate solution of the equation for the electron term is the hydrogen molecule. This calculation is of great theoretical importance.

If the energy is measured with respect to the energy of the separated motionless atoms, then the energy values of the electron terms of the stable molecule have negative values. The energy (negative) of the molecule is a measure of the chemical binding of the constituent atoms. Thus the calculation of the electron terms of the molecule represents at the same time a quantitative theory of chemical binding between atoms.

The establishment of the nature of chemical binding is one of the fundamental results of quantum mechanics.

Before the appearance of quantum mechanics there were no substantiated concepts of the nature of chemical binding, in particular of the nature of homopolar molecules. We recall that homopolar molecules are molecules made up of neutral atoms. For example, molecules containing identical atoms are of this type. We shall try to account for certain characteristic features of the theory of chemical binding in the example of the hydrogen molecule.

The Schrödinger equation for the electron terms of the hydrogen molecule is of the form

$$\left(-\frac{\hbar^2}{2m} \nabla_1^2 - \frac{\hbar^2}{2m} \nabla_2^2 + \frac{e^2}{R} - \frac{e^2}{r_{a_1}} - \frac{e^2}{r_{a_2}} - \frac{e^2}{r_{b_1}} - \frac{e^2}{r_{b_2}} + \frac{e^2}{r_{12}} \right) \psi = E\psi . \qquad (79.1)$$

Here R is the distance between the nuclei of the hydrogen atoms; the quantities $r_{a_1}, r_{a_2}, r_{b_1}, r_{b_2}$ are respectively the distances between nucleus a and the first electron, nucleus a and the second electron, nucleus b and the first electron, and nucleus b and the second electron; and r_{12} is the distance between the electrons.

If the atoms forming the molecule are placed an infinitely large distance apart, then it can be said that one of the electrons, for example N_1, will be bound to nucleus a, and the other (electron N_2) to nucleus b. By virtue of the identity of electrons, such a statement makes no sense when the atoms are brought together.

An exact solution of the Schrödinger equation (79.1) involves great mathematical difficulties. Hence a number of approximate calculations have been carried out. We shall make use of perturbation theory, which allows the basic properties of the system to be elucidated relatively simply. The question as to the degree of accuracy of the calculation will be discussed later. We choose as the wave function of zero order approximation the wave function of the system with infinitely distant nuclei. For an infinite distance between the nuclei ($R \to \infty$) the wave function of the two electrons and two nuclei has the form

$$\varphi_1^0 = \psi_a(\mathbf{r}_{a_1})\psi_b(\mathbf{r}_{b_2}) , \qquad (79.2)$$

where $\psi_a(\mathbf{r}_{a_1})$ and $\psi_b(\mathbf{r}_{b_2})$ are the wave functions of the hydrogen atom in which respectively the first electron is near nucleus a and the second electron is near nucleus b.

Evidently, these functions satisfy the equations

$$\left(-\frac{\hbar^2}{2m}\nabla_1^2 - \frac{e^2}{r_{a_1}}\right)\psi_a(\mathbf{r}_{a_1}) = E_0\psi_a(\mathbf{r}_{a_1}),$$

$$\left(-\frac{\hbar^2}{2m}\nabla_2^2 - \frac{e^2}{r_{b_2}}\right)\psi_b(\mathbf{r}_{b_2}) = E_0\psi_b(\mathbf{r}_{b_2}).$$

(79.3)

We are interested in the normal state of the hydrogen molecule. Therefore E_0 must be understood to be the lowest energy level of the hydrogen atom. Indeed, the same energy is possessed by the state

$$\varphi_2^0 = \psi_a(\mathbf{r}_{a_2})\psi_b(\mathbf{r}_{b_1}),$$

(79.4)

which differs from the first state by electron exchange. We stress that φ_1^0 and φ_2^0 are eigenfunctions of different operators and are not orthogonal to each other.

We write the wave functions of the zero order approximation in the form of symmetrized combinations of the functions φ_1^0 and φ_2^0, i.e. as

$$\psi_s^0 = A_1[\psi_a(\mathbf{r}_{a_1})\psi_b(\mathbf{r}_{b_2}) + \psi_a(\mathbf{r}_{a_2})\psi_b(\mathbf{r}_{b_1})],$$

$$\psi_a^0 = A_2[\psi_a(\mathbf{r}_{a_1})\psi_b(\mathbf{r}_{b_2}) - \psi_a(\mathbf{r}_{a_2})\psi_b(\mathbf{r}_{b_1})].$$

(79.5)

The constants A_1 and A_2 are defined by the normalization condition

$$\int |\psi_s^0|^2\, dV_1 dV_2 = \int |\psi_a^0|^2\, dV_1 dV_2 = 1.$$

They are equal to

$$A_1 = [2(1+s^2)]^{-\frac{1}{2}}, \qquad A_2 = [2(1-s^2)]^{-\frac{1}{2}},$$

where the quantity s represents the degree of non-orthogonality of the functions φ_1^0 and φ_2^0 and is equal to

$$s^2 = \int \varphi_1^0 \varphi_2^0\, dV_1 dV_2.$$

(79.6)

For such a choice of unperturbed functions, which represent the symmetrized wave functions of individual atoms, the perturbation operator is expressed by

$$\hat{H}' = \frac{e^2}{R} - \frac{e^2}{r_{a_2}} - \frac{e^2}{r_{b_1}} + \frac{e^2}{r_{12}}.$$

Direct application of perturbation theory is inadmissible: the zero order wave functions are not orthogonal to each other. Therefore it is necessary to modify the perturbation theory somewhat. We write the perturbed wave functions and the energy of the perturbed system in the form

$$\psi_s = \psi_s^0 + \psi_s',$$
$$\psi_a = \psi_a^0 + \psi_a', \qquad\qquad (79.7)$$
$$E = E_0 + \epsilon.$$

Then eq. (79.1) for the function ψ_s' is written in the form

$$\frac{\hbar^2}{2m}(\nabla_1^2+\nabla_2^2)A_1[\psi_a(\mathbf{r}_{a_1})\psi_b(\mathbf{r}_{b_2}) + \psi_a(\mathbf{r}_{a_2})\psi_b(\mathbf{r}_{b_1})] + \frac{\hbar^2}{2m}(\nabla_1^2+\nabla_2^2)\psi_s' +$$

$$+\left[E_0 + \epsilon - \frac{e^2}{R} - \frac{e^2}{r_{12}} + \frac{e^2}{r_{a_1}} + \frac{e^2}{r_{a_2}} + \frac{e^2}{r_{b_1}} + \frac{e^2}{r_{b_2}}\right] \times$$

$$\times A_1[\psi_a(\mathbf{r}_{a_1})\psi_b(\mathbf{r}_{b_2}) + \psi_a(\mathbf{r}_{a_2})\psi_b(\mathbf{r}_{b_1})] +$$

$$+\left[E_0 + \epsilon - \frac{e^2}{R} - \frac{e^2}{r_{12}} + \frac{e^2}{r_{a_1}} + \frac{e^2}{r_{a_2}} + \frac{e^2}{r_{b_1}} + \frac{e^2}{r_{b_2}}\right]\psi_s' = 0.$$

Using eq. (79.3) and dropping small terms containing the product of the perturbation operator with the perturbed wave function, we obtain

$$\left\{\frac{\hbar^2}{2m}(\nabla_1^2+\nabla_2^2) + \frac{e^2}{r_{a_1}} + \frac{e^2}{r_{b_2}} + E_0\right\}\psi_s' =$$

$$= -\left\{A_1\left[\epsilon - \frac{e^2}{R} - \frac{e^2}{r_{12}}\right]\psi_s^0 + A_1\left(\frac{e^2}{r_{b_1}} + \frac{e^2}{r_{a_2}}\right)\varphi_1^0 +$$

$$+ A_1\left(\frac{e^2}{r_{a_1}} + \frac{e^2}{r_{b_2}}\right)\varphi_2^0\right\}. \qquad\qquad (79.8)$$

An analogous expression is obtained upon substituting ψ_a.

For further calculations we shall make use of the following general theorem. In order that the solution of the equation with the right-hand side

$$(\hat{H}_0 - E_k^0)\psi = \varphi$$

may exist it is necessary that the right-hand side φ be orthogonal to the function ψ_k^0 satisfying a homogeneous equation.

(The proof is particularly easy for the case of a non-degenerate spectrum. We write the equation in the form

$$(\hat{H}_0 - E_n^0)\psi = \varphi , \tag{79.9}$$

where $\hat{H}_0$ is a linear operator having the non-degenerate spectrum of eigenvalues

$$\hat{H}_0 \psi_k^0 = E_k^0 \psi_k^0 .$$

We expand the function ψ in terms of the functions ψ_k^0:

$$\psi = \sum a_k \psi_k^0 .$$

Substituting into (79.9), we have

$$\sum_{k \neq n} a_k (E_k^0 - E_n^0)\psi_k^0 = \varphi .$$

Calculating the integral $\int \psi_n^0 \varphi \, dx$, we find that it is equal to zero. This proves our statement.)

Applying this theorem to (79.8), we require that the right-hand side of (79.8) be orthogonal to the solution of the homogeneous equation, i.e. to the unperturbed eigenfunction φ_1^0. Multiplying the right-hand side of (79.8) by φ_1^0 and integrating, we find

$$\int \left(\epsilon - \frac{e^2}{r_{12}} - \frac{e^2}{R} \right) [\varphi_1^0 \pm \varphi_2^0] \varphi_1^0 \, dV +$$

$$+ \int \left(\frac{e^2}{r_{b_1}} + \frac{e^2}{r_{a_2}} \right) (\varphi_1^0)^2 \, dV \pm \int \left(\frac{e^2}{r_{a_1}} + \frac{e^2}{r_{b_2}} \right) \varphi_1^0 \varphi_2^0 \, dV = 0 ,$$

where the plus sign refers to the symmetric function. Analogous calculations lead to a formula with minus sign in the case of the antisymmetric function.

Taking into account the normalization condition, we have

$$\epsilon(1 \pm s^2) = \int \left(\frac{e^2}{R} + \frac{e^2}{r_{12}} - \frac{e^2}{r_{b_1}} - \frac{e^2}{r_{a_2}} \right) (\varphi_1^0)^2 \, dV \pm$$

$$\pm \int \left(\frac{e^2}{R} + \frac{e^2}{r_{12}} - \frac{e^2}{r_{a_1}} - \frac{e^2}{r_{b_2}} \right) \varphi_1^0 \varphi_2^0 \, dV$$

Solving this equation for ϵ, we find the correction to the energy

$$\epsilon = \frac{J \pm K}{1 \pm s^2} , \tag{79.10}$$

where

$$J(R) = e^2 \int \psi_a^2(\mathbf{r}_{a_1}) \psi_b^2(\mathbf{r}_{b_2}) \left(\frac{1}{R} - \frac{1}{r_{a_2}} - \frac{1}{r_{b_1}} + \frac{1}{r_{12}} \right) dV_1 dV_2 ,$$

$$K(R) = e^2 \int \psi_a(\mathbf{r}_{a_1}) \psi_b(\mathbf{r}_{b_1}) \psi_a(\mathbf{r}_{a_2}) \psi_b(\mathbf{r}_{b_2}) \times \qquad (79.11)$$

$$\times \left(\frac{1}{R} - \frac{1}{r_{a_1}} - \frac{1}{r_{b_2}} + \frac{1}{r_{12}} \right) dV_1 dV_2 .$$

In formula (79.10) the plus sign corresponds to the symmetric state characterized by the wave function ψ_s^0, and the minus sign corresponds to the antisymmetric state (the wave function ψ_a^0). If ψ_s^0 and ψ_a^0 were mutually orthogonal, then we would find that $s = 0$. In this case (79.10) would be the same as the ordinary formula of perturbation theory for the correction to the energy.

The expressions found for J and K are analogous to the integrals obtained in §67. The integral J defines the Coulomb interaction of the nuclei and electrons with each other. The quantity K represents the exchange energy. The total wave function of the hydrogen molecule represents the product of the spatial functions and the spin function. Since the total wave function of the system must be antisymmetric, the spatial function symmetric in coordinates must be multiplied by the antisymmetric spin function and vice versa. Hence the state (ψ_s^0) is a state with zero spin for the electrons (singlet state). Analogous reasoning shows that the wave function ψ_a^0 describes a state of the molecule with a spin of one. To find out which of these two states is the bound state (molecule), it is necessary to find the dependence of the quantity ϵ on the radius R. This can be done if the wave functions of the normal state of the hydrogen atom are substituted for ψ_a and ψ_b in integral (79.11). The results of the calculation are conveniently presented in a graph (fig. V.23). Here E_1 and E_2 are the energies of the molecule corresponding respectively to the singlet and triplet states. We see that two hydrogen atoms having the total electron spin equal to one cannot form a bound state, since E_2 has no minimum. The bound state can only be the singlet state. Knowing the form of $E_1(R)$, it is possible to determine the binding energy as well as the effective size of the molecule. The minimum of the potential energy lies at $R_0 = 0.79$Å. The binding energy is not in very good agreement with experimental values. This is due to the fact that the operator chosen as the perturbation does not contain a small parameter and is not small in comparison with the unperturbed operator. Therefore quantitative application of the perturbation theory is inapplicable. Somewhat better results are obtained by other approximate methods. However, the general qualitative conclusions on the

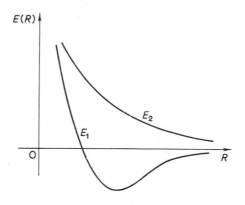

Fig. V.23

nature of chemical binding with the formation of a stable molecule from two atoms is correct. The stability of the molecule is wholly determined by the magnitude and sign of the exchange integral. Namely, for a stable chemical compound to be formed it is necessary (although not sufficient) that the spins of the atoms be antiparallel. This is often expressed by the not strictly correct statement that 'the forces of chemical binding are exchange forces'. In §67 we have discussed in detail the quantum-mechanical theory of exchange forces and the meaning of such formulations. Anyhow, it is beyond doubt that the forces responsible for the formation of homopolar chemical compounds are of a specific quantum-mechanical character. It is also often said that 'antiparallel spins are coupled'. The preceding calculation clearly shows the large degree of relativity of such a terminology. In the example of the hydrogen molecule one can show not only the quantum-mechanical nature of the forces of chemical binding but also the difficulties arising in calculating the formation of molecular systems.

The above calculation allows one to draw the general conclusion that to each valence in a chemical compound there corresponds a pair of electrons with antiparallel spins, bound to each other by the exchange interaction.

Hence it follows that unpaired outer electrons, which are in open shells, are responsible for the chemical properties of atoms. The valence of the atom is determined by the number of such unpaired electrons. When a homopolar chemical compound is formed these electrons get 'collectivized', i.e. can no longer be considered to belong to a given atom. At the same time the configuration of the unfilled state is modified in such a way that it approaches a filled structure. In other words, the electrons get paired in such a way that the spins of all the electrons in the molecule tend to compensate for each other. Stable homopolar molecules tend to have all their electrons paired. In the most stable molecules all electrons are paired, the total spin of the molecule is $S = 0$ and the multiplicity is equal to one.

These qualitative results are in good agreement with experimental data for most molecules. We should also mention some facts which are important for understanding the formation of molecules. The first of these is the property (mentioned in §73) of atoms to enter into chemical binding in an excited state rather than in the normal state. In the light of the theory just discussed this property becomes comprehensible. When two atoms interact, the action of the perturbation can cause one of them to make a transition into an excited state. If the energy gained in forming a compound with an atom in an excited state is larger than that for a compound with the atom in the normal state, then the former will correspond to a more stable molecular configuration.

Thus, for example, we have seen in §73 that the stable term of the carbon atom is $^3P_0$. Carbon atoms have two unpaired electrons in the 2p-state, and the valence in the normal state is equal to two. However, the carbon atom has an excited state with the configuration $1s^2 2s 2p^3$, in which the atom is in the state ^5S. This state is higher than the normal state by 4.2 eV. In the ^5S-state carbon has four unpaired electrons and its valence is equal to four. In compounds for which an energy larger than 4.2 eV is gained, carbon has valence four.

Another property of molecules following from the general theory is their geometric form. Formation of a chemical compound is related to the values of the Coulomb and exchange integrals which contain the products of the wave functions. If among these wave functions there are wave functions of electrons in the p-state, which have anisotropy in space (see §38), then the largest overlap of the wave functions is reached in selected spatial directions. As an example we can mention the molecule NH_3. The nitrogen atom, having the configuration $1s^2 2s^2 2p^3 - {}^4S$, has three p-electrons responsible for chemical binding. The wave functions of these electrons have their largest value in three mutually perpendicular directions. In the molecule NH_3 the angles between the bonds N–H are close (although not exactly equal) to $90°$.

The last fact important for understanding the structure of molecules is that the wave function of an excited state represents a linear combination of the wave functions of electrons. For example, in the excited state of the carbon atom mentioned above the wave function is a linear combination of the wave functions of one 2s-electron and three 2p-electrons. 2s- and 2p-states may be involved with different weights in forming the wave function.

This fact, called the hybridization of states, makes it possible, for instance, to understand why all four valences of the carbon atom are completely identical to each other.

Referring the reader to the specialist literature* for details, we stress that very great mathematical difficulties arise in attempting quantitative calculations of the forces of chemical binding and of the structure of the molecules formed. These difficulties are associated with the fact that the interaction responsible for the formation of a chemical bond cannot be considered as a small perturbation.

§80. The interaction of atoms at large distances

In addition to the forces of chemical character which in certain cases bind atoms into molecules, there is a weak interaction between atoms which are at relatively large distances from each other. Let us consider two widely separated atoms. Because of their spherical symmetry, the atoms have no mean dipole moments. However, the non-diagonal matrix elements of the dipole moments are different from zero.

One can imagine dipole moments of atoms arising in an obvious way as a result of the quantum-mechanical motion of the electrons, with the appearance and disappearance of a dipole moment, equal to zero only on the average. As a result instantaneous dipole moments are induced in both atoms. As an obvious illustration, one often considers the interaction of two oscillators in which the induced dipole moments are directly expressed in terms of their zero-point oscillations.

The interaction energy of atoms can be calculated by perturbation theory. As the perturbation operator one takes the energy of interaction of two dipoles given by formula (17.12) of Part I. The first order correction to the energy is equal to

$$E_1 = H'_{nn} = \int \psi^* \left\{ \frac{d_1 \cdot d_2 - 3(d_1 \cdot n)(d_2 \cdot n)}{R^3} \right\} \psi \, dV =$$
$$= \frac{\bar{d}_1 \cdot \bar{d}_2 - 3(\bar{d}_1 \cdot n)(\bar{d}_2 \cdot n)}{R^3} = 0$$

---

* See, for example, H.Eyring, I.Walter and G.Kimbal, *Quantum chemistry* (Wiley, New York, 1944, 1958); U.Kosman, *Introduction to quantum chemistry* (Academic Press, New York, 1957).

by virtue of $\bar{d}_1 = \bar{d}_2 = 0$. The second order correction can be written in the form

$$E_2 = \frac{1}{R^6} \sum_m \frac{[d_1 \cdot d_2 - 3(d_1 \cdot n)(d_2 \cdot n)]^2_{mn}}{E_n^{(0)} - E_m^{(0)}} = \frac{A}{R^6} \, .$$

According to the results of §53, the quantity

$$A = \sum_m \frac{[d_1 \cdot d_2 - 3(d_1 \cdot n)(d_2 \cdot n)]^2_{mn}}{E_n^{(0)} - \dot{E}_m^{(0)}} \, ,$$

is always negative. Hence finally

$$U_{vdW} = E_2 = - \frac{const}{R^6}$$

This formula expresses the law of the van der Waals interaction. This interaction has no specific character, in the sense that it corresponds to attractive forces which decrease as R^{-7} for all atoms irrespective of their nature. The value of the constant can be expressed in terms of the polarizability of the atoms and varies for different atoms. Thus the van der Waals interaction between atoms represents the same specific quantum-mechanical effect as the chemical interaction. It cannot be understood on the basis of classical concepts, since atoms do not have a 'permanent' dipole moment.

The van der Waals forces, in contrast to forces leading to the formation of a chemical bond, possess additivity. If the interaction involves not two but three and more atoms, then the energy of interaction of the system, as any other weak perturbation, is obtained by the addition of the energies of pair interactions.

This result is of a general character, since we have not used a particular wave function.

However, if the atoms are not in S-states, then they can have a mean quadrupole moment different from zero. In this case, in addition to the van der Waals interaction a quadrupole–quadrupole interaction $\sim R^{-5}$ will exist between the atoms.

As distinct from atoms, molecules may have a mean dipole moment. If, however, its value as well as the value of the quadrupole moment is small, then the formula for U_{vdW} also applies to the molecules.

A particular situation arises when two identical atoms in different states interact, say, when an excited and a non-excited atom of one and the same element interact. In this case an additional degeneracy, associated with the possibility of excitation exchange between the atoms, arises in the system.

The perturbation operator in this case is also the dipole–dipole interaction operator. However, the interaction energy is defined not by the mean value of this operator but by the solution of the corresponding secular equation (see §54). If the given atoms have non-zero matrix elements of the transition between the ground state and the excited state considered, then the interaction energy already turns out to be different from zero in the first approximation of perturbation theory. In this case the dipole–dipole interaction with resonant transfer of excitation takes place between the atoms. As can easily be seen, the energy of this interaction decreases only in inverse proportion to the cube of the distance between the atoms, $U \sim R^{-3}$.

Suppose, for example, that one of the atoms is in the ground $^1S_0$ state, while the other is in the excited $^1P_1$ state. We denote the wave functions corresponding to these states respectively by φ and ψ_m (the subscript m characterizes the angular momentum component in the state $^1P_1$, $m = -1, 0, 1$). Thus a system of two non-interacting atoms turns out to be six-fold degenerate. It is described by unperturbed functions of the form $\varphi(1)\psi_m(2)$ and $\varphi(2)\psi_m(1)$. The matrix elements of the interaction operator $\hat{H}'$

$$\hat{H}' = \frac{\mathbf{d}_1 \cdot \mathbf{d}_2}{R^3} - \frac{3(\mathbf{d}_1 \cdot \mathbf{R})(\mathbf{d}_2 \cdot \mathbf{R})}{R^5}$$

with respect to these wave functions are not equal to zero for transitions between states differing by excitation transference. The calculations are conveniently carried out in a system of coordinates with the z-axis direct along the vector $\mathbf{R}$. Solving a secular equation of the form (54.4), we find the expression for the interaction energy

$$U_1 = \pm \frac{g^2}{R^3}, \qquad U_2 = \pm \frac{2g^2}{R^3}, \qquad (80.1)$$

where g denotes a matrix element of the form

$$g = \int \psi_0^* d_z \varphi \, dV.$$

The upper signs in formulae (80.1) refer to symmetric, and the lower signs to antisymmetric, excitations states. Energies U_1 correspond to states with $\Lambda = 1$, while energies U_2 correspond to states with $\Lambda = 0$.

In gaseous systems in which there is a considerable concentration of excited atoms the dipole–dipole interaction with resonant transfer of excitation can play a more important role than the van der Waals interaction. It does not disappear in correct averaging over the orientations of the dipole moment of the atom and gives a basic contribution to the thermodynamic functions of the system*. The resonant dipole–dipole interaction is not additive.

§81. The comparison of molecular terms with atomic terms

The states of a molecule formed from two atoms can be related to the states of the atoms if the process of formation of the molecule is imagined as a result of their infinitely slow approach to each other.

The angular momentum component along the axis joining the two nuclei is conserved in the course of the process. On the other hand, as we have seen before, the component of the total angular momentum Λ along this axis will also be conserved for this molecule (see §78). We shall determine the possible values of Λ, as well as the number of energy states of the molecule formed.

Let the atoms be characterized by total angular momenta L_1 and L_2, respectively. We assume that $L_1 > L_2$. The components of the angular momenta of the atoms can take on, respectively, the following values:

$$M_1 = L_1, \quad L_1 - 1, \quad L_1 - 2, \quad \dots, \quad -L_1,$$

$$M_2 = L_2, \quad L_2 - 1, \quad L_2 - 2, \quad \dots, \quad -L_2,$$

In accordance with the definition of the quantity Λ (see §78), there corresponds to the maximum value $\Lambda = L_1 + L_2$ the only state in which the components of the angular momenta of the atoms are equal to $M_1 = L_1$, $M_2 = L_2$. The next possible value of Λ is equal to $\Lambda = L_1 + L_2 - 1$. To this value of Λ there correspond two terms arising respectively from two states; in the first $M_1 = L_1$, $M_2 = L_2 - 1$, and the second $M_1 = L_1 - 1$, $M_2 = L_2$. Analogously, to the value $\Lambda = L_1 + L_2 - 2$ there correspond 3 terms arising from the states: $M_1 = L_1, M_2 = L_2 - 2; M_1 = L_1 - 1, M_2 = L_2 - 1; M_1 = L_1 - 2$, $M_2 = L_2$. The results obtained are conveniently expressed in the table:

for $\quad \Lambda = L_1 + L_2 \qquad$ 1 term is possible ,

for $\quad \Lambda = L_1 + L_2 - 1 \qquad$ 2 terms are possible ,

for $\quad \Lambda = L_1 + L_2 - 2 \qquad$ 3 terms are possible ,

$\quad \cdot$

$\quad \cdot$

$\quad \cdot$

for $\quad \Lambda = L_1 - L_2 \qquad 2L_2 + 1$ terms are possible .

* V.I.Malnev and S.I.Pekar, Soviet Physics JETP 24 (1967) 1220; 31 (1970) 597; Yu.A.Vdovin, Soviet Physics JETP 27 (1968) 242.

We can see by a simple calculation that the number of terms for $\Lambda < L_1 - L_2$ is equal to $2L_2 + 1$ and does not depend on Λ. In determining all possible states of the system account must be taken of the fact that each energy level with $\Lambda \neq 0$ is degenerate, since the energy of the system cannot depend on the orientation of the angular momentum in space.

The Σ term requires particular consideration.

A molecule turns out to be in the Σ state if $M_1 = -M_2$. This condition is fulfilled in L_2 cases where we have for the angular momentum components $M_1 > 0$ and $M_2 < 0$ and also in L_2 cases where $M_1 < 0$ and $M_2 > 0$. Furthermore, M_1 and M_2 can be equal to zero. Consequently, in the Σ state the molecule can also be formed from $2L_2 + 1$ energy states.

In §78 we have pointed out that Σ terms are divided into Σ^+ and Σ^- terms, depending on the symmetry properties of the system. The symmetry properties of the system do not change when the atoms are put an infinite distance apart. Hence the wave functions of the system for the states $|M_1| = |M_2|$ can be written in the form of symmetric or antisymmetric combinations

$$\psi_s = \psi_M^{(1)} \psi_{-M}^{(2)} + \psi_{-M}^{(1)} \psi_M^{(2)} , \tag{81.1}$$

$$\psi_a = \psi_M^{(1)} \psi_{-M}^{(2)} - \psi_{-M}^{(1)} \psi_M^{(2)} . \tag{81.2}$$

The Σ state corresponding to the values $M_1 = M_2 = 0$ is determined by the behaviour of the function $\psi = \psi_0^{(1)} \psi_0^{(2)}$ under reflection in a plane joining the nuclei of the atoms. Depending on the actual properties of the wave functions $\psi_0^{(1)}$ and $\psi_0^{(2)}$ there arise Σ^+ or Σ^- terms. Thus in L_2 cases a molecule in the Σ^+ state is formed, while in another L_2 cases a molecule in the Σ^- state is formed. One more Σ^+ or Σ^- term arises depending on the form of the function $\psi_0^{(1)} \psi_0^{(2)}$.

So far we have considered molecules formed from two different atoms. If the molecule is made up of identical atoms, then the calculation of its possible states is somewhat modified. Two cases are possible; either the separated atoms are in different states or they are in identical states. In the first case the number of possible terms must be doubled in comparison with the number of terms of a molecule consisting of different atoms, since the state of a molecule made up of identical atoms is invariant under the inversion transformation, and even and odd terms can be formed. If the atoms are in identical states, then the total number of states remains the same as for a molecule with different atoms. The problem of the parity of these states is rather complex*.

* See L.D.Landau and E.M.Lifshitz, *Quantum mechanics* (Pergamon Press, Oxford, 1965) or E.Wigner and E.Witmer, Z.f.Phys. 51 (1958) 859.

§82. Rotation and vibration of diatomic molecules

We can now turn to the quantitative consideration of the motion of the nuclei in diatomic molecules. We shall not be interested in the translational motion of the molecule as a whole.

The motion of the nuclei in a molecule depends only on the distance between the nuclei. In the adiabatic approximation, according to eq. (78.3), the wave function satisfies the Schrödinger equation, which in spherical coordinates has the form

$$\left[-\frac{\hbar^2}{2\mu} \frac{1}{R^2} \frac{\partial}{\partial R} \left(R^2 \frac{\partial}{\partial R} \right) + E_n(R) + \frac{\hbar^2 \hat{K}^2}{2\mu R^2} \right] \alpha_n = 0 , \qquad (82.1)$$

where $E_n(R)$ is the electron energy, and $\hat{K}$ is the angular momentum operator of the motion of the nuclei. We assume the electron energy to be fixed and consider the nuclei for a given E_n. Then the motion of the nuclei amounts to rotations and vibrations about an equilibrium position. The angular momentum operator of the nuclei must be expressed in terms of the total angular momentum operator of the molecule, which can be written in the form $\hat{J} = \hat{K} + \hat{L}$, where $\hat{L}$ is the angular momentum of the system of electrons.

It is clear that the molecule is in a state with a definite value of the total angular momentum. The angular momentum of the nuclei can run over a sequence of values corresponding to different rotational states of the electrons. Hence we shall be interested only in the mean value of the quantity

$$\overline{(\hat{K})^2} = \overline{(\hat{J}-\hat{L})^2} = J^2 + \overline{(\hat{L})^2} - 2\overline{\hat{J} \cdot \hat{L}}$$

since $\hat{J}^2$ has a definite value and is conserved.

According to what was said in §78, the component of the angular momentum of the electrons along the axis of the molecule, $L_z = \Lambda$, is conserved. The two other components are on the average equal to zero: $\overline{L_x} = 0$, $\overline{L_y} = 0$. Hence we have $\overline{L^2} = \Lambda^2$. Furthermore, since the direction of the vector n (the axis of the molecule) is the only specified direction, the following equality holds:

$$\overline{L} = n\Lambda .$$

The vector of the mean angular momentum of the nuclei in a diatomic molecule is perpendicular to n, i.e.

$$\hat{K} \cdot n = (\hat{J} - \hat{L}) \cdot n = 0 .$$

(This statement follows directly from the fact, known from classical mechanics, that the angular momentum vector in a two-body system is perpendicular to the axis joining to two bodies.) Hence it follows that

$$\hat{\mathbf{J}} \cdot \mathbf{n} = \hat{\mathbf{L}} \cdot \mathbf{n} = \Lambda . \qquad (82.2)$$

From (82.2) we obtain

$$(\hat{\mathbf{J}} \cdot \hat{\mathbf{L}}) = \Lambda^2 .$$

Finally, we find

$$(\hat{K})^2 = \hat{J}^2 + \overline{L^2} - 2\Lambda^2 = J(J+1) + \overline{L^2} - 2\Lambda^2 . \qquad (82.3)$$

where the quantum number J runs over a sequence of integers $J \geqslant \Lambda$.

In formula (82.3) the last two terms depend only on the state of the system of electrons, whereas the first term characterizes the rotation of the molecule as a whole.

The Schrödinger equation assumes the form

$$\left\{ -\frac{\hbar^2}{2\mu} \frac{1}{R^2} \frac{\partial}{\partial R} \left(R^2 \frac{\partial}{\partial R} \right) + E_n(R) + \frac{\hbar^2 (\overline{\hat{L}^2 - 2\Lambda})^2}{2\mu R^2} + \frac{\hbar^2 J(J+1)}{2\mu R^2} \right\} \alpha_n = 0 . \qquad (82.4)$$

Denoting

$$U(R) = E_n(R) + \frac{\hbar^2 (\overline{\hat{L}^2 - 2\Lambda^2})}{2\mu R^2} \qquad (82.5)$$

we see that $U(R)$ plays the role of an effective potential energy. We shall consider those states of the nuclei for which the distance between the nuclei remains close to the equilibrium distance.

The effective potential energy can be written in the form

$$U(R) \cong U(R_0) + \frac{d^2 U}{dR^2}\bigg|_{R=R_0} \tfrac{1}{2}(R-R_0)^2 = U(R_0) + \tfrac{1}{2}\mu\omega_0^2(R-R_0)^2 ,$$

where ω_0 is the frequency of vibration. Eq. (82.4) finally assumes the form

$$\left[-\frac{\hbar^2}{2\mu} \frac{\partial}{\partial R} \left(R^2 \frac{\partial}{\partial R} \right) + U(R_0) + \tfrac{1}{2}\mu\omega_0^2(R-R_0)^2 + \frac{\hbar^2}{2\mu R_0^2} J(J+1) \right] \alpha = 0 . \qquad (82.6)$$

We see that in the adiabatic approximation used, the motion of the molecule for a given electron state amounts to its rotation as a whole together with harmonic vibrations.

The total energy of the molecule is given by the formula

$$E = E^{\text{el}} + \frac{\hbar^2}{2\mu R_0^2} J(J+1) + \hbar\omega_0(v+\tfrac{1}{2}) , \qquad (82.7)$$

where v is the vibrational quantum number.

Let us also estimate the degree of accuracy of the adiabatic approximation. According to (57.6) the parameter of non-adiabaticity is in order of magnitude

$$\hat{C}\alpha \sim \frac{\hbar^2}{\mu}\nabla\alpha\int\varphi^*\,\nabla\varphi\,dV. \tag{82.8}$$

We are interested in the dependence of this expression on the reduced mass μ. We estimate in order of magnitude the derivative $\nabla\alpha$ in the vibrational ground state. Evidently, $\nabla\alpha \sim \alpha[\overline{(R-R_0)^2}]^{-\frac{1}{2}}$. In order to estimate the mean displacement $[\overline{(R-R_0)^2}]^{\frac{1}{2}}$ we note that in the ground state the mean potential energy is equal to one half of the total energy, i.e.

$$\tfrac{1}{2}\mu\omega_0^2\overline{(R-R_0)^2} = \tfrac{1}{2}\hbar\omega_0 .$$

Hence $[\overline{(R-R_0)^2}]^{\frac{1}{2}} \sim (\omega_0\mu)^{-\frac{1}{2}}$. Thus $\nabla\alpha \sim (\omega_0\mu)^{\frac{1}{2}}\alpha$. But by definition

$$\omega_0 = [\mu^{-1}(d^2U/dR^2)_{R=R_0}]^{\frac{1}{2}} \sim \mu^{-\frac{1}{2}} .$$

The integral in (82.8) depends only on the electron part of the system and does not depend on μ. Hence, finally, $C \sim \mu^{-\frac{3}{4}}$. From dimensionality considerations it follows that

$$C \sim (m/\mu)^{\frac{3}{4}} . \tag{82.9}$$

Indeed, the total Schrödinger equation describing the motion of all particles in the molecule involves only two quantities of the dimensionality of mass; μ and the electron mass m. One cannot construct any other quantities of the dimensionality of mass from the quantities involved in the Schrödinger equation. Thus the parameter C is very small even for the hydrogen molecule. The quantity $(m/\mu)^{\frac{3}{4}}$ is the basic small parameter of the theory of molecules.

The spacings between electron energy levels ΔE^{el} for molecules do not depend on the mass of the nuclei and are of the same order of magnitude as for atoms (i.e. are of the order of a few eV).

The spacing between vibrational levels is

$$\Delta E^{\mathrm{vib}} = \hbar\omega_0 \sim \mu^{-\frac{1}{2}} \ll \Delta E^{\mathrm{el}} . \tag{82.10}$$

This amounts to a few tenths of an electronvolt.

Finally, the spacing between rotational levels is

$$\Delta E^{\mathrm{rot}} = \frac{\hbar^2}{2\mu R_0^2}[J'(J'+1)-J(J+1)] \tag{82.11}$$

where $\Delta J = J' - J = \pm 1$.

Since $\Delta E^{\mathrm{rot}} \sim \mu^{-1}$, this spacing is much smaller than that between vibrational levels and amounts to few meV.

Knowing the distribution of levels, one can find the emission (or absorption) spectra of molecules, which differ strongly in character from atomic line spectra. However, one then has to take into account an important fact. In ch. 12, devoted to radiation theory, it will be shown that transitions between levels are limited by so-called selection rules. It turns out that transitions are possible only between levels for which a change in the quantum numbers defined by the conditions

$$
J' \to J \qquad
\begin{array}{c}
\nearrow \ J+1 \\
\\
\searrow \ J-1
\end{array}
\qquad v' = v \pm 1
$$

takes place (except for the transition $J' = J = 0$, which is forbidden).

Taking into account the selection rules, the proof of which will be given in ch. 12, it is possible to find the frequencies emitted or absorbed. In the case of transitions for which the electron state of the molecule does not change, we obtain (82.7)

$$
\hbar\omega = \hbar\omega_n(v'-v'') + B[J'(J'+1) - J''(J''+1)] , \tag{82.12}
$$

where $B = \hbar/2\mu R^2$. Taking into account the selection rules, we obtain two frequency branches for a given difference $v' - v''$. For $J'' = J' + 1$ we have the first branch of frequencies:

$$
\hbar\omega_1 = \hbar\omega_n(v'-v'') - 2B(J'+1) , \qquad J' = 0, 1, 2, \dots . \tag{82.13}
$$

For $J'' = J' - 1$ we find the second branch of frequencies:

$$
\hbar\omega_2 = \hbar\omega_n(v'-v'') + 2BJ' , \qquad J' = 1, 2, 3, 4, \dots . \tag{82.14}
$$

We note that J' cannot be equal to zero, since this would correspond to $J'' = -1$.

Let us consider the order and distribution of these frequencies for a given difference $v'-v''$. The frequency ω_1 decreases beginning with $\omega_1 = \omega_n(v'-v'')-2B$, and the frequency ω_2 increases from the lowest value equal to $\omega_n(v'-v'')+2B$. The spacing between the lines in each branch is equal to $2B$, while the spacing between the branches is equal to $4B$. The frequency $\omega_n(v'-v'')$ lying between the bands is not observed. The set of lines ω_1 is also called the P-branch of frequencies, and the set of frequencies ω_2 is the R-branch. These frequencies lie in the infrared part of the spectrum.

We now turn to the study of frequencies which arise in transitions associated with a change in the electron state. The character of such a spectrum differs fundamentally from the infrared spectrum considered. Radiation frequencies are defined in this case by the formula

$$\hbar\omega = E_0 + \hbar\omega_n(v'+\tfrac{1}{2}) - \hbar\omega_m(v''+\tfrac{1}{2}) + B_n(J'+1)J' - B_mJ''(J''+1) . \qquad (82.15)$$

Here it should be stressed that $\omega_n \neq \omega_m$, and $B_n \neq B_m$. As a matter of fact, the vibration frequencies ω and the quantities B are determined by the electron state of the molecule and, consequently, when this state changes these quantities change substantially.

Since the change in the energy of the molecule in transitions associated with a change in the electron state is rather large, the frequencies observed in this case lie in the visible part of the spectrum. The set of lines corresponding to the chosen pair of quantum numbers v'_l and v''_l is called a band. The band is in its turn made up of three branches. These branches are obtained in the following way. In accordance with the selection rules the quantum number J'' can be equal to $J'' = J'-1$, $J'' = J'+1$ and $J'' = J'$. To the first case there corresponds the R-branch, whose frequencies are defined by the relation

$$\omega_1 = A + \bar{B}J'^2 + CJ' , \qquad (82.16)$$

where

$$A = E_0 + \hbar\omega_n(v'+\tfrac{1}{2}) - \hbar\omega_n(v''+\tfrac{1}{2}) ,$$

$$\bar{B} = B_n - B_m , \qquad C = B_n + B_m .$$

Transitions $J'' = J' + 1$ constitute the Q-branch, and frequencies are in this case defined by the formula

$$\hbar\omega_2 = A + \bar{B}J'(J'+1) . \qquad (82.17)$$

and, finally, for the P-branch we obtain

$$\hbar\omega_3 = A - 2B_m + \bar{B}J'^2 + (B_n - 3B_m)J' \qquad (82.18)$$

In all three cases ω is a quadratic function of the quantum number J. To examine the distribution of frequencies it is convenient to refer to the diagram of fig. V.24. Here a parabola corresponding to eq. (82.18) is shown for $\bar{B} > 0$. The quantum number J is plotted on the vertical axis, and the frequency ω on the horizontal axis. Experimentally observed frequencies can easily be obtained by means of this diagram. If the points of intersection of horizontal lines drawn through integer values of J' and the parabola are projected on the horizontal axis, then we obtain the observed values of frequencies. These radiation frequencies are shown below the horizontal axis.

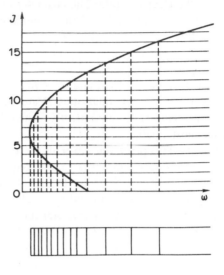

Fig. V.24

We see that the frequencies of the spectrum are not equally spaced, as was the case in the infrared part of the spectrum, but become denser in a certain part of it; as ω increases the spacings between the observed frequencies become larger. The place of convergence of the lines is called the band edge. In the case given the band edge lies on the low frequency side of the spectrum. If $B_n < B_m$, then the parabola is curved in the opposite way. In this case the band edge lies on the high frequency side. However, it should be stressed that many exceptions to the above rule are observed. For example, in the case where electron terms are degenerate more than three bands are observed; sometimes, for example, the Q-branch is not observed and so on*.

---

* For a detailed exposition of these problems see L.D.Landau and E.M.Lifshitz, *Quantum mechanics* (Pergamon Press, Oxford, 1965).

Scattering Theory

§83. Scattering amplitude and cross section

We mean by a scattering process the deflection of particles from their initial directions of motion caused by their interaction with a system which we shall call the scatterer.

The investigation of scattering processes of charged and neutral particles is one of the basic experimental methods of studying the structure of atoms, atomic nuclei and elementary particles.

Indeed, the very existence of the atomic nucleus was established in Rutherford's experiments on the scattering of α-particles. The analysis of the results of experiments on the scattering of neutrons by nuclei enabled Bohr to formulate modern concepts on the structure of the nucleus. The study of the laws of scattering of fast particles is the main source of information about nuclear forces and of the properties of elementary particles.

From these examples, although they are far from being complete, it is easy to evaluate the significance of scattering theory, one of the most important branches of quantum mechanics.

The scattering of a flux of particles is characterized by the differential scattering cross section. This quantity is defined as the ratio of the number of particles dN_{scat} scattered per unit time into the solid angle $d\Omega$ to the flux density of the incident particles, j_{inc}, i.e. the differential cross section is defined by the relation

$$d\sigma(\theta,\varphi) = \frac{dN_{scat}(\theta,\varphi)}{j_{inc}} \, ,$$

where the angles θ and φ define the direction of motion of the scattered particles. The z-axis is taken along the direction of motion of the incident particles.

For our purposes it is convenient to write dN_{scat} in the form

$$dN_{scat}(\theta,\varphi) = j_{scat}(\theta,\varphi) \, ds \, ,$$

where j_{scat} is the flux density of the scattered particles at large distances from the scattering centre, and ds is an element of area perpendicular to the radius vector drawn from the scattering centre at the angles θ, φ. The quantity ds is related to the solid angle element $d\Omega$ by the equality

$$ds = r^2 d\Omega \, .$$

Thus the differential cross section is defined by the formula

$$d\sigma = \frac{j_{scat}}{j_{inc}} \, ds \, . \tag{83.1}$$

This definition of the cross section is the same as that introduced in §43 of Part I. In quantum mechanics j_{scat} and j_{inc} are probability current densities defined in §7.

In the mutual scattering of two quantum-mechanical systems, for example the scattering of an electron by an atom, a neutron by a nucleus, an atom by an atom and so on, one has to distinguish between elastic and inelastic scattering. In elastic scattering the internal state of both the scatterer and the scattered system remains unchanged. For example, in the elastic scattering of electrons by atoms the state of the latter remains unchanged. In inelastic scattering the internal state of one or both systems changes. For example, the scattering of electrons by atoms is inelastic if in the process of scattering the atoms make a transition into an excited state.

In inelastic scattering a part of the kinetic energy goes over into the internal energy or, conversely, the internal energy goes over into kinetic energy. Collisions of this latter type are called collisions of the second kind.

We shall begin the exposition of scattering theory with the simpler case of elastic scattering. In this case one need not be concerned with the internal state of the systems and can call the interacting systems particles (although these systems may have a complex internal structure, for example they may be atoms, molecules or nuclei).

In the scattering process there is an interaction between two particles; the scattered particle and the scatterer. In this case the potential energy of the interaction very often has the form $U(r)$. Then the problem of the motion of two interacting particles can always be reduced to the study of the motion of one particle (with reduced mass μ) in the field of the motionless centre of force at the centre of mass of the system.

In practice it is always necessary to know how the process appears in the laboratory system of coordinates. Therefore if the problem of motion of one particle in the field of external forces is solved, then in the final formulae one has to transform to the laboratory system. This can easily be done, knowing that the cross section (83.1) is invariant under transformation from one Galilean system to another, and that the angles transform by means of the relations (see §43 of Part I)

$$\tan\theta_1 = \frac{m_1 \sin\theta}{m_1 + m_2 \cos\theta}; \qquad \theta_2 = \frac{\pi-\theta}{2}. \tag{83.2}$$

Here θ is the scattering angle of the two particles in the centre-of-mass system; θ_1 and θ_2 are the scattering angles of the first and second particles in the laboratory system of coordinates, in which the second particle was at rest before the collision.

We now consider the wave function of the particle scattered by the force centre. We shall not, as yet, make any assumptions about the concrete form of the potential energy of interaction.

We let the motionless scattering centre be at the origin and take the direction of the incident particle flux to be the z-axis. At a distance from the scattering centre the incident particle moves as a free particle, and its wave function has the form of a plane wave e^{ikz}. Near the force centre the particle undergoes scattering and the form of its wave function is different.

However, when the scattered particle goes sufficiently far from the force centre, it will again move as a free particle. Since the scattered particle flux at a large distance will always be directed from the scattering centre, the motion of the scattered particles must be described by a diverging wave $f(\theta,\varphi)\, e^{ikr}/r$.

The total wave function describing the motion of the incident and scattered particles at large distances from the scattering centre can be written in the form

$$\psi = e^{ikz} + f(\theta,\varphi)\frac{e^{ikr}}{r}, \tag{83.3}$$

where the first term describes the motion of the incident particles, and the second term the motion of the scattered particles.

The amplitude of the diverging wave $f(\theta,\varphi)$, called the scattering amplitude, depends, generally speaking, on the angles θ and φ. According to (83.1), one has to calculate the incident and scattered particle flux densities. In accord with formula (7.6), the flux density in the plane wave e^{ikz} incident on the scattering centre is equal to $p/m = v$, where v is the velocity of the particles. Analogously, the flux density in the diverging wave is given by the expression

$$\frac{|f(\theta,\varphi)|^2 v}{r^2} . \tag{83.4}$$

Taking the ratio of the incident and scattered fluxes we obtain, in correspondence with formula (83.1), the differential cross section

$$d\sigma = |f(\theta,\varphi)|^2 \, d\Omega . \tag{83.5}$$

Thus we see that the cross section is completely determined by the value of the scattering amplitude. The calculation of the latter is usually carried out in the following way. One finds the solution of the Schrödinger equation for the motion of the particle in the field of the scattering centre, which at large distances from the centre has the form (83.3). Then the coefficient of the factor e^{ikr}/r gives the scattering amplitude to be determined.

The wave function describing the motion of the particle at a distance from the scattering centre was written in the form (83.3), i.e. as the sum of incident and diverging waves, on the basis of simple and obvious physical considerations.

However, it can be shown rigorously that at a large distance from the fixed scattering centre $U(r)$ the solution of the Schrödinger equation has indeed the form (83.3). For this we write the Schrödinger equation in the form

$$(\nabla^2 + k^2)\psi = \frac{2mU}{\hbar^2} \psi ,$$

where $k^2 = 2mE/\hbar^2$, and m and E are respectively the mass and energy of the scattered particle. By means of a Green's function the solution can (see §15) be written in the form

$$\psi = \psi_0 + \int G(\mathbf{r},\mathbf{r}') \frac{2m}{\hbar^2} U(\mathbf{r}')\psi(\mathbf{r}') \, dV' , \tag{83.6}$$

where the function ψ_0 satisfies the equation

$$(\nabla^2 + k^2)\psi_0 = 0 .$$

The solution of this equation evidently has the form of a plane wave e^{ikz}.

The Green's function satisfies the equation

$$(\nabla^2 + k^2) G(\mathbf{r}, \mathbf{r}') = \delta(\mathbf{r} - \mathbf{r}') \; .$$

This equation is formally identical with eq. (24.20) of Part I, if ω^2/c^2 in it is replaced by k^2 and $-4\pi \mathbf{j}_0/c$ by $\delta(\mathbf{r}-\mathbf{r}')$. Without reproducing the calculations of § 24 of Part I, we make use of a formula similar to eq. (24.19) of Part I and write the solution for $G(\mathbf{r}, \mathbf{r}')$ in the form

$$G(\mathbf{r}, \mathbf{r}') = \frac{1}{4\pi} \int \frac{\delta(\mathbf{r}'' - \mathbf{r}') \, e^{ik|\mathbf{r} - \mathbf{r}''|} \, dV''}{|\mathbf{r} - \mathbf{r}''|} \; .$$

Carrying out the integration with respect to dV'', we obtain

$$G(\mathbf{r}, \mathbf{r}') = -\frac{1}{4\pi} \frac{e^{ik|\mathbf{r} - \mathbf{r}'|}}{|\mathbf{r} - \mathbf{r}'|} \; .$$

Substituting the values of ψ_0 and G into (83.6), we arrive at the integral equation

$$\psi = e^{ikz} - \frac{m}{2\pi\hbar^2} \int \frac{e^{ik|\mathbf{r} - \mathbf{r}'|} U(\mathbf{r}') \psi(\mathbf{r}') dV'}{|\mathbf{r} - \mathbf{r}'|} \; . \tag{83.7}$$

Further, we consider the integral involved in formula (83.7) and determine its values at large distances $\mathbf{r}$. We define large distances in the following way. Let the range of values of $\mathbf{r}'$ in which the integrand differs considerably from zero and which gives the basic contribution to the value of the integral be R. We shall call large distances those distances $|\mathbf{r}|$ for which the inequality

$$|\mathbf{r}| \gg R \tag{83.8}$$

is fulfilled; such distances always exist when $U(r)$ decreases sufficiently rapidly. In calculating integral (83.7) at large distances it can be assumed that $|\mathbf{r}| \gg |\mathbf{r}'|$.

Expanding $|\mathbf{r} - \mathbf{r}'|$ in a series, we have

$$|\mathbf{r} - \mathbf{r}'| = [(\mathbf{r} - \mathbf{r}')^2]^{\frac{1}{2}} = (r^2 - 2\mathbf{r} \cdot \mathbf{r}')^{\frac{1}{2}} = r - \frac{\mathbf{r} \cdot \mathbf{r}'}{r} \; .$$

Substituting this expansion into (83.7), we find

$$\psi = e^{ikz} - \frac{m e^{ikr}}{2\pi\hbar^2 r} \int U(\mathbf{r}') \, e^{-i\mathbf{k} \cdot \mathbf{r}'} \psi(\mathbf{r}') \, dV' \; . \tag{83.9}$$

Here $\mathbf{k} = k\mathbf{r}/r$. The wave vector $\mathbf{k}$ is evidently directed along the radius vector. It characterizes the direction of propagation of the diverging spherical wave. Comparing (83.9) with (83.3) we see that this last expression is of a general character. The scattering amplitude is equal to

$$f(\theta,\varphi) = - \frac{m}{2\pi\hbar^2} \int U(\mathbf{r}')\psi(\mathbf{r}')\, e^{-i\mathbf{k}\cdot\mathbf{r}'}\, dV' \, . \qquad (83.10)$$

We shall need formulae (83.9) and (83.10) for what follows.

§84. The Born approximation

Although we could find the asymptotic expression for the wave function, the problem of obtaining the concrete form of the scattering amplitude is still far from being solved. Indeed, according to formula (83.10) the scattering amplitude is expressed in terms of the unknown wave function ψ. The exact solution of the Schrödinger equation and the determination of $f(\theta,\varphi)$ in most problems of practical interest is associated with very great mathematical difficulties. Therefore approximate methods are widely used in scattering theory. The most important of these is the Born approximation. This method is based on the assumption that the potential energy of interaction of the scattered particle with the force centre is small, so that it can be considered as a small perturbation.

If the potential energy is a small perturbation, then it can be assumed that the initial motion of the particle is changed only slightly. Then the integral equation (83.9) can easily be solved by a method of successive approximation. In the zero order approximation the small term containing the potential energy can be dropped. Then

$$\psi_0 = e^{ikz} = e^{i\mathbf{k}_0\cdot\mathbf{r}} \, , \qquad (84.1)$$

where $\mathbf{k}_0$ is a vector equal to $\mathbf{k}_0 = k\mathbf{n}_0$, and $\mathbf{n}_0$ is the unit vector along the z-axis. In the first approximation, in place of the wave function on the right-hand side of (83.9), one has to substitute the value of its zero order approximation (84.1). We obtain

$$\psi = e^{ikz} - \frac{m}{2\pi\hbar^2 r} e^{ikr} \int U(\mathbf{r}')\, e^{ikz'-i\mathbf{k}\cdot\mathbf{r}'}\, dV' \, . \qquad (84.2)$$

In this approximation the scattering amplitude is equal to

$$f(\theta,\varphi) = - \frac{m}{2\pi\hbar^2} \int U(\mathbf{r}')\, e^{i\mathbf{K}\cdot\mathbf{r}'}\, dV' \, . \qquad (84.3)$$

Here we have introduced the notation

$$\mathbf{K} = \mathbf{k}_0 - \mathbf{k} \, . \qquad (84.4)$$

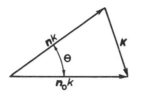

Fig. V.25

where, in correspondence with fig. V.25, the modulus of the vector $\mathbf{K}$ is defined by the relation

$$K = k|\mathbf{n}-\mathbf{n}_0| = 2k \sin \tfrac{1}{2}\theta = \frac{2mv}{\hbar} \sin \tfrac{1}{2}\theta \ . \tag{84.5}$$

The vector $\mathbf{K}$ is often called the collision vector. Correspondingly the vector $\mathbf{P} = \hbar\mathbf{K}$ is called the momentum transfer vector. If the potential energy does not depend on angles, $U = U(|\mathbf{r}|)$, then in (84.3) one can carry out the integration over angles

$$f(\theta) = -\frac{m}{2\pi\hbar^2} \int_0^\infty U(|\mathbf{r}'|)r'^2 \, dr' \int_0^\pi e^{iKr' \cos\theta} \sin\theta \, d\theta \int_0^{2\pi} d\varphi =$$

$$= -\frac{2m}{\hbar^2} \int_0^\infty U(|\mathbf{r}'|) \frac{\sin Kr'}{Kr'} r'^2 \, dr' \ . \tag{84.6}$$

In the first approximation the scattering amplitude is determined by the potential energy to the first power. Then, if (84.3) is substituted into definition (83.5), we find

$$d\sigma = |f(\theta)|^2 \, d\Omega = \frac{m^2}{4\pi^2\hbar^4} \left| \int_0^\infty U(|\mathbf{r}'|) e^{i\mathbf{K}\cdot\mathbf{r}'} \, dV' \right|^2 d\Omega =$$

$$= \frac{4m^2}{\hbar^4} \left| \int_0^\infty U(|\mathbf{r}'|) \frac{\sin Kr'}{Kr'} r'^2 \, dr' \right|^2 d\Omega \ . \tag{84.7}$$

Expression (84.7) is called the Born approximation. It is widely used in nuclear physics.

Continuing with successive approximations, i.e. substituting the wave function ψ from (84.2) into (83.9), it would be possible to find the wave function and scattering amplitude in the second approximation. The addition to the scattering amplitude in the second approximation would be determined by the integral of the square of the potential energy of interaction. Corrections of higher orders can be found in an analogous way.

For small values of the scattering angle we have from (84.7)

$$d\sigma = \frac{4m^2}{\hbar^4} \left| \int_0^\infty U(|\mathbf{r}'|)r'^2 \, dr' \right|^2 d\Omega \, ,$$

i.e. the cross section turns out to be independent of the velocity of the particle. In the next section we shall give an example of a concrete calculation of the cross section by the Born approximation. We now turn to a discussion of the applicability of the Born approximation.

For a rapid convergence of the series of successive approximations it is necessary that the correction to the wave function of the first approximation ψ_1 be small in comparison with the wave function of the zero order approximation ψ_0, i.e. the following condition must be fulfilled:

$$|\psi_1| \ll |\psi_0| \, . \tag{84.8}$$

By means of (83.7) one can find the value of the function ψ_1 which is valid for arbitrary values of r; then (84.8) is written in the form

$$\frac{m}{2\pi\hbar^2} \left| \int \frac{e^{ikz'} e^{ik|\mathbf{r}-\mathbf{r}'|} U(\mathbf{r}') \, dV'}{|\mathbf{r}-\mathbf{r}'|} \right| \ll 1 \, . \tag{84.9}$$

Since $\psi_1(\mathbf{r}')$ decreases with distance from the scattering centre, then condition (84.9) will be fulfilled if it is fulfilled at the origin. Hence condition (84.9) can be replaced by the inequality

$$|\psi_1(0)| = \frac{m}{2\pi\hbar^2} \left| \int \frac{e^{ik(r'+z')} U(r') \, dV'}{r'} \right| \ll 1 \, . \tag{84.10}$$

Further estimates of the integral can be carried out in two limiting cases:
1. When the relation $kR \ll 1$ is satisfied, where R is the effective range of interaction. This corresponds to small energies of the particles

$$E \ll \frac{\hbar^2}{mR^2} \, .$$

2. When the inverse inequality $kR \gg 1$ is fulfilled. This corresponds to the condition

$$E \gg \frac{\hbar^2}{mR^2} \, .$$

In the first case, in estimating the integral one can set $e^{ik(r'+z')}$ in (84.10) equal to unity.

Then (84.10) gives in order of magnitude

$$\frac{m}{\hbar^2} \int \frac{|U(r')|\, dV'}{r'} \approx \frac{m}{\hbar^2}\, |U_0|R^2 \ll 1 \;.$$

Here U_0 is a mean value of the interaction energy in the range R.

We write this last relation in the form

$$U_0 \ll \frac{\hbar^2}{mR^2} \;. \tag{84.11}$$

According to (37.9), the expression $\hbar^2/mR^2$ is equal in order of magnitude to the minimum depth of the potential well of radius R for which the level arises. We see that the condition of applicability of the Born approximation for the scattering of slow particles has a simple meaning. Namely, the mean interaction energy must be small in comparison with the minimum potential energy of the particle in the well for which the bound state is formed.

In the case of a large energy of the particle the region of applicability of the Born approximation is considerably enlarged. The exponential factor in formula (84.10) oscillates very rapidly, leading to a decrease in the total value of the integral.

In calculating the integral one can take the slowly varying factors out, writing

$$|\psi_1(0)| = \frac{m}{2\pi\hbar^2}\, |U_0| \left| \iint \frac{e^{ikz'}\, e^{ikr'}\, dV'}{r'} \right| =$$

$$= \frac{m|U_0|}{\hbar^2} \left| \int_0^R \int_0^\pi e^{ikr'(1+\cos\vartheta)} \sin\vartheta\, d\vartheta r'\, dr' \right| =$$

$$= \frac{m|U_0|}{\hbar^2 k} \left| \int_0^R (1-e^{2ikr'})\, dr' \right| \approx \frac{m|U_0|R}{\hbar^2 k} \ll 1 \;.$$

Here we have dropped the integral of the rapidly oscillating quantity $e^{2ikr'}$ since it is small in comparison with the integral retained.

Rewriting this last inequality in the form

$$\frac{|U_0|R}{\hbar v} \ll 1 \;, \tag{84.12}$$

we see that the Born approximation becomes valid for particles with larger energy the larger the product $U_0 R$ determined by the properties of the scattering centre.

In the important case of the Coulomb field the potential Ze^2/r decreases so slowly that the concept of an effective range of interaction R cannot be introduced.

However, we note that for $U_0 = Ze^2/R$ the product $U_0 R$ contained in the inequality (84.12) does not depend on R.

Hence for the Coulomb field inequality (84.12) assumes the form

$$\frac{Ze^2}{\hbar v} \ll 1 . \tag{84.13}$$

This has an obvious meaning: if the velocity of the electron in the first Bohr orbit of a hydrogen-like atom with nuclear charge Ze (the quantity $v_k = Ze^2/\hbar$) is introduced, then formula (84.13) assumes the form

$$\frac{v_k}{v} \ll 1 , \tag{84.14}$$

i.e. the velocity of the particle must be large in comparison with the velocity of the electron in the first Bohr orbit.

For the Born approximation to be applicable inequality (84.13) requires a larger energy the larger the charge of the scattering nucleus.

§85. The scattering of fast charged particles by atoms

Let us apply the Born approximation to the calculation of the cross section for the scattering of fast charged particles by atoms.

We shall assume that the nucleus of the atom with charge Ze is at the origin, and that the charge of the atomic shell is distributed in space with density $n(\mathbf{r})$. We shall disregard the size of the nucleus treating it as a point.

The differential scattering cross section is given by formula (84.6), which for $U = e\varphi$, where φ is the potential of the electric field acting on the particle to be scattered and e is its charge, assumes the form

$$d\sigma = \frac{m^2 e^2}{4\pi^2 \hbar^4} \left| \int \varphi(\mathbf{r}') \, e^{i\mathbf{K}\cdot\mathbf{r}'} \, dV' \right|^2 d\Omega . \tag{85.1}$$

The integral in formula (85.1) is conveniently expressed in terms of the charge density distribution in the atom.

For this we note that $\int \varphi(\mathbf{r}') \, e^{i\mathbf{K}\cdot\mathbf{r}'} \, dV'$ represents the Fourier component of the potential. It can be expressed in terms of the Fourier component of the charge density analogously to formula (24.25) of Part I which relates the Fourier component of the current density to the Fourier component of the potential.

We then have

$$d\sigma = \frac{4m^2e^2}{K^4\hbar^4} \left| \int \rho(\mathbf{r}') e^{i\mathbf{K}\cdot\mathbf{r}'} dV' \right|^2 d\Omega . \tag{85.2}$$

The charge density in the atom can be written in the form

$$\rho(\mathbf{r}) = Ze\delta(\mathbf{r}) - en(\mathbf{r}) . \tag{85.3}$$

For the differential cross section we finally obtain

$$d\sigma = \frac{4m^2e^4}{\hbar^4 K^4} \left| \int Z e^{i\mathbf{K}\cdot\mathbf{r}'} \delta(\mathbf{r}') dV' - \int n(\mathbf{r}') e^{i\mathbf{K}\cdot\mathbf{r}'} dV' \right|^2 d\Omega =$$

$$= \frac{4m^2e^4}{\hbar^4 K^4} |Z - F(K)|^2 d\Omega , \tag{85.4}$$

where

$$F(K) = \int n(\mathbf{r}') e^{i\mathbf{K}\cdot\mathbf{r}'} dV' . \tag{85.5}$$

The quantity F is called the atomic form factor. Its value is determined by the electron charge density distribution.

Substituting into (85.5) the value of the collision vector $\mathbf{K}$ according to (84.5), we rewrite the differential cross section in the form

$$d\sigma = \left(\frac{e^2}{2mv^2} \right)^2 |Z - F(K)|^2 \frac{d\Omega}{\sin^4 \frac{1}{2}\theta} . \tag{85.6}$$

Let us first consider a particular case of formula (85.6). If the scattering takes place on a point nucleus without an electron shell, $n = 0$, then, consequently, $F = 0$. We then obtain for the differential cross section

$$d\sigma = \left(\frac{Ze^2}{2mv^2} \right)^2 \frac{d\Omega}{\sin^4 \frac{1}{2}\theta} . \tag{85.7}$$

This is the well-known Rutherford formula, which is obtained in classical mechanics. The Rutherford formula in the case given was obtained by means of the approximate Born method. However, it is interesting to note that the same expression is obtained in an exact solution of the problem*. Since the

* See, for example, N.F.Mott and H.S.W.Massey, *The theory of atomic collisions* (Clarendon Press, Oxford, 1965).

scattering cross section in the exact solution does not contain Planck's constant $\hbar$, the results given by classical and quantum physics must naturally be the same.

The fact that the cross section becomes infinity for scattering at infinitely small angles is associated with the slow change of the Coulomb potential. Hence particles are scattered no matter how far away they pass from the scattering centre. However, as we shall see later, in practice the screening effect of the electron shell ensures a finite value of the scattering cross section.

Let us now consider the atomic form factor (85.5). The effective range of integration in it has a size of the order of the atomic size a. Outside this range $n(\mathbf{r})$ reduces to zero. Hence for small angles θ, for which $Ka \ll 1$, the exponent in integral (85.5) can be expanded in a series. We then have

$$Z - F(K) = Z - Z - i\mathbf{K} \int n(\mathbf{r}')\mathbf{r}'\, dV' + \frac{1}{2}\int n(\mathbf{r}')(\mathbf{K} \cdot \mathbf{r}')^2\, dV' . \qquad (85.8)$$

In formula (85.8) the first two terms mutually cancel, since the charge of the electron shell of the atom is equal to the charge of the nucleus. The third term represents the dipole moment of the atom, which, as we have seen (see §72), is equal to zero. In the last term, integrating over angles we obtain

$$Z - F = \frac{2\pi K^2}{3} \int\limits_0^\infty n(|\mathbf{r}|)\, r^4 dr .$$

The differential cross section in the limiting case $Ka \ll 1$ will have the form

$$d\sigma = \left(\frac{4\pi m e^2}{3\hbar^2}\right)^2 \left|\int n(r)\, r^4 dr\right|^2 d\Omega .$$

Thus owing to the screening by the charge of the electron shell the differential cross section for small scattering angles turns out to be a finite and constant (independent of angle) quantity. On the contrary, for large scattering angles, when the inverse inequality $Ka \gg 1$ is fulfilled, the exponential in integral (85.5) begins to oscillate rapidly and the form factor turns out to be small. Neglecting it in comparison with Z, we arrive at (85.7). The screening of the nuclear charge is not manifested for large scattering angles.

As an example let us calculate the form factor for the hydrogen atom. According to §38, the charge density in the hydrogen atom in the ground state is equal to

$$n(r) = |\psi(r)|^2 = \frac{1}{\pi a^3} e^{-2r/a} , \qquad a = \frac{\hbar^2}{me^2} .$$

Consequently, the form factor is defined by the integral

$$F(K) = \frac{1}{\pi a^3} \int e^{-2r/a} \, e^{i\mathbf{K}\cdot\mathbf{r}} \, r^2 \, dr \, \sin\vartheta \, d\vartheta \, d\varphi . \qquad (85.9)$$

Directing the z-axis along the vector $\mathbf{K}$, we have

$$F(K) = \frac{1}{\pi a^3} \int e^{-2r/a} \, e^{iKr \cos\vartheta} \, r^2 \, dr \, \sin\vartheta \, d\vartheta \, d\varphi .$$

Carrying out the integration, we finally find

$$F(K) = \frac{16}{(4+K^2 a^2)^2} .$$

Then the differential cross section for the hydrogen atom can be written in the form

$$d\sigma = \left(\frac{e^2}{2mv^2}\right)^2 \left[1 - \frac{16}{(4+K^2 a^2)^2}\right]^2 \frac{d\Omega}{\sin^4 \frac{1}{2}\theta} .$$

The total cross section is obtained by integration over all values of the scattering angle.

For other atoms of the periodic system of elements the charge density and the potential of interaction with the scattered particle can be calculated by means of Hartree or Thomas–Fermi approximate methods. After this the calculation of the form factor can be carried out in accordance with formula (85.5).

§86. Partial wave scattering theory

In the preceding sections we have considered one form of approximate scattering theory.

In addition to approximate theories it is possible to develop an exact scattering theory, which is often called partial wave theory.

The general scheme of partial wave scattering theory does not differ from that assumed in §83. We consider the motion of a particle in the field of a scattering centre. We assume that the scattering field is spherically symmetric and that at a distance from the centre the incident particle is described by the plane wave e^{ikz} and the scattered particle by a diverging spherical wave. Let the general solution of the Schrödinger equation in the centrally symmetric field be found. At a distance from the scattering centre this solution must be written in the form (83.3), i.e. in the form of an incident plane wave and a diverging spherical wave. As we know, the amplitude of the latter determines the scattering cross section which is of interest to us.

According to (35.31), the general solution of the Schrödinger equation in a centrally symmetric field independent of the angle φ can be represented by the expansion

$$\psi = \sum_{l=0}^{\infty} A_l R_l(r) P_l(\cos\theta) . \tag{86.1}$$

We shall call a given term of the series (86.1) the lth partial wave. At a large distance from the force centre the asymptotic form of the radial functions R_l is given by formulae (35.26) and (35.27)

$$R_l = B_l \frac{\sin(kr+\delta_l-\tfrac{1}{2}\pi l)}{kr} =$$

$$= B_l \frac{\exp i(kr+\delta_l-\tfrac{1}{2}\pi l) - \exp -i(kr+\delta_l-\tfrac{1}{2}\pi l)}{2ikr} . \tag{86.2}$$

We recall (see §36) that if the potential energy $U(r)$ is equal to zero over all space, then the set of phase shifts δ_l reduces to zero. The asymptotic expression which we need for ψ for the motion of a particle in the potential field $U(r)$ can be written in the following form:

$$\psi = \sum_{l=0}^{\infty} C_l P_l(\cos\theta) \frac{\exp i(kr+\delta_l-\tfrac{1}{2}\pi l) - \exp -i(kr+\delta_l-\tfrac{1}{2}\pi l)}{2ikr} . \tag{86.3}$$

We now have to write expression (83.3) in the form (86.3). This will allow us to relate the coefficients C_l and the phase shifts δ_l to the scattering amplitude $f(\theta)$. Expression (83.3) is most simply brought into the form (86.3) by expanding (83.3) in a series in terms of Legendre polynomials. We need the expansion of the plane wave e^{ikz} only for large distances, which can be found very simply. We write the plane wave in the form

$$e^{ikz} = e^{ikr\cos\theta} = \sum_{l=0}^{\infty} i^l(2l+1) \bar{P}_l(\cos\theta) G_l(r) , \tag{86.4}$$

where $G_l(r)$ is an unknown function of the radius. Multiplying this equation by $P_{l'}(\cos\theta)\sin\theta$ and integrating with respect to θ, we find

$$\frac{1}{2i^l}\int_{-1}^{+1} e^{ikrx}P_l(x)\,dx = G_l(r) .\tag{86.5}$$

We have made use of the conditions of orthogonality and normalization of the Legendre polynomials

$$\int_{-1}^{+1} P_l^2(x)\,dx = \frac{2}{2l+1} .$$

Integrating the left-hand side of (86.5) by parts, we have

$$G_l(r) = \frac{i^{-l}}{2ikr} e^{ikrx}P_l(x)\Big|_{x=-1}^{x=1} + \text{terms of the order of } r^{-2} .$$

Finally, using the known property of the Legendre polynomials $P_l(1) = 1$, $P_l(-1) = (-1)^l$, we obtain for the function $G_l(r)$ at large distances

$$G_l(r) = \frac{\sin(kr-\frac{1}{2}\pi l)}{kr} .$$

Thus the expansion of the plane wave at large distances is written in the form

$$e^{ikz} = \sum_l i^l(2l+1)P_l(\cos\theta)\,\frac{\sin(kr-\frac{1}{2}\pi l)}{kr} .\tag{86.6}$$

We also expand $f(\theta)$ in a series in terms of the Legendre polynomials

$$f(\theta) = \sum_{l=0}^{\infty} D_l P_l(\cos\theta) .\tag{86.7}$$

Substituting series (86.6) and (86.7) into (83.3) and equating the expression found and the asymptotic expression (86.3), we have

$$\sum_l C_l \frac{P_l(\cos\theta)}{2ikr} [\exp i(kr - \tfrac{1}{2}\pi l + \delta_l) - \exp -i(kr - \tfrac{1}{2}\pi l + \delta_l)] =$$

$$= \sum_l \left[\frac{i^l(2l+1)}{2ikr} [\exp i(kr - \tfrac{1}{2}\pi l) - \exp -i(kr - \tfrac{1}{2}\pi l)] + D_l \frac{e^{ikr}}{r} \right] P_l(\cos\theta) .$$

(86.8)

For eq. (86.8) to be fulfilled for arbitrary values of the angle θ it is necessary that the coefficients of the polynomials P_l on each side be equal to each other. Equating these coeffcients we find

$$\frac{C_l}{2ikr} [\exp i(kr - \tfrac{1}{2}\pi l + \delta_l) - \exp -i(kr - \tfrac{1}{2}\pi l + \delta_l)] =$$

$$= \frac{i^l(2l+1)}{2ikr} [\exp i(kr - \tfrac{1}{2}\pi l) - \exp - i(kr - \tfrac{1}{2}\pi l)] + D_l \frac{e^{ikr}}{r} .$$

(86.9)

This relation must be fulfilled for any arbitrary value of the radius r. This means that the coefficients of exponentials with the same indices must be equal to each other. Hence we find the following relation between the coefficients:

$$C_l = i^l(2l+1) \exp(i\delta_l) ,$$

$$i^l(2l+1) + 2ikD_l \exp(\tfrac{1}{2}i\pi l) = C_l \exp(i\delta_l) .$$

(86.10)

Finding D_l from this and substituting it into expansion (86.7), we find for the scattering amplitude the expression

$$f(\theta) = \frac{1}{2ik} \sum_{l=0}^{\infty} (2l+1)[e^{2i\delta_l} - 1] P_l(\cos\theta) .$$

(86.11)

Consequently, the differential cross section will be equal to

$$d\sigma = \frac{1}{4k^2} \left| \sum_{l=0}^{\infty} (2l+1)(e^{2i\delta_l} - 1) P_l(\cos\theta) \right|^2 d\Omega .$$

(86.12)

We find the total cross section by integrating (86.12) and taking into account the orthogonality relations for the Legendre polynomials. A simple calculation gives

$$\sigma = \sum_{l=0}^{\infty} \frac{4\pi}{k^2} (2l+1) \sin^2 \delta_l .$$ (86.13)

We see that the differential cross section and the total cross section for the scattering of a particle in a given force field is expressed in terms of the set of phase shifts δ_l. Hence it follows that for the calculation of scattering cross sections it is necessary to find the solution of the Schrödinger equation (35.8) for a particle moving in the given force field. Defining the form of the solution for large distances and comparing it with (86.2), we find δ_l.

The exact solution of the Schrödinger equation makes it possible to find the infinite set of phase shifts δ_l and, consequently, the value of the scattering cross section. The exact or partial wave theory of scattering was first developed by Rayleigh, who studied the scattering of sound waves. Faxen and Holtsmark were the first to use Rayleigh's method for solving the problems of quantum mechanics.

From (86.13) it is seen that the total cross section can be written in the form of a sum of the so-called partial cross sections

$$\sigma = \sum_{l=0} \sigma_l , \qquad \sigma_l = \frac{4\pi}{k^2} (2l+1) \sin^2 \delta_l .$$

Each of the partial cross sections corresponds to taking into account one of the terms of the series (86.2)

$$B_l P_l(\cos\theta) \frac{\sin(kr - \frac{1}{2}\pi l + \delta_l)}{kr} .$$

It is clear that it describes a state of the particle with definite angular momentum $L^2 = \hbar^2 l(l+1)$. For this reason a notation analogous to the notation of atomic terms is adopted in scattering theory. For example, to $l = 0$ there corresponds s-wave scattering, which is characterized by the partial cross section σ_0; to $l = 1$ there corresponds p-wave scattering with the partial cross section σ_1 and so on.

The total particle flux through an arbitrary surface surrounding the scattering centre for a particle in the state with angular momentum L, is equal to zero. It could be calculated by the general formula (7.3). However, this can also be seen without carrying out the calculation, on the basis of the general theorem presented in §7. There it was pointed out that the total flux is always equal to zero in the case of a real wave function. In our case this is so, since the wave function is expressed by formula (86.2).

The equality to zero of the total flux of scattered particles has an obvious meaning; it means the law of conservation of the number of particles in the process of scattering. It is important to note that the conservation law holds for particles of each value of l separately. We shall come back to the discussion of this fact in §91.

Finding the values of all the phase shifts δ_l is, as a rule, a very complex problem. Furthermore, the practical value of formulae written in the form of a series is not great if the series does not have a sufficiently rapid convergence. We cannot dwell here on the problems of convergence of the series (86.12) and (86.13) and shall give only the final result*.

For the convergence of series (86.13) it is necessary that the potential energy $U(r)$ should decrease at large distances more rapidly than r^{-n}, where $n > 2$. Further, the series for the differential cross section diverges for $\theta = 0$, if $U(r)$ has the form r^{-n}, where $n \leqslant 3$ at large distances. When $r \to 0$, $U(r)$ must increase more slowly than r^{-2}.

The practical value of formulae (86.12) and (86.13) for the scattering cross section becomes greater, the smaller the number of terms of the series which play an essential role. Simple reasoning shows that as the energy of the particle increases the number of phase shifts δ_l which must be taken into account in series (86.12) and (86.13) increases.

Indeed, let R be the radius of the sphere in which the interaction energy is substantially different from zero. For a sufficiently rapid decrease of $U(r)$ the introduction of such a quantity is always possible. The wave function R_l has its first maximum at the distance r defined by the relation $kr \sim l$. At the next maximum R_l has a considerably smaller value because of the decrease of the factor r^{-1}.

For small values of r the wave function is also small. Thus the wave function R_l has its basic value for $r \sim l/k$. If $r \sim l/k > R$, then the wave function is small in the interaction sphere. But in this case the scattering amplitude will also be small. Thus only those particles for which $l/k \leqslant R$ undergo effective scattering.

The angular momentum l of the effectively scattered particles increases with increasing energy of the particle. For small energies the number of terms which must be taken into account in the series (86.12) and (86.13) is relatively small. Therefore partial wave scattering theory is particularly important for the study of the scattering of slow particles. This qualitative reasoning can be replaced by a quantitative rule, which we shall give without proof.

* L.D.Landau and E.M.Lifshitz, *Quantum mechanics* (Pergamon Press, Oxford, 1965).

If a classical particle having momentum p and impact parameter

$$\rho_l = \frac{\hbar \, [l(l+1)]^{\frac{1}{2}}}{p} = \frac{[l(l+1)]^{\frac{1}{2}}}{k} , \qquad (86.14)$$

in moving does not penetrate the region where the potential energy of inter-action differs considerably from zero, then the phase δ_l corresponding to the angular momentum $\hbar^2 l(l+1)$ is small[*].

We apply this rule to the investigation of the scattering of a slow particle. Let the scattering centre produce a field effective to the range R. By slow particles we shall mean particles with quantum number k for which $kR \ll 1$.

In this case

$$\rho_l > R . \qquad (86.15)$$

For all values $l > 0$ all phase shifts, except δ_0, are small. We thus see that only s-wave scattering is important for the scattering of slow particles.

The differential cross section is then equal to

$$d\sigma = \frac{1}{4k^2} \, |e^{2i\delta_0} - 1|^2 \, d\Omega = \frac{\sin^2 \delta_0}{k^2} \, d\Omega , \qquad (86.16)$$

since $P_0(\theta) = 1$.

The cross section for s-wave scattering does not depend on the scattering angle. This means that the scattering is spherically symmetric. As the energy of the particle increases phase shifts of higher order begin to play a role and the scattering progressively assumes an ever more asymmetric character.

For large energies the cross section becomes substantially different from zero only for very small angles θ. This can best be seen by means of the Born approximation (84.3). For large energies the vector $\mathbf{K}$ is large, the integral rapidly oscillates, and hence the cross section is small. For $\theta = 0$ the vector $\mathbf{K}$ is equal to zero, there is no oscillation, and the cross section is large. Finally, we note that partial wave scattering theory in the form in which it has been described here is inapplicable to scattering in the Coulomb field. The wave function in this case does not have the asymptotic form (83.3). This fact is associated with the very slow decrease of the Coulomb potential as a function of the distance. This case requires particular consideration[**].

---

[*] The derivation of this statement is given in the book: N.F.Mott and H.S.W.Massey, *The theory of atomic collisions* (Clarendon Press, Oxford, 1965).

[**] For the exact solution of this problem see, for example, L.D.Landau and E.M. Lifshitz, *Quantum mechanics* (Pergamon Press, Oxford, 1965).

§87. Scattering by a spherical potential well (the concept of resonance scattering)

As an example of the use of partial wave theory we shall consider the scattering of a particle in a potential field which we define in the following way:

$$U = -U_0 \quad \text{for} \quad r < R \,,$$
$$U = 0 \qquad \text{for} \quad r > R \,. \tag{87.1}$$

For simplicity we confine ourselves to the case where the scattered particle has a small energy, i.e. $kR \ll 1$. In this case, as we know, s-wave scattering is important, and we need only determine the phase shift δ_0. In the case of the potential field given by formula (87.1) the solution of the problem offers no difficulty. By means of the relations already found we can also illustrate a very interesting phenomenon occurring in the scattering process, so-called resonance scattering. It consists in the fact that the scattering cross section under certain conditions turns out to be very large. This effect occurs when there exists an energy level in the potential field close to zero and the energy of the particle to be scattered is sufficiently small. We write the wave function in the form $\psi = A_0 R_0(r) = \chi/r$. The function $\chi(r)$ satisfies the equation

$$\frac{d^2\chi}{dr^2} + k^2\chi = 0 \qquad \text{for} \quad r > R \,, \tag{87.2}$$

$$\frac{d^2\chi}{dr^2} + \beta^2\chi(r) = 0 \quad \text{for} \quad r < R \,, \tag{87.3}$$

where

$$\beta^2 = \frac{2m(E+U_0)}{\hbar^2} \,.$$

The form of the function χ for $r > R$ is easily obtained from the solution of eq. (87.2)

$$\chi = C \sin(kr + \delta_0) \,. \tag{87.4}$$

In the general case the function χ has the form (87.4) only for large distances (see formula (86.2)). However, in our case, by virtue of the sharp boundary of the potential energy, the function R_0 has the form (86.2) for all distances $r > R$.

For $r < R$ we obtain

$$\chi = A \sin \beta r + B \cos \beta r \,,$$

where A and B are constants. The function R_0 must remain finite for $r \to 0$. Hence the coefficient B must be set equal to zero. Thus we get

$$\chi = A \sin \beta r \quad \text{for} \quad r < R \,.$$

The function ψ and its first derivative must be continuous at the point $r = R$. These two relations are conveniently replaced by the equality of the logarithmic derivatives. We then find

$$\beta \cotan \beta R = k \cotan (kR + \delta_0) \,. \tag{87.5}$$

We have obtained a transcendental equation for the phase shift δ_0. We first assume that δ_0 is small. Then $\cotan (kR + \delta_0)$ can be expanded in a series in terms of the small argument $kR + \delta_0$. As a result we have

$$\beta \cotan \beta R = \frac{k}{kR + \delta_0} \,,$$

whence we can find the phase shift δ_0 which is equal to

$$\delta_0 = \frac{k}{\beta \cotan \beta R} - kR \,. \tag{87.6}$$

We see from relation (87.6) that the phase shift will indeed be considerably smaller than unity if the following relation is fulfilled:

$$\frac{k}{\beta \cotan \beta R} \ll 1 \,. \tag{87.7}$$

The differential cross section can easily be found by making use of formula (86.16) and recalling that $\delta_0 \ll 1$. It has the form

$$d\sigma = \frac{\delta_0^2}{k^2} d\Omega = \frac{1}{k^2} \left(\frac{k}{\beta \cotan \beta R} - kR \right)^2 d\Omega \,.$$

However, a form of the potential well for which $\beta \cotan \beta R$ approaches zero is possible. In this case inequality (87.7) is violated, and the phase shift δ_0 is large. To find the conditions under which δ_0 is large, we shall establish the relation between the quantity $\beta \cotan \beta R$ involved in formula (87.5) and the energy level of the particle in a bound state. In §37 we have obtained for the energy levels of a particle in a potential well formula (37.6)

$$\frac{2m(U_0 - \epsilon)}{\hbar^2} \cotan^2 \left(\frac{2m(U_0 - \epsilon)R^2}{\hbar^2} \right)^{\frac{1}{2}} = \frac{2m\epsilon}{\hbar^2} \,. \tag{87.8}$$

ϵ is the energy level of the particle in the well. If the energy level of the particle in the well is close to zero, i.e. if $\epsilon \ll U_0$, then relation (87.8) may be rewritten in the form

$$\frac{2mU_0}{\hbar^2} \cotan^2 \left(\frac{2mU_0 R^2}{\hbar^2} \right)^{\frac{1}{2}} = \frac{2m\epsilon}{\hbar^2}. \tag{87.9}$$

In the case being considered the energy of the scattered particle is also small ($E \ll U_0$), hence relation (87.9) can be written in the form

$$\beta^2 \cotan^2 \beta R = \frac{2m\epsilon}{\hbar^2}. \tag{87.10}$$

Thus we see that the increase of δ_0 is associated with the presence of an energy level ϵ close to zero.

We now turn to the calculation of the phase shift in the case where relation (87.7) is violated and δ_0 is large. We denote this value of the phase shift by δ_{0r}. We find δ_{0r} again from relations (87.5). For this we expand $\cotan(kR+\delta_{0r})$ in a series in terms of the small parameter kR and restrict ourselves to the zero order term of the expansion. Then we obtain

$$\beta \cotan \beta R = k \cotan \delta_{0r}.$$

Squaring this relation and making use of formula (87.10), we find

$$\cotan^2 \delta_{0r} = \frac{\epsilon}{E}. \tag{87.11}$$

We now see that the phase shift δ_{0r} will not be a small quantity if $\epsilon < E$.

We find the scattering cross section by means of the general formula (86.16). In this case we have

$$\sigma_r = \frac{4\pi \sin^2 \delta_{0r}}{k^2} = \frac{2\pi \hbar^2}{m(E+\epsilon)}. \tag{87.12}$$

This expression is called the Wigner formula. It is easily seen that the cross section in the case of a resonance is considerably larger than in the absence of a resonance. The ratio of the cross section is equal to

$$\frac{\sigma_r}{\sigma} = \frac{\sin^2 \delta_{0r}}{\delta_0^2}.$$

Since $\delta_0 \ll 1$, and $\sin \delta_{0r}$, as is seen from formula (87.11), for $\epsilon \approx E$ is close to one, then it is evident that

$$\frac{\sigma_r}{\sigma} \gg 1 \ .$$

We have obtained formula (87.12) for a particular form of the potential energy. It should be stressed, however, that the dependence of the cross section on ϵ (87.12) is general and is not related to the actual form of the potential energy*.

Resonance scattering also occurs in the case where the system does not have a real level close to zero but the configuration of the field is similar to that for which such a level appears. In such a situation the function $\cotan \beta R$ is positive, whereas for the real level we necessarily have $\cotan \beta R < 0$ (see §37). Relation (87.10) contains $\cotan^2 \beta R$ and hence is fulfilled independently of the sign of the function $\cotan \beta R$. In the case where $\cotan \beta R > 0$, scattering takes place at the virtual level, not at the real level.

By means of the relations obtained above one can also easily find the differential cross section for scattering by a potential barrier, i.e. by a potential field having the following form:

$$U = 0 \qquad \text{for} \quad r > R \ ,$$

$$U = |U_0| \qquad \text{for} \quad r < R \ .$$

For this it is sufficient to carry out the replacement $\beta \to i\beta$. Then for the differential cross section we obtain

$$d\sigma = \frac{1}{|\beta|^2} \, (\tanh |\beta| R - |\beta| R)^2 \, d\Omega \ . \tag{87.13}$$

Formula (87.13) is simplified in the case of an infinitely high potential barrier $U_0 \to \infty$. In this case we find the following expression for the total cross section:

$$\sigma = 4\pi R^2 \ . \tag{87.14}$$

It is interesting to note that the scattering cross section is larger in this case than the geometric size of the scatterer, by a factor of four.

---

* For a more general derivation of the formula for resonance scattering see L.D. Landau and E.M.Lifshitz, *Quantum mechanics* (Pergamon Press, Oxford, 1965).

§88. The elastic scattering of identical particles

Up till now we have assumed that the scattered particle and the target are different particles. We now consider the case where the scattered and target particles are identical. As we shall now see, the identity of the particles has an essential effect on the scattering process. We shall begin with the consideration of particles of zero spin. We first suppose that the identical particles move towards each other with equal velocities. In this case the centre of mass of the system is at rest and the wave function will, in correspondence with (14.14), have the form

$$\psi = \psi(x,y,z)$$

and will depend only on the relative coordinates. The wave function $\psi_0(x,y,z)$ satisfies eq. (14.11)

$$\left[-\frac{\hbar^2}{2\mu}\nabla^2 + U(r)\right]\psi = E\psi(x,y,z) . \tag{88.1}$$

The reduced mass of two identical particles is equal to $\mu = \frac{1}{2}m$. We cannot for our case write the wave function in the form

$$\psi = e^{ikz} + \frac{f(\theta)}{r} e^{ikr} ,$$

since this function does not satisfy the symmetry requirements.

As a matter of fact according to (14.6) there corresponds to the exchange of two particles (i.e. to the replacement $x_1 \to x_2, y_1 \to y_2, z_1 \to z_2$) the transformation $\mathbf{r} \to -\mathbf{r}$. Here the modulus of the vector $\mathbf{r}$ does not change, and the angle θ is replaced by $\pi - \theta$. Taking into account this last transformation, it is easily found that the symmetrized wave function must have the form

$$\psi_s = e^{ikz} + e^{-ikz} + \frac{e^{ikr}}{r} [f(\theta) + f(\pi-\theta)] . \tag{88.2}$$

A diverging wave again describes the scattered particles. The differential scattering cross section is now given by the expression

$$d\sigma = |f(\theta) + f(\pi-\theta)|^2 \, d\Omega = |f(\theta) + f(\pi-\theta)|^2 \sin\theta \, d\theta \, d\varphi . \tag{88.3}$$

Thus we have found the differential cross section for the process in which one of the colliding identical particles is scattered at an angle θ with respect to the direction of its initial flight.

From formula (88.3) it follows that the number of particles scattered at angle θ and at angle $\pi - \theta$ is the same. If one of the particles was at rest before the collision, then the differential cross section in this system of coordinates can be found in the following way. In the system of coordinates in which the centre of mass is at rest the differential cross section is given by expression (88.3). The transition to the laboratory system is carried out by means of formulae (83.2). In the case given the mass of the particles is the same, and we obtain

$$\tan \vartheta_1 = \frac{\sin \theta}{1 + \cos \theta} = \tan \tfrac{1}{2}\theta$$

and, correspondingly, $\vartheta_1 = \tfrac{1}{2}\theta$.

Expressing the differential cross section as a function of the angle ϑ_1, we find

$$d\sigma = |f(2\vartheta_1) + f(\pi - 2\vartheta_1)|^2 \; 4 \cos \vartheta_1 \, \sin \vartheta_1 \, d\vartheta_1 d\varphi_1 =$$

$$= |f(2\vartheta_1) + f(\pi - 2\vartheta_1)|^2 \; 4 \cos \vartheta_1 \, d\Omega_1 \, , \qquad (88.4)$$

where $d\Omega_1$ is a solid angle element in the laboratory system.

Expression (88.4) gives the differential cross section for the process in which one of the particles is scattered into the solid angle element $d\Omega_1$. Since the two particles are identical, the question as to which one of the particles entered $d\Omega_1$, that which was initially moving or that which was initially at rest, makes no physical sense.

As an example of the application of formula (88.4) let us consider the collision of two identical particles in which the interacting energy has the simple form

$$U = U_0 \quad \text{for} \quad r < R \, ,$$

$$U = 0 \quad \text{for} \quad r > R \, .$$

We suppose that before the collision one of the particles was at rest while the other was moving sufficiently slowly that the relation $kR \ll 1$ was fulfilled. In this case $\delta_l \ll 1$ and in accord with (86.11) the scattering amplitude can be written in the form $f(\theta) = \delta_0/k$.

For the differential scattering cross section in the laboratory system we obtain

$$d\sigma = |f(2\vartheta_1) + f(\pi - 2\vartheta_1)|^2 \; 4 \cos \vartheta_1 \, d\Omega_1 = \frac{16 \delta_0^2}{k^2} \cos \vartheta_1 \, d\Omega_1 \, .$$

Thus we see that if in the centre-of-mass system the scattering is spherically symmetric, then in the laboratory system the differential cross section is proportional to the cosine of the scattering angle.

The theory of the scattering of identical particles with spin different from zero is constructed according to the same scheme as for spinless particles. For concreteness we assume that both colliding particles have spin $\frac{1}{2}$. The generalization of the theory to the case of arbitrary spin offers no difficulty.

We consider the collision of two identical particles in the centre-of-mass system in the case where the total spin of the system is equal to zero (i.e. the spins of the particles are antiparallel). Then the spin part of the wave function must be antisymmetric and, consequently, the coordinate part must be symmetric. In other words, the coordinate part of the wave function can, as in the case of spinless particles, be written in the form

$$\psi_s = e^{ikz} + e^{-ikz} + \frac{e^{ikr}}{r} \left[f(\theta) + f(\pi-\theta) \right] . \tag{88.5}$$

Correspondingly we have for the differential scattering cross section

$$d\sigma_s = |f(\theta) + f(\pi-\theta)|^2 \, d\Omega . \tag{88.6}$$

If the total spin is equal to one (i.e. the spins are parallel), then the spin part of the wave function is symmetric, and the coordinate part is antisymmetric. Hence for this case we can write the following asymptotic expression:

$$\psi_a = e^{ikz} - e^{-ikz} + \frac{e^{ikr}}{r} \left[f(\theta) - f(\pi-\theta) \right] . \tag{88.7}$$

Then for the differential cross section we obtain

$$d\sigma_a = |f(\theta) - f(\pi-\theta)|^2 \, d\Omega . \tag{88.8}$$

We have considered above processes in which the scattered particles had a definite spin orientation. However, in scattering, particles are often in a state with indefinite spin. In this case one is usually interested in the mean cross section which is obtained by averaging over all possible spin states. The mean cross section for particles with spin $\frac{1}{2}$ can easily be found as follows. The colliding particles can be in four states: in one state with spin 0 and in three states with spin 1 (three possible projections on the z-axis). Since all these states are equally probable, the state with spin 0 has a statistical weight equal to $\frac{1}{4}$, and the weight of the state with spin 1 is equal to $\frac{3}{4}$. Hence the mean differential cross section can be written in the form

$$d\sigma = \tfrac{1}{4} d\sigma_s + \tfrac{3}{4} d\sigma_a . \tag{88.9}$$

As an example let us consider the scattering of two slow identical particles with spin $\frac{1}{2}$, for which the interaction energy can be written in the form

$$U = U_0 \quad \text{for} \quad r < R ,$$

$$U = 0 \quad \text{for} \quad r > R .$$

In the case of parallel spins the scattering cross section given by expression (88.8) turns out to be equal to zero

$$d\sigma_a = |f(\theta) - f(\pi - \theta)|^2 \, d\Omega = 0 .$$

Consequently, the scattering of particles with parallel spins is associated with effects of higher orders, i.e. with p-wave, d-wave etc. scattering. The cross section for the scattering of particles with antiparallel spins at small energies is the same as for particles with spin zero,

$$d\sigma_s = \frac{4\delta_0^2}{k^2} \, d\Omega .$$

The mean cross section according to formula (88.9) is given by the expression

$$d\sigma = \frac{\delta_0^2}{k^2} \, d\Omega .$$

Thus we see that taking into account the identity of the particles leads to the appearance of a basic dependence of the scattering cross section on the mutual orientation of their spins.

The transition from cross sections calculated in the centre-of-mass system to cross sections calculated in the laboratory system is carried out in the same way as for spinless particles.

§89. The effect of polarization in scattering processes

All the results obtained up to now apply to the scattering of beams in which all the particles are in one and the same state, i.e. are described by one and the same wave function. However, the particles of a beam can be in different spin states. We shall now confine ourselves to considering beams made up of particles with spin $\frac{1}{2}$ scattered by non-polarized targets. As is known, each of the particles of the beam is described by a two-component spinor.

In §61 it was shown that an arbitrary state of a particle is at the same time a state with a definite projection of the spin on a certain direction in space. In other words, for a state with an indefinite z-component of the spin one can always find some z'-axis with respect to which the given state will be a state with a definite spin projection. Consequently, we see that if the beam consists of particles which are in the same state, then it will be fully polarized along a certain direction. If the beam is partially polarized, then the particles are described by different spinors. In this case the beam cannot be described by means of a wave function, and we have a mixture of states (see §23). Nevertheless, for the description of the spin properties of the particles of the beam one can introduce a function φ defined by the formula*

$$\varphi = c_1 \varphi_1 \epsilon_1 + c_2 \varphi_2 \epsilon_2 + \ldots .$$

The summation is carried out over the spin states of the particles of the beam. We denote the spinor which describes the group of particles in the kth spin state by φ_k. The coefficient c_k determines the weight of this state. It is proportional to the number of particles in a given group. The quantities ϵ_i and ϵ_k, satisfying the condition $\epsilon_i^2 = 1$ and $\epsilon_i \epsilon_k = 0$, are introduced in order to eliminate the interference between wave functions of particles which are in different spin states in the quadratic expressions defining the mean values. We define the polarization vector as the spin vector averaged over the beam:

$$\mathbf{P} = \bar{\boldsymbol{\sigma}} = \frac{\varphi^\dagger \boldsymbol{\sigma} \varphi}{\varphi^\dagger \varphi} = \frac{\sum_n |c_n|^2 \varphi_n^\dagger \boldsymbol{\sigma} \varphi_n}{\sum_n |c_n|^2} = \frac{\sum \varphi_n'^\dagger \boldsymbol{\sigma} \varphi_n'}{\sum \varphi_n'^\dagger \varphi_n'} , \tag{89.1}$$

where $\varphi_n' = c_n \varphi_n$.

Formula (89.1) has a simple meaning: $(\varphi_n^\dagger \boldsymbol{\sigma} \varphi_n)$ represents the mean value of the spin vector in the nth state, and the ratio $|c_n|^2 / \Sigma_n |c_n|^2$ determines the probability of realization of the nth state in the beam. This probability is equal to N_n/N, where N_n is the number of particles in the nth state, and N is the total number of particles in the beam.

We write $c_n \varphi_n$ in the form

$$c_n \varphi_n = \begin{pmatrix} u_n \\ v_n \end{pmatrix} . \tag{89.2}$$

Substituting expression (89.2) into (89.1), we easily find the components of the polarization vector

* L.Wolfenstein, Phys. Rev. 75 (1943) 1664.

$$P_x = \frac{2\,\mathrm{Re}\,\sum_n u_n^* v_n}{\sum_n (|u_n|^2 + |v_n|^2)},$$

$$P_y = \frac{2\,\mathrm{Im}\,\sum_n u_n^* v_n}{\sum_n (|u_n|^2 + |v_n|^2)}, \qquad (89.3)$$

$$P_z = \frac{\sum_n (|u_n|^2 - |v_n|^2)}{\sum_n (|u_n|^2 + |v_n|^2)}.$$

If one half of the particles constituting the beam is polarized in some direction, for example in the positive direction of the z-axis, and the other half is polarized in the opposite direction, then the polarization vector $\mathbf{P}$ of the beam will be equal to zero. Indeed, one group of particles is described by the spin functions.

$$c_1 \varphi_1 = \begin{pmatrix} c_1 \\ 0 \end{pmatrix},$$

while the other group has the spin functions

$$c_2 \varphi_2 = \begin{pmatrix} 0 \\ c_2 \end{pmatrix}, \qquad |c_1|^2 = |c_2|^2 .$$

Substituting these values into (89.3), we find that the polarization vector is equal to zero.

We now turn to the case of the scattering of particles with spin $\frac{1}{2}$ by a target with spin 0. Then the wave function ψ, describing the process of elastic scattering, at large distances has the form

$$\psi = \varphi\, e^{ikz} + \frac{e^{ikr}}{r} f\varphi . \qquad (89.4)$$

Here φ is the spinor characterizing the state of the incident particle, and f is a certain two-row matrix depending on the scattering angles. Let us establish the general form of this matrix. First of all we note that any two-row matrix can be expressed in terms of a unit matrix and the Pauli matrices $\sigma_x, \sigma_y, \sigma_z$, since the matrices mentioned make up a complete system (see §60). Correspondingly we have

$$f = g(\theta)I + \mathbf{h}(\theta) \cdot \boldsymbol{\sigma} . \tag{89.5}$$

The further form of the functions g and $\mathbf{h}$ can be obtained from the following considerations. The laws of transformation of the first and second terms of formula (89.4) must be the same under spatial rotations and reflections. Since the first term transforms as a spinor, the second term in this formula must also have the character of a spinor. Hence it follows that the function g must be a scalar. Since the operator $\boldsymbol{\sigma}$ transforms as a pseudovector, $\mathbf{h}$ also must be a pseudovector. On the other hand, the pseudovector $\mathbf{h}$ depends on the quantities which characterize the scattering process, and can be defined by only two vectors $\mathbf{k}_0$ and $\mathbf{k}_1$ (the wave vectors of the particle before and after scattering). From these two vectors one can construct the single unit pseudovector

$$\mathbf{n} = \frac{\mathbf{k}_0 \times \mathbf{k}_1}{|\mathbf{k}_0 \times \mathbf{k}_1|} .$$

Hence $\mathbf{h} = h(\theta)\mathbf{n}$ where $h(\theta)$ is a scalar.

Finally we obtain

$$f = g(\theta)I + \mathbf{n} \cdot \boldsymbol{\sigma} h(\theta) . \tag{89.6}$$

Correspondingly, the elastic scattering cross section has the form

$$\frac{d\sigma}{d\Omega} = \varphi^\dagger f^\dagger f \varphi = |g|^2 + |h|^2 + 2\,\mathrm{Re}\,(g^*h)\boldsymbol{\mu} \cdot \mathbf{n} . \tag{89.7}$$

where $\boldsymbol{\mu} = \varphi^\dagger \boldsymbol{\sigma} \varphi$.

We average expression (89.7) over the spin states of the particles of the incident beam. Then, making use of (89.1), we find

$$\frac{d\sigma}{d\Omega} = |g|^2 + |h|^2 + 2\,\mathrm{Re}\,(g^*h)\mathbf{P}_{\mathrm{inc}} \cdot \mathbf{n} =$$

$$= (|g|^2 + |h|^2) \left(1 + \frac{2\,\mathrm{Re}\,(g^*h)\mathbf{P}_{\mathrm{inc}} \cdot \mathbf{n}}{|g|^2 + |h|^2} \right) . \tag{89.8}$$

where $\mathbf{P}_{\mathrm{inc}}$ is the polarization vector of the incident beam.

If the incident beam is not polarized ($\mathbf{P}_{\mathrm{inc}} = 0$), then the differential cross section is equal to

$$\frac{d\sigma}{d\Omega} = |g|^2 + |h|^2 \tag{89.9}$$

We now turn to the investigation of the state of the scattered beam. We stress that a polarization of the beam can arise after scattering even in the case where the incident beam was not polarized. From general considerations it is easy to indicate the direction of polarization of the scattered beam. As a matter of fact, the polarization is described by the pseudovector $\mathbf{P}$, which can be oriented only in the direction of the pseudovector $\mathbf{n}$. Consequently, for the scattered beam, which was not polarized before scattering, we have

$$\mathbf{P}_{scat} = P_{scat}\mathbf{n} \; . \tag{89.10}$$

We shall determine the value of the polarization of the scattered beam. Based on definition (89.1), we have

$$\mathbf{P}_{scat} = \frac{\sum_n (f\varphi_n)^\dagger \boldsymbol{\sigma}(f\varphi_n)}{\sum_n (f\varphi_n)^\dagger (f\varphi_n)} = \frac{\sum_n \varphi_n^\dagger f^\dagger \boldsymbol{\sigma} f \varphi_n}{\sum_n \varphi_n^\dagger f^\dagger f \varphi_n} \; . \tag{89.11}$$

Since by assumption the incident beam is not polarized, it can be represented in the form of two beams consisting of the same number of particles but with oppositely directed spins. Then the summation over n reduces to the summation over two states characterized by oppositely directed spins.

Consequently, we have

$$\mathbf{P}_{scat} = \frac{\displaystyle\sum_{i=1}^{2} \varphi_i^\dagger f^\dagger \boldsymbol{\sigma} f \varphi_i}{\displaystyle\sum_{i=1}^{2} \varphi_i^\dagger f^\dagger f \varphi_i} \; .$$

We see that to calculate the polarization it is necessary to find the sums of the diagonal elements (traces) of the matrices of certain operators. In the notation of §45 this last formula can be rewritten in the form

$$\mathbf{P}_{scat} = \frac{\mathrm{Tr} f^\dagger \boldsymbol{\sigma} f}{\mathrm{Tr} f^\dagger f} \; . \tag{89.12}$$

We calculate first $\mathrm{Tr} f^\dagger \boldsymbol{\sigma} f$. Making use of expression (89.6), we have

$$\mathrm{Tr} f^\dagger \boldsymbol{\sigma} f = \mathrm{Tr} \left\{ [g^* I + h^* (\mathbf{n} \cdot \boldsymbol{\sigma})] \, \boldsymbol{\sigma} \, [g I + h(\mathbf{n} \cdot \boldsymbol{\sigma})] \right\} \; .$$

It is easily seen from formulae (60.15) and (60.16) that $\mathrm{Tr}\,\sigma_i = 0$ $(i=1,2,3)$. From relations of the type $\sigma_x\sigma_y = i\sigma_z$ it follows that

$$\mathrm{Tr}\,\sigma_i\sigma_k = 0 \qquad (i \neq k) .$$

Since $\sigma_i^2 = I$, then $\mathrm{Tr}\,\sigma_i^2 = 2$ $(i=1,2,3)$.

Using these relations, we obtain

$$\mathrm{Tr}\,f^\dagger\boldsymbol{\sigma}f = \mathrm{Tr}\,[\,g^*h\boldsymbol{\sigma}(\mathbf{n}\cdot\boldsymbol{\sigma}) + h^*g(\mathbf{n}\cdot\boldsymbol{\sigma})\boldsymbol{\sigma}] =$$

$$= \mathrm{Tr}\,(g^*h+h^*g)(\sigma_x^2 n_x\mathbf{i}+\sigma_y^2 n_y\mathbf{j}+\sigma_z^2 n_z\mathbf{k}) = 4n\,\mathrm{Re}\,(g^*h) .$$

By means of analogous calculations we find

$$\mathrm{Tr}\,f^\dagger f = 2(|h|^2+|g|^2) .$$

Thus the polarization vector of the scattered beam has the form

$$\mathbf{P}_{\mathrm{scat}} = \frac{2\,\mathrm{Re}\,(g^*h)}{|g|^2 + |h|^2}\,\mathbf{n}\,; \qquad \mathbf{n} = \frac{\mathbf{k}\times\mathbf{k}_1}{|\mathbf{k}\times\mathbf{k}_1|}. \qquad (89.13)$$

Making use of (89.13) and (89.8), we express the scattering cross section in terms of the polarization vector $\mathbf{P}_{\mathrm{scat}}$:

$$\frac{d\sigma}{d\Omega} = (|g|^2+|h|^2)(1+\mathbf{P}_{\mathrm{inc}}\cdot\mathbf{P}_{\mathrm{scat}}) , \qquad (89.14)$$

where $\mathbf{P}_{\mathrm{scat}}$ is the polarization vector of the beam of scattered particles in the case where the beam was not polarized before scattering.

Thus we see that the scattering cross section depends on the polarization of the incident and scattered beams. Experimentally such dependences can be observed in experiments on double scattering. An unpolarized beam of particles (fig. V.26) becomes polarized after scattering. Then the polarized beam of particles falls on the second scatterer. In this case the cross section for scattering to the left (vector $\mathbf{k}_2$) turns out to be different from the cross section for scattering to the right (vector $\mathbf{k}_2'$).

For simplicity we assume that all the vectors $\mathbf{k}, \mathbf{k}_1, \mathbf{k}_2$ and $\mathbf{k}_2'$ lie in one plane. The vector $\mathbf{n}$, characterizing the polarization after the first scattering, is directed upwards perpendicular to the plane of the drawing. The vectors $\mathbf{P}_{\mathrm{scat}}$ involved in formula (89.14) have opposite directions for beams scattered a second time to the left and to the right because of the different directions of the vectors $\mathbf{k}_2$ and $\mathbf{k}_2'$. Thus the cross section for the beam scattered to the left is equal to

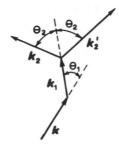

Fig. V.26

$$\frac{d\sigma}{d\Omega} = (|g|^2 + |h|^2)[1 + P_{\text{scat}}(\theta_1) \, P_{\text{scat}}(\theta_2)] \ . \tag{89.15}$$

Correspondingly, for the beam scattered to the right we have

$$\frac{d\sigma}{d\Omega} = (|g|^2 + |h|^2)[1 - P_{\text{scat}}(\theta_1) \, P_{\text{scat}}(\theta_2)] \ . \tag{89.16}$$

We see that the ratio of the number of particles scattered to the left and to the right is determined by the polarization P_{scat}. We have

$$R = \frac{1 + P_{\text{scat}}(\theta_1) \, P_{\text{scat}}(\theta_2)}{1 - P_{\text{scat}}(\theta_1) \, P_{\text{scat}}(\theta_2)} \ . \tag{89.17}$$

As an example let us consider the scattering of a neutron by a nucleus taking into account the spin—orbit interaction between them. The concept of this interaction was first introduced by Fermi to account for the phenomenon of polarization of fast neutrons. It has the form

$$\hat{H}' = V(r) + W(r)\boldsymbol{\sigma} \cdot \hat{\mathbf{l}} \ . \tag{89.18}$$

Here $V(r)$ and $W(r)$ are functions depending only on the radius, and $\mathbf{l}$ is the neutron orbital angular momentum operator.

From experiment it follows that parity is conserved in nuclear interactions. Operator (89.18) is constructed in such a way that it automatically satisfies this conservation law. For what follows it is convenient to write the function $W(r)$ in the form

$$W(r) = \frac{1}{r} \frac{d}{dr} \, Y(r) \ .$$

Let us find the functions g and h by making use of the Born approximation. As was shown in §84, the amplitude f in this approximation is equal to

$$f = -\frac{m}{2\pi\hbar^2} \int e^{-i\mathbf{k}_1\cdot\mathbf{r}} \hat{H}'(\mathbf{r}) e^{i\mathbf{k}_0\cdot\mathbf{r}} dV =$$

$$= -\frac{m}{2\pi\hbar^2} \left[V_{\mathbf{k}_0-\mathbf{k}_1} + \frac{\hbar}{i}\boldsymbol{\sigma} \int e^{-i\mathbf{k}_1\cdot\mathbf{r}} \frac{dY}{r\,dr} [\mathbf{r}\times\boldsymbol{\nabla}] e^{i\mathbf{k}_0\cdot\mathbf{r}} dV \right],$$

where $V_{\mathbf{k}-\mathbf{k}'}$ is the Fourier component of the function V. By means of elementary transformations we find

$$f = -\frac{m}{2\pi\hbar^2} \left\{ V_{\mathbf{k}_0-\mathbf{k}_1} - \hbar\boldsymbol{\sigma}\cdot[\mathbf{k}_0\times\int e^{i(\mathbf{k}_0-\mathbf{k}_1)\cdot\mathbf{r}}\boldsymbol{\nabla}Y\,dV] \right\}.$$

Integrating by parts we obtain

$$f = -\frac{m}{2\pi\hbar^2} \left\{ V_{\mathbf{k}_0-\mathbf{k}_1} + \hbar\boldsymbol{\sigma}\cdot[\mathbf{k}_0\times\int Y\boldsymbol{\nabla}\,e^{i(\mathbf{k}_0-\mathbf{k}_1)\cdot\mathbf{r}}\,dV] \right\} =$$

$$= -\frac{m}{2\pi\hbar^2} \left\{ V_{\mathbf{k}_0-\mathbf{k}_1} - i\hbar\boldsymbol{\sigma}\cdot[\mathbf{k}_0\times\mathbf{k}_1] Y_{\mathbf{k}_0-\mathbf{k}_1} \right\}. \qquad (89.19)$$

Comparing (89.19) and (89.6) we find the functions h and g to be

$$g = -\frac{m}{2\pi\hbar^2} V_{\mathbf{k}_0-\mathbf{k}_1},$$

$$h = \frac{imk^2\sin\theta}{2\pi\hbar} Y_{\mathbf{k}_0-\mathbf{k}_1}. \qquad (89.20)$$

We note that in the first approximation of the perturbation theory considered there is no polarization of the scattered particles. Indeed, substituting relation (89.20) into formula (89.13) we obtain $\mathbf{P}_{scat} = 0$. However, in a more accurate calculation $\mathbf{P}_{scat} \neq 0$.

A more general formalism, suitable for the treatment of the scattering of particles by polarized targets, may be found, for example, in the book by Davydov*.

§90. The transition to the classical limit in the quantum scattering formulae

We first of all transform the exact formula for the scattering amplitude into a form convenient for transition to the classical limit.

* A.S.Davydov, *Theorie des Atomkerns* (Deutscher Verlag der Wissenschaften, Berlin, 1963).

If we make use of the expansion of the δ-function in terms of the Legendre polynomials (III.11), then the scattering amplitude (86.11) can be written in the form

$$f(\theta) = \frac{1}{2ik} \sum_{l=0} (2l+1) P_l(\cos\theta) e^{2i\delta_l} - \frac{1}{ik} \delta(1-\cos\theta).$$ (90.1)

For all angles $\theta \neq 0$ formula (90.1) assumes the form

$$f(\theta) = \frac{1}{2ik} \sum_i (2l+1) P_l(\cos\theta) e^{2i\delta_l}.$$ (90.2)

In the quasi-classical approximation the radial part of the wave function has the form (43.2)

$$R_l = \frac{A_l}{rp_r^{\frac{1}{2}}} \sin\left[\frac{1}{\hbar} \int_a^r \left(2m[E-U(r)] - \frac{\hbar^2(l+\frac{1}{2})^2}{r^2} \right)^{\frac{1}{2}} dr + \tfrac{1}{4}\pi \right].$$

The expression for R_l must be understood as an asymptotic expression, i.e. it must be assumed that $r \to \infty$; a denotes the coordinate of the turning point, where the total energy E is equal to the sum of the potential and centrifugal energies, i.e.

$$E = U(a) + \frac{\hbar^2(l+\frac{1}{2})^2}{2ma^2}.$$

In §43 the condition for definition of the turning point did not involve the centrifugal energy, since the motion was assumed to be one-dimensional.

Comparing the expression for R_l with formula (86.2) we see that the scattering phase shift can be written in the form

$$\delta_l = \frac{1}{\hbar} \int_a^r \left(2m[E-U(r)] - \frac{\hbar^2(l+\frac{1}{2})^2}{r^2} - k \right)^{\frac{1}{2}} dr + \tfrac{1}{2}\pi(l+\tfrac{1}{2}) - ka.$$ (90.3)

In (90.3) it is necessary to assume that $r \to \infty$, $l \gg 1$. Then the values of the phase shifts δ_l are very large in absolute magnitude. The formula for the scattering amplitude (90.2) can be simplified by taking into account that in the quasi-classical approximation l must be assumed $\gg 1$. Then for the Legendre polynomials $P_l(\cos\theta)$ one can write an asymptotic expression for $l \gg 1$. They have the form*

---

* N.N. Lebedev, *Special functions and their application* (Prentice Hall, Englewood Cliffs, N.J., 1965).

$$P_l(\cos\theta) = \frac{1}{i(2\pi l \sin\theta)^{\frac{1}{2}}} \, [e^{i(l+\frac{1}{2})\theta + \frac{1}{4}i\pi} - e^{-i(l+\frac{1}{2})\theta - \frac{1}{4}i\pi}] \ .$$

Then for the scattering amplitude we obtain

$$f \equiv \frac{1}{ik} \sum_{l \geqslant 1} l P_l(\cos\theta) \, e^{2i\delta_l} = \frac{1}{k} \sum B(l) \, (e^{i\alpha(l)} - e^{i\beta(l)}) \,, \qquad (90.4)$$

where

$$B(l) = -\left(\frac{l}{2\pi \sin\theta}\right)^{\frac{1}{2}} \,,$$

$$\alpha(l) = 2\delta_l + (l+\tfrac{1}{2})\theta + \tfrac{1}{4}\pi \,,$$

$$\beta(l) = 2\delta_l - (l+\tfrac{1}{2})\theta - \tfrac{1}{4}\pi \,.$$

To obtain $f(\theta)$ it is necessary to sum the series

$$\sum_l B(l) \, e^{i\alpha(l)} \quad \text{and} \quad \sum_l B(l) \, e^{i\beta(l)} \,.$$

We shall consider one of these series, since, as will be clear from what follows, only one of the series has a sum different from zero in a given force field (repulsive or attractive). The quantities $\alpha(l)$, as can be seen from their definition, are large for large l. Hence the terms of the series $\Sigma \, B(l) \, e^{i\alpha(l)}$, containing rapidly oscillating factors, are mutually cancelled. An exception is possible in the case where for a certain value $l = l_0$ the quantity $\alpha(l_0)$ has an extremum, i.e.

$$\left(\frac{d\alpha(l)}{dl}\right)_{l=l_0} = 0 \,. \qquad (90.5)$$

Near the extremum the function $\alpha(l)$ changes slowly and the sum of the series reduces to a sum of terms with values of l close to l_0.

    In this case, to carry out the summation the sum can be replaced by an integral. In the integral the integrand is substantially different from zero only for $l \approx l_0$, and the integral can be calculated by the method of steepest descent (see Part III, §20). Thus one can write

$$\sum_l B(l) \, e^{i\alpha(l)} = B(l_0) \, e^{i\alpha(l_0)} \int_{-\infty}^{+\infty} e^{i\gamma(l-l_0)^2} dl = e^{i\alpha(l_0)} B(l_0) \int_{-\infty}^{+\infty} e^{-c(l-l_0)^2} \, dl \,, \qquad (90.6)$$

where

$$\gamma = \frac{1}{2}\left(\frac{d^2\alpha}{dl^2}\right)_{l=l_0}, \qquad c = -i\gamma .$$

The calculation of the integral in (90.6) is carried out directly, and we obtain

$$\sum_l B(l) e^{i\alpha l} = B(l_0) e^{i\alpha(l_0)} \left(\frac{i\pi}{\gamma}\right)^{\frac{1}{2}} . \tag{90.7}$$

By means of this relation the scattering amplitude can be written in the form

$$f(\theta) = \frac{B(l_0)}{k} e^{i\alpha(l_0)} \left(\frac{i\pi}{\gamma}\right)^{\frac{1}{2}} . \tag{90.8}$$

We shall later find the quantity $\alpha(l_0)$, but for the present we shall consider the physical meaning of eq. (90.5). For this we define the derivative $(d\alpha/dl)_{l=l_0}$. By means of relations (90.4) we have

$$\left(\frac{d\alpha}{dl}\right)_{l=l_0} = 2\left(\frac{d\delta_l}{dl}\right)_{l=l_0} + \theta = 0 . \tag{90.9}$$

In differentiating it should be recalled that the angle θ is given, and that we determine the cross section for a definite value of the angle. If we differentiate with respect to l and make use of formula (90.3), we then obtain

$$\left(\frac{d\delta_l}{dl}\right)_{l=0} = -\int_a^\infty \frac{\hbar(l_0+\frac{1}{2})dr}{r^2[2m(E-U)-\hbar^2(l_0+\frac{1}{2})^2/r^2]^{\frac{1}{2}}} -$$

$$- \left[2m(E-U)-\hbar^2(l+\frac{1}{2})^2/r^2\right]_{r=a}^{\frac{1}{2}} \left(\frac{da}{dl}\right)_{l=l_0} + k\frac{da}{dl} - k\frac{da}{dl} + \frac{\pi}{2} =$$

$$= -\int_a^\infty \frac{\hbar(l_0+\frac{1}{2})dr}{r^2[2m(E-U)-\hbar^2(l_0+\frac{1}{2})^2/r^2]^{\frac{1}{2}}} + \frac{1}{2}\pi ,$$

since at the turning point $r = a$ the square root reduces to zero. Condition (90.9) assumes the form

$$-\int_a^\infty \frac{\hbar(l_0+\frac{1}{2})dr}{r^2[2m(E-U)-\hbar^2(l_0+\frac{1}{2})^2/r^2]^{\frac{1}{2}}} + \frac{1}{2}\pi \pm \frac{1}{2}\theta = 0 . \tag{90.10}$$

If we carried out the corresponding calculations for the second sum, then lower sign in (90.10) would correspond to the extremum $\beta(l)$. For brevity the two conditions are combined. Formula (90.10) defines the value of l_0.

The quantity $\hbar(l_0+\frac{1}{2}) = L$ represents the angular momentum. After introducing the angular momentum L formula (90.10) can be transformed into the form

$$\int_a^\infty \frac{L\,dr}{r^2\,[2m(E-U)-L^2/r^2]^{\frac{1}{2}}} = \tfrac{1}{2}(\pi\pm\theta)\,. \tag{90.11}$$

In classical mechanics the angular momentum can be connected with the impact parameter ρ by means of the following relation:

$$L = m\rho\upsilon\,,$$

where υ is the velocity of the particle at infinity. Substituting this value for the angular momentum into formula (90.11), we obtain an expression which is exactly the same as the classical relation connecting the impact parameter with the scattering angle θ*

$$\int_{r=a}^\infty \frac{m\upsilon\rho\,dr}{r^2\,[2m(E-U)-(m\upsilon\rho/r)^2]^{\frac{1}{2}}} = \tfrac{1}{2}(\pi\pm\theta)\,. \tag{90.12}$$

The values of the impact parameter ρ is determined by the positive root of eq. (90.12). It is known from mechanics that in a repulsive force field the positive root of this equation exists only for a negative θ. On the contrary, in an attractive force field this root exists for a positive θ.

Let us consider the case of repulsive forces. Then condition (90.9) can be fulfilled only for $\alpha(l)$ but not for $\beta(l)$. Correspondingly only the first of the series in (90.4) has a sum different from zero.

We now go on to the calculation of the cross section. According to formulae (83.5) and (90.8), it is defined by the expression

$$d\sigma = |f(\theta)|^2\,d\Omega = \frac{1}{k^2}\,|B(l_0)|^2\,\frac{\pi}{|\gamma|}\,d\Omega\,.$$

The quantity γ is defined by expression (90.6). By means of (90.4) and (90.3) we obtain

---

* See, for example, L.D.Landau and E.M.Lifshitz, *Mechanics* (Pergamon Press, Oxford, 1960).

$$\gamma = \hbar \frac{\partial^2}{\partial L^2} \int_a^\infty [2m(E-U) - L^2/r^2]^{\frac{1}{2}} \, dr =$$

$$= -\frac{\partial}{\partial L} \int_a^\infty \frac{(L\hbar/r^2) \, dr}{[2m(E-U) - L^2/r^2]^{\frac{1}{2}}}. \qquad (90.13)$$

Making use of (90.11) we transform the expression for γ into the form

$$\gamma = \pm \frac{\hbar}{2} \frac{\partial \theta}{\partial L}.$$

If the value of B from (90.4) is substituted and the value found for γ is used, then the differential cross section takes the form

$$d\sigma = |f(\theta)|^2 \, d\Omega = \frac{L}{m^2 v^2 \sin \theta} \left| \frac{\partial L}{\partial \theta} \right| d\Omega. \qquad (90.14)$$

Replacing the quantity L in (90.14) by its classical value, we obtain

$$d\sigma = \frac{\rho}{\sin \theta} \left| \frac{\partial \rho}{\partial \theta} \right| d\Omega. \qquad (90.15)$$

Expression (90.15) represents the ordinary scattering formula given by classical mechanics.

We examine the limits of applicability of the formulae (90.15) for the scattering cross section. They can be established from the following obvious considerations.

One can speak of the motion of a particle in a trajectory in the case where the corresponding wavelength is small in comparison with the size of the system. In the given case the wavelength must be small in comparison with the size of the region in which a considerable interaction takes place. If the size of this region is denoted by R, then this requirement can be written in the form

$$\lambda \ll R, \qquad (90.16)$$

where λ is the de Broglie wavelength.

Substituting the value of λ into formula (90.16) we find

$$R \gg \frac{2\pi\hbar}{mv}. \qquad (90.17)$$

In order that the behaviour of the particle may be characterized by classical concepts, it is necessary that the quantum-mechanical uncertainties be small. In other words, it is necessary that the following relations be fulfilled:

$$\frac{\Delta\theta}{\theta} \ll 1 , \qquad \frac{\Delta\rho}{\rho} \ll 1 , \qquad\qquad\qquad (90.18)$$

where ρ is the classical impact parameter, and $\Delta\theta$ and $\Delta\rho$ are respectively the quantum-mechanical uncertainties for the scattering angle θ and the impact parameter ρ.

For the quantity $\Delta\theta$ one can write an expression valid in order of magnitude

$$\Delta\theta \sim \frac{\Delta p}{p} , \qquad\qquad\qquad (90.19)$$

where Δp is the uncertainty in the transverse component of the momentum. Making use of the uncertainty relation for the coordinate and momentum

$$\Delta p \cdot \Delta\rho \sim \hbar$$

and eliminating the quantity Δp from (90.19), and then using (90.18), we obtain

$$\theta \gg \Delta\theta \sim \frac{\hbar}{\Delta\rho p} \gg \frac{\hbar}{\rho p} . \qquad\qquad\qquad (90.20)$$

This condition assumes a considerably simpler form if the scattering angles are small. Namely, in this case the scattering angle θ can be found in a simple way. It is equal to the ratio of the value of the transverse momentum acquired by the scattered particle in traversing the field of the scatterer to the longitudinal momentum. The transverse momentum is equal to the force $U'(\rho)$ acting on the particle multiplied by the time τ for which this force acts: $\tau = \rho/v$. Thus the scattering angle θ is, in order of magnitude, equal to

$$\theta \approx |U'(\rho)| \frac{\rho}{vp} . \qquad\qquad\qquad (90.21)$$

Or, substituting (90.21) into (90.20), we find the condition of applicability of the theory

$$|U'(\rho)|\rho^2 \gg \hbar v .$$

But if the derivative $U'(\rho)$ is replaced by $U(\rho)/\rho$, then the condition of applicability can be rewritten in the form

$$U(\rho) \gg \frac{\hbar v}{\rho} . \qquad\qquad\qquad (90.22)$$

Comparing (90.22) with the condition of applicability of the Born approximation (84.12), we see that the conditions are opposite to each other. Thus these methods supplement each other to a considerable degree.

§91. The general theory of inelastic scattering and the absorption of particles

So far we have confined ourselves to consideration of the elastic scattering process. We now turn to the more general case where inelastic scattering is also possible.

Any process in which the internal state of the particles changes is said to be inelastic. Thus, for example, collisions accompanied by an excitation (for instance, an excitation of the atom or nucleus), by a decay or by the production of new particles, for example, are inelastic. Each of the possible processes is called a reaction channel. If the process is compatible with conservation laws, the channel is said to be open. In what follows we shall consider processes for which the inelastic and elastic reaction channels are open. We shall begin with a generalization of partial wave scattering theory. This will allow us to cover, at the same time, processes of elastic and inelastic scattering and absorption. For a formal description of any scattering process we shall surround the scattering centre by a fictitious sphere of sufficiently large radius R_0.

Let us consider the character of the lth partial wave for $r > R_0$ in three cases:

(1) at the origin there is no scattering centre,

(2) at the origin there is a scattering centre at which the particle undergoes only elastic scattering,

(3) at the origin there is a scattering centre at which the particle undergoes inelastic scattering.

In the first case the radial function of the lth partial wave can be written (see (36.10)) in the form of a superposition of two waves

$$R_l = a_l \frac{e^{i(kr-\frac{1}{2}\pi l)}}{2ikr} - a_l \frac{e^{-i(kr-\frac{1}{2}\pi l)}}{2ikr} = a_l \frac{\sin (kr-\frac{1}{2}\pi l)}{kr}.$$

The second term represents a converging wave, and the first term a diverging wave. Here we make use of asymptotic expressions, since by assumption R_0 is sufficiently large. The amplitudes and phases of the two waves are the same and the wave function R_l is the product of a real function and a constant factor. Hence the flux through a closed surface is equal to zero:

$$j_l = \frac{\hbar}{2mi} \int \left(\psi_l^* \frac{\partial \psi_l}{\partial r} - \psi_l \frac{\partial \psi_l^*}{\partial r} \right) r^2 \, d\Omega = 0 \,,$$

where

$$\psi_l = P_l(\cos\theta) R_l(r) \,.$$

In the second case the radial function of the lth partial wave is written, according to (86.2), in the form

$$R_l = B_l \frac{\sin(kr + \delta_l - \frac{1}{2}\pi l)}{kr} = F_l \frac{e^{2i\delta_l} \, e^{i(kr - \frac{1}{2}\pi l)} - e^{-i(kr - \frac{1}{2}\pi l)}}{2ikr} \,. \quad (91.1)$$

The amplitudes of the converging and diverging waves differ from each other by the phase factor $e^{2i\delta_l}$, where $|e^{2i\delta_l}| = 1$. In this case the total partial flux through the surface of the sphere is also equal to zero (the partial wave function depending on l is real). Hence it follows that the diverging and converging fluxes of the lth partial wave are equal to each other. The fact that the converging and diverging waves have different coefficients, $e^{2i\delta_l}$ and unity, does not contradict this equality, since $|e^{2i\delta_l}| = 1$.

In the third case, where the particles undergo inelastic scattering, it is impossible to write a general expression for the radial function taking into account all possible inelastic processes. We can, however, simplify the problem if we consider elastic scattering separately from all possible forms of inelastic scattering.

In this case we can write the following formal expression for the radial function of the lth partial wave describing the elastic scattering of a particle:

$$R_l = b_l \frac{S_l \, e^{i(kr - \frac{1}{2}\pi l)} - e^{-i(kr - \frac{1}{2}\pi l)}}{2ikr} \quad (91.2)$$

This expression is constructed according to the same principle as (91.1), but it takes into account the particular process, in which inelastic processes or absorption may exist along with elastic scattering. The coefficient S_l introduced is in magnitude less than one. This expresses the fact that in the presence of absorption or inelastic scattering the converging flux of elastically scattered particles is larger than the diverging flux. Then the wave function is written in the form

$$\psi = \sum_l b_l \frac{S_l \, e^{i(kr - \frac{1}{2}\pi l)} - e^{-i(kr - \frac{1}{2}\pi l)}}{2ikr} P_l(\cos\theta) = \sum_l \psi_l \,.$$

The coefficients b_l are again defined by the requirement that the wave function ψ be the same as (83.3). On carrying out calculations analogous to those for elastic scattering, we find the wave function in the form

$$\psi = \sum_{l=0}^{\infty} \frac{i^l}{2ikr} (2l+1)[S_l \, e^{i(kr-\frac{1}{2}\pi l)} - e^{-i(kr-\frac{1}{2}\pi l)}] P_l(\cos\theta) . \quad (91.2')$$

It is easily shown that the flux of elastically scattered particles with given angular momentum through a sphere of radius $r \gg R_0$ is different from zero. Indeed, we have

$$\frac{\partial \psi_l}{\partial r} = \frac{i^l(2l+1)P_l(\cos\theta)}{2ikr} [ikS_l e^{i(kr-\frac{1}{2}\pi l)} + ik \, e^{-i(kr-\frac{1}{2}\pi l)}] P_l(\cos\theta) .$$

Here only terms proportional to r^{-1} are retained in the expression for $\partial\psi_l/\partial r$. Terms proportional to r^{-2} are dropped, since we desire to find the flux through a sphere of large radius. We further calculate the total flux of particles through a sphere of radius $r \gg R_0$. It is equal to

$$j_l = \frac{i\hbar}{2m} r^2 \int \left(\psi_l^* \frac{\partial\psi_l}{\partial r} - \psi_l \frac{\partial\psi_l^*}{\partial r} \right) d\Omega .$$

Substituting the functions ψ_l and $\partial\psi_l/\partial r$ into the expression for the flux and taking into account the conditions of normalization of the Legendre polynomials $P_l(\cos\theta)$, we obtain

$$j_l = - \frac{\pi\hbar}{mk} (2l+1)(1-|S_l|^2) . \tag{91.3}$$

Since $|S_l| \ll 1$, the flux is negative. This means that the total flux is directed inwards through the sphere.

It is easy to understand the meaning of this result: the flux of particles incident on the centre with angular momentum l turns out to be larger than the flux of elastically scattered particles. The particles undergo inelastic scattering or absorption, and the intensity of the beam of elastically scattered particles is reduced. It is clear that on dividing the flux j_l by the flux density of incident particles we find, by definition, the partial inelastic scattering cross section. Here inelastic scattering is understood to be all processes reducing the intensity of elastic scattering. Since the flux density of incident particles is equal to v, then for the lth partial inelastic scattering cross section we obtain

$$\sigma_{l\text{inel}} = \frac{\pi}{k^2} (2l+1)(1-|S_l|^2) . \tag{91.4}$$

As for the elastic scattering amplitude, we can, without reproducing the calculations of §86, write for it the expression

$$f(\theta) = \frac{1}{2ik} \sum_{l=0}^{\infty} (2l+1)(S_l-1) P_l(\cos \theta) ,$$ (91.5)

since formula (91.1) differs from (91.2) by the substitution of S_l for $e^{2i\delta_l}$.

The set of complex quantities S_l defines the cross section for inelastic as well as elastic scattering. In particular, if $S_l = e^{2i\delta_l}$, where δ_l is real, then the inelastic scattering cross section reduces to zero, and the elastic scattering amplitude is the same as expression (86.11).

Besides the lth partial cross section for elastic and inelastic scattering processes one can also write the total cross sections for the processes.

The total inelastic scattering cross section is evidently equal to

$$\sigma_{\text{inel}} = \frac{\pi}{k^2} \sum_{l=0}^{\infty} (2l+1)(1-|S_l|^2) = \sum_{l} \sigma_l ,$$ (91.6)

and the total elastic scattering cross section is equal to

$$\sigma_{\text{el}} = \int |f(\theta)|^2 \, d\Omega = \frac{\pi}{k^2} \sum_{l=0}^{\infty} (2l+1)[|1-S_l|^2 .$$ (91.7)

We now turn to the consideration of formula (91.6).

Each cross section σ_l can be pictured as a characteristic of the process of inelastic scattering or absorption of particles with angular momentum l. Since the quantity $|S_l|^2 < 1$, it can be stated that the partial cross section σ_l has the upper limit $\sigma_{l\,\text{max}} = \pi k^{-2}(2l+1)$.

The structure of formula (91.6) and the physical meaning of the coefficient $1-|S_l|^2$ can easily be understood by means of the following reasoning based on the quasi-classical approximation.

The collision parameter of a particle can be (see (86.14)) written in the form

$$\rho_l = \frac{\hbar}{p} [l(l+1)]^{\frac{1}{2}} .$$ (91.8)

For large l we obtain

$$\rho_l = \frac{\hbar}{p} l \,.$$

The area of the annulus lying between two circles of radii ρ_l and ρ_{l+1} is equal to

$$2\pi\rho_l \frac{\hbar}{p} \approx \frac{\pi\hbar^2}{p^2} (2l+1) \,.$$

The number of particles passing through this annulus oriented perpendicularly to the incident flux can easily be found. If the flux density of incident particles is equal to unity, then the number of particles crossing the ring is numerically equal to $\pi\hbar^2 p^{-2}(2l+1)$.

We introduce the so-called absorption coefficient ξ_l, which by definition represents the ratio of the number of absorbed particles incident on a given surface to the total number of particles incident on this surface. The number of particles absorbed by the surface of the annulus defined by radii ρ_l and ρ_{l+1} is defined by the expression $\pi\hbar^2 p^{-2}(2l+1)\xi_l$, and, correspondingly, the absorption cross section will have the form

$$\sigma_{l_{\text{inel}}} = \frac{\pi}{k^2} (2l+1)\xi_l \,. \tag{91.9}$$

Comparing formulae (91.6) and (91.9) we see that

$$1 - |S_l|^2 = \xi_l \,, \tag{91.10}$$

i.e. the quantity $1-|S_l|^2$ is the absorption coefficient.

Finally, we also obtain the formula relating elastic and inelastic scattering cross sections. It turns out that the following equality holds:

$$\frac{4\pi}{k} \operatorname{Im} f(0) = \sigma_{\text{inel}} + \sigma_{\text{el}} \,. \tag{91.11}$$

To obtain this relation we shall calculate the sum of elastic and inelastic cross sections. By means of (91.6) and (91.7) we have

$$\sigma_{\text{inel}} + \sigma_{\text{el}} = \frac{\pi}{k^2} \sum_{l=0}^{\infty} (2l+1)(2-S_l-S_l^*) = \frac{\pi}{k^2} \sum_{l=0}^{\infty} (2l+1)(2-2\operatorname{Re} S_l) \,. \tag{91.12}$$

On the other hand, since the Legendre polynomials for $\theta = 0$ are equal to one, we have for the scattering amplitude

$$f(0) = \frac{1}{2ik} \sum_{l=0}^{\infty} (2l+1)(S_l - 1) \, ,$$

and the imaginary part of the scattering amplitude is equal to

$$\text{Im} f(0) = \frac{1}{2k} \sum_{l=0}^{\infty} (2l+1)(1 - \text{Re } S_l) \, .$$

Comparing the expressions obtained, we see that eq. (91.11) is valid. Thus we have shown that the sum of inelastic and elastic scattering cross sections is proportional to the imaginary part of the scattering amplitude taken for the value of the angle $\theta = 0$. Formula (91.11) is called the optical theorem.

In conclusion we note that the absorption of particles can be described by introducing a complex potential $U = V_1 - iV_2$, where V_1 and V_2 are real functions. The imaginary part of the potential characterizes the absorption or emission of particles. As a matter of fact, in this case the Schrödinger equation has the form

$$i\hbar \frac{\partial \psi}{\partial t} = \left(-\frac{\hbar^2}{2m} \nabla^2 + V_1 - iV_2 \right) \psi \, . \tag{91.13}$$

Carrying out calculations analogous to those of §7, we obtain

$$\frac{\partial \rho}{\partial t} + \nabla \cdot \mathbf{j} - \frac{2V_2 \psi^* \psi}{\hbar} = 0 \, , \tag{91.14}$$

where

$$\rho = \psi^* \psi \, , \qquad \mathbf{j} = \frac{\hbar}{2mi} [\psi \nabla \psi^* - \psi^* \nabla \psi] \, .$$

In the stationary case for V_2 equal to zero, $\nabla \cdot \mathbf{j} = 0$, which corresponds to the absence of absorption or emission of particles. If $V_2 \neq 0$, then we obtain

$$\nabla \cdot \mathbf{j} = 2V_2 \rho / \hbar \, .$$

Depending on the sign of V_2 this formula describes the absorption or emission of particles.

§92. The diffraction of fast neutrons by nuclei

The study of the interaction of fast neutrons with nuclei shows that in the region of neutron energies above a few tens of MeV for light nuclei and a few hundreds of MeV for heavy nuclei a very intense capture of neutrons takes place.

The intense absorption of neutrons is also accompanied by their elastic scattering. In describing the strong absorption of fast neutrons the following optical analogy turns out to be very useful. The nucleus behaves with respect to the neutrons as a perfectly absorbing (black) sphere on which a light wave is incident. The absorption of the light wave by the black sphere is accompanied by its perturbation in the region of space near the absorber. This means that in addition to absorption, light scattering occurs. Analogously to this the absorption of neutrons by a nucleus will perturb their wave function and the neutrons will undergo elastic scattering.

To calculate the neutron elastic scattering cross section we shall make use of the analogy with optical phenomena. In §36 of Part IV we have seen that diffraction phenomena occur when the wavelength of the light is less than the radius of the scattering sphere. In this case the intensity of light scattered by a black sphere of radius R into solid angle $d\Omega$ is given by expression (36.13) of Part IV

$$ dI = \frac{I}{\pi} \left| \frac{J_1(kR\theta)}{\theta} \right|^2 d\Omega , \qquad (92.1) $$

where θ is the angle between the direction of motion of the scattered light and the initial direction of its incidence, I is the total intensity of light incident on the screen, and J_1 is the Bessel function of first order.

Simple estimation shows that the wavelength of neutrons of an energy of the order of 1 MeV is smaller by a factor of several hundred than the nuclear size. Therefore the optical formula (92.1) can be applied to the scattering of neutrons by the absorbing nucleus. To obtain the neutron differential scattering cross section the flux of neutrons scattered into angle $d\Omega$ must be divided by the incident neutron flux density $I/\pi R^2$. We then have

$$ d\sigma = R^2 \left| \frac{J_1(kR\theta)}{\theta} \right|^2 d\Omega . \qquad (92.2) $$

This expression, of course, can also be obtained from the general formula (91.5).

From the condition of 'blackness' of the nucleus it follows that the absorption coefficient ξ_l is equal to one for those l for which $\rho < R$ and $\xi_l = 0$, if $\rho_l > R$. Since $\rho_l \sim \hbar l/p$ (see §91), then

$$S_l = \begin{cases} 0 & l < pR/\hbar = kR , \\ \\ 1 & l > pR/\hbar = kR , \end{cases}$$

where $kR \gg 1$. Substituting these values of S_l into (91.5) we find the elastic scattering amplitude

$$f(\vartheta) = -\frac{1}{2ik} \sum_{l=0}^{kR} (2l+1) P_l(\cos \vartheta) .$$

The dominant part in the sum is played by terms with large l. Hence we can disregard unity in comparison with $2l$, and use for the Legendre polynomial $P_l(\cos \vartheta)$ the approximate expression valid for $\vartheta \ll 1$*.

$$P_l(\cos \vartheta) = J_0 [(l+\tfrac{1}{2})\vartheta] \cong J_0(l\vartheta)$$

and pass from a summation over l to an integration

$$f(\vartheta) = \frac{i}{k} \int_0^{kR} l J_0(l\vartheta) \, dl = \frac{iR}{\vartheta} J_1(kR\vartheta) .$$

From which we immediately obtain expression (92.2) for the cross section.

Let us consider the dependence of the differential cross section (92.2) on the scattering angle θ in more detail. The differential cross section does not depend on the azimuthal angle. Evidently we have

$$d\sigma = 2\pi R^2 \left| \frac{J_1(kR\theta)}{\theta} \right|^2 \sin \theta \, d\theta . \tag{92.3}$$

For small angles $kR\theta < 1$, we expand the Bessel function in a series and find $J_1(kR\theta) \approx \tfrac{1}{2} kR\theta$. Consequently, for small angles the cross section assumes the form

$$d\sigma = \tfrac{1}{4} k^2 R^4 \, d\Omega . \tag{92.4}$$

This is independent of the scattering angle θ.

---

* See, for example, L.D.Landau and E.M.Lifshitz, *Quantum mechanics* (Pergamon Press, Oxford, 1965).

For larger angles up to values lying in the interval $1 \gg \theta \gg 1/kR$ one can write the asymptotic expression for the Bessel function

$$J_1(kR\theta) \approx \left(\frac{2}{\pi kR\theta}\right)^2 \sin(kR\theta - \tfrac{1}{4}\pi) .$$

In this range of angles the cross section, which oscillates, decreases rapidly with increasing θ. The value of the cross section at the maxima decreases in proportion to θ^{-3}.

Thus the cross section has a sharp maximum for scattering at an angle $\theta \approx 0$, i.e. for forward scattering in directions close to the direction of the incident beam.

The total scattering cross section σ can be found by integrating (92.3) over the entire solid angle,

$$\sigma = 2\pi R^2 \int \frac{J_1^2}{\theta^2} \sin\theta \, d\theta .$$

In view of the rapid convergence of the integral the contribution to it by large values of θ is small, and the upper limit of the integral can approximately be replaced by infinity. Then, making use of the formula

$$\int_0^\infty \frac{J_1^2(x)}{x} \, dx = \tfrac{1}{2} ,$$

we find finally

$$\sigma = \pi R^2 . \tag{92.5}$$

The total cross section for the scattering of neutrons with $\lambda \leqslant R$ is the same as the geometric cross section of the nucleus.

Let us also define the total cross section for the absorption of neutrons by a nucleus. Making use of the expression for S_l and substituting it into formula (91.6) we obtain

$$\sigma_{\text{inel}} = \frac{\pi}{k^2} \sum_{l=0}^{kR} (2l+1) = \pi R^2 . \tag{92.6}$$

Consequently, the cross section for the absorption of neutrons by a black nucleus is also the same as the geometric cross section of the nucleus.

From relations (92.5) and (92.6) it follows that the total cross section for the interaction of neutrons with a nucleus is equal to twice the geometric cross section of the nucleus,

$$\sigma_{\text{inel}} + \sigma_{\text{el}} = 2\pi R^2 .$$ (92.7)

By analogous methods one can also calculate the cross sections for scattering by nuclei which only partially absorb the neutrons incident on them, as well as the diffraction of charged particles by nuclei*.

§93. The scattering of slow particles. The threshold approximation

We shall apply the partial wave scattering formulae already obtained to find the cross sections for the elastic and inelastic scattering of slow particles. As in §86, we shall understand slow particles to be those whose de Broglie wavelength λ is large compared with the size of the region of interaction. We shall restrict ourselves to the case where the interaction energy decreases sufficiently rapidly with increasing distance, that an effective radius R of the interaction sphere can be introduced.

As we have seen in §86, only s-wave scattering is significant at small energies. The radial part of the wave function corresponding to the angular momentum $l = 0$ satisfies eq. (35.8)

$$-\frac{\hbar^2}{2m}\frac{1}{r^2}\frac{\partial}{\partial r}\left(r^2\frac{\partial R_0}{\partial r}\right) + U(r)R_0 = ER_0 .$$

Introducing the wave number $k = (2mE/\hbar^2)^{\frac{1}{2}}$, the above equation can be rewritten in the form

$$R_0'' + \frac{2}{r}R_0' + k^2R_0 - \frac{2m}{\hbar^2}U(r)R_0 = 0 .$$ (93.1)

For $r > a$, the potential energy is equal to zero outside the interaction region and the equation for the function R_0 assumes the form

$$R_0'' + \frac{2}{r}R_0' + k^2R_0 = 0 .$$ (93.2)

It is clear that the potential energy does not reduce to zero sharply at a certain limit, but changes over a transitional region according to a complex and usually unknown law.

Therefore at first sight the finding of the wave function over all space seems to be an extremely complex problem. In reality, however, this is not so.

---

* For more details see A.I.Akhiezer and I.Ya.Pomeranchuk, *Nekotorye voprosy teorii yadra (Some problems in nuclear theory)* (Gostekhizdat, Moscow, 1950).

It turns out that use can be made of the large value of λ or, what is the same, of the small k, essentially to simplify the problem. Namely, in the region $r < a$, i.e. in the region of effective interaction, the term k^2 can be neglected since it is small in comparison with $2m\hbar^{-2}U(r)$. Then we have

$$R_0'' + \frac{2}{r}R_0' - \frac{2m}{\hbar^2}U(r)R_0 = 0 . \tag{93.3}$$

The solution of eq. (93.3) for a given function $U(r)$ can be written in the form $R_0(r,c_1,c_2)$, where c_1 and c_2 are two arbitrary constants. Since (93.3) does not involve the quantity k, we shall assume the wave function to be independent of k in the region $r < a$.

The solution of eq. (93.2) is

$$R_0 = \frac{1}{2ikr}(c_3 e^{ikr} + c_4 e^{-ikr}) . \tag{93.4}$$

Here c_3 and c_4 are two constants of integration which do not depend on r but, generally speaking, are functions of k.

The equation for the wave function cannot be written for the transitional region, since the behaviour of the potential energy here is unknown. However, the width of the intermediate region is small in comparison with the size of the interaction region and is very small in comparison with the wavelength λ.

Nevertheless, a substantial change of the wave function takes place over a distance λ. Hence one can disregard the change of the wave function in the transition region and replace it by a sharp boundary at $r = a$. At this surface the two solutions must join smoothly.

It is clear, however, that two functions, one of which depends on k as a parameter and the other of which does not depend on k at all, can only be joined when in the neighbourhood of the boundary of the region the function (93.4) also becomes independent of k.

For $r \sim a$ the quantity ka is small by definition. Hence expanding the exponentials (93.4) in a series in powers of kr and retaining the first two terms of the expansion, we obtain

$$R_0 = \frac{c_3(k)(1+ikr) + c_4(k)(1-ikr)}{2ikr} . \tag{93.5}$$

This expression will not depend on k when the following relations are fulfilled

$$c_4(k) + c_3(k) = 2ika_2 \,,$$
$$c_3(k) - c_4(k) = 2a_1 \,,$$

$$(93.6)$$

where a_1 and a_2 are constant quantities independent of k.

Solving the system of equations (93.6), we find for c_3 and c_4

$$c_4 = -a_1 + ika_2 \,,$$
$$c_3 = ika_2 + a_1 \,.$$

$$(93.7)$$

Comparing expression (93.4) with (93.3), we find the quantity S_0 to have the form

$$S_0 = -\frac{c_3}{c_4} = -\frac{a_1 + ika_2}{ika_2 - a_1} \,.$$

$$(93.8)$$

Expanding in a series in small values of k, we obtain

$$S_0 = 1 + 2ika_2/a_1 \,.$$

Making use of formulae (91.6) and (91.7) we find the expression for the cross sections for elastic and inelastic processes:

$$\sigma_{el} = \frac{\pi}{k^2} |1 - S_0|^2 = 4\pi \left|\frac{a_2}{a_1}\right|^2 \,,$$

$$\sigma_{inel} = \frac{\pi}{k^2} (1 - |S_0|^2) = \frac{4\pi}{k} \operatorname{Im} \frac{a_2}{a_1} \,.$$

$$(93.9)$$

From these formulae it follows that the elastic scattering cross section in the case considered does not depend on the energy of the scattered particle. The inelastic scattering cross section is inversely proportional to the wave number k, i.e. inversely proportional to the velocity v of the particle.

The method which we have used of neglecting the width of the transition zone and the replacement of eq. (93.1) by eq. (93.2) in the internal region is of a very general character and is called the threshold approximation. This approximation can be applied successfully in all cases where the wavelength can be considered to be large in comparison with the width of the transition region.

We shall encounter the use of the threshold approximation later.

§94. The Breit–Wigner formula

In the preceding sections we have considered the laws of elastic scattering as well as of the absorption of particles. We now turn to the study of some phenomena occurring in nuclear reactions of the type

$$A + a \rightarrow B + b . \tag{94.1}$$

Here A and B are the initial and final nuclei, a is the incident particle and b is the particle emerging as a result of the reaction. In order to avoid complications associated with the effect of the nuclear electric field, we shall confine ourselves to the case where the incident and outgoing particles are neutrons.

The study of reactions caused by neutrons of relatively small energies showed that the cross sections for the reactions as a function of the energy of the incident neutrons display maxima at definite energy values. The phenomenon is of a pronounced resonant character and the maxima correspond to very narrow neutron energy intervals.

To account for the resonant character of nuclear reactions Bohr proposed the following general scheme of nuclear reactions. Neutron a, penetrating the nucleus, strongly interacts with the nuclear particles and transfers its excess energy to them. This latter energy is evidently equal to the sum of its kinetic energy and the binding energy of the particle in the nucleus U_0. The energy brought by the neutron is rapidly distributed among all the nucleons in the nucleus, since they interact strongly with each other. As a result a new, so-called compound nucleus C arises from nucleus A and the neutron. The compound nucleus is not a stable system, since its energy is higher than the energy of the normal state by an amount $E + U_0$. After a certain lapse of a time the compound nucleus will make a transition to the normal state. At small excitation energies this transition may proceed in one of two ways.

First, as a result of a fluctuation all the excitation energy may be concentrated on one of the nuclear particles. This particle (for simplicity of the argument, a neutron) then has the possibility of escaping from the nucleus, with energy E. Evidently this mode of reaction corresponds to the elastic scattering of the neutron by the nucleus.

Secondly, only part of the excitation energy may be carried away by the escaping neutron. The rest of the excitation energy is emitted by the system of nuclear particles in the form of a γ-quantum. In this case inelastic scattering of the neutron occurs. A particular case of this reaction is the radiative capture of the neutron, in which the entire excitation energy is carried away by a γ-quantum and the neutron remains in the nucleus.

For the cross sections for the elastic and inelastic scattering use can be made of formulae (91.6) and (91.7). We shall restrict ourselves to the case of slow neutrons, described by an s-wave, and shall consider the nucleus to be a sphere of radius R. Although the nucleus cannot be considered to have sharp geometric bounds, its diffuseness is very small in comparison with the wavelength of the incident neutron $\lambda \gg R$.

The neutron inside the nucleus must be in a state to which there corresponds a wavelength λ_{int}. At the surface $r = R$ the wave functions describing the neutron outside and inside the nucleus must join for which the following conditions are to be fulfilled:

$$\psi = \psi_{int}, \qquad \frac{d\psi}{dr} = \frac{d\psi_{int}}{dr}.$$

It follows from the second condition that the orders of magnitude of the amplitudes of the wave functions of the external and internal motions are in the ratio $\sim\lambda_{int}/\lambda$. This means that the probability for the particle to get inside the nucleus is $\sim(\lambda_{int}/\lambda)^2$, i.e. is very small.

The corresponding energy is determined by the value of the normal derivative of the wave function at the surface of the nucleus.

We denote by $f(E)$ the quantity

$$f(E) = R \left(\frac{d(r\psi)/dr}{r\psi} \right)_{r=R}. \tag{94.2}$$

The quantity $f(E)$ is related directly to the normal derivative $(d\psi/dr)_{r=R}$ and depends on the neutron energy E. The quantity S_0, defining the cross section for the elastic and inelastic scattering of the s-wave, can easily be expressed in terms of f.

Substituting the value of ψ from (91.2) into (94.2), we find

$$f = -i \frac{kR \cdot e^{-ikR} + kRS_0 e^{ikR}}{e^{-ikR} - S_0 e^{ikR}}.$$

Hence it follows that

$$S_0 = -e^{-2ikR} \frac{kR - if(E)}{kR + if(E)}. \tag{94.3}$$

Since $f(E)$ is, generally speaking, a complex quantity, one can write

$$f(E) = f_1(E) - if_2(E), \tag{94.4}$$

where $f_1(E)$ and $f_2(E)$ are real functions. Since $|S_0|$ is always $\leqslant 1$, then the function $f_2(E) \geqslant 0$.

Taking into account (94.4) we have for S_0

$$S_0 = -e^{-2ikR} \frac{kR - if_1(E) - f_2(E)}{kR + if_1(E) + f_2(E)}. \tag{94.5}$$

Substituting this value of S_0 into (91.6), we find

$$\sigma_{\text{inel}} = \frac{\pi}{k^2}(1-|S_0|^2) = \frac{4\pi}{k^2}\frac{kRf_2}{(kR+f_2)^2 + f_1^2}. \tag{94.6}$$

Analogously from (91.7) it follows that

$$\sigma_{\text{el}} = \frac{\pi}{k^2}|1-S_0|^2 = \frac{\pi}{k^2}\left|1+e^{-2ikR}\frac{kR-if_1-f_2}{kR+if_1+f_2}\right|^2 =$$

$$= \frac{4\pi}{k^2}\left|e^{-ikR}\frac{kR\cos kR - f_1\sin kR + if_2\sin kR}{kR+if_1+f_2}\right|^2 =$$

$$= \frac{4\pi}{k^2}\left|\frac{kR}{i(kR+f_2)-f_1} + e^{ikR}\sin kR\right|^2. \tag{94.7}$$

Let us first discuss the formula for σ_{inel}. Since $f_2 > 0$, the cross section has a maximum for $f_1(E_0) = 0$. When the neutron has an energy equal to E_0 it has a relatively high probability of penetrating the nucleus. Accordingly the energy E_0 corresponds to a resonance value of the energy of the nucleus. Near the resonance energy we can expand the function $f_1(E)$ in a series in powers of $E-E_0$ and restrict ourselves to the first term of the expansion

$$f_1(E) = f'(E_0)(E - E_0).$$

It can be shown* that the quantity $f'(E_0) < 0$. We introduce the notation

$$\Gamma_e = -\frac{2kR}{f'(E_0)}; \qquad \Gamma_r = -\frac{2f_2}{f'(E_0)}; \qquad \Gamma = \Gamma_e + \Gamma_r. \tag{94.8}$$

Then we find

$$\sigma_{\text{inel}} = \frac{\pi}{k^2}\frac{\Gamma_e\Gamma_r}{(E-E_0)^2 + \frac{1}{4}\Gamma^2}. \tag{94.9}$$

In formula (94.7) for σ_{el} for $E \approx E_0$ the first term is usually large in comparison with the second, and one can write

$$(\sigma_e)_{\text{el}} \approx \frac{\pi}{k^2}\frac{\Gamma_e^2}{(E-E_0)^2 + \frac{1}{4}\Gamma^2}. \tag{94.10}$$

The formulae for the cross sections for elastic and inelastic scattering of slow neutrons are called the Breit–Wigner formulae. To explain the physical meaning of the quantities Γ_e, Γ_r and Γ which have been introduced it is useful to

---

* See A.I.Akhiezer and I.Ya.Pomeranchuk, *Nekotorye voprosy teorii yadra* (*Some problems in nuclear theory*) (Gostekhizdat, Moscow, 1950).

·compare the Breit–Wigner formulae with the dispersion formulae of the theory of light scattering (§ 108). We see that the general structure of the formulae is the same. This is quite natural, since the Breit–Wigner formulae could be obtained by considering the reaction as the transition of the system (nucleus + neutron) from the initial into the final state via a compound nucleus as an intermediate state (i.e. according to the same scheme as in the scattering of photons). Direct application of perturbation theory leads to the Breit–Wigner formulae. However this way of obtaining the Breit–Wigner formulae cannot be substantiated, since the perturbation of the state of the neutron is not weak.

Nevertheless, such an obvious although not rigorous calculation shows that the quantities Γ_e and Γ_r characterize transition probabilities. Namely, Γ_e is proportional to the matrix element of the transition of the system from the intermediate state (nucleus C) to the final state (nucleus A and the neutron with energy E). Hence the quantity Γ_e, which is called the partial width of the resonance level E corresponding to elastic scattering, determines the probability of decay of the nucleus C with elastic scattering of the neutron. The quantity Γ_r is called the partial width of the resonance level with respect to the reaction. It determines the probability of decay of the nucleus with inelastic scattering and neutron capture. In the case of slow neutrons the probability of inelastic scattering is small and the reaction reduces to the capture of the neutron. Finally, Γ determines the total probability of decay of nucleus C. It is equal to the energy half-width of the resonance maximum of the cross section.

Fig. V.27 shows the energy dependence of the cross sections for the elastic scattering and radiative capture of slow neutrons. We have introduced the Breit–Wigner formulae in the particular case where the energy of the neutron is close to one of the resonance levels E_0 of the nucleus. They can be generalized to the case of many levels. They may also take into account the spin states of the nucleus and of light particles. Finally, the Breit–Wigner formulae can be generalized to the case of charged particles and particles with an angular momentum. We discuss certain properties of the widths of the resonance levels of a nucleus. The reaction width Γ_r for slow neutrons reduces to the radiative capture width Γ_γ, since no inelastic scattering occurs at small energies. The value of Γ_γ amounts to about 10^{-1} eV and does not depend on the velocity of the neutron. The width $\Gamma_e \sim k \sim v$, where v is the velocity of the neutron, and at small energies for heavy and medium nuclei $\Gamma_e \ll \Gamma_r$. This means that neutron capture predominates over elastic scattering. In the case of light nuclei the situation is the reverse: resonance scattering predominates over capture.

The Bohr concept of the formation of a compound nucleus is valid for

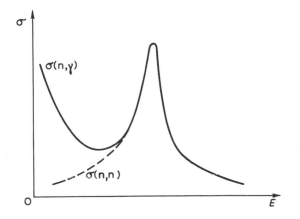

Fig. V.27

nuclear reactions proceeding at not too large energies. As the energy of the incident particles increases their cross section for scattering by individual nuclear nucleons decreases sharply. Hence at energies $E > 50$–100 MeV the interaction of the particles with the nucleus reduces to an interaction with an individual nucleon. The Breit–Wigner formulae turn out to be no longer applicable.

§95. The scattering matrix (S-matrix)

The mathematical technique of scattering theory described above is associated with an explicit form of the interaction potential distribution over all space. However, in a number of important cases there is no potential energy (independent of the velocity). Hence in modern scattering theory an important role is played by a more general statement of the problem. Let the wave function $\psi_a(t \to -\infty)$ of a system of particles be given in the initial state before the interaction. The general problem of scattering theory is to find the wave function of the system a long time after the interaction, $\psi(t \to \infty)$. The wave function $\psi(t \to \infty)$ can be expressed in terms of the initial function $\psi_a(t \to -\infty)$ by means of the operator $\hat{V}(t,t_0)$ introduced in §49 and describing the development of the wave function in time. By the scattering matrix S we shall mean the limiting expression of the operator $\hat{V}(t,t_0)$ (in the interaction representation) describing the development of the process in time (see §49).

$$\hat{S} = \lim_{\substack{t_0 \to -\infty \\ t \to \infty}} \hat{V}(t,t_0) \, . \tag{95.1}$$

Thus the scattering matrix $\hat{S}$ carries out the transformation of the initial state $\psi_a(-\infty)$ into the final state $\psi(\infty)$,

$$\psi(\infty) = \hat{S}\psi_a(-\infty) \, . \tag{95.2}$$

The index a denotes the complete set of quantum numbers defining the state of the system before scattering. It is assumed that the particles in the initial as well as in final state are separated by sufficiently large distances from each other that the interactions between them need not be taken into account (the so-called adiabatic hypothesis).

We expand the function ψ in a series in terms of a certain complete system of functions ψ_b, where b denotes the corresponding set of quantum numbers

$$\psi = \sum_b c_b \psi_b \, .$$

Here the symbol $\sum_b$ denotes summation over a discrete sequence of quantum numbers and integration over quantum numbers changing continuously.

As follows from (95.2), the expansion coefficients c_b are expressed in terms of the matrix elements of the operator $\hat{S}$

$$c_b = (\psi_b, \psi) = (\psi_b, \hat{S}\psi_a) = \langle b|S|a\rangle = S_{ba} \, . \tag{95.3}$$

The square of the amplitude c_b gives the total probability of the transition of the system in scattering from state a into state b

$$W'_{ba} = |S_{ba}|^2 \, . \tag{95.4}$$

Thus the matrix elements of the operator $\hat{S}$ are directly related to the corresponding transition probabilities.

Since the operator $\hat{V}(t,t_0)$ is unitary (see §96), its limiting value is also unitary, i.e. for the operator $\hat{S}$ we can write

$$\hat{S}\hat{S}^\dagger = \hat{S}^\dagger\hat{S} = \hat{I} \, , \tag{95.5}$$

where $\hat{I}$ denotes the unit operator.

Taking the diagonal matrix elements of one of the relations (95.5), we obtain the obvious result

$$\sum_b S_{ab}^\dagger S_{ba} = \sum_b |S_{ba}|^2 = 1 \, , \tag{95.6}$$

i.e. the sum of the probabilities of all possible transitions is equal to one. Using relation (95.4), one can express the cross section for the process in terms of the matrix elements of the operator $\hat{S}$. However, it is first necessary to obtain the expression for the transition probability per unit time.

We assume that the initial state ψ_a is characterized by a definite energy value E_a. The total energy of the system is conserved in time. Hence the matrix S_{ba} can be written in the form

$$S_{ba} = S_{ba}^E \delta(E_a - E_b) \, .$$

The matrix S_{ba}^E is said to be given on the energy surface. Then the total transition probability (95.4) is written in the form

$$W_{ba}' = |S_{ba}^E|^2 \delta^2(E_a - E_b) \, . \tag{95.7}$$

This probability is proportional to the square of the δ-function. We write one of the δ-functions in the form (see Vol. 1, Appendix III)

$$\delta(E_a - E_b) = \lim_{T \to \infty} \int_{-\frac{1}{2}T}^{\frac{1}{2}T} \frac{1}{2\pi\hbar} \exp\left((i/\hbar)(E_b - E_a)t\right) dt \, .$$

Substituting this expression into (95.7) and integrating the transition probability over the energy of the final state, we obtain the transition probability in time T

$$\bar{W}_{ba} = \int W_{ba}' dE_b = \frac{1}{2\pi\hbar} |S_{ba}^E|^2 T \, . \tag{95.8}$$

Hence we find for the transition probability per unit time

$$W = \frac{1}{2\pi\hbar} |S_{ba}^E|^2 \, . \tag{95.9}$$

To find the cross section for the process we have to divide the transition probability by the incident particle flux density.

In the initial state there are two particles. As usual we consider the scattering process in the centre-of-mass system. The wave function of the initial state ψ_a describes states with given energy of relative motion E_a and a direction of the momentum of relative motion $n_a = p_a/p_a$, and is normalized by the condition

$$\int \psi_{E_a \mathbf{n}}^* \psi_{E_a' \mathbf{n}'} \, dV = \delta(E_a - E_a') \delta(\mathbf{n}_a - \mathbf{n}_a') = p_a^2 \frac{dp}{dE} \delta(\mathbf{p}_a - \mathbf{p}_a') . \qquad (95.10)$$

Then

$$\psi_{E_a \mathbf{n}} = |E_a, \mathbf{n}\rangle = p_a \left(\frac{dp}{dE}\right)^{\frac{1}{2}} \psi_{\mathbf{p}_a} = \frac{p_a}{v_a^{\frac{1}{2}}} |\mathbf{p}_a\rangle . \qquad (95.11)$$

The incident particle flux density is equal to

$$\mathbf{j}_0 = \frac{p_a^2}{(2\pi\hbar)^3} \mathbf{n}_a . \qquad (95.12)$$

As always when we deal with a continuously changing quantity we have to introduce the differential transition probability dW_{ba} and, consequently, the differential cross section $d\sigma_{ba}$. Denoting the solid angle interval in which the vector $\mathbf{n}_b$ lies by $d\Omega_b$, we obtain from (95.9) and (95.12)

$$d\sigma_{ba} = \frac{4\pi^2}{k_a^2} |\langle b, E, \mathbf{n}_b | S^E | a, E, \mathbf{n}_a \rangle|^2 \, d\Omega_b , \qquad (95.13)$$

where $k_a = \hbar^{-1} p_a$.

Let us now consider the case where elastic and different forms of inelastic scattering may occur as a result of the interaction of two particles, i.e.

$$
A + B \to C + D
\begin{cases}
\nearrow A + B \\
\\
\searrow C' + D' .
\end{cases}
$$

We shall call each form of transformation a reaction channel. Formula (95.13) for $b \neq a$ corresponds to an inelastic reaction channel. The cross section taking into account elastic and inelastic channels can be written in the form

$$d\sigma_{ba} = \frac{4\pi^2}{k_b^2} |\langle b, E, \mathbf{n}_b | S^E - I | a, E, \mathbf{n}_a \rangle|^2 \, d\Omega_b , \qquad (95.14)$$

where I is the unit matrix. Since in the matrix I only diagonal elements are different from zero, for $b \neq a$ the cross section (95.14) is the same as (95.13).

Expressions (95.13) and (95.14) can be written in a form analogous to (86.12) if the initial state $|a, E, \mathbf{n}_a\rangle$ is expanded in terms of partial waves

$$|a, E, \mathbf{n}_a\rangle = |a, E, l, m\rangle \langle l, m | \mathbf{n}_a\rangle . \qquad (95.15)$$

The transformation functions $\langle l, m | \mathbf{n}_a\rangle$ were found in §48:

$$\langle l,m|\mathbf{n}_a \rangle = Y_{lm}^*(\mathbf{n}_a) \ . \tag{95.16}$$

Choosing the z-axis to be along the direction of the vector $\mathbf{n}_a$, we obtain

$$Y_{lm}(\mathbf{n}_a) = Y_{lm}(0) = \left(\frac{2l+1}{4\pi}\right)^{\frac{1}{2}} \delta_{m0} \ . \tag{95.17}$$

In substituting expressions (95.16), (95.17) into (95.13) and (95.14) matrix elements of the following form arise:

$$\langle b,E,\mathbf{n}_b|S^E|a,E,l,0 \rangle \ .$$

In motion in a central field angular momentum is conserved. Hence the S-matrix is diagonal with respect to the quantum numbers l, m, and one can write

$$\langle b,E,\mathbf{n}_b|S^E|a,E,l,0 \rangle = \langle \mathbf{n}_b|l,0 \rangle \langle b,E,l,0|S^E|a,E,l,0 \rangle =$$

$$= Y_{l0}(\mathbf{n}_b) \langle b,E,l,0|S^E|a,E,l,0 \rangle = P_l(\cos\theta_b) \left(\frac{2l+1}{4\pi}\right)^{\frac{1}{2}} S_{ba}^l \ . \tag{95.18}$$

Correspondingly for the differential cross section for scattering into solid angle $d\Omega_b$ we obtain

$$d\sigma_{ba} = \frac{1}{4k_a^2} \left| \sum_l (2l+1)(S_{ba}^l - \delta_{ba}) P_l(\cos\theta) \right|^2 d\Omega_b \ . \tag{95.19}$$

Integrating this expression over all directions of the vector $\mathbf{n}_b$, we obtain the cross section for the scattering $a \rightarrow b$

$$\sigma_{ba} = \frac{\pi}{k_a^2} \sum_l (2l+1) \ |S_{ba}^l - \delta_{ba}|^2 \ . \tag{95.20}$$

From this formula it follows that the total elastic scattering cross section has the form

$$\sigma_{aa} = \frac{\pi}{k_a^2} \sum_l (2l+1) \ |S_{aa}^l - 1|^2 \ . \tag{95.21}$$

We can also write down the expression for the total cross section for all inelastic processes σ_{inel}, which is obtained by summing σ_{ba} over all channels $b \neq a$

$$\sigma_{\text{inel}} = \sum_{b \neq a} \sigma_{ba} = \frac{\pi}{k_a^2} \sum_{b \neq a} \sum_l (2l+1) \ |S_{ba}|^2 \ .$$

This expression can be transformed by making use of the unitarity of the S-matrix. Namely, we have

$$\sum_{b\neq a} |S_{ba}^l|^2 = 1 - |S_{aa}^l|^2 .$$ (95.22)

Correspondingly for σ_{inel} we obtain

$$\sigma_{\text{inel}} = \frac{\pi}{k_a^2} \sum_l (2l+1)(1-|S_{aa}^l|^2) .$$ (95.23)

Formulae (95.23) and (95.21) are the same as formulae (91.6) and (91.7) of partial wave scattering theory. We see that the quantities S_l introduced in §91 are the diagonal matrix elements of the scattering matrix S. If inelastic processes are impossible, i.e. $S_{ba}^l = 0$ for $b \neq a$, then from unitarity relation (95.22) it follows that $|S_{aa}^l|^2 = 1$, i.e. that

$$S_{aa}^l = e^{2i\delta_l} .$$ (95.24)

Then expressions (95.19)–(95.21) are the same as the expressions for the elastic scattering cross section obtained in §86.

From relations (95.19), (95.21) and (95.22) it follows that the cross section for a process is determined by the matrix elements of the operator $\hat{F}$, $i\hat{F} = \hat{S}-\hat{I}$ (the factor i is introduced for convenience). The unitarity of the S-matrix leads to the following relation:

$$\hat{S}^\dagger \hat{S} = (\hat{I}-i\hat{F}^\dagger)(\hat{I}+i\hat{F}) = \hat{I} ,$$

or

$$-i\hat{F} + i\hat{F}^\dagger = \hat{F}^\dagger \hat{F} .$$

Taking the matrix elements of the left-hand and right-hand sides of this relation with respect to the wave functions (95.11), we obtain

$$\hat{F}_{ba} - \hat{F}_{ba}^\dagger = i \sum_c \hat{F}_{bc}^\dagger \hat{F}_{cb} ,$$ (95.25)

where Σ_c denotes the summation over the discrete and integration over the continuous states of the system of two particles after collision. In fact, the system (95.25) is a system of integral equations expressing the property of unitarity of the S-matrix.

The system of equations (95.25) is substantially simplified if only elastic scattering is possible. As is seen from comparison of (95.19), (86.11) and (86.12), in this case the matrix elements of the operator $\hat{F}$ are to within a factor the same as the elastic scattering amplitude $f(\mathbf{n}',\mathbf{n})$

$$\frac{2\pi}{k} \hat{F}_{\mathbf{n}',\mathbf{n}} = f(\mathbf{n}',\mathbf{n}) , \tag{95.26}$$

where $\mathbf{n}$ and $\mathbf{n}'$ are the unit vectors characterizing the direction of the momentum vector of the relative motion of the incident and scattered particles. From (95.25) we obtain

$$f(\mathbf{n}',\mathbf{n}) - f^*(\mathbf{n},\mathbf{n}') = \frac{ik}{2\pi} \int f^*(\mathbf{n}'',\mathbf{n}')f(\mathbf{n}'',\mathbf{n})\,d\Omega'' . \tag{95.27}$$

Relation (95.27) expresses the condition of unitarity for elastic scattering. For scattering in a central field the amplitude f depends only on the angle ϑ between the vectors $\mathbf{n}$ and $\mathbf{n}'$, and relation (95.27) can be rewritten in the form

$$\operatorname{Im} f(\mathbf{n}',\mathbf{n}) = \frac{k}{4\pi} \int f^*(\mathbf{n}'',\mathbf{n}')f(\mathbf{n}'',\mathbf{n})\,d\Omega'' . \tag{95.28}$$

For $\mathbf{n} = \mathbf{n}'$ we obtain the relation connecting the imaginary part of the amplitude of the scattering at zero angle with the total cross section (optical theorem; see §91).

We note that eq. (95.28) makes it possible, in principle, to find the scattering amplitude if its modulus, which is defined by the scattering law, is known. Setting

$$f(\vartheta) = \left(\frac{d\sigma}{d\Omega}\right)^{\frac{1}{2}} e^{i\alpha(\vartheta)}$$

and substituting this expression into (95.28), we obtain the integral equation for the phase $\alpha(\vartheta)$. Thus knowing the scattering cross section $d\sigma/d\Omega$ we can, in principle, also determine the scattering amplitude $f(\vartheta)$. We note, however, that eq. (95.28) does not change under the replacement $\alpha \to \pi - \alpha$, i.e. it determines the scattering amplitude with an accuracy to within the transformation $f(\vartheta) \to -f^*(\vartheta)$.

Let us consider the effect of this uncertainty on the values of the scattering phase shifts. For this we calculate the integral

$$\int |f(\theta)| \, e^{i\alpha(\theta)} \, P_{l'}(\cos\theta) \sin\theta \, d\theta =$$

$$= \frac{1}{2ik} \sum_{l=0}^{\infty} (2l+1)(e^{2i\delta_l}-1) \, P_l P_{l'} \sin\theta \, d\theta \ .$$

Making use of the properties of orthogonality of the Legendre polynomials, we obtain

$$\int |f(\theta)| \, e^{i\alpha(\theta)} \, P_{l'}(\cos\theta) \sin\theta \, d\theta = \frac{1}{ik}(e^{2i\delta_{l'}}-1) \ . \tag{95.29}$$

Equating the real parts of relation (95.29), we find

$$|f(\theta)| \cos\alpha(\theta) \, P_l(\cos\theta) \, d\cos\theta = \frac{\sin 2\delta_l}{k} \ . \tag{95.30}$$

From formula (95.30) it is clear that the replacement of α by $\pi - \alpha$ leads to a change of sign of the left-hand side. To conserve the equality it is necessary to reverse the sign of all the phase shifts δ_l. Thus the uncertainty in the quantity α leads to an uncertainty in the sign of all the phase shifts.

If the sign of only one of the phase shifts is determined in an independent way, then the relation of δ_l to all the other phase shifts becomes unambiguous. The sign of one phase shift (namely, the s-wave one) can be established, for example, from the study of the scattering and interference of slow particles. It should be pointed out that although the calculations show us the possibility of determining the scattering amplitude, the solution of the integral equation (95.28) is a difficult problem.

§96. S-matrix and perturbation theory

If the total Hamiltonian can be written in the form of a sum $\hat{H} = \hat{H}_0 + \hat{H}'$ where $\hat{H}_0$ describes the behaviour of non-interacting particles and $\hat{H}'$ their interaction, then to find the explicit form of the S-matrix it is convenient to make use of the interaction representation. The wave function in this representation is defined by eq. (49.21). The operator $\hat{V}(t,t_0)$ defined by formula (49.1), which transforms the wave function at a given instant of time t_0 into the wave function at the instant of time t, can also be defined in the interaction representation. That is, writing

$$\varphi(t) = \hat{V}(t,t_0)\, \varphi(t_0) \tag{96.1}$$

and substituting into (49.21), we find

$$i\hbar\, \frac{\partial \hat{V}(t,t_0)}{\partial t} = \hat{H}'_{int}(t)\, \hat{V}(t,t_0)\,, \tag{96.2}$$

$$\hat{V}(t_0,t_0) = 1\,. \tag{96.3}$$

System (96.2) and (96.3) can be compared with the integral equation

$$\hat{V}(t,t_0) = 1 - \frac{i}{\hbar}\, \int\limits_{t_0}^{t} dt'\hat{H}'_{int}(t')\hat{V}(t',t_0)\,. \tag{96.4}$$

Integral equation (96.4) can be solved by a method of successive approximation

$$V(t,-\infty) = 1 - \frac{i}{\hbar}\, \int\limits_{-\infty}^{t} dt_1\hat{H}'_{int}(t_1) +$$

$$+ \left(-\frac{i}{\hbar}\right)^2\, \int\limits_{-\infty}^{t} dt_1 \int\limits_{-\infty}^{t_1} dt_2 \hat{H}'_{int}(t_1)\hat{H}'_{int}(t_2) + \dots\,. \tag{96.5}$$

The general term of the series is of the form

$$\hat{V}^{(n)} = \left(-\frac{i}{\hbar}\right)^n\, \int\limits_{-\infty}^{t} dt_1 \int\limits_{-\infty}^{t_1} dt_2 \dots \int\limits_{-\infty}^{t_{n-1}} \hat{H}'_{int}(t_1)\, \hat{H}'_{int}(t_2) \dots \hat{H}'_{int}(t_n)\, dt_n\,. \tag{96.6}$$

It is evident that the range of integration over the variables $t_1, t_2, \dots t_n$ has the order

$$t_1 > t_2 > \dots > t_n\,. \tag{96.7}$$

In order to simplify the notation and so that one need not follow the order of carrying out the integration, it is convenient to symmetrize formula (96.6). In the case of a function symmetric with respect to its variables use can be made of the formula

$$\int\limits_{a}^{b} dt_1 \int\limits_{a}^{t_1} dt_2 \dots \int\limits_{a}^{t_{n-1}} dt_n f(t_1,\dots,t_n) =$$

$$= \frac{1}{n!} \int\limits_{a}^{b} dt_1 \int\limits_{a}^{b} dt_2 \dots \int\limits_{a}^{b} dt_n f(t_1,\dots,t_n)\,. \tag{96.8}$$

For the purpose mentioned we introduce the so-called chronological operator $\hat{P}$, which by definition arranges time-dependent operators in chronological sequence, i.e. in order of decreasing times (96.7):

$$\hat{P}\hat{L}(t_1)\hat{M}(t_2) = \begin{cases} \hat{L}(t_1)\hat{M}(t_2) & \text{for} \quad t_1 > t_2 \\ \hat{M}(t_2)\hat{L}(t_1) & \text{for} \quad t_2 > t_1 . \end{cases} \tag{96.9}$$

A representation of this operator could, for example, be the expression

$$\hat{P} = \frac{1 + \epsilon(t_1 - t_2)}{2} + \frac{1 - \epsilon(t_2 - t_1)}{2} ,$$

where $\epsilon(x)$ is the so-called sign function

$$\epsilon(x) = \frac{|x|}{x} = \begin{cases} 1 & \text{for} \quad x > 0 , \\ -1 & \text{for} \quad x < 0 . \end{cases}$$

By means of the chronological operator we can write

$$\int_{-\infty}^{t} dt_1 \int_{-\infty}^{t_1} dt_2 \cdots \int_{-\infty}^{t_{n-1}} dt_n \hat{H}'_{\text{int}}(t_1) \ldots \hat{H}'_{\text{int}}(t_n) =$$

$$= \frac{1}{n!} \int_{-\infty}^{t} dt_1 \int_{-\infty}^{t} dt_2 \cdots \int_{-\infty}^{t} dt_n \, \hat{P}\{\hat{H}'_{\text{int}}(t_1 \ldots \hat{H}'_{\text{int}}(t_n)\} . \tag{96.10}$$

Hence for $\hat{V}(t, -\infty)$ we find

$$\hat{V}(t, -\infty) = 1 + \sum_{n=1}^{\infty} \left(-\frac{i}{\hbar}\right)^n \frac{1}{n!} \int_{-\infty}^{t} dt_1 \cdots \int_{-\infty}^{t} dt_n \, \hat{P}\{\hat{H}'_{\text{int}}(t_1) \ldots \hat{H}'_{\text{int}}(t_n)\} =$$

$$= P \exp\left(-\frac{i}{\hbar} \int_{-\infty}^{t} \hat{H}'_{\text{int}}(t) \, dt\right) .$$

In accordance with the definition of the S-matrix

$$S = \lim_{\substack{t \to \infty \\ t_0 \to -\infty}} \hat{V}(t, t_0) = \lim_{t \to \infty} \hat{V}(t, -\infty) =$$

$$= \hat{P} \exp\left(-\frac{i}{\hbar} \int_{-\infty}^{\infty} \hat{H}'_{\text{int}}(t) \, dt\right) . \tag{96.11}$$

The formula obtained, called Dyson's formula, allows one to relate the S-matrix to the interaction energy $\hat{H}'$ (if this last exists). It is accurate in the sense that the summation of the entire perturbation series is carried out in it.

It is easily seen that the first terms of the expansion of the general formula for the S-matrix lead to ordinary perturbation theory.

For ease of calculation we restrict ourselves to first order perturbation theory, writing

$$S^{(1)} = 1 - \frac{i}{\hbar} \int_{-\infty}^{\infty} dt \, \hat{H}'_{int}(t) \, . \tag{96.12}$$

The operator $\hat{P}$ is in this case identically equal to one. Taking the matrix element with respect to states $a \neq b$, which are eigenstates of the Hamiltonian H_0, we have

$$S_{ba} = -\frac{i}{\hbar} \int_{-\infty}^{\infty} dt \, (\hat{H}'_{int})_{ba} \, .$$

Passing to the Schrödinger representation and making use of definition (49.19), we obtain

$$S_{ba}^{(1)} = -\frac{i}{\hbar} \int_{-\infty}^{\infty} dt \, \langle b | e^{(i/\hbar)\hat{H}_0 t} \hat{H}' \, e^{-(i/\hbar)\hat{H}_0 t} | a \rangle =$$

$$= -\frac{i}{\hbar} \langle b | \hat{H}' | a \rangle \int_{-\infty}^{\infty} \exp\left[\frac{i}{\hbar}(E_b - E_a)t\right] dt = -2\pi i \hat{H}'_{ba} \delta(E_b - E_a) \, .$$

We see that $S_{ba}^{(1)}$ is the same as the transition amplitude in the first approximation of perturbation theory.

Analogous, although more cumbersome calculations allow $S_{ba}^{(2)}$ to be identified with the transition amplitude in the second order of perturbation theory.

In spite of the convenience of the notation of Dyson's formula, which is often used in intermediate calculations, for the actual calculation of the S-matrix one has to carry out an expansion in a series and integration by terms.

An important feature of Dyson's formula is the fact that it can easily be transformed into a relativistically invariant form. Hence it is of particular importance in calculating relativistic effects.

§97. Analytic properties of the S-matrix

As we have already stressed, a number of important results of scattering theory which are not associated with the use of a particular form of the interaction potential can be obtained by means of the S-matrix technique. This is associated, in particular, with the analytic properties of the S-matrix*. For simplicity, we shall in what follows restrict ourselves to the case of elastic scattering. Then the elements of the S-matrix are given by formula (95.24).

As was shown in §35, the asymptotic expression for the radial component of the wave function of a particle of energy $E = \hbar^2 k^2 / 2m$ and angular momentum l, regular at the origin, has the form

$$\chi_{kl} = r R_{kl} = a_l(k)\, e^{i(kr - \frac{1}{2}l\pi)} + b_l(k)\, e^{-i(kr - \frac{1}{2}l\pi)} . \tag{97.1}$$

In deriving (97.1) it was assumed that the potential decreases more rapidly than r^{-1} at large distances.

Comparing expression (97.1) with (86.2) and taking into account (95.24), the matrix elements $S_{aa}^l \equiv S_l$ can be expressed in terms of the constants $a_l(k)$ and $b_l(k)$

$$S_l(k) = -\frac{a_l(k)}{b_l(k)} . \tag{97.2}$$

We shall now formally consider the wave function χ_{kl} and, correspondingly, the function $S_l(k)$ to be functions of the complex variable k. We shall show, first of all, that the function of a complex variable $S_l(k)$ should be given only in one quadrant and not in the entire plane of the complex variable k. Indeed, since the Schrödinger equation does not change under the replacement of k by $-k$, the function χ_{-kl}, by virtue of the uniqueness of the solution, describes the same state as the function χ_{kl}. These two functions can differ only by a constant factor. Replacing k by $-k$ in (97.1), we obtain

$$\frac{a_l(k)}{b_l(k)} = \frac{b_l(-k)}{a_l(-k)} .$$

Hence it follows that

$$S_l(k) = S_l^{-1}(-k) . \tag{97.3}$$

---

* For a more detailed consideration of the problems touched upon in this section and for a bibliography see A.I.Baz, L.B.Zeldovich and A.M.Perelomov, *Rasseyanie, reaktsii i raspady v nerelyativistskoi kvantovoi mekhanike* (*Scattering, reactions and decays in nonrelativistic quantum mechanics*) (Nauka, Moscow, 1966).

Further, we note that, since the Schrödinger equation is real, the function χ_{kl}^* also must be the same to within a constant as the function χ_{kl}. Hence it is again easily found that

$$S_l(k) = (S_l^*(k))^{-1} . \qquad (97.4)$$

Formula (97.4) is obtained for real k. Carrying out analytic continuation to the entire plane of the complex variables k, we have

$$S_l(k) = (S_l^*(k^*))^{-1} . \qquad (97.5)$$

Relations (97.3) and (97.5) connect the values of the function $S_l(k)$ given in one of the quadrants of the plane of the complex variable k with its values at the corresponding points of the remaining three quadrants. From relation (97.4) it follows that for real k, $|S_l(k)|^2 = 1$, i.e. the phase shift δ_l is real $(S_l = e^{2i\delta_l})$. On the contrary, as is seen from (97.5) and (97.3), the function $S_l(k)$ is real on the imaginary axis, so that the phase shift δ_l is imaginary.

Let us consider the position of the singularities of the function $S_l(k)$. We assume that there corresponds to the potential $U(r)$ a bound state of the particle with energy $-E_0$. The bound state is described by the wave function χ_{k_0l} regular at the origin and falling off at large distances as $e^{-|k_0|r}$, where $k_0 = i(2m\hbar^{-2}|E_0|)^{\frac{1}{2}}$. Consequently, the function χ_{kl} analytically continued to the complex plane must fall off for $k = k_0$ as $e^{-|k_0|r}$. Hence the relation $b_l(k_0) = 0$ must be satisfied at the point $k = k_0$. In accordance with formula (97.2), the function $S_l(k)$ has a pole at the point $k = k_0$. As follows from (97.3), the function $S_l(k)$ reduces to zero at the symmetric point lying in the lower half-plane, i.e. at $k = -k_0$. Thus we arrive at the conclusion that to each bound state there corresponds a pole of the function $S_l(k)$ lying at the corresponding point of the upper imaginary semiaxis in the plane of the complex variable k. It should be noted that so-called 'false' poles, which do not correspond to any stationary state, may also arise on the imaginary semiaxis. It can be shown (see ref. on p. 412) that 'false' poles do not arise when the so-called cut-off radius R is introduced, i.e. when the condition $U(r) = 0$ is introduced for $r > R$, where the radius R may be as large as one wants.

We note that the function $S_l(k)$ cannot have poles in the upper half-plane lying anywhere off the imaginary axis. Indeed, to such a pole there would correspond a complex value of the energy of the bound state, which is impossible.

The function $S_l(k)$ may also have poles in the lower half-plane, and there they may also lie off the imaginary semiaxis. As follows immediately from relations (97.3) and (97.5), these poles must be situated in pairs symmetric with respect to the imaginary semiaxis. In the upper half-plane there correspond to these poles zeros of the function $S_l(k)$. It is easily seen that to poles lying in the lower half-plane there correspond wave functions exponentially increasing at large distances. Such wave functions cannot, of course, correspond to a bound state. It can be shown that to poles in the lower half-plane there correspond quasi-stationary states of the system, i.e. states which decay in the course of a certain finite time T.

We find the residue of the function $S_l(k)$ with respect to the pole to which there corresponds a bound state with energy $E = -E_0$ or the value $k = k_0 = i(2m\hbar^{-2}|E_0|)^{\frac{1}{2}}$. Denoting this residue by c_l, we write the function S_l in the neighbourhood of the point $k = k_0$ in the form

$$S_l = \frac{c_l}{k - k_0} . \qquad (97.6)$$

The quantity c_l is connected by a simple relation with the amplitude of the wave function corresponding to the stationary state with energy $E = -E_0$. In order to establish this relation we write down equations satisfied by the function χ_{kl} and by its derivative with respect to energy

$$\chi_{kl}'' + \frac{2m}{\hbar^2}\left(E - U - \frac{\hbar l(l+1)}{2mr^2}\right)\chi_{kl} = 0$$

$$\left(\frac{\partial \chi_{kl}}{\partial E}\right)'' + \frac{2m}{\hbar^2}\left(E - U - \frac{\hbar^2 l(l+1)}{2mr^2}\right)\frac{\partial \chi_{kl}}{\partial E} = -\frac{2m}{\hbar^2}\chi_{kl} .$$

We shall assume the function χ_{kl} to be normalized by the condition

$$\int_0^\infty |\chi_{kl}|^2 \, dr = 1 .$$

Multiplying the first equation by $\partial\chi_{kl}/\partial E$ and the second by χ_{kl}, subtracting one from the other and integrating with respect to dr, we obtain

$$\chi_{kl}'\frac{\partial \chi_{kl}}{\partial E} - \chi_{kl}\left(\frac{\partial \chi_{kl}}{\partial E}\right)' = \frac{2m}{\hbar^2}\int_0^r \chi_{kl}^2 \, dr . \qquad (97.7)$$

We apply this relation for $E = -E_0$ and $r \to \infty$. Expanding the functions $a_l(k)$ and $b_l(k)$ near the point $k = k_0$ in a series and rename the constant

$$a_l(k) = a_l(k_0) = A_l i^{-l} , \qquad b_l(k) = \beta_l(k - k_0) . \qquad (97.8)$$

Making use of these expansions and of relations (97.7) and (97.1), we obtain

$$\beta_l = -\frac{i}{a_l} = -\frac{i^{l+1}}{A_l}. \tag{97.9}$$

Substituting expression (97.9) into (97.2), the residue c_l at the point $k = k_0$ is

$$c_l = i A_l^2 (-1)^{l+1}. \tag{97.10}$$

Thus we have related the value of the residue c_l to the amplitude A_l in the asymptotic expression of the wave function $\chi_{kl} = A_l e^{-|k_0|r}$ of the bound state.

The study of the behaviour of the scattering phase shifts $\delta_l(k)$ and, consequently, also of the function $S_l(k) = e^{2i\delta_l(k)}$, and the extrapolation of these results to the complex region make it possible, on the basis of (97.10), also to draw definite conclusions concerning the wave function of the bound state.

The analytic properties of the quantities $S_l(k)$ make it possible to obtain important relations which must be satisfied by the scattering amplitude. These relations are called dispersion relations. Dispersion relations for the amplitude of the scattering at zero angle $f(0, k)$ are the simplest and at the same time the most important. Dispersion relations establish the connection between the real and imaginary parts of the scattering amplitude $f(\vartheta, k)$

$$f(\vartheta, k) = \text{Re } f(\vartheta, k) + i \text{ Im } f(\vartheta, k)$$

and are based on the use of the Cauchy formula in the theory of analytic functions.

Suppose that $F(k)$ is a certain function which is analytic in the upper half-plane of the complex variable k and has simple poles on the upper imaginary axis. Let us consider the integral

$$\int_C \frac{F(k') \, dk'}{k' - k}$$

taken over the contour shown in fig. V.28.

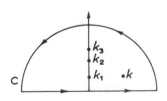

Fig. V.28

The integral is determined by the sum of the residues of the integrand. These residues are taken at the point $k' = k$ and at the points $k' = k_1, k_2, ...,$ where the poles of the function $F(k)$ lie. If the function $F(k)$ tends to zero sufficiently rapidly as $|k| \to \infty$, then the integral over the upper semicircle is equal to zero. We then have

$$\int_{-\infty}^{\infty} \frac{F(k')\,dk'}{k' - k} = 2\pi i \left(F(k) + \sum_n \frac{\text{Res}\, F(k_n)}{k_n - k} \right). \tag{97.11}$$

Here $\text{Res}\, F(k_n)$ denotes the residue of the function F at the point $k' = k_n$. Now let the imaginary part of k tend to zero, so that k tends to the point k_0 lying on the real axis. In this case

$$\int_{-\infty}^{\infty} \frac{F(k')\,dk'}{k' - k_0} = P \int_{-\infty}^{\infty} \frac{F(k')\,dk'}{k' - k_0} + i\pi F(k_0). \tag{97.12}$$

Here P denotes that the integral is understood in the sense of the principal value

$$P \int_{-\infty}^{\infty} \frac{F(k')\,dk'}{k' - k_0} = \lim_{\epsilon \to 0} \left[\int_{-\infty}^{k_0-\epsilon} \frac{F(k')\,dk'}{k' - k_0} + \int_{k_0+\epsilon}^{\infty} \frac{F(k')\,dk'}{k' - k_0} \right],$$

and the second term on the right-hand side of (97.12) arises from the integration over a small semicircle around the point $k = k_0$.

Based on the results (97.11) and (97.12) we obtain the dispersion relation for the amplitude of the scattering at zero angle $f(0, k)$. This amplitude is connected with the matrix elements S_l (95.24) by relation (86.11)

$$f(0,k) = \frac{1}{2ik} \sum_l (2l+1)(S_l-1). \tag{97.13}$$

From this expression it follows that the poles of the function $S_l(k)$ are also the poles of the function $f(0, k)$, and the function $f(0, k)$ has no other poles. The point $k = 0$ is not a pole at all, since for $k \to 0$, $\delta \to 0$, $S_l \to 1$ (see §86). Thus the function $f(0, k)$ is analytic in the upper half-plane of the complex variable k and has poles on the upper imaginary semiaxis. Dispersion relations for this function are easily obtained if one substitutes into relations (97.11) and (97.12) the function $F(k)$ in the form

$$F(k) = f(0, k) - f(0, \infty). \tag{97.14}$$

From the amplitude $f(0, k)$ one subtracts its value for $k \to \infty$ in order to reduce the integral over the large semicircle to zero (fig. V.28). For $k \to \infty$ the term with the potential $U(r)$ can be neglected in the Schrödinger equation. The solution of such an equation has the form of a plane wave. Substituting such a solution into (83.10), we obtain

$$f(0,\infty) = - \frac{m}{2\pi\hbar^2} \int U(r) \, dV \,. \tag{97.15}$$

Expression (97.15) represents the scattering amplitude in the Born approximation (see §84), i.e. $f(0,\infty) = f_B$. Substituting (97.14) into relations (97.11) and (97.12) and taking into account that the integral with the Born amplitude reduces to zero, we obtain

$$f(0,k) = f_B + \frac{1}{i\pi} P \int_{-\infty}^{\infty} \frac{f(0,k') \, dk'}{k' - k} - 2 \sum_{n,l} \frac{\text{Res}\, f(0,k_{nl})}{k_{nl} - k} \,. \tag{97.16}$$

In this relation k is assumed to be real, and the index zero is dropped. The points $k = k_{nl}$ lie on the upper imaginary semiaxis and correspond to the poles of the function $S_l(k)$. The summation in (97.16) is carried out over all bound states. The expression for the residue of the function S_l in terms of the amplitude of the corresponding bound state is given by formula (97.10). Taking into account (97.13), we have

$$\text{Res}\, f(0, k_{nl}) = \frac{1}{2k_{nl}} A_{nl}^2 (-1)^{l+1} (2l+1) \,. \tag{97.17}$$

Relation (97.16) can be rewritten in a somewhat different from, if it is taken into account, according to (97.3) and (97.4), that for real k $S_l(-k) = S_l^*(k)$ and, correspondingly (see (97.13)), that $f(0,-k) = f^*(0, k)$. Hence the integration in (97.16) can be carried out only over positive values of k, having physical meaning. Equating the real parts on the left and the right in (97.16), we have finally

$$\text{Re}\, f(0,k) = f_B + \frac{2}{\pi} P \int_0^{\infty} \frac{\text{Im}\, f(0,k')k' \, dk'}{k'^2 - k^2} - \text{Re} \sum_{n,l} \frac{A_{nl}^2(-1)^{l+1}(2l+1)}{k_{nl}(k_{nl}-k)} \,. \tag{97.18}$$

The imaginary part of the amplitude $\text{Im}\, f(0,k)$ involved in the right-hand side of the equation can be expressed, according to the optical theorem (see (91.11)), in terms of a physically observed quantity; the total scattering cross section $\sigma(k)$. Hence also the real part, $\text{Re}\, f(0, k)$, according to (97.18) can be expressed in terms of physically observable quantities. Dispersion relations are at present widely used. In particular, by means of them one can immediately remove the ambiguity (noted in §95) in the choice of phase shifts for a known law of scattering, i.e. for a known cross section. We stress that dispersion relations are based on as general a property of the S-matrix as its analyticity which results from the causality principle.

§98. Time reversal and the principle of detailed balance

Let us consider the properties of the S-matrix associated with the symmetry of the Schrödinger equation with respect to time reversal. We have already touched upon this question in §6 and shall now consider it in more detail.

The symmetry with respect to time reversal means that there exists a solution $\psi_{rev}(x,t)$ of the 'reversed' Schrödinger equation expressed in terms of the function $\psi(x,-t)$. If the operator $\hat{H}$ does not depend explicitly on time, then

$$i\hbar \frac{\partial \psi^*(x,-t)}{\partial t} = \hat{H}^* \psi^*(x,-t) . \tag{98.1}$$

For $\hat{H}^* = \hat{H}$ eq. (98.1) is the same as the initial equation (27.7), and the function $\psi^*(x,-t)$ describes the process reversed in time (see (6.9)). In the more general case (a charged particle in a magnetic field) we have to set

$$\psi_{rev}(x,t) = \hat{V}\psi^*(x,-t) , \tag{98.2}$$

where $\hat{V}$ is a certain operator. Operating on eq. (98.1) from the left with the operator $\hat{V}$, we obtain the equation for the function ψ_{rev}

$$i\hbar \frac{\partial \psi_{rev}(x,t)}{\partial t} = \hat{V}\hat{H}^* \hat{V}^{-1} \psi_{rev}(x,t) . \tag{98.3}$$

This equation is the same as the initial Schrödinger equation (27.7) under the condition

$$\hat{V}\hat{H}^* = \hat{H}\hat{V} . \tag{98.4}$$

From the Hermitian property of the operator $\hat{H}$ it follows that the operator $\hat{V}$ must be unitary, i.e. $\hat{V}^{-1} = \hat{V}^\dagger$. To the law of transformation of wave functions (98.2) there corresponds a definite law of transformation of arbitrary operators $\hat{F}$. This law can be found by the usual methods (see §46, 48,49).

A lack of generality arises in the case given only in connection with the fact that the operator $\hat{V}$ operates not on the function ψ but on the function ψ^*. We shall find the operator $\hat{F}_{rev}$ (reversed in time) proceeding from the requirement that the matrix element of the operator $\hat{F}$ taken with respect to the functions ψ_{rev} must be the same as the matrix elements of the operator $\hat{F}_{rev}$ taken with respect to the functions $\psi(x, -t)$

$$\langle \psi_{rev} | \hat{F} | \psi_{rev} \rangle = \langle \psi(-t) | \hat{F}_{rev} | \psi(-t) \rangle . \tag{98.5}$$

Making use of relation (98.2), we obtain

$$\langle \psi_{rev} | \hat{F} | \psi_{rev} \rangle = \langle \hat{V} \psi^*(-t) | \hat{F} | \hat{V} \psi^*(-t) \rangle = \langle \psi^*(-t) | \hat{V}^\dagger \hat{F} \hat{V} | \psi^* -t) \rangle .$$

Hence it follows (see (17.3)) that

$$\tilde{\hat{F}}_{rev} = \hat{V}^\dagger \hat{F} \hat{V} , \tag{98.6}$$

where $\tilde{\hat{F}}_{rev}$ denotes the transpose of the operator $\hat{F}_{rev}$. As is easily seen from (98.4) and (98.6), the operator $\hat{H}$ is invariant under time reversal, i.e. $\hat{H}_{rev} = \hat{H}$. Here we have made use of the condition of hermiticity of the Hamiltonian $\hat{H} = \hat{H}^*$. Relation (98.6) can serve as a basis for finding the operator $\hat{V}$. Indeed, it is natural to require that the quantum operators transform under time reversal in the same way as the corresponding classical quantities. Quantities such as energy, coordinates, electric field strength and so on are invariant under time reversal. The corresponding operators also must be invariant. Velocity, momentum, angular momentum, magnetic field strength and so on change sign under time reversal. The corresponding operators must have the same property. For example, the relations

$$\hat{\mathbf{r}}_{rev} = \hat{\mathbf{r}} , \qquad \hat{\mathbf{p}}_{rev} = -\hat{\mathbf{p}} , \qquad \hat{\mathbf{L}}_{rev} = -\hat{\mathbf{L}} \tag{98.7}$$

must be fulfilled. The spin transforms as the angular momentum, i.e. the following relation must be fulfilled:

$$\hat{\mathbf{s}}_{rev} = -\hat{\mathbf{s}} . \tag{98.8}$$

Let us consider, for example, a particle with spin $\frac{1}{2}$. Proceeding from relations (98.8) it is easy to find the operator $\hat{V}_s$ operating on the spin variables under time reversal. Making use of expression (98.6) and taking into account the form of the spin operators (60.15) and (60.16), we have

$$\hat{V}_s^\dagger \hat{s}_x \hat{V}_s = -\hat{s}_x , \qquad \hat{V}_s^\dagger \hat{s}_y \hat{V}_s = \hat{s}_y , \qquad \hat{V}_s^\dagger \hat{s}_z \hat{V}_s = -\hat{s}_z . \tag{98.9}$$

From these relations by means of (60.12) we easily find

$$\hat{V}_s = i\sigma_y . \tag{98.10}$$

(We have chosen the phase factor in such a way that the operator $\hat{V}_s$ is real.)

For the motion of a particle in a magnetic field the operator $\hat{V}$ must involve changing the direction of the magnetic field (or of the vector potential $\mathbf{A}$) to the opposite direction. Taking this fact into account, relation (98.4) has the form

$$\sigma_y \hat{H}^*(-\mathbf{A}) = \hat{H}(\mathbf{A})\sigma_y \;. \tag{98.11}$$

It is easily verified that the Hamiltonian $\hat{H}$ (see (63.3)) satisfies this relation. The invariance of the Schrödinger equation under time reversal means that one can always find an operator $\hat{V}$ satisfying condition (98.4) (for more details see the reference below*). However, the discovery in 1964 of the anomaly in the decay of K-mesons shows that under certain conditions the principle of time reversal may apparently be violated.

From the invariance of the Hamiltonian $\hat{H}$ under the replacement $t \rightarrow -t$ there results the invariance of the S-matrix, i.e. (see (98.6)) the following relation holds:

$$\hat{V}^\dagger \hat{S} \hat{V} = \tilde{\hat{S}} \;. \tag{98.12}$$

The validity of this relation is easily checked, taking into account (98.4), for the operator $\hat{V}(t, t_0)$ (see (96.5)). Since the operator $\hat{S}$ is defined as the limit of the operator $\hat{V}(t, t_0)$ (see (96.11)), it also satisfies relation (98.12).

Based on relation (98.12), it is easy to establish the relation directly between the matrix elements of the S-matrix for the direct and inverse reactions. We denote by ψ_a and ψ_b the wave functions of the initial and final states of the system. Then, taking into account (17.3), (98.2) and (98.12), we have

$$\langle \psi_b | \hat{S} | \psi_a \rangle = \langle \psi_a^* | \tilde{\hat{S}} | \psi_b^* \rangle = \langle \psi_a^* | \hat{V}^\dagger \hat{S} \hat{V} | \psi_b^* \rangle = \langle \hat{V} \psi_a^* | \hat{S} | \hat{V} \psi_b^* \rangle = \langle \psi_{a*} | \hat{S} | \psi_{b*} \rangle \;, \tag{98.13}$$

where ψ_{a*} and ψ_{b*} denote the 'reversed' wave functions of the states a and b. Thus the following equality is fulfilled:

$$S_{ba} = S_{a*b*} \;. \tag{98.14}$$

Relation (98.14) establishes the connection between the matrix elements of the S-matrix of the direct and 'reversed' processes. The states ψ_{a*} and ψ_{b*} differ from the states ψ_a and ψ_b by the sign of quantities such as velocities, momenta, angular momentum components, spin components and so on. Relation (98.14) or the equivalent relation (98.13) is called the reciprocity theorem. On the basis of this theorem a relation can be established between

* A.M.Baldin, V.I.Goldanskii and I.L.Rozenthal, *Kinematics of nuclear reactions* (Pergamon Press, Oxford, 1961).

the cross sections for direct and inverse reactions (principle of detailed balance).

Let us consider the reaction

$$a + A \rightleftharpoons b + B .$$

We denote by j_a, m_a, j_A, m_A, j_b, m_b, j_B, m_B respectively the total angular momenta and their components of the particles taking part in the reaction. According to (95.14), the cross sections for the direct and inverse reactions expressed in terms of the matrix elements of the S-matrix have the form

$$\frac{d\sigma_{ba}}{d\Omega_b} = \frac{4\pi^2}{k_a^2} |\langle j_b, m_b, j_B, m_B; -\mathbf{n}_b | \hat{S} | j_a, m_a, j_A, m_A; \mathbf{n}_a \rangle|^2 , \qquad (98.15)$$

$$\frac{d\sigma_{ab}}{d\Omega_a} = \frac{4\pi^2}{k_b^2} |\langle j_a, m_a, j_A, m_A; -\mathbf{n}_a | \hat{S} | j_b, m_b, j_B, m_B; \mathbf{n}_b \rangle|^2 . \qquad (98.16)$$

Since the momentum vector of the relative motion of the particles in the final state is directed away from the centre of mass, it is assigned a minus sign.

The relation between these cross sections cannot be written directly, since the reciprocity theorem relates the cross section for the direct process to that of the 'reversed' process which differs from (98.16) by the change of the signs of the angular momentum components m_a, m_A, m_b, m_B into the opposite signs. However, one can write the relation between averaged cross sections, i.e. cross sections summed over the components of the angular momenta of the final states and averaged over the components of the angular momenta of the initial states. Such cross sections no longer depend on the angular momentum components, and for them the reciprocity theorem (98.14) gives

$$\frac{1}{k_b^2} (2j_a+1)(2j_A+1) \frac{\overline{d\sigma_{ba}}}{d\Omega_b} = \frac{1}{k_a^2} (2j_b+1)(2j_B+1) \frac{\overline{d\sigma_{ab}}}{d\Omega_a} , \qquad (98.17)$$

where

$$\frac{\overline{d\sigma_{ba}}}{d\Omega_b} = \frac{1}{(2j_a+1)(2j_A+1)} \sum_{\substack{m_a, m_A \\ m_b, m_B}} \frac{d\sigma_{ba}}{d\Omega_b} \qquad (98.18)$$

and

$$\frac{\overline{d\sigma_{ab}}}{d\Omega_a} = \frac{1}{(2j_b+1)(2j_B+1)} \sum_{\substack{m_a, m_A \\ m_b, m_B}} \frac{d\sigma_{ab}}{d\Omega_a} . \tag{98.19}$$

A relation analogous to (98.17) can also be written for the total cross sections

$$k_a^2(2j_a+1)(2j_A+1)\bar{\sigma}_{ba} = k_b^2(2j_b+1)(2j_B+1)\bar{\sigma}_{ab} . \tag{98.20}$$

We note also that the relation between non-averaged cross sections for the direct and inverse reactions can be established within the framework of applicability of perturbation theory:

$$\frac{1}{k_b^2} \frac{d\sigma_{ba}}{d\Omega_b} = \frac{1}{k_a^2} \frac{d\sigma_{ab}}{d\Omega_a} . \tag{98.21}$$

Indeed, in this case the transition probability and, consequently, also the cross section for the process, is determined by the square of the modulus of the matrix element of the perturbation Hamiltonian H'_{ba}, for which, by virtue of hermiticity, the relation $|H'_{ba}|^2 = |H'_{ab}|^2$ is fulfilled. From this equality there results relation (98.21).

12

The Method of Second Quantization and Radiation Theory

§99. Second quantization for systems of bosons and fermions*

One of the important formal mathematical methods often used in the quantum mechanics of a system of many particles is the so-called second-quantization method.

In this method a transition from the coordinate representation of the wave function to new variables is carried out. As new variables the numbers of particles in a given quantum state are chosen. Thus the system of particles is now characterized not by defining the wave function $\psi(\xi_1,\xi_2,...,\xi_N,t)$ but by defining a new function $c(n_1,n_2,...,t)$, where n_1, n_2, ... are the numbers of particles in the 1st, 2nd and so on states. We shall call the quantities $n_1, n_2, ...$ the occupation numbers.

The quantity

$$|c(n_1,n_2,...,n_k,...,t)|^2 \tag{99.1}$$

gives the probability that at instant of time t there are n_1 particles in the first state, n_2 particles in the second state and so on. The second-quantization method turns out to be very convenient for those systems in which the

---

* In this section we follow L.D.Landau and E.M.Lifshitz, *Quantum mechanics* (Pergamon Press, Oxford, 1965).

number of particles in a given state changes, and the production and disappearance of particles of a given kind occurs (for example, in the emission and absorption of photons, or in the β-decay of nuclei). The transition from the ordinary description to second quantization is an example of a transformation from one representation to another.

Let us formally consider a system of non-interacting identical particles. We shall first assume that the particles obey Bose statistics.

We denote by $\psi_1(\xi)$, $\psi_2(\xi)$, ..., $\psi_k(\xi)$ the whole set of orthogonal and normalized wave functions of an individual particle forming a complete system of functions chosen in an arbitrary way. The index k denotes the set of four quantum numbers characterizing the state of the particle. We pass to the representation in which the occupation numbers n_k and not the coordinates, ξ_i, of the particles are chosen as independent variables.

In the new representation the basis functions (see §65) are the symmetrized and normalized products of the wave functions $\psi_k(\xi_i)$ of the individual particles. Formula (65.5) for the general case where n_1 particles are in state ψ_1, n_2 particles in state ψ_2 and so on assumes the form

$$\psi_{n_1,n_2,n_3,...}(\xi_1,\xi_2,...,\xi_N) =$$

$$= \left(\frac{n_1!n_2!n_3!...}{N!}\right)^{\frac{1}{2}} \sum \psi_{k_1}(\xi_1)\psi_{k_2}(\xi_2)...\psi_{k_N}(\xi_N) . \qquad (99.2)$$

The summation is carried out only over all permutations of different indices $k_1, k_2, ...$.

We introduce the operators $\hat{a}_k^{\dagger}$ and $\hat{a}_k$ which act on the new variables, the occupation numbers in state k. We define these operators by the formulae

$$\hat{a}_k \psi_{n_1,...,n_k,...} = (n_k)^{\frac{1}{2}} \psi_{n_1,...,n_k-1,...} , \qquad (99.3)$$

$$\hat{a}_k^{\dagger} \psi_{n_1,...,n_k,...} = (n_k+1)^{\frac{1}{2}} \psi_{n_1,...,n_k+1,...} . \qquad (99.4)$$

The operator $\hat{a}_k$ reduces the number of particles in state k by one, i.e. it replaces n_k by n_k-1. The operator $\hat{a}_k^{\dagger}$ increases this number by one, i.e. it replaces n_k-1 by n_k. It is obvious that the consecutive application of the operators $\hat{a}_k$ and $\hat{a}_k^{\dagger}$ does not change the number of particles in the kth state, i.e.

$$\hat{a}_k^{\dagger}\hat{a}_k \psi_{n_1,...,n_k,...} = n_k \psi_{n_1,...,n_k,...} . \qquad (99.5)$$

The matrix elements of the operators $\hat{a}_k$ and $\hat{a}_k^{\dagger}$ are of the form

$$\langle n_1,n_2,...,n_k-1,...|\hat{a}_k|n_1,n_2,...,n_k,...\rangle = (a_k)_{n_k-1,n_k} = n_k^{\frac{1}{2}} , \qquad (99.6)$$

$$\langle n_1, n_2, ..., n_k+1, ... | \hat{a}_k^\dagger | n_1, n_2, ..., n_k, ... \rangle = (a_k^\dagger)_{n_k+1, n_k} = (n_k+1)^{\frac{1}{2}} , \qquad (99.7)$$

$$(a_k^\dagger a_k)_{n'_k, n_k} = n_k \delta_{n'_k, n_k} . \qquad (99.8)$$

In accordance with their meaning, the operators $\hat{a}_k$ and $\hat{a}_k^\dagger$ are called respectively the annihilation and creation operators of a particle in the kth state. The operator $\hat{a}_k^\dagger \hat{a}_k$ is called the operator of the number of particles n_k in the state k.

We have already encountered operators similar to the operators $\hat{a}_k$ and $\hat{a}_k^\dagger$ in §50 in considering the problem of the harmonic oscillator. It is easily seen that the operators $\hat{a}_k$ and $\hat{a}_k^\dagger$ satisfy the commutation relations

$$\hat{a}_k \hat{a}_l^\dagger - \hat{a}_l^\dagger \hat{a}_k = \delta_{kl} ,$$

$$\hat{a}_k \hat{a}_l - \hat{a}_l \hat{a}_k = 0 , \qquad (99.9)$$

$$\hat{a}_k^\dagger \hat{a}_l^\dagger - \hat{a}_l^\dagger \hat{a}_k^\dagger = 0 .$$

We shall show how the ordinary operators acting on a wave function in the coordinate representation can be expressed in terms of the creation and annihilation operators of particles, i.e. in the second-quantization representation.

Let us consider the operator $\hat{L}(\xi_i)$ acting on the coordinates of the ith particle. The coordinates are understood to include the spin coordinates. Since all the particles are equivalent, we introduce the operator $\hat{L}_1 = \Sigma_{i=1}^N \hat{L}(\xi_i)$. Let us find the expression for it in the second-quantization representation. We obtain the matrix elements of $\hat{L}_1$ by means of the basis functions (99.2).

We have by definition

$$\langle n'_1, ..., n'_k, ... | \hat{L}_1 | n_1, ..., n_k, ... \rangle =$$

$$= \left\langle n'_1, ..., n'_k, ... \left| \sum_{i=1}^N \hat{L}(\xi_i) \right| n_1, ..., n_k, ... \right\rangle . \qquad (99.10)$$

Let us consider one term of the sum over the particles

$$\langle n'_1, ..., n'_k, ... | \hat{L}(\xi_i) | n_1, ..., n_k, ... \rangle =$$

$$= \int \psi^*_{n'_1, ..., n'_k, ...} \hat{L}(\xi_i) \psi_{n_1, ..., n_k, ...} \, d\xi_1 ... d\xi_N . \qquad (99.11)$$

(Summation over spin variables is implied.) The operator $\hat{L}(\xi_i)$ acts only on the variables of the ith particle. Hence we can write

$$\hat{L}(\xi_i)\psi_{n_1,\ldots,n_k,\ldots} = \left(\frac{n_1!\ldots n_k!\ldots}{N!}\right)^{\frac{1}{2}} \sum \psi_{k_1}(\xi_1)\ldots\psi_{k_N}(\xi_N)\hat{L}(\xi_i)\psi_{k_i}(\xi_i) . \quad (99.12)$$

Multiplying (99.12) by the function $\psi^*_{n_1\ldots}$ and integrating we note, first of all, that the integrals over all variables except ξ_i contain only the products of wave functions.

By virtue of orthogonality of the latter all integrals involving factors of the form $\psi^*_1(\xi_1)\psi_2(\xi_1)$, i.e. containing the products of the wave functions of the particles (except of the ith particle) referring to different states, will reduce to zero.

In the double sum over permutations (99.11) only those terms which contain the products of the wave functions of the particles (except the ith particle) referring to the same states differ from zero. The integral over the variables ξ_i is of the form

$$(\hat{L}(\xi_i))_{lk} = \int \psi^*_l(\xi_i)\hat{L}(\xi_i)\psi_k(\xi_i)\,d\xi_i .$$

This means that for $l \neq k$ a transition of the particle takes place from the kth state into the lth state. Consequently, the number of particles in the kth state decreases by one, and in the lth state increases by one. We denote the corresponding matrix element by

$$\langle n_k-1,n_l|\hat{L}(\xi_i)|n_k,n_l-1\rangle \qquad\qquad (99.13)$$

(the operator is diagonal with respect to other occupation numbers and we do not write them down). The functions involved in the matrix element are of the form

$$\psi^*_{n_1,\ldots,n_k-1,\ldots} = \left(\frac{n_1!\ldots(n_k-1)!\ldots n_l!\ldots}{N!}\right)^{\frac{1}{2}} \sum \psi^*_{k_1}(\xi_1)\ldots\psi^*_{k_N}(\xi_N) ,$$

$$\psi_{n_1,\ldots,n_l-1,\ldots} = \left(\frac{n_1!\ldots n_k!\ldots(n_l-1)!\ldots}{N!}\right)^{\frac{1}{2}} \sum \psi_{k_1}(\xi_1)\ldots\psi_{k_N}(\xi_N) .$$

By virtue of the orthogonality of the wave functions integration over the coordinates of all particles gives (taking into account the permutations of $N-1$ particles excluding the ith particle)

$$\langle n_k-1,n_l|\hat{L}(\xi_i)|n_k,n_l-1\rangle = \left(\frac{n_1!...(n_k-1)!...n_l!...}{N!}\right)^{\frac{1}{2}} \left(\frac{n_1!...n_k!...(n_l-1)!...}{N!}\right)^{\frac{1}{2}} \times$$

$$\times \frac{(N-1)!}{n_1!...(n_k-1)!...(n_l-1)!...} (\hat{L}(\xi_i))_{lk}$$

$$= \frac{(n_k n_l)^{\frac{1}{2}}}{N} (\hat{L}(\xi_i))_{lk} .$$

Since the operators $\hat{L}(\xi_1), \hat{L}(\xi_2), ...$ differ from each other only in the number of particles on whose coordinates they act, all matrix elements differing in the number of particles are equal to each other. Hence for the matrix element (99.10) of the operator $\hat{L}$ we can finally write

$$(n_k-1,n_l|\hat{L}_1|n_k,n_l-1) = \left\langle n_k-1,n_l \left| \sum_{i=1}^{N} \hat{L}(\xi_i) \right| n_k,n_l-1 \right\rangle =$$

$$= N\langle n_k-1,n_l|\hat{L}(\xi_i)|n_k,n_l-1\rangle = (n_k n_l)^{\frac{1}{2}} (L(\xi))_{lk} . \qquad (99.14)$$

In the case where the diagonal matrix element is considered, i.e. where the distribution of the number of particles over states does not change, we have analogously

$$\langle n_1,n_2,...|\hat{L}_1|n_1,n_2,...\rangle = \sum_k n_k (L(\xi))_{kk} . \qquad (99.15)$$

We now introduce the operators $\hat{a}^\dagger$ and $\hat{a}$ into formulae (99.14) and (99.15). Then the operator $\hat{L}_1$ can be written in the form

$$\hat{L}_1 = \sum_{k,l} (L(\xi))_{lk} \hat{a}_l^\dagger \hat{a}_k . \qquad (99.16)$$

Indeed, the matrix elements of this operator are, by virtue of (99.6) and (99.8), the same as the matrix elements (99.14) and (99.15).

An analogous result can be obtained in the same way for operators which act on the coordinates of two particles ξ_i and ξ_j.

The operator

$$\hat{L}_2 = \sum_{i,j \neq 1} \hat{L}(\xi_i,\xi_j)$$

is expressed in the second-quantization representation by the formula

$$\hat{L}_2 = \sum_{k,p,l,m} \langle l,m|\hat{L}(\xi,\xi')|k,p\rangle \hat{a}_l^\dagger \hat{a}_m^\dagger \hat{a}_k \hat{a}_p \ , \tag{99.17}$$

where the matrix elements are equal to

$$\langle l,m|\hat{L}(\xi,\xi')|k,p\rangle = \int \psi_l^*(\xi)\psi_m^*(\xi')\hat{L}(\xi,\xi')\psi_k(\xi)\psi_p(\xi')\,d\xi\,d\xi' \ . \tag{99.18}$$

By means of the general formulae (99.16) and (99.17) one can write the Hamiltonian of a system of particles in the second-quantization representation. In the case of a system of non-interacting particles in a given external field we have

$$\hat{H} = \sum_{i=1}^{N} \hat{H}_i = \sum_{i=1}^{N} (\hat{T}_i + U(\xi_i)) = \sum_{i=1}^{N} \left(-\frac{\hbar^2}{2m} \nabla_i^2 + U(\xi_i) \right) , \tag{99.19}$$

where $U(\xi_i)$ is the potential energy of the ith particle in the external field, and $\hat{T}_i$ is its kinetic energy operator. Operator (99.19) is evidently a particular case of the operator $\hat{L}_1$. Correspondingly we can immediately write operator (99.19) in the second-quantization representation

$$\hat{H} = \sum_{k,l} (\hat{H}_i)_{lk} \hat{a}_l^\dagger \hat{a}_k \ . \tag{99.20}$$

Choosing as ψ_k the eigenfunctions of the Hamiltonian $\hat{H}_i$ of an individual particle, we have

$$(H_i)_{lk} = \int \psi_l^*(\xi)\hat{H}_i(\xi)\psi_k(\xi)\,d\xi = E_k \delta_{lk} \ ,$$

where E_k is the energy of the particle in the kth state.
    Hence, finally,

$$\hat{H} = \sum_k E_k \hat{a}_k^\dagger \hat{a}_k \ . \tag{99.21}$$

The energy of a system of particles is, by virtue of (99.8), equal to

$$E = \sum_k E_k n_k .$$

(99.22)

If the eigenfunctions of the operator $\hat{T}_i$ corresponding to the eigenvalues ϵ_k are chosen as ψ_k, then (99.20) is rewritten in the form

$$\hat{H} = \sum_k \epsilon_k \hat{a}_k^\dagger \hat{a}_k + \sum_{k,l} \hat{a}_l^\dagger \hat{a}_k \int \psi_l^* U(\xi) \psi_k \, d\xi .$$

(99.23)

In the case of a system of particles between which there is a pair interaction the interaction energy operator has the form $\frac{1}{2}\sum_{i \neq j} W(\xi_i, \xi_j)$. Making use of (99.17) we write the Hamiltonian in the second-quantization representation

$$\hat{H} = \sum_{k,l} (\hat{H}_i)_{lk} \hat{a}_l^\dagger \hat{a}_k + \frac{1}{2} \sum_{k,p,l,m} \langle lm|W|kp\rangle \hat{a}_l^\dagger \hat{a}_m^\dagger \hat{a}_k \hat{a}_p ,$$

(99.24)

or, taking as the functions ψ_k the eigenfunctions of the operator $\hat{H}_i$

$$\hat{H} = \sum_k E_k \hat{a}_k^\dagger \hat{a}_k + \frac{1}{2} \sum_{k,p,l,m} \langle lm|W|kp\rangle \hat{a}_l^\dagger \hat{a}_m^\dagger \hat{a}_k \hat{a}_p .$$

(99.25)

We note that the pair interaction (the last term of formula (99.24)) has an obvious interpretation. The interaction can be treated as the collision of two particles which are in the pth and kth states. After the interaction they make a transition to the lth and mth states.

It is useful to note that formula (99.20) can be obtained by means of the following formal method. In the expression for the mean energy (65.9) we replace the wave function by the operator in the space of occupation numbers defined as

$$\psi(\xi) \rightarrow \hat{\psi}(\xi) = \sum_k \hat{a}_k \psi_k(\xi)$$

(99.26)

and correspondingly

$$\psi^*(\xi) \rightarrow \hat{\psi}^\dagger(\xi) = \sum_l \hat{a}_l^\dagger \psi_l^*(\xi) .$$

(99.26′)

Then on the right-hand side of (65.9) we have

$$\int \psi^*(\xi)\hat{H}_i\psi(\xi)\,d\xi \rightarrow \sum_{k,l} \int \hat{a}_l^\dagger \psi_l^*(\xi)\hat{H}_i\hat{a}_k\psi_k(\xi)\,d\xi = \sum_{k,l} \hat{a}_l^\dagger \hat{a}_k(\hat{H}_i)_{lk}\,. \tag{99.27}$$

Comparing (99.27) and (99.20) we see that when the ordinary wave function is replaced by the operator the right-hand side of (65.9) is the same as (99.20). This means that in this case $\bar{H}$ can formally be replaced by the operator $\hat{H}$ in the second-quantization representation.

The name second quantization is due to the replacement of the wave function ψ by the operator $\hat{\psi}$. In second quantization not only are all mechanical quantities replaced by quantum operators (ordinary quantization) but the wave function itself is also quantized, i.e. replaced by an operator. Although second quantization is a formal method, it turns out to be very useful in a number of cases.

The Hamiltonian of a system of particles interacting in pairs can also be obtained easily in an analogous way. For this we again replace the functions ψ and ψ^* in formula (65.8) by operators (99.26). Then, in correspondence with what was said above, we make the replacement $\bar{H} \rightarrow \hat{H}$, where $\hat{H}$ is the Hamiltonian in the second-quantization representation.

After the replacement we obtain formula (99.24).

All the results obtained so far apply to bosons. It can be shown* that formulae (99.20) and (99.24) remain valid also for a system of fermions. However, the operators $\hat{a}_k$ and $\hat{a}_k^\dagger$ can then no longer satisfy relations (99.9). Indeed, for the operators $\hat{a}_k$ and $\hat{a}_k^\dagger$ defined by formulae (99.9) the eigenvalues of the product $\hat{a}_k^\dagger\hat{a}_k$ are equal to arbitrary positive integers n_k. For a system of fermions the occupation numbers can be equal only to zero or one in accordance with the Pauli principle. The operators $\hat{a}_k$ and $\hat{a}_k^\dagger$ must now be defined in such a way that the eigenvalues of the operator $\hat{a}_k^\dagger\hat{a}_k$ are equal either to zero or to one, i.e.

$$(\hat{a}_k^\dagger\hat{a}_k)_{n_kn_k} = n_k = \left\{ {0 \atop 1} \right.. \tag{99.28}$$

We shall show that conditions (99.28) are fulfilled if the operators $\hat{a}_k$ and $\hat{a}_k^\dagger$ satisfy the following anticommutation rules:

$$\hat{a}_k\hat{a}_l^\dagger + \hat{a}_l^\dagger\hat{a}_k = \delta_{kl}\,, \tag{99.29}$$

* See L.D.Landau and E.M.Lifshitz, *Quantum mechanics* (Pergamon Press, Oxford, 1965).

$$\hat{a}_k\hat{a}_l + \hat{a}_l\hat{a}_k = \hat{a}_k^\dagger\hat{a}_l^\dagger + \hat{a}_l^\dagger\hat{a}_k^\dagger = 0 \ . \tag{99.30}$$

For this we convince ourselves of the fact that

$$(\hat{a}_k^\dagger\hat{a}_k)^2 = \hat{a}_k^\dagger\hat{a}_k \ . \tag{99.31}$$

Indeed, we evaluate the left-hand side and making use of (99.29) we obtain

$$(\hat{a}_k^\dagger\hat{a}_k)^2 = \hat{a}_k^\dagger\hat{a}_k\hat{a}_k^\dagger\hat{a}_k = \hat{a}_k^\dagger\hat{a}_k(1-\hat{a}_k\hat{a}_k^\dagger) = \hat{a}_k^\dagger\hat{a}_k - \hat{a}_k^\dagger\hat{a}_k\hat{a}_k\hat{a}_k^\dagger = \hat{a}_k^\dagger\hat{a}_k \ ,$$

since $\hat{a}_k^2 = 0$, which follows from (99.30).

Taking the diagonal matrix elements of relation (99.31), we find $n_k^2 = n_k$. This equality can be fulfilled only for $n_k = 0$ and $n_k = 1$. One can find the explicit form of the matrices $\hat{a}_k$ based on relations (99.30). Since the numbers n_k take on only two values 0 and 1, the operators $\hat{a}_k$ and $\hat{a}_k^\dagger$ are two-row matrices with respect to these variables. We shall present the corresponding matrix elements without derivation. They are

$$(a_k)_{01} = (a_k^\dagger)_{10} = \prod_{l=1}^{k-1} (1-2n_l) \ . \tag{99.32}$$

All other matrix elements are equal to zero. As a result of the multiplication of the quantities $1-2n_l$, where $l = 1, 2, ..., k-1$, either +1 or −1 is obtained, depending on the value of the occupation numbers of states preceding the given state.

Hence it is clear that the numbering of states $1, 2, ..., k$, chosen initially, must not be changed.

The Schrödinger equation in the occupation number representation, where the Hamiltonian is given by formula (99.24), involves the law of conservation of the total number of particles (see §7). However, the introduction of the operators $\hat{a}_k^\dagger$ and $\hat{a}_k$ describing the absorption and production of particles allows one in a corresponding generalization also to investigate processes in which the number of particles of a given kind is not conserved.

§ 100. The quantum mechanics of the photon

The experimental establishment of the quantum or corpuscular nature of light was a spur to the creation of quantum theory as a whole.

On the other hand, the construction, as a consequence, of the quantum theory of the electromagnetic field has been one of the most notable successes of quantum theory.

Light quanta or photons are elementary particles whose distinctive property is the fact that their rest mass is equal to zero. Hence they always move with the velocity c in vacuum. This fact leads to certain important features of the method of describing their behaviour. Namely, the relation between the energy and momentum of the photon is given by the general formula

$$\epsilon = cp = \hbar ck .\tag{100.1}$$

If the momentum of the photon is replaced by its momentum operator, then the energy operator in the momentum representation has the form

$$\hat{H} = c\hat{p} = \hbar c\hat{k} .\tag{100.2}$$

Correspondingly the Schrödinger equation can be written in the momentum representation as

$$i\hbar \frac{\partial \psi_p}{\partial t} = \hat{H}\psi_p ,\tag{100.3}$$

where ψ_p is the wave function of the photon in the momentum representation.

The operator $\hat{H}$ is related to the photon energy ϵ by the general formula

$$\epsilon = \int \psi_p^* \hat{H} \psi_p \, d\mathbf{p} = \hbar c \int \psi_p^* \hat{k}\psi_p \, d\mathbf{p} .\tag{100.4}$$

On the other hand it can be assumed that there corresponds to a photon an electromagnetic field over all space. Its energy is

$$\epsilon = \int \frac{E^2 + H^2}{8\pi} \, dV = \frac{1}{4\pi} \int E^2 \, dV .\tag{100.5}$$

It is natural to identify the energy of the photon with the energy of the electromagnetic field. Both field vectors satisfy Maxwell's equations, which in a vacuum are reduced to the form

$$\mathbf{V}^2 \mathbf{E} - \frac{1}{c^2} \frac{\partial^2 \mathbf{E}}{\partial t^2} = 0$$

and analogously for the vector $\mathbf{H}$.

Expanding $\mathbf{E}$ in a Fourier integral

$$\mathbf{E}(\mathbf{r},t) = \int \mathbf{E}(\mathbf{k},t) \, e^{i\mathbf{k}\cdot\mathbf{r}} \, d\mathbf{k}$$

we have

$$\frac{\partial^2 \mathbf{E}(\mathbf{k},t)}{\partial t^2} + k^2 \mathbf{E}(\mathbf{k},t) = 0 ,$$

or

$$\left[\frac{\partial E(k,t)}{\partial t} - ikE(k,t)\right]\left[\frac{\partial E(k,t)}{\partial t} + ikE(k,t)\right] = 0 .$$ (100.6)

By virtue of the fact that the field is real the following condition must be fulfilled:

$$E(k) = E(-k) .$$ (100.7)

In place of the Fourier component $E(k,t)$ we introduce the new function $f(k,t)$ defined by the relations

$$E(k,t) = N(k)[f(k,t)+f^*(-k,t)] ,$$
$$\dot{E}(k,t) = -ikN(k)[f(k,t)-f^*(-k,t)]$$ (100.8)

where N is a factor of proportionality. The dot denotes differentiation with respect to time.

It is easily seen that in such a representation of $E(k,t)$ the condition (100.7) is automatically fulfilled.

Substituting the values of $E(k,t)$ and $\dot{E}(k,t)$ into (100.6) we arrive at two equations

$$i\frac{\partial f}{\partial t} = kf , \qquad -i\frac{\partial f^*}{\partial t} = kf^* .$$ (100.9)

We stress that eqs. (100.9) only represent another form of notation of Maxwell's equations. Multiplying (100.9) by $\hbar$ we obtain

$$i\hbar\frac{\partial f}{\partial t} = pf , \qquad -i\hbar\frac{\partial f^*}{\partial t} = pf^* .$$ (100.10)

We see that the function $f(k,t)$ satisfies an equation which is in form identical with the Schrödinger equation. If p is replaced by the operator $\hat{H}$, then the function $f(k,t)$ must be identified with the wave function of the photon in the k-representation.

The factor of proportionality N, which has so far remained arbitrary, can be defined from the comparison of (100.4) and (100.5).

Substituting expressions (100.8) into (100.5), we have

$$\epsilon = \frac{1}{4\pi} \int E(k,t) \cdot E(k',t) \, e^{i(k+k')\cdot r} \, dk \, dk' \, dV =$$

$$= \frac{1}{4\pi} \int E(k,t) \cdot E(k',t) \, dk \, dk' \int e^{i(k+k')\cdot r} \, dV =$$

$$= \frac{(2\pi)^3}{4\pi} \int E(k,t) \cdot E(k',t) \delta(k+k') \, dk \, dk' =$$

$$= 2\pi^2 \int E(k,t) \cdot E(-k,t) \, dk = 4\pi^2 \int N^2(k) \, f(k) \cdot f^*(k) \, dk \;.$$

For $N = (ck/4\pi^2)^{\frac{1}{2}}$ the energy of the electromagnetic field and the energy of the photon turn out to be identical. Thus in the k-representation the photon is described by the wave function

$$\psi(k,t) = f(k,t) \;.$$

Then the following condition is fulfilled:

$$\int f^* f \, dk = 1 \;.$$

In this case the Maxwell equations for the electromagnetic field of a monochromatic wave turn out to be identical with the Schrödinger equation for an individual photon. Introducing the explicit dependence on time, we can write

$$\psi(k,t) = f_0(k) \, e^{-i\omega t} = f_0(k) \, e^{-(i/\hbar)et} \;.$$

By virtue of Maxwell's equation $\nabla \cdot E = 0$, the amplitude in k-space satisfies the condition $k f_0(k) = 0$. We shall not dwell on the problems of normalization of the wave function and on the calculation of other quantum-mechanical quantities of photons, for example spin angular momentum, parity and so on: we refer the reader to the monograph of Akhiezer and Berestetskii*.

We confine ourselves only to some remarks of theoretical importance. We stress, first of all, that since Maxwell's equations are relativistically invariant so is the Schrödinger equation for the photon.

This is natural, since the photon always moves with the velocity of light.

We have found the wave function of the photon in the k-representation (or, what is the same, in the p-representation). This wave function has the usual probabilistic meaning. However, the wave function of the photon in the x-representation, which would allow one to establish the probability of localization of the photon at a given point of space, does not exist.

* A.I.Akhiezer and V.B.Berestetskii, *Quantum electrodynamics* (Interscience Publishers, New York, 1965).

For free particles of rest mass m_0 different from zero the wave function in the x-representation is obtained from the wave function in the p-representation by means of the Fourier transformation.

In our case the Fourier transformation gives

$$\mathbf{f}(\mathbf{r},t) = \int \mathbf{f}(\mathbf{k},t)\, e^{i\mathbf{k}\cdot\mathbf{r}}\, d\mathbf{k}\,.$$

However, and here lies the fundamental difference between photons and particles with $m_0 \neq 0$, the position of a photon can be determined only as a result of interaction with charged particles, for example with electrons.

This interaction is determined by the value of the field vectors $\mathbf{E}$ and $\mathbf{H}$ at the point at which the electron is localized. The strength of the field at a certain point is defined by the Fourier inversion transformation, i.e.

$$\mathbf{E}(\mathbf{r},t) = \int \mathbf{E}(\mathbf{k},t)\, e^{i\mathbf{k}\cdot\mathbf{r}}\, d\mathbf{k} = \frac{1}{2\pi}\int (ck)^{\frac{1}{2}}\, [\mathbf{f}(\mathbf{k},t)+\mathbf{f}^*(\mathbf{k},t)]\, e^{i\mathbf{k}\cdot\mathbf{r}}\, d\mathbf{k}\,.$$

This formula shows that the field strength is not expressed in terms of $\mathbf{f}(\mathbf{r},t)$, i.e. is not determined by the value of any wave function at the same point of space. On the contrary, $\mathbf{E}(\mathbf{r},t)$ is determined by the distribution of $\mathbf{f}(\mathbf{r},t)$ in space.

Photons have a spin equal to one. However, the definition of spin as the intrinsic angular momentum of the particle at rest makes no sense in the case of photons. Hence the division of the total angular momentum of the photon into an orbital part and a spin part is to a certain degree arbitrary.

This important last remark is associated with the description of a system of photons.

Photons do not interact directly with each other. The very weak interaction existing between photons is due to their interaction with electrons of the background. Hence the wave function of a system of photons is the wave function of a system of non-interacting particles. Photons as particles with integer spin obey **Bose—Einstein statistics**.

When photons interact with other particles the number of photons changes in the processes of emission and absorption. Photons are absorbed and emitted one at a time. The interaction of photons with charges can be described by means of their wave function (see the monograph of Akhiezer and Berestetskii cited above). However, this interaction is described in a much more effective and simple way by means of the second-quantization representation. We note that the method of second quantization was itself devised by Dirac for just this purpose.

§101. The quantization of the radiation field

As is well known, the development of quantum theory began with the establishment of the quantum properties of the electromagnetic field and the creation of a semiempirical theory of light quanta. Hence it is natural to try to apply the mathematical apparatus of quantum mechanics to the electromagnetic field. It turns out, however, that the electromagnetic field has a number of features which make this a complex problem. The modern quantum theory of the electromagnetic field, commenced by the studies of Dirac, is based on special methods, in particular on the method of second quantization*.

We recall that in the classical theory of the electromagnetic field in vacuum it was shown that a charge-free electromagnetic field can formally be compared to a mechanical system with an infinitely large number of degrees of freedom.

Expanding the vector potential, $\mathbf{A}$, of the electromagnetic field in terms of plane waves and taking the infinite set of amplitudes of the expansion q_i as generalized coordinates, it was possible to compare the electromagnetic field with a certain mechanical system; a set of field oscillators (see §38 of Part I). To each of the Fourier components of the expansion $\mathbf{A}$ there corresponds one of the oscillators. Hence the complete set of field oscillators includes an infinitely large number and, consequently, the electromagnetic field could be compared to a mechanical system with an infinitely large number of degrees of freedom.

We write the Hamiltonian of this system as follows:

$$H = \sum \tfrac{1}{2}(p_\lambda^2 + \omega_\lambda^2 q_\lambda^2) = \sum H_\lambda \,, \tag{101.1}$$

where H_λ is the Hamiltonian of the λth oscillator, p_λ is the generalized momentum corresponding to the coordinate q_λ, and ω_λ is the corresponding frequency. The summation is carried out over all values of frequencies and polarizations.

The quantum theory of the electromagnetic field is based on the assumption that this analogy can be given a direct physical content. Namely, it is assumed that a real electromagnetic field represents a quantum system which obeys the ordinary laws of quantum mechanics. The Hamiltonian H is ob-

---

* A more detailed exposition of the quantum theory of radiation may be found in the book of W.Heitler, *The quantum theory of radiation* (Clarendon Press, Oxford, 1954).

tained from the classical Hamiltonian (101.1) by means of the usual replacement of mechanical quantities, generalized coordinates and momenta, by
corresponding quantum operators. That is, we replace q_λ and p_λ by operators
satisfying the commutation relations:

$$\hat{p}_\lambda \hat{q}_\mu - \hat{q}_\mu \hat{p}_\lambda = \frac{\hbar}{i} \delta_{\lambda\mu} , \qquad \hat{q}_\lambda \hat{q}_\mu - \hat{q}_\mu \hat{q}_\lambda = 0 , \qquad \hat{p}_\lambda \hat{p}_\mu - \hat{p}_\mu \hat{p}_\lambda = 0 .$$

Since different field oscillators are independent, the operators $\hat{p}_\lambda$ and $\hat{q}_\lambda$
referring to different oscillators commute with each other. Then $\hat{H}$ will represent the Hamiltonian of a quantum system. It is advisable, however, to carry
out the canonical transformation to new variables (see formulae (50.11)).
Namely, we write

$$\hat{a}_\lambda = \frac{1}{\sqrt{2}} \left[\left(\frac{\omega_\lambda}{\hbar}\right)^{\frac{1}{2}} \hat{q}_\lambda + \frac{i\hat{p}_\lambda}{(\omega_\lambda \hbar)^{\frac{1}{2}}} \right] ,$$

$$\hat{a}_\lambda^\dagger = \frac{1}{\sqrt{2}} \left[\left(\frac{\omega_\lambda}{\hbar}\right)^{\frac{1}{2}} \hat{q}_\lambda - \frac{i\hat{p}_\lambda}{(\omega_\lambda \hbar)^{\frac{1}{2}}} \right] .$$

(101.2)

In the new representation

$$\hat{p}_\lambda^2 + \omega_\lambda^2 \hat{q}_\lambda^2 = \hbar\omega_\lambda (\hat{a}_\lambda \hat{a}_\lambda^\dagger + \hat{a}_\lambda^\dagger \hat{a}_\lambda) ,$$

so that

$$\hat{H} = \frac{1}{2} \sum_\lambda \hbar\omega_\lambda (\hat{a}_\lambda \hat{a}_\lambda^\dagger + \hat{a}_\lambda^\dagger \hat{a}_\lambda) .$$

To the operators $\hat{a}_\lambda$ and $\hat{a}_\lambda^\dagger$ there correspond the commutation relations

$$\hat{a}_\lambda \hat{a}_\mu^\dagger - \hat{a}_\mu^\dagger \hat{a}_\lambda = \delta_{\lambda\mu} ,$$

$$\hat{a}_\lambda \hat{a}_\mu - \hat{a}_\mu \hat{a}_\lambda = 0 ,$$

$$\hat{a}_\lambda^\dagger \hat{a}_\mu^\dagger - \hat{a}_\mu^\dagger \hat{a}_\lambda^\dagger = 0 ,$$

(101.3)

which follow immediately from the definition and commutation relations for
$\hat{p}_\lambda$ and $\hat{q}_\lambda$.

The Hamiltonian can be transformed by means of (101.3), writing

$$\hat{a}_\lambda \hat{a}_\lambda^\dagger = 1 + \hat{a}_\lambda^\dagger \hat{a}_\lambda .$$

Then

$$\hat{H} = \sum_\lambda \hbar\omega_\lambda (\hat{a}_\lambda^\dagger \hat{a}_\lambda + \tfrac{1}{2}) .$$

(101.4)

Comparing the expression (101.4) for $\hat{H}$ and the commutation relations
(101.3) for the operators $\hat{a}$ and $\hat{a}^\dagger$ with the corresponding expressions (99.9)
and (99.21), we see that they are completely analogous. This means that a

free electromagnetic field represents a system of bosons which are usually called photons or light quanta.

To each plane wave in expansion (38.19) of Part I there corresponds a photon. The energy of each photon, according to formula (101.4), is equal to $\hbar\omega_\lambda$. The total energy of the electromagnetic field correspondingly has the form

$$E = \sum E_\lambda n_\lambda + \sum \tfrac{1}{2}\hbar\omega_\lambda = \sum E_\lambda n_\lambda + E_0 , \qquad (101.5)$$

where $E_\lambda = \hbar\omega_\lambda$, and n_λ is the number of photons with energy E_λ.

The second term of formula (101.5), denoted by E_0, is called the energy of the zero-point oscillations of the electromagnetic field. Formula (101.5) shows that if all $n_\lambda = 0$, i.e. if there are no photons in the field, then the energy of the electromagnetic field is equal to E_0. Moreover, the quantity E_0 itself is infinitely large, since the sum for E_0 involves an infinitely large number of positive terms $\hbar\omega_\lambda$.

The presence of the infinitely large constant term in the energy of the electromagnetic field has no effect on the processes of interaction of the field with matter (the emission, absorption and scattering of light) which will be considered in this chapter. In these processes changes occur in the state of the electromagnetic field for which only the difference between the energies of two states is important.

When an energy difference is formed the zero-point energy is cancelled. Hence, until recently, it was assumed that the zero-point energy could be taken as the zero of energy and that it could be formally omitted in all expressions. However, as quantum electrodynamics developed it turned out that this is not so and that the appearance of the term E_0 in the formula for the energy of the electromagnetic field has a profound meaning.

From the point of view of modern electrodynamics the 'emptiness', the absence of particles and of photons, is not 'nothing' but is a definite state of the field, called the vacuum. The existence of the vacuum state and of zero-point oscillations with frequencies ω_λ is important in certain interactions between the electromagnetic field and electrons and leads to a number of observed effects.

We shall touch briefly upon the problem of the vacuum in §116 and §128. In the meanwhile we shall not consider the zero-point energy.

Let us now find the momentum of a charge-free electromagnetic field. According to (38.25) of Part I, we have for the momentum of a plane wave

$$\mathbf{p}_\lambda = \frac{\mathbf{k}_\lambda}{k_\lambda c} E_\lambda , \qquad (101.6)$$

where $\mathbf{k}_\lambda$ and E_λ are respectively the wave vector and the energy of the wave.

If we pass to quantized expressions and replace E_λ by its eigenvalue, then we easily obtain

$$\mathbf{p}_\lambda = \hbar \mathbf{k}_\lambda \ .$$

Just as $\hbar\omega_\lambda$ represents the energy of an individual photon, $\hbar\mathbf{k}_\lambda$ is its momentum. We see that between the energy and momentum of the photon there is the relation found from the analysis of experimental data even before the creation of quantum mechanics

$$|\mathbf{p}_\lambda| = \frac{E_\lambda}{c} \ .$$

From (101.6) it follows, in particular, that the rest mass of the photon is equal to zero (see §14 of Part II). The total momentum of the electromagnetic field is equal to

$$\mathbf{P} = \sum \hbar \mathbf{k}_\lambda n_\lambda \ . \tag{101.7}$$

It is determined by the occupation numbers n_λ.

We now turn to the formulation of the Schrödinger equation for the electromagnetic field. It has the usual form

$$i\hbar = \frac{\partial \psi}{\partial t} = \hat{H}\psi \ .$$

The wave function of the electromagnetic field is usually called the amplitude of the state of the field. If use is made of the Hamiltonian in the representation of occupation numbers, then the amplitude of the state of the electromagnetic field will also be a function of the occupation numbers n_λ

$$\psi = \psi(n_1, n_2, \dots n_\lambda, \dots t) \ .$$

According to the conclusions of §99, the operators $\hat{a}_\lambda^\dagger$ and $\hat{a}_\lambda$ represent the photon creation and annihilation operators. When they act on the wave function they respectively increase and reduce by one the number of photons of frequency ω_λ. The matrix elements of these operators are given by formulae (99.6) and (99.7).

§102. The interaction of an electron with radiation

Having carried out the quantization of a free electromagnetic field, we can turn to the consideration of a system consisting of an electromagnetic field and particles. We shall assume that there is one electron in the radiation field and shall find the interaction between the electron and the electromagnetic field. In this chapter we shall suppose that the electron has a velocity small in comparison with the velocity of light and that it is described by a non-relativistic Hamiltonian. We write the Hamiltonian of the system (radiation field + electron) in the form

$$\hat{H} = \frac{1}{2m} \left(\hat{\mathbf{p}} - \frac{e}{c} \hat{\mathbf{A}} \right)^2 + \hat{H}_{rad} \ .$$

We assume that the scalar potential φ is chosen to be equal to zero, and that the gauge condition (see (10.5) of Part I) of the vector potential $\mathbf{A}$ is of the form $\nabla \cdot \mathbf{A} = 0$. From this relation it follows that the momentum operator $\hat{\mathbf{p}}$ commutes with the vector $\mathbf{A}$, and hence the Hamiltonian $\hat{H}$ can be rewritten as

$$\hat{H} = \frac{\hat{\mathbf{p}}^2}{2m} - \frac{e}{mc} \ (\hat{\mathbf{p}} \cdot \hat{\mathbf{A}}) + \frac{e^2}{2mc^2} \ \hat{\mathbf{A}}^2 + \hat{H}_{rad} \ . \tag{102.1}$$

The first term of (102.1) represents the Hamiltonian of the free particle, and the last term the Hamiltonian of the free radiation field. The Hamiltonian of the interaction of the electron with the radiation field, responsible for all processes of emission and absorption of photons by the electron, is of the form

$$\hat{H}' = - \frac{e}{mc} \ (\hat{\mathbf{p}} \cdot \hat{\mathbf{A}}) + \frac{e^2}{2mc^2} \ \hat{\mathbf{A}}^2 \ . \tag{102.2}$$

We shall formally assume the electron charge to be the small parameter in terms of which the perturbation theory expansion is carried out. In what follows we shall in fact see that the expansion is carried out in powers of the small quantity $e^2/\hbar c = \frac{1}{137}$ which figures in the corresponding matrix elements and is called the interaction constant. We shall confine ourselves to the consideration of some simple processes in the first non-vanishing approximation of perturbation theory. We have, in §56, obtained the general expressions for the probabilities of different processes, and our problem reduces to the calculation of the matrix elements of the interaction operator $\hat{H}'$ considered as the perturbation operator. The expansion of the vector potential is conveniently written in the form (38.19) of Part I:

$$\mathbf{A} = \sum_\lambda \ (b_\lambda \mathbf{A}_\lambda + b_\lambda^* \mathbf{A}_\lambda^*) \ ,$$

where

$$A_\lambda = e_\lambda (4\pi c^2/V)^{\frac{1}{2}} e^{ik_\lambda \cdot r} . \tag{102.3}$$

We pass to the quantum operators

$$\hat{A} = \sum_\lambda (\hat{b}_\lambda A_\lambda + \hat{b}_\lambda^\dagger A_\lambda^*) . \tag{102.3'}$$

Making use of relations (38.20) of Part I, we express the operators $\hat{b}_\lambda$ and $\hat{b}_\lambda^\dagger$ in terms of the operators $\hat{q}_\lambda$ and $\hat{p}_\lambda$

$$\hat{b}_\lambda = \frac{1}{2\omega_\lambda} (\omega_\lambda \hat{q}_\lambda + i\hat{p}_\lambda) , \qquad \hat{b}_\lambda^\dagger = \frac{1}{2\omega_\lambda} (\omega_\lambda \hat{q}_\lambda - i\hat{p}_\lambda) .$$

Using formulae (101.2) we introduce the operators $\hat{a}_\lambda$ and $\hat{a}_\lambda^\dagger$. We then obtain

$$\hat{b}_\lambda = (\hbar/2\omega_\lambda)^{\frac{1}{2}} \hat{a}_\lambda , \qquad \hat{b}_\lambda^\dagger = (\hbar/2\omega_\lambda)^{\frac{1}{2}} \hat{a}_\lambda^\dagger . \tag{102.4}$$

Comparing with (99.6) and (99.7) we find that the operators $\hat{b}_\lambda$ and $\hat{b}_\lambda^\dagger$ have the following matrix elements different from zero:

$$\langle n_1,...n_\lambda,...|\hat{b}_\lambda|n_1,...n_\lambda+1,...\rangle = [\hbar(n_\lambda+1)/2\omega_\lambda]^{\frac{1}{2}} ,$$
$$\langle n_1,...n_\lambda,...|\hat{b}_\lambda^\dagger|n_1,...n_\lambda-1,...\rangle = [\hbar n_\lambda/2\omega_\lambda]^{\frac{1}{2}} . \tag{102.5}$$

Thus the matrix elements of the vector potential differ from zero only for the processes of emission and absorption of one photon. For the operator $\hat{A}^2$ involved in (102.2) we have

$$\hat{A}^2 = \sum_{\lambda,\lambda'} [\hat{b}_\lambda \hat{b}_{\lambda'}(A_\lambda A_{\lambda'}) + \hat{b}_\lambda \hat{b}_{\lambda'}^\dagger(A_\lambda A_{\lambda'}^*) + \hat{b}_\lambda^\dagger \hat{b}_{\lambda'}(A_\lambda^* A_{\lambda'}) + \hat{b}_\lambda^\dagger \hat{b}_{\lambda'}^\dagger(A_\lambda^* A_{\lambda'}^*)] . \tag{102.6}$$

From this expression it is seen that the matrix elements of the operator $\hat{A}^2$ differ from zero for two-photon transitions, i.e. for the emission or absorption of two photons or the emission of one photon and the absorption of another.

The term containing the operator $\hat{A}^2$ as well as the term with operator $-(e/mc)(\hat{p}\cdot\hat{A})$ gives a contribution to processes involving two photons, the latter term being taken into account in the second approximation of perturbation theory.

The vector potential (102.3) describes the state of a photon with given momentum. One can also introduce the concept of the state of a photon with given angular momentum. In order to find the expression for the vector potential describing the state of a photon with angular momentum and its component along the z-axis we should carry out the expansion of the vector potential $\mathbf{A}$ not in terms of plane waves, but in terms of spherical waves. The amplitudes of the expansion must be considered as operators in the space of occupation numbers satisfying commutation relations of the same type as (101.3). In a state with given momentum the angular momentum of the photon does not have a definite value. This corresponds to the fact that the plane wave can be written in the form of an expansion in terms of an infinite sequence of spherical waves.

The photon possesses definite 'internal' degrees of freedom, since in describing its state it is necessary to take into account different possible polarizations.

The 'internal' state of a system is usually associated with its spin. However, the definition of the spin of a system as its 'intrinsic' angular momentum, i.e. the angular momentum at rest, is inapplicable to the photon. The photon in any reference frame moves with velocity c.

Nevertheless, it sometimes appears to be convenient to introduce the concept of spin also for the photon, writing the total angular momentum operator in the form of a superposition of the orbital angular momentum operator and the spin operator. In this case it turns out that the spin of the photon must be considered to be equal to one. In correspondence with three possible spin components $s_z = 0, \pm 1$, one would think that the photon may be in three different states with different polarization. However, the condition of the transverse nature of electromagnetic waves leads to the fact that actually only two spin components are possible, which correspond to the two independent polarization states of the photon. The reader may find a detailed consideration of the problems which are touched upon here in the monograph of Akhiezer and Berestetskii*.

§103. The absorption and emission of light

Let us consider the probability of a one-photon transition; i.e. the process of absorption and emission. We shall first of all write down the matrix elements corresponding to the absorption and emission of a photon of frequency

---

* A.I.Akhiezer and V.B.Berestetskii, *Quantum electrodynamics* (Interscience Publishers, New York, 1965).

ω_λ. Suppose the electron was in the initial state ψ_1 before absorption and in the state ψ_2 after absorption. The transition $1 \to 2$ proceeds with the absorption, and the transition $2 \to 1$ with the emission of a photon of frequency ω_λ. The matrix element of the perturbation operator (102.2) for the transition with the absorption of a photon is of the form

$$\langle 2, n_\lambda - 1 | \hat{H}' | 1, n_\lambda \rangle =$$

$$= -\frac{e}{mc} \int \psi_2^* (\hat{p} \cdot e_\lambda) \left(\frac{4\pi c^2}{V} \right)^{\frac{1}{2}} e^{ik_\lambda \cdot r} (b_\lambda)_{n_\lambda - 1, n_\lambda} \psi_1 \, dV =$$

$$= -\frac{e}{m} \left(\frac{2\pi \hbar n_\lambda}{V \omega_\lambda} \right)^{\frac{1}{2}} \int \psi_2^* (\hat{p} \cdot e_\lambda) e^{ik_\lambda \cdot r} \psi_1 \, dV . \tag{103.1}$$

Analogously for the process of emission of a photon we have

$$\langle 1, n_\lambda + 1 | \hat{H}' | 2, n_\lambda \rangle = -\frac{e}{m} \left(\frac{2\pi \hbar (n_\lambda + 1)}{V \omega_\lambda} \right)^{\frac{1}{2}} \int \psi_1^* (\hat{p} \cdot e_\lambda) e^{-ik_\lambda \cdot r} \psi_2 \, dV . \tag{103.2}$$

The probability per unit time of the transition with the absorption of a photon is given by the formula (see §56)

$$dW = \frac{2\pi}{\hbar} |\langle 2, n_\lambda - 1 | \hat{H}' | 1, n_\lambda \rangle|^2 \rho(\omega) \, d\Omega . \tag{103.3}$$

Here $d\Omega$ is the solid angle element corresponding to the direction of propagation of the photon before absorption. We shall assume that states 1 and 2 of the electron belong to a discrete spectrum. In this case the final state of the system with energy E_2 belongs to a discrete spectrum, while the initial state with energy $E_1 + \hbar\omega$ belongs to a continuous spectrum (since the frequency ω changes in a continuous way). Then the photon absorbed may belong to any of the oscillators in the interval of states $d\omega d\Omega$ in volume V. The number of such oscillators for given polarization per unit volume is given by formula (38.23) of Part I. Passing to a continuous distribution of frequencies we shall omit the index λ where this cannot lead to misunderstanding, or replace it by the index **k**.

By $\rho(\omega)$ in expression (103.3) is meant the number of oscillators in volume V corresponding to unit energy and angular intervals for a given polarization:

$$\rho(\omega) = \frac{\omega^2 V}{(2\pi c)^3 \hbar} . \tag{103.4}$$

For the transition probability per unit time taking into account (103.1) we obtain

$$dW = \frac{e^2\omega}{m^2 2\pi\hbar c^3} |((\hat{p}\cdot e)e^{ik\cdot r})_{21}|^2 n_k \, d\Omega \, .$$
(103.5)

The absorption probability is equal to zero for all energies except those which satisfy the conservation law

$$E_2 = E_1 + \hbar\omega \, .$$
(103.6)

Let us determine the intensity $J_0(\omega)$ of the incident radiation corresponding to the frequency interval $d\omega$ and angular interval $d\Omega$. Since to one oscillator there correspond n_k photons with given polarization, we have

$$J_0(\omega) \, d\omega \, d\Omega = n_k \hbar\omega c\rho\hbar \, d\omega \, d\Omega = n_k \hbar \frac{\omega^3 d\omega \, d\Omega}{(2\pi)^3 c^2} \, .$$

The total probability is proportional to the intensity of the incident radiation. The probability of emission of a photon by the electron is easily calculated in a completely analogous way.

The probability per unit time of the transition with the emission of a photon with momentum $\hbar k$ and polarization e is given by a formula of the type of (103.3)

$$dW = \frac{e^2\omega}{m^2 2\pi\hbar c^3} |((\hat{p}\cdot e)e^{-ik\cdot r})_{12}|^2 (n_k + 1) \, d\Omega \, .$$
(103.7)

The emission probability is different from zero if the frequency of the emitted quantum is equal to

$$\hbar\omega = E_2 - E_1 \, .$$
(103.8)

We see further that the probability of the transition $2 \rightarrow 1$ with the emission of a photon, given by formula (103.7), consists of two terms. One of these is proportional to the intensity of radiation (to the number of photons n_k) existing before the emission. The initially existing electromagnetic field acts on the electron, favouring its transition into a new state with the emission of an additional photon. This is called stimulated emission. The existence of stimulated emission was first pointed out by Einstein before the creation of the modern quantum theory of radiation. The second term of formula (103.7) does not depend on the intensity of the initial radiation and also ensures the possibility of emission in the case where before the emission the electromagnetic field was not excited (the number of photons $n_k = 0$). Emission of such a type is called spontaneous emission.

From the comparison of formulae (103.5) and (103.7), taking into account the hermitian property of the matrix elements, it follows that for the ratio of the probabilities of emission and absorption of a photon one can write

$$\frac{dW_{emiss}}{dW_{abs}} = \frac{n_k + 1}{n_k} . \tag{103.9}$$

We shall see in §12 of Part VI (Volume 4) that it is easy to obtain from (103.9) the Planck formula for the intensity distribution in black-body radiation.

We shall now show that only electrons in bound states can absorb and emit photons. For this we calculate the integrals involved in the matrix elements for the transition probabilities, assuming the electron to be free. The wave functions ψ_1 and ψ_2 are written in the form of plane waves

$$\psi_1 = C e^{(i/\hbar)(\mathbf{p}_1 \cdot \mathbf{r})} , \qquad \psi_2 = C e^{(i/\hbar)(\mathbf{p}_2 \cdot \mathbf{r})} ,$$

where C is the normalization constant. Substituting these wave functions into (103.2), we easily find

$$\int \psi_1^* \left(\frac{\hbar}{i} \nabla \cdot \mathbf{e} \right) e^{-i\mathbf{k} \cdot \mathbf{r}} \psi_2 \, dV =$$
$$= |C|^2 \int e^{(-i/\hbar)(\mathbf{p}_1 \cdot \mathbf{r})} \left(\frac{\hbar}{i} \mathbf{e} \cdot \nabla \right) e^{(i/\hbar)(\mathbf{p}_2 - \hbar\mathbf{k}) \cdot \mathbf{r}} \, dV \sim \delta(\mathbf{p}_2 - \hbar\mathbf{k} - \mathbf{p}_1) . \tag{103.10}$$

Formula (103.10) expresses the momentum conservation law in the interaction of a photon with a free electron. Furthermore, the energy conservation law holds in the transition. Thus the following equalities must be fulfilled simultaneously

$$\mathbf{p}_2 = \mathbf{p}_1 + \hbar\mathbf{k} , \tag{103.11}$$

$$E_2 = E_1 + \hbar\omega . \tag{103.12}$$

It is easily seen that eqs. (103.11) and (103.12) are inconsistent. An analogous conclusion, of course, also applies to the case of absorption.

For the laws of conservation of energy and momentum to hold simultaneously it is necessary that a third body, to which the excess momentum is transferred, be involved. In the case of atomic electrons such a body can be the nucleus of the atom.

§104. Dipole transitions in atomic systems

The matrix element for the process of emission of a photon (103.2) can in most cases be substantially simplified. Usually the wavelength of the photon emitted is considerably larger than the linear size of the region of space in which the wave functions of the electron ψ_1 and ψ_2 are considerably different from zero.

For example, let the electron move in an atom whose effective radius is equal to a. Then the wave functions of the initial and final states are very small outside the range a. The energy of the electron in the field of the nucleus with effective charge Z^* is in order of magnitude equal to $Z^* e^2/a$. The change ΔE in the energy of the atom in the transition and, consequently, the energy of the emitted photon is of the same order of magnitude. Then the length of the emitted wave is $\lambda \approx c/\omega \approx \hbar c/\hbar\omega \approx \hbar c a/Z^* e^2$. The ratio of the atomic size to the wavelength is of the order of

$$\frac{a}{\lambda} \approx \frac{Z^* e^2}{hc} \approx \frac{Z^*}{137} .$$

For the outer electrons $Z^* \approx 1$ and the wavelength is substantially larger than the atomic size. In the case of X-radiation arising in transitions in the K-shell of heavy atoms this approximation turns out to be inadequate. For $\lambda \gg a$ the index of the exponential function inside the integral in (103.2) is very small within the limits of the effective range of integration, and hence the factor $e^{-i\mathbf{k} \cdot \mathbf{r}}$ can be replaced by unity.

The probability of the transition with emission, (103.7), is then rewritten in the form

$$dW = \frac{e^2 \omega}{m^2 2\pi\hbar c^3} |(p_e)_{12}|^2 (n_\mathbf{k}+1) d\Omega . \tag{104.1}$$

Here $\hat{p}_e$ is the operator of the component of the momentum of the particle along the direction of polarization of the emitted quantum.

The matrix element of the momentum operator can be expressed in terms of the matrix element of the coordinate. According to (31.7) and (49.5) we have

$$\mathbf{p}_{12} = m\mathbf{v}_{12} = m\dot{\mathbf{r}}_{12} = \frac{im}{\hbar} (E_1 - E_2) \mathbf{r}_{12} = -\frac{im}{e} \omega \mathbf{d}_{12} , \tag{104.2}$$

where $\mathbf{d}$ is the dipole moment of the particle. Substituting (104.2) into (104.1) we obtain

$$dW = \frac{\omega^3}{2\pi\hbar c^3} |(d_e)_{12}|^2 (n_k+1) \, d\Omega .$$ (104.3)

Here d_e is the component of the dipole moment vector of the particle along the direction of polarization. We see that the transition probability (104.3) depends on the matrix element of the dipole moment of the particle and hence such transitions are called dipole transitions and the radiation is called dipole radiation. If the angle between $(\mathbf{d})_{12}$ and the direction of polarization of the radiation is denoted by θ, then expression (104.3) can be rewritten as

$$dW = \frac{\omega^3}{2\pi\hbar c^3} |\mathbf{d}_{12}|^2 (n_k+1) \cos^2 \theta \, d\Omega .$$ (104.4)

We sum this expression over the polarizations of the quantum. As independent directions of polarization we choose the polarization in the plane $(\mathbf{d},\mathbf{k})$ and the polarization in the direction perpendicular to this plane. Expression (104.4) is then brought into the form

$$dW = \frac{\omega^3}{2\pi\hbar c^3} |\mathbf{d}_{12}|^2 (n_k+1) \sin^2 \vartheta \, d\Omega ,$$ (104.5)

where ϑ is the angle between the vector $\mathbf{d}_{12}$ and the direction of propagation of the radiation $\mathbf{k}$.

The intensity of emission per unit time into the element of solid angle $d\Omega$ is obtained by multiplying (104.5) by the energy of the photon $\hbar\omega$. For spontaneous emission we have

$$J \, d\Omega = \frac{\omega^4}{2\pi c^3} |\mathbf{d}_{12}|^2 \sin^2 \vartheta \, d\Omega .$$ (104.6)

Integrating over angles we find the total spontaneous emission per unit time

$$\frac{dE}{dt} = \frac{4\omega^4}{3c^3} |\mathbf{d}_{12}|^2 .$$ (104.7)

This expression is very similar to the classical formula for the intensity of dipole radiation (see (27.9) of Part I). The difference between the classical and quantum formulae lies only in the fact that the averaged square of the dipole moment $\overline{\mathbf{d}^2}$ involved in the classical expression must be replaced by the corresponding matrix element (doubled) $2|\mathbf{d}_{12}|^2$.

Dipole transitions in the absorption of light can be considered in an analogous way. Setting $e^{i\mathbf{k}\cdot\mathbf{r}} = 1$ in (103.1) and taking into account (104.2), we obtain for the transition probability per unit time

$$dW = \frac{\omega^3}{2\pi\hbar c^3} |d_{21}|^2 n_k \cos^2\theta \, d\Omega \, . \tag{104.8}$$

Averaging this expression over all orientations of the vector $\mathbf{d}$ with respect to the direction of incident radiation, we find

$$\overline{\cos^2\theta} = \frac{1}{4\pi} \int \cos^2\theta \, d\Omega = \tfrac{1}{3} \, . \tag{104.9}$$

Expressing n_k in terms of the intensity of incident radiation $J_0(\omega)$ and multiplying (104.8) by $\hbar\omega$, we find the energy absorbed per unit time

$$J \, d\Omega = \frac{4\pi^2}{3} \frac{e^2}{\hbar c} \omega |r_{21}|^2 J_0(\omega) d\Omega \, . \tag{104.10}$$

So far we have considered the absorption and emission of a photon by one electron. If the absorbing or emitting system contains several electrons, then, disregarding the interaction between them, it can be assumed that formulae (104.10) and (104.5) will remain valid, provided that the dipole moment of the electron in them is replaced by the sum of the dipole moments of all the electrons.

§105. Quadrupole and magnetic dipole radiation

The matrix elements of a dipole transition are obtained from the general expression (103.2) when the exponential function $e^{-i\mathbf{k}\cdot\mathbf{r}}$ is replaced by unity. It may turn out, however, that the matrix element of the dipole transition reduces to zero, whereas the precise matrix element (103.2) differs from zero. In this case one has to expand the exponential $e^{-i\mathbf{k}\cdot\mathbf{r}}$ in a series, writing out the higher terms of the expansion. Then the emission probability will be different from zero, although substantially lower than the probability of dipole radiation. For this reason such transitions are called forbidden. The emission probability determined by the following terms of the expansion will have the form

$$dW = \frac{e^2\omega^3}{2\pi\hbar c^3} |(r_e(\mathbf{k}\cdot\mathbf{r}))_{21}|^2 (n_k+1) \, d\Omega \, . \tag{105.1}$$

The intensity of spontaneous emission in such a transition, analogously to (104.6), will be equal to

$$J \, d\Omega = \frac{e^2 \omega^4}{2\pi c^3} \, |(\mathbf{r}(\mathbf{k \cdot r}))_{21}|^2 \sin^2 \vartheta \, d\Omega \; . \tag{105.2}$$

Comparing this formula with the classical expressions (see Vol. 1, §31 of Part I), we see that (105.2) represents the magnetic dipole and quadrupole radiation. The probability of the forbidden radiation (magnetic dipole and quadrupole) is related to the probability of allowed dipole radiation as a^2/λ^2 $(k \approx \lambda^{-1}, r \approx a)$. If for some reason or other the matrix elements (105.1) are equal to zero, then the probability of radiation of higher order can be found in an analogous way.

§106. Selection rules

We see that the character of radiation from atomic and nuclear systems is determined by the matrix element $\mathbf{d}_{21} = e \mathbf{r}_{21}$. Let us now establish when this matrix element can be different from zero, i.e. between which states of the system transitions accompanied by dipole radiation are possible. The set of requirements which must be satisfied by the wave functions of the initial and final states of the system in order that the matrix element of the dipole transition $\mathbf{r}_{21}$ may not reduce to zero are called the selection rules for dipole radiation. The selection rules can easily be formulated in the general form if the wave functions ψ_1 and ψ_2 describe the state of a particle moving in a centrally symmetric field. In this case the dependence of ψ_1 and ψ_2 on angles is characterized by spherical functions (see §35). For dipole transitions to be possible in the system the matrix element of the projection of the radius vector on the direction of polarization of the quantum e must be different from zero. Let us first consider a quantum polarized along the z-axis. In this case $r_e = z = r \cos \vartheta$. The matrix element of the dipole transition will be proportional to the integral

$$\int_0^\pi \int_0^{2\pi} Y^*_{l_2 m_2} \cos \vartheta \, Y_{l_1 m_1} \sin \vartheta \, d\vartheta \, d\varphi \; . \tag{106.1}$$

Here l_1, m_1 and l_2, m_2 are the quantum numbers of the states of the system before and after the emission of the quantum. Taking into account the definition of spherical functions (30.16), the integral (106.1) can be rewritten in the form

$$\int_0^\pi P_{l_2}^{m_2}(\cos\vartheta)P_{l_1}^{m_1}(\cos\vartheta)\cos\vartheta\,\sin\vartheta\,d\vartheta\int_0^{2\pi}e^{i(m_1-m_2)\varphi}\,d\varphi\ . \tag{106.2}$$

The integral over the angle φ is different from zero only for $m_1 = m_2$. The integral over the angle ϑ then has the form

$$\int_{-1}^1 P_{l_2}^m(x)\,xP_{l_1}^m(x)\,dx\ . \tag{106.3}$$

It can be shown that the following relation is valid for associated Legendre polynomials*:

$$xP_l^m(x)=\frac{l+|m|}{2l+1}P_{l-1}^m(x)+\frac{l-|m|+1}{2l+1}P_{l+1}^m(x)\ . \tag{106.4}$$

Substituting this expression into (106.3) and taking into account the conditions of orthogonality of associated Legendre polynomials, we find that integral (106.3) is different from zero only for $l_2 = l_1 \pm 1$.

Thus we see that if the radiation is polarized along the z-axis, then the matrix element of the dipole transition differs from zero only for transitions with $m_2 = m_1, l_2 = l_1 \pm 1$.

Let us now define analogous selection rules for the quantum numbers l, m in the case where the quantum is emitted in the direction of the z-axis and, consequently, is polarized in the (x,y)-plane. We consider the case of circular polarization with a phase shift equal to $\frac{1}{2}\pi$. Then the transition probability is determined by the matrix element of the quantity $x \pm iy$

$$(x\pm iy)_{21} = (r\sin\vartheta\,e^{\pm i\varphi})_{21}\ . \tag{106.5}$$

Separating the integral over the angle φ, we obtain

$$\int_0^{2\pi}e^{i(m_1-m_2\pm 1)\varphi}\,d\varphi\ . \tag{106.6}$$

This integral is different from zero under the condition

$$m_2 = m_1 \pm 1\ . \tag{106.7}$$

The corresponding integral over the angle ϑ is different from zero if $l_2 =$

---

* See, for example, N.N.Lebedev, *Special functions and their applications* (Prentice Hall, Englewood Cliffs, N.Y., 1965).

$l_1 \pm 1$. Thus the selection rules obtained for the quantum numbers l and m for dipole transitions can finally be formulated in the form

$$\Delta m = 0, \pm 1 \; ; \qquad \Delta l = \pm 1 \; . \tag{106.8}$$

It can easily be seen that the selection rules given by relations (106.8) express the angular momentum conservation law. The fact that l may change by one shows that in a dipole transition the emitted quantum carries away an angular momentum equal to one. At first sight this conclusion may seem to be strange. As a matter of fact, we have considered (formula (103.7)) transitions with the emission of a photon of given momentum. But in a state with given momentum and polarization the angular momentum of the photon does not have a sharp value. If, however, the wavelength of the photon is large in comparison with the size of the system, then it is possible to expand the function $e^{i\mathbf{k}\cdot\mathbf{r}}$ in a series. Carrying out this expansion and retaining the first non-vanishing term, i.e. the dominant term determining the value of the matrix element, we in fact separate photons with given total angular momentum. To dipole radiation there correspond photons with angular momentum one, to quadrupole radiation photons with angular momentum two and so on. Direct calculation, on which we cannot dwell, confirms this conclusion.

Selection rules (106.8) automatically satisfy the requirements of the parity conservation law. Since the operator $\mathbf{r}$ is odd, the functions ψ_2 and ψ_1 must have different parity. Then the entire matrix element remains invariant under the replacement $\mathbf{r} \rightarrow (-\mathbf{r})$.

In deriving relations (106.1) the spin states of the electron were not taken into account, i.e. it was assumed that the spin state is not related to the orbital motion. In this case conditions (106.8) must be supplemented by the relation $\Delta s = 0$ which expresses spin conservation in the dipole transition. However, if the spin–orbit interaction cannot be disregarded, as, for example, in the case of heavy atoms and nuclei, then it is necessary to formulate selection rules for the total angular momentum J. Taking into account that in a dipole transition the quantum carries away an angular momentum equal to one, then, according to the rule of addition of angular momenta in quantum mechanics we obtain

$$\Delta j = 0, \pm 1 \quad (\text{excluding } 0 \rightarrow 0 \text{ transitions}) \; . \tag{106.9}$$

In this case transitions with $\Delta j = 0$ are not forbidden, since the total angular momentum is not directly related to the parity of the state. The transition from the state $j_1 = 0$ into the state $j_2 = 0$ is forbidden, since in this case the total angular momentum conservation law cannot be satisfied.

In the case of magnetic dipole radiation the quantum also carried away an angular momentum equal to one. However, the magnetic dipole quantum has a parity opposite to the parity of the electric dipole quantum. This is associated with the fact that the magnetic moment operator does not change sign under the inversion of the system of coordinates, since the magnetic moment is a pseudovector. Consequently, the matrix elements of the magnetic moment operator are different from zero only for transitions between states of the same parity.

The electric quadrupole quantum carries away an angular momentum equal to two. In correspondence with this the total angular momentum selection rules are of the form

$$\Delta j = 0, \pm 1, \pm 2 . \tag{106.10}$$

Transitions with the angular momenta

$$0 \to 0 ; \qquad \tfrac{1}{2} \to \tfrac{1}{2} ; \qquad 0 \rightleftharpoons 1$$

are forbidden. The change in angular momentum in the emission given by relations (106.9) and (106.10) refers either to one particle, if only its state is changed, or to the system as a whole, for example to an atom or nucleus.

If the system is in a certain excited state and dipole transition to a lower energy state is forbidden, then the lifetime of the system in this excited state can be rather large. States of such a type are said to be metastable. In gases which are not very rarefied a metastable atom usually transfers its excitation energy in collisions with other atoms without emission.

Transitions associated with the angular momentum change $\Delta j \approx 4, 5$, which are strongly forbidden, are observed in nuclei. The lifetime of the nucleus with respect to such a transition for small excitation energies may reach several months. Such nuclei are said to be isomeric. They were first observed by Kurchatov and Rusinov.

§ 107. The photoelectric effect

The process of absorption of a photon by a bound particle when the energy of the photon exceeds the binding energy of the particle is called the photoelectric effect. In particular, in the photoelectric effect in an atom an electron in a state belonging to a discrete spectrum absorbs the photon and makes a transition to the continuous spectrum. The kinetic energy T of the electron knocked out of the atom is defined by the Einstein relation

$$T = \hbar\omega - I , \tag{107.1}$$

where I is the ionization energy of the atom.

The momentum excess arising when the photon is absorbed is transferred to the nucleus. The more strongly the electron is bound in the atom, the more easily the momentum is transformed to the nucleus. Hence it is to be expected that the probability of occurrence of the photoelectric effect will have a maximum value for the most strongly bound electrons, the electrons of the K-shell.

In what follows we shall restrict ourselves to the consideration of this case. The matrix element of the transition with the absorption of one quantum has the form (103.1). The wave functions ψ_1 and ψ_2 in the matrix element correspond respectively to the ground state of the electron in the atom and to a state belonging to the continuous spectrum. Since we do not take into account relativistic effects, it is obvious that the energy of the photon must in any case be small in comparison with the rest energy of the electron $\hbar\omega \ll mc^2$.

On the other hand, we exclude the region close to the threshold of the photoelectric effect, and assume that the energy of the photon is large in comparison with the ionization energy of the atom. Taking into account (107.1) and (38.17), these requirements lead to the inequality

$$T = \frac{p^2}{2m} \gg I = \frac{Z^2 e^4 m}{2\hbar^2} \quad \text{or} \quad \frac{Ze^2}{\hbar v} \ll 1 . \tag{107.2}$$

According to the results of §84, the fulfillment of inequality (107.2) means that the Coulomb field acting on the electron can be considered as a small perturbation. Consequently, one can take a plane wave for the wave function ψ_2 of the free particle in the zero order approximation (disregarding the action of the Coulomb field on the free electron). The wave function of an electron in the K-shell can be written in the form of the hydrogen function with the effective nuclear charge Z (the action of other electrons on K-electrons is small). We then have

$$\psi_1 = \left(\frac{Z^3}{\pi a^3}\right)^{\frac{1}{2}} e^{-Zr/a} ; \qquad \psi_2 = \frac{1}{(2\pi\hbar)^{\frac{3}{2}}} e^{(i/\hbar)(\mathbf{p}\cdot\mathbf{r})} . \tag{107.3}$$

We normalize the wave function of the final state ψ_2 belonging to the continuous spectrum to the δ-function in momentum space. The transition probability per unit time, according to (56.8′), is equal to

$$dW = \frac{2\pi}{\hbar} |H'_{21}|^2 \delta(E_2 - E_1 - \hbar\omega) p^2 \, dp \, d\Omega . \tag{107.4}$$

Here $d\Omega$ is the element of solid angle characterizing the direction of the momentum $\mathbf{p}$ of the emitted photoelectron, $E_1 = -I$, and $E_2 = p^2/2m$.

We integrate (107.4) over the energies of the final state

$$dW = \frac{2\pi m}{\hbar} |H'_{21}|^2 p \, d\Omega .$$

(107.5)

The value of the momentum p is determined by relation (107.1). The integral in the matrix element (107.5) is easily calculated after the substitution of the wave functions given above. For this we write it in the form

$$H'_{21} = -\frac{e}{m(2\pi\hbar)} \left(\frac{Z^3}{\pi a^3 V \omega}\right)^{\frac{1}{2}} \int e^{(i/\hbar)(\mathbf{q}\cdot\mathbf{r})} \frac{\hbar}{i} (\mathbf{e}\cdot\nabla) \, e^{-Zr/a} \, dV ,$$

where $\mathbf{e}$ is the polarization vector of the photon.

Here we have introduced the vector $\mathbf{q} = \hbar\mathbf{k} - \mathbf{p}$ which represents the momentum transferred to the nucleus, and used the transverse property of electromagnetic waves $\mathbf{e}\cdot\mathbf{k} = 0$. Integrating by parts we obtain

$$H'_{21} = -\frac{e}{m(2\pi\hbar)} \left(\frac{Z^3}{\pi a^3 V \omega}\right)^{\frac{1}{2}} (\mathbf{e}\cdot\mathbf{p}) \int e^{-Zr/a} \, e^{(i/\hbar)(\mathbf{q}\cdot\mathbf{r})} \, dV .$$

Passing to spherical coordinates with the polar axis directed along the vector $\mathbf{q}$, we find

$$\int e^{-Zr/a} \, e^{(i/\hbar)(\mathbf{q}\cdot\mathbf{r})} \, dV = \frac{2\pi\hbar}{iq} \int_0^\infty (e^{(i/\hbar)(\mathbf{q}\cdot\mathbf{r})} - e^{-(i/\hbar)(\mathbf{q}\cdot\mathbf{r})}) e^{-Zr/a} r \, dr =$$

$$= \frac{8\pi a^3}{Z^3 (1 + q^2 a^2/Z^2 \hbar^2)^2} .$$

Consequently, the matrix element H'_{21} is of the form

$$H'_{21} = -\frac{4e}{m\hbar} \left(\frac{a^3}{\pi Z^3 V \omega}\right)^{\frac{1}{2}} (\mathbf{e}\cdot\mathbf{p}) \frac{1}{(1 + q^2 a^2/\hbar^2 Z^2)^2} .$$

(107.6)

We obtain the differential cross section for the photoelectric effect if we divide the transition probability per unit time (107.5) by the incident photon flux density. Since one quantum is absorbed and the process is normalized in such a way that in volume V there is one photon, then the incident photon flux density is equal to c/V. In correspondence with this we have

$$d\sigma = \frac{32 \times 137^4}{Z^3} \frac{p(\mathbf{p}\cdot\mathbf{e})^2 c^3}{(mc^2)^2 \hbar\omega} \left(\frac{e^2}{mc^2}\right)^2 \frac{d\Omega}{(1 + q^2 a^2/\hbar^2 Z^2)^4} .$$

(107.7)

The constant $r_0 = e^2/mc^2 \sim 10^{-13}$ cm, as is well known, is called the classical radius of the electron. Expression (107.7) can be simplified. First of all, we denote the angle between the direction of momentum of the incident quantum and that of the emitted photoelectron by ϑ, i.e. the angle between the vectors $\mathbf{k}$ and $\mathbf{p}$, and the angle between the $(\mathbf{p},\mathbf{k})$-plane and $(\mathbf{e},\mathbf{k})$-plane by φ. Then, writing $\hbar k = \kappa$, we have

$$\mathbf{p} \cdot \mathbf{e} = p \sin \vartheta \cos \varphi ,$$

$$q^2 = p^2 + \kappa^2 - 2p\kappa \cos \vartheta . \tag{107.8}$$

The expression $1 + q^2 a^2/\hbar^2 Z^2$ involved in (107.7) can also be rewritten in a simpler form:

$$1 + \frac{q^2 a^2}{\hbar^2 Z^2} = \frac{a^2}{\hbar^2 Z^2} \left(\frac{Z^2 \hbar^2}{a^2} + q^2 \right) = \frac{a^2}{\hbar^2 Z^2} \left(\frac{Z^2 m^2 e^4}{\hbar^2} + p^2 + \kappa^2 - 2p\kappa \cos \vartheta \right) .$$

Taking into account that $I = Z^2 e^4 m/2\hbar^2$, it follows from relation (107.1) that

$$\frac{Z^2 m^2 e^4}{\hbar^2} + p^2 = 2m\hbar\omega = 2m\kappa c .$$

Then the preceding relation is rewritten as

$$1 + \frac{q^2 a^2}{\hbar^2 Z^2} = \frac{a^2}{\hbar^2 Z^2} \kappa (2mc + \kappa - 2p \cos \vartheta) = \frac{a^2}{\hbar^2 Z^2} 2m\hbar\omega(1 - \beta \cos \vartheta) , \tag{107.9}$$

where $\beta = v/c$.

Here we have made use of the condition $\kappa c = \hbar\omega \ll mc^2$. The absolute value of the momentum $\mathbf{p}$ involved in (107.5) can be replaced, according to (107.1) and (107.2), by the quantity $(2m\hbar\omega)^{\frac{1}{2}}$. Taking into account relations (107.9) and (107.8), we obtain for the differential cross section (107.7) the following expression:

$$d\sigma = 4\sqrt{2} \, \frac{Z^5}{137^4} \, r_0^2 \left(\frac{mc^2}{\hbar\omega} \right)^{\frac{7}{2}} \frac{\sin^2 \vartheta \cos^2 \varphi}{(1 - \beta \cos \vartheta)^4} \, d\Omega . \tag{107.10}$$

Since expression (107.10) is obtained in the non-relativistic approximation, it makes sense only to within terms of the first order with respect to β. Hence we have finally

$$d\sigma = 4\sqrt{2}\,\frac{Z^5}{137^4}\,r_0^2\left(\frac{mc^2}{\hbar\omega}\right)^{\frac{7}{2}}\sin^2\vartheta\cos^2\varphi\,(1+4\beta\cos\vartheta)\,d\Omega\;.\tag{107.11}$$

From expression (107.11) it follows that photoelectrons are emitted mainly in the direction of polarization of the photon $\vartheta = \frac{1}{2}\pi$, $\varphi = 0$. Photoelectrons are not emitted in the direction of propagation of the quantum ($\vartheta = 0$). As the energy of the quantum increases the maximum is considerably displaced in the forward direction. In order to obtain the total cross section for the photoelectric effect on the K-shell it is necessary to integrate (107.11) over all angles ϑ, φ and in addition to introduce the factor 2, since there are two electrons in the K-shell

$$\sigma = \frac{32\sqrt{2}}{3}\,\pi\,\frac{Z^5}{137^4}\,r_0^2\left(\frac{mc^2}{\hbar\omega}\right)^{\frac{7}{2}}\;.\tag{107.12}$$

We see that the total cross section increases rapidly with increasing charge of the nucleus (as Z^5) and decreases with increasing frequency of the quantum (as $\omega^{-\frac{7}{2}}$).

§ 108. The scattering of light by atoms

As an important example of a process involving two photons we shall consider the quantum theory of the scattering of light by atoms. Let a photon with wave vector $\mathbf{k}_1$ be incident on an atom, and let a photon with wave vector $\mathbf{k}_2$ be emitted. We denote the corresponding frequencies by ω_1 and ω_2, and the polarization vectors by $\mathbf{e}_1$ and $\mathbf{e}_2$. If the frequency of the incident photon is equal to the frequency of the emitted photon, i.e. $\omega_1 = \omega_2$, then after scattering the atom returns to its initial state. Such scattering without change in the frequency is called coherent. We have seen in Part I that from the point of view of classical radiation theory only coherent scattering is possible. From the quantum-mechanical point of view scattering with a change in the frequency is as natural as coherent scattering, and was observed experimentally by Raman and independently by Mandelshtam and Landsberg. It is called Raman scattering.

For generality we shall assume that the state of the atom changes in the act of scattering. We shall suppose that the energy of the incident photon is lower than the binding energy of the electron in the atom, which corresponds to the visible region of the spectrum. For photon energies large in comparison with this quantity the electron can be considered as free. However, we shall defer the consideration of the scattering of a photon by a free electron (the Compton effect) to § 127.

If in the initial state the energy of the atom is E_1 and in the final state E_2, then the energy conservation law gives

$$\hbar\omega_2 = \hbar\omega_1 + E_1 - E_2 . \tag{108.1}$$

Let us write down the matrix elements of the interaction operator $\hat{H}'$ (102.2) for the scattering process. Since the process involves two photons, it is necessary to take into account the contribution of the operator $\hat{A}^2$ to the matrix element. We denote the operator $-(e/mc)(\hat{p}\cdot\hat{A})$ by $\hat{H}'_1$, and the operator $(e^2/2mc^2)\hat{A}^2$ by $\hat{H}'_2$. Then the total perturbation operator is

$$\hat{H}' = \hat{H}'_1 + \hat{H}'_2 . \tag{108.2}$$

Taking into account (102.5), it follows from (102.4) that the operator $\hat{A}^2$ gives a contribution to the process studied in the first approximation of perturbation theory

$$(\hat{H}'_2)_{21} = \frac{2\pi e^2}{mV} \frac{\hbar}{(\omega_1\omega_2)^{\frac{1}{2}}} \int \psi_2^* e^{i(\mathbf{k}_1-\mathbf{k}_2)\cdot\mathbf{r}} (\mathbf{e}_1\cdot\mathbf{e}_2)\psi_1 \, dV . \tag{108.3}$$

For the frequencies considered the wavelength of the light is considerably larger than the size of the atom. In correspondence with this the exponential function in (108.3) can be set equal to unity. Also taking into account the orthogonality of the functions ψ_1 and ψ_2, we have

$$(\hat{H}'_2)_{21} = \frac{2\pi e^2}{mV} \frac{\hbar}{(\omega_1\omega_2)^{\frac{1}{2}}} (\mathbf{e}_1\cdot\mathbf{e}_2)\delta_{12} . \tag{108.4}$$

The operator $\hat{H}'_1$ has matrix elements different from zero only for processes involving one photon. In the case of scattering the operator $\hat{H}'_1$ can give a contribution to the transition probability only in the second order approximation of perturbation theory. In §56 it was shown that in order to define the probability of a process in the second order approximation it is necessary to find the matrix elements of the operator $\hat{H}'_1$ corresponding to transitions into intermediate states.

Two types of intermediate states, over which the summation should be carried out, are possible in the process of scattering. (1) In the transition from the initial to the intermediate state of the first type the photon k_1 is absorbed and the atom makes a transition into a certain state which we shall characterize by the index i (energy E_i). In the subsequent transition from the intermediate into the final state the photon k_2 is emitted. According to (103.1) and (103.2) the matrix elements for the transition from the initial into the intermediate state and from the intermediate into the final state are of the form

$$(\hat{H}'_1)_{\text{I}1} = -\frac{e}{m}\left(\frac{2\pi\hbar}{V\omega_1}\right)^{\frac{1}{2}}\int\psi_i^*(\hat{\mathbf{p}}\cdot\mathbf{e}_1)e^{i\mathbf{k}_1\cdot\mathbf{r}}\psi_1\,dV,$$

$$(\hat{H}'_1)_{2\text{I}} = -\frac{e}{m}\left(\frac{2\pi\hbar}{V\omega_2}\right)^{\frac{1}{2}}\int\psi_2^*(\hat{\mathbf{p}}\cdot\mathbf{e}_2)e^{-i\mathbf{k}_2\cdot\mathbf{r}}\psi_i\,dV. \tag{108.5}$$

(2) In the transition to the intermediate state of the second type the emission of the photon k_2 first occurs. Then the photon k_1 is absorbed and the atoms makes a transition from the intermediate into the final state. We recall that the energy conservation law holds only for the initial and final states. The matrix elements of the transition via the second intermediate state are written as

$$(\hat{H}'_1)_{\text{II}1} = -\frac{e}{m}\left(\frac{2\pi\hbar}{V\omega_2}\right)^{\frac{1}{2}}\int\psi_i^*(\hat{\mathbf{p}}\cdot\mathbf{e}_2)e^{-i\mathbf{k}_2\cdot\mathbf{r}}\psi_1\,dV,$$

$$(\hat{H}'_1)_{2\text{II}} = -\frac{e}{m}\left(\frac{2\pi\hbar}{V\omega_1}\right)^{\frac{1}{2}}\int\psi_2^*(\hat{\mathbf{p}}\cdot\mathbf{e}_1)e^{i\mathbf{k}_1\cdot\mathbf{r}}\psi_i\,dV. \tag{108.6}$$

The constitutive matrix element Λ is given by formula (56.19)

$$\Lambda_{21} = \sum_i\left(\frac{(\hat{H}'_1)_{21}(\hat{H}'_1)_{11}}{E_{\text{init}}-E_i} + \frac{(\hat{H}'_1)_{2\text{II}}(\hat{H}'_1)_{\text{II}1}}{E_{\text{init}}-(E_i+\hbar\omega_1+\hbar\omega_2)}\right). \tag{108.7}$$

The energy of the initial state is made up of the energy of the atom E_1 and the energy of the incident quantum $\hbar\omega_1$, i.e.

$$E_{\text{init}} = E_1 + \hbar\omega_1. \tag{108.8}$$

The energy of the intermediate state of the second type involves, in addition to the energy of the atom, the total energy of the two photons.

Substituting expressions (108.5) and (108.6) into (108.7) we obtain

$$\Lambda_{21} = \frac{e^2 2\pi\hbar}{m^2 V(\omega_1\omega_2)^{\frac{1}{2}}}\sum_i\left(\frac{(\mathbf{e}_2\cdot\hat{\mathbf{p}})_{2i}(\mathbf{e}_1\cdot\hat{\mathbf{p}})_{i1}}{E_1-E_i+\hbar\omega_1} + \frac{(\mathbf{e}_1\cdot\hat{\mathbf{p}})_{2i}(\mathbf{e}_2\cdot\hat{\mathbf{p}})_{i1}}{E_1-E_i-\hbar\omega_2}\right). \tag{108.9}$$

Here we have substituted unity for the exponential expressions $e^{i\mathbf{k}_1\cdot\mathbf{r}}$ and $e^{-i\mathbf{k}_2\cdot\mathbf{r}}$. The summation over the energy states of the atom must also involve integration over states belonging to the continuous spectrum. The total matrix element for the process considered is obtained by adding to (108.9) the matrix element (108.4)

$$M_{21} = \frac{2\pi e^2}{mV} \frac{n}{(\omega_1\omega_2)^{\frac{1}{2}}} \left[\frac{1}{m} \sum_i \left(\frac{(e_2 \cdot \hat{p})_{2i}(e_1 \cdot \hat{p})_{i1}}{E_1 - E_i + \hbar\omega_1} + \right. \right.$$

$$\left. \left. + \frac{(e_1 \cdot \hat{p})_{2i}(e_2 \cdot \hat{p})_{i1}}{E_1 - E_i - \hbar\omega_2} \right) + \delta_{12}(e_1 \cdot e_2) \right] . \tag{108.10}$$

The transition probability per unit time is given, as always, by the formula

$$dW = \frac{2\pi}{\hbar} |M_{21}|^2 \rho(\omega_2) d\Omega . \tag{108.11}$$

Here $\rho(\omega_2)$, as in §103, denotes the number of field oscillators in volume V corresponding to unit energy interval (see (103.4)). The solid angle element $d\Omega$ characterizes the direction of the momentum of the scattered photon.

Dividing the transition probability per unit time (108.11) by the incident photon flux density which, as in the preceding section, is equal to c/V, we obtain the expression for the differential cross section for the process

$$d\sigma = r_0^2 \frac{\omega_2}{\omega_1} \left| \frac{1}{m} \sum_i \frac{(e_2 \cdot \hat{p})_{2i}(e_1 \cdot \hat{p})_{i1}}{E_1 - E_i + \hbar\omega_1} + \frac{(e_1 \cdot \hat{p})_{2i}(e_2 \cdot \hat{p})_{i1}}{E_1 - E_i - \hbar\omega_2} + \delta_{12}(e_1 \cdot e_2) \right|^2 d\Omega . \tag{108.12}$$

It is clear that formula (108.12) is inapplicable in the case of perfect resonance $\hbar\omega_1 = E_1 - E_i$.

Formula (108.12) characterizes the scattering capacity of the atom as a function of the frequency of the incident light, and hence is called, as in classical electrodynamics, a dispersion formula. The last term of (108.12) is different from zero only for coherent scattering $\omega_1 = \omega_2$, for which the initial and final states of the atom are the same. If the initial and final states of the atom are not identical, then the frequency of the scattered radiation is shifted with respect to the frequency of the incident radiation by an amount corresponding to the difference between the energy states of the atom (108.1). Scattering of such a type is called the Raman effect.

Formula (108.12) can be rewritten in a somewhat different form if the matrix elements of the momentum are expressed in terms of the matrix elements of the coordinates of the electron. First of all, it is convenient to write the last term in the bracket in (108.12) in the same form as the two preceding terms.

For this we note* that, as a result of the commutation of the correspond-

---

* See the monograph of A.I.Akhiezer and V.B.Berestetskii, *Quantum electrodynamics* (Interscience Publishers, New York, 1965).

ing components of the operators $\hat{\mathbf{p}}$ and $\hat{\mathbf{r}}$, the scalar product $\mathbf{e}_1 \cdot \mathbf{e}_2$ can be expressed as

$$(\mathbf{e}_1 \cdot \mathbf{e}_2) = \frac{i}{\hbar} [(\mathbf{e}_1 \cdot \hat{\mathbf{p}})(\mathbf{e}_2 \cdot \mathbf{r}) - (\mathbf{e}_2 \cdot \mathbf{r})(\mathbf{e}_1 \cdot \hat{\mathbf{p}})] \ . \tag{108.13}$$

We shall take into account the fact that the scalar product $\mathbf{e}_1 \cdot \mathbf{e}_2$ is involved in (108.12) only for coherent scattering if we take from (108.13) the matrix element of the transition from the initial into the final state. This matrix element is, by virtue of the orthogonality of the corresponding wave functions, different from zero only under the condition that the initial state of the atom is the same as the final state

$$\delta_{21}(\mathbf{e}_1 \cdot \mathbf{e}_2) = \frac{i}{\hbar} [(\mathbf{e}_1 \cdot \hat{\mathbf{p}})(\mathbf{e}_2 \cdot \mathbf{r}) - (\mathbf{e}_2 \cdot \mathbf{r})(\mathbf{e}_1 \cdot \hat{\mathbf{p}})]_{21} =$$

$$= \frac{i}{\hbar} \sum_i ((\mathbf{e}_1 \cdot \hat{\mathbf{p}})_{2i}(\mathbf{e}_2 \cdot \mathbf{r})_{i1} - (\mathbf{e}_2 \cdot \mathbf{r})_{2i}(\mathbf{e}_1 \cdot \mathbf{p})_{i1}) \ . \tag{108.14}$$

We substitute the expression obtained into (108.12). The matrix elements of the momentum, according to (49.5), are expressed in terms of the matrix elements of the coordinate

$$(\mathbf{p})_{21} = \frac{i}{\hbar} m(E_2 - E_1)(\mathbf{r})_{21} \ . \tag{108.15}$$

Adding up in (108.12) the corresponding matrix elements in brackets and taking into account the energy conservation law (108.1), we obtain the following expression for the bracket involved in (108.12):

$$[\ldots] = \frac{m}{\hbar} \sum_i \left(\frac{\omega_2(E_i - E_1)(\mathbf{e}_2 \cdot \mathbf{r})_{2i}(\mathbf{e}_1 \cdot \mathbf{r})_{i1}}{E_1 - E_i + \hbar\omega_1} + \frac{\omega_2(E_2 - E_i)(\mathbf{e}_1 \cdot \mathbf{r})_{2i}(\mathbf{e}_2 \cdot \mathbf{r})_{i1}}{E_1 - E_i - \hbar\omega_2} \right) \ . \tag{108.16}$$

This expression can be simplified if the following term, equal to zero, is added to it:

$$\frac{m\omega_2}{\hbar} [(\mathbf{e}_1 \cdot \mathbf{r})(\mathbf{e}_2 \cdot \mathbf{r}) - (\mathbf{e}_2 \cdot \mathbf{r})(\mathbf{e}_1 \cdot \mathbf{r})]_{21} =$$

$$= \frac{m\omega_2}{\hbar} \sum_i ((\mathbf{e}_2 \cdot \mathbf{r})_{2i}(\mathbf{e}_1 \cdot \mathbf{r})_{i1} - (\mathbf{e}_1 \cdot \mathbf{r})_{2i}(\mathbf{e}_2 \cdot \mathbf{r})_{i1}) \ .$$

We then obtain

$$[...] = m\omega_1\omega_2 \sum_i \left(\frac{(e_2 \cdot r)_{2i}(e_1 \cdot r)_{i1}}{E_1 - E_i + \hbar\omega_1} + \frac{(e_1 \cdot r)_{2i}(e_2 \cdot r)_{i1}}{E_1 - E_i - \hbar\omega_2} \right) . \tag{108.17}$$

Substituting (108.17) into (108.12) we find the final expression for the differential scattering cross section

$$d\sigma = \frac{e^4}{c^4} \omega_1 \omega_2^3 \left| \sum_i \left(\frac{(e_2 \cdot r)_{2i}(e_1 \cdot r)_{i1}}{E_1 - E_i + \hbar\omega_1} + \frac{(e_1 \cdot r)_{2i}(e_2 \cdot r)_{i1}}{E_1 - E_i - \hbar\omega_2} \right) \right|^2 d\Omega . \tag{108.18}$$

A remarkable feature of the formulae obtained is the fact that for co-herent scattering they are the same as the classical formulae in §36 of Part I. Formula (108.18) is widely used in practice, since the study of Raman scatter-ing turned out to be a very effective method of investigating the energy levels and other properties of complex molecules.

We note also that the matrix elements of the atomic dipole moment induced by light can be expressed on the basis of the relations obtained. In turn, as was shown in §34 of Part IV, it is easy to find the relation between the dielectric constant and the induced dipole moment for a rarefied gas. Hence formula (108.18) is the basis for the quantum-mechanical calculation of the dielectric constant ϵ and correspondingly of the refractive index $n = \epsilon^{\frac{1}{2}}$. In particular, for coherent scattering the following expression is obtained for the quantity n^2:

$$n^2 = 1 + \frac{4\pi e^2 N}{m} \sum_i \frac{f_i}{\omega_{i1}^2 - \omega^2} , \tag{108.19}$$

where the quantity $f_i = 2m\hbar^{-1}\omega_{i1}|x_{i1}|^2$ is called the oscillator strength, $\hbar\omega_{i1} = E_i - E_1$, and N is the number of atoms per unit volume. We have chosen the direction of polarization of the photons as the x-axis. The following relation holds for the oscillator strengths f_i:[*]

$$\sum_i f_i = 1 .$$

As was pointed out earlier the expressions obtained make no sense near a

---

[*] See, for example, H. Bethe and E. Salpeter, *Quantum mechanics of one and two electron systems*, Handbuch der Physik, Bd. 35 (Springer-Verlag, Berlin, 1957).

resonance. Investigation of this phenomenon, called resonance fluorescence, is impossible without introducing the concept of linewidth, to which the next section is devoted. There the formulae of scattering theory taking into account the linewidth will be presented.

§109. The theory of the natural linewidth

In §103–105 we have found the total probability of emission of a photon by an atomic system. A more precise investigation of the corresponding equations makes it possible to find also the frequency distribution of the intensity of the radiation, i.e. to determine the form of the spectral line of the radiation. We recall that such a problem can also be solved within the framework of classical concepts (see Part I). In this case in order to obtain the natural form of a line it was necessary to take into account the damping of the amplitude of the radiation oscillator.

Although the interaction of charged particles with the electromagnetic field is weak, we cannot confine ourselves to the usual approximation of perturbation theory in considering the form of a spectral line. As a matter of fact, finding the form of a line is associated with the necessity of taking into account the decay of the initial state of the atomic system. Such a decay takes place over a rather long time t and, naturally, cannot be taken into account by the methods of perturbation theory (see §56).

We shall proceed from the general system of eqs. (55.5) for the amplitudes c_m of unperturbed states (atom and radiation field). We shall denote the amplitude of the initial state of the system by $\varphi'(t)$. In this state there are no photons, and the atom is in an excited state with energy E_2. In the system of eqs. (55.5) it is sufficient to take into account only those states whose energy is approximately the same as the energy of the initial state. Only these states play an important role. We shall assume for simplicity that the state with energy E_2 is the first excited state of the atom, and that the energy of the ground state is equal to E_1

$$E_2 - E_1 = \hbar\omega_0 .$$

The emitted photon has a frequency ω close to the frequency ω_0. We shall denote the amplitude of the state arising in the transition $2 \to 1$ by $f'(t)$ (the atom makes a transition into the ground state and emits a photon with wave vector $\mathbf{k}$, frequency ω_λ and polarization $\mathbf{e}$).

In the case given the system of eqs. (55.5) will have the form

$$i\hbar \dot{f}'_\lambda(t) = \langle 1, 1_\lambda | H' | 2, 0 \rangle e^{-i(\omega_\lambda - \omega_0)t} \varphi'(t) ,$$

$$i\hbar \dot{\varphi}'(t) = \sum_\lambda \langle 2, 0 | H' | 1, 1_\lambda \rangle e^{i(\omega_\lambda - \omega_0)t} f'_\lambda(t) .$$

(109.1)

From the initial conditions it follows that $f'_\lambda(0) = 0$, $\varphi'(0) = 1$. We introduce the notation

$$f_\lambda(t) = f'_\lambda(t) e^{-i\omega_\lambda t} , \qquad \langle 1, 1_\lambda | H' | 2, 0 \rangle = H'_{12} ,$$

$$\varphi(t) = \varphi'(t) e^{-i\omega_0 t} , \qquad \langle 2, 0 | H' | 1, 1_\lambda \rangle = H'_{21} = H'^{*}_{12} .$$

(109.2)

The Hamiltonian of the interaction of charged particles with the electromagnetic field, $\hat{H}'$, is given by expression (102.2). The form of the line is determined by the value of $|f'_\lambda(t)|^2 = |f_\lambda(t)|^2$ for $t \to \infty$.

Substituting expression (109.2) into (109.1) we obtain the system of equations of the amplitudes $f_\lambda(t)$ and $\varphi(t)$

$$i\hbar \dot{f}_\lambda(t) = H'_{12}\varphi(t) + \hbar\omega_\lambda f_\lambda(t) ,$$

$$i\hbar \dot{\varphi}'(t) = \hbar\omega_0 \varphi(t) + \sum_\lambda H'^{*}_{12} f_\lambda(t) .$$

(109.3)

The system of eqs. (109.3) is conveniently solved by means of a Laplace transformation. We denote the amplitudes in the Laplace representation by $f_\lambda(p)$

$$f_\lambda(p) = \int_0^\infty f_\lambda(t) e^{-pt} \, dt ,$$

(109.4)

$$f_\lambda(t) = \frac{1}{2\pi i} \int_{-i\infty+\delta}^{i\infty+\delta} f_\lambda(p) e^{pt} \, dp .$$

Taking into account the initial conditions, we have

$$ipf_\lambda(p) = \frac{1}{\hbar}H'_{12}\varphi(p) + \omega_\lambda f_\lambda(p),$$

(109.5)

$$ip\varphi(p) = i + \omega_0\varphi(p) + \frac{1}{\hbar}\sum_\lambda H'^*_{12} f_\lambda(p).$$

Eliminating the amplitude $\varphi(p)$ from this system, we obtain the equation for the function $f_\lambda(p)$

$$(ip-\omega_\lambda)(ip-\omega_0)f_\lambda(p) = \frac{i}{\hbar}H'_{12} + \frac{H'_{12}}{\hbar^2}\sum_\lambda H'^*_{12}f_\lambda(p).$$

(109.6)

Multiplying the left-hand and right-hand sides of the equation by the function H'^*_{12} and summing over λ, we find the expression for the sum

$$\frac{1}{\hbar}\sum_\lambda H'^*_{12}f_\lambda(p) = \frac{\gamma}{2(ip-\omega_0+\frac{1}{2}i\gamma)},$$

(109.7)

where γ is defined by the relation

$$\frac{1}{\hbar^2}\sum_\lambda \frac{|H'_{12}|^2}{ip-\omega_\lambda} = -\frac{1}{2}i\gamma.$$

(109.8)

We pass from a discrete to a continuous distribution of frequencies and replace the summation by integration. The denominator of the left-hand side of expression (109.8) contains an infinitely small imaginary positive addition. The integration over the frequency ω amounts to taking the integral in the sense of the principal value and the semiresidue at the point $\omega_\lambda = ip$

$$\int \frac{F(\omega)\,d\omega}{ip-\omega} = P\int \frac{F(\omega)\,d\omega}{ip-\omega} - i\pi F(ip),$$

(109.9)

where F is an arbitrary function.

We disregard the integral taken in the sense of the principal value (it gives a small real shift of the spectrum of radiation frequencies). Since in integrating over the Laplace variable p the basic contribution is given by the region where $ip \approx \omega_0$, in finding the value of γ we can immediately take the residue at this point.

We have finally

$$\gamma = \frac{2\pi}{\hbar^2} \sum \int d\Omega |H'_{12}|^2 \frac{\omega_0^2 V}{(2\pi c)^3}.$$ (109.10)

On the right-hand side the summation is carried out over the polarizations and the integration over the directions of the vector $\mathbf{k}$ of the emitted photon. From comparison of (109.10) with the expressions obtained in § 103 it follows that the quantity γ is the total probability of emission of a photon by the atom per unit time.

Substituting expression (109.7) into (109.6) we find the amplitude $f_\lambda(p)$

$$f_\lambda(p) = \frac{iH'_{12}}{\hbar(ip-\omega_0+\frac{1}{2}i\gamma)(ip-\omega_\lambda)}.$$ (109.11)

From the transform (109.11) we find the original, the function $f_\lambda(t)$ (see (109.4)). Closing the integration contour in the left half-plane of the complex variable p and determining the residues, we find

$$f_\lambda(t) = \frac{H'_{12} e^{-i\omega_\lambda t}}{(\omega_\lambda-\omega_0+\frac{1}{2}i\gamma)} [e^{i(\omega_\lambda-\omega_0)t-\frac{1}{2}\gamma t}-1].$$ (109.12)

The expression $|f_\lambda(t)|^2$, taken after the lapse of a sufficiently large time $t \gg \gamma^{-1}$, defines the probability of emission by the atom of a photon with given polarization and given wave vector $\mathbf{k}$, i.e. defines the form of the emission line

$$|f_\lambda(\infty)|^2 = \frac{|H'_{12}|^2}{\hbar^2} \frac{1}{(\omega_\lambda-\omega_0)^2 + \frac{1}{4}\gamma^2}.$$ (109.13)

The intensity of emission by the atom of a photon with given frequency ω, $J(\omega)$, is obtained from (109.13) multiplying $|f_\lambda(\infty)|^2$ by $\hbar\omega\rho(\omega)$ (see (103.4)) and summing and integrating over the polarizations and the directions of the vector $\mathbf{k}$ of the emitted photon. Taking into account (109.10) we have

$$J(\omega) d\omega = \frac{\gamma}{2\pi} \frac{\hbar\omega \, d\omega}{(\omega-\omega_0)^2 + \frac{1}{4}\gamma^2}.$$ (109.14)

The intensity distribution of the radiation is of a dispersion character, and the distribution width γ is equal to the total probability of emission per unit time. The relation obtained between the distribution width and the transition probability is in correspondence with the general Heisenberg uncertainty relation for time and energy $\Delta E \Delta t > \hbar$ (see § 34). Here ΔE is the uncertainty in the energy of the excited state, $\Delta E \sim \hbar\Delta\omega$, $\Delta\omega$ is the distribution width and Δt is the mean lifetime of the atom in the excited state; $\Delta t \sim \gamma^{-1}$, since the emission proceeds in a time of the order of γ^{-1}. Hence it follows that $\Delta \omega \sim \gamma$ in correspondence with the result obtained.

In the case where the transition takes place not between the first excited and ground levels but between arbitrary ith and kth levels, the width of the transition line γ_{ik} is equal to the sum of the widths γ_i and γ_k of the levels

$$\gamma_{ik} = \gamma_i + \gamma_k . \tag{109.15}$$

Each of the level widths γ_i and γ_k is equal to the sum of the probabilities of transition from the given level to all lower levels.

The width considered here is called 'natural', since it is determined by the process of emission itself, by the radiative reaction. In addition there are other mechanisms of spectral-line broadening, which usually lead to more noticeable effects. Thus, for example, in a gaseous system the collision broadening and the Doppler broadening are considerable*. The collision broadening is due to collisions between the molecules. Indeed, collisions interrupt the process of emission. Hence, if τ is the lifetime of the atom with respect to collisions (the mean time between collisions), then, as follows from the uncertainty relation for time and energy, the linewidth is of the order of τ^{-1}.

By taking into account the linewidth the results obtained in the preceding section can be extended to the region near resonance, where $\hbar\omega \approx E_n - E_1$, and where E_n is the energy of one of the atomic levels. In this case the basic contribution in the summation is given only by states with the 'resonance' energy E_n. Thus for the differential coherent scattering cross section we have, instead of (108.18),

$$d\sigma = \left(\frac{e\omega}{c}\right)^4 \frac{\sum_i (\mathbf{e}_2 \cdot \mathbf{r})_{2i}(\mathbf{e}_1 \cdot \mathbf{r})_{i1}}{(E_1 - E_n + \hbar\omega)^2 + \frac{1}{4}\Gamma_n^2} d\Omega , \tag{109.16}$$

where Γ_n is the total width of the nth level. The summation is carried out over all states with energy E_n.

Carrying out the summation and averaging over the initial states and summing over the final states of the system, we obtain

$$\sigma = \pi\lambda^2 \frac{2j_n + 1}{2j_1 + 1} \frac{\gamma_n^2}{(E_1 - E_n + \hbar\omega)^2 + \frac{1}{4}\Gamma_n^2} , \tag{109.17}$$

---

* This problem is considered in detail, for example, in the book of I.I.Sobelman, *Vvedenie v teoriyu atomnykh spektrov* (*Introduction to the theory of atomic spectra*) (Nauka, Moscow, 1963).

where γ_n is the natural width of the nth level, $\lambdabar = c/\omega$, and j_1 and j_n are the total angular momenta of the initial and the nth atomic states. The cross section reaches a maximum value equal to $4\pi\lambdabar^2(2j_n+1)/(2j_1+1)$ for perfect resonance and $\gamma_n = \Gamma_n$.

13

Relativistic Quantum Mechanics

§110. The relativistic wave equation for a particle of zero spin

So far we have restricted ourselves to the study of the properties of particles moving with velocities small in comparison with the velocity of light.

Indeed, in obtaining the Schrödinger equation we have written the non-relativistic Hamiltonian of a particle in an external potential field

$$H = \frac{\mathbf{p}^2}{2m} + U(\mathbf{r})$$

and have replaced the corresponding quantities in it by operators. To obtain the relativistic theory, use should be made of the same scheme as that developed in §27. Namely, to construct the wave equation one has to use the relativistic expression for the Hamiltonian. For generality we shall immediately assume that the particle moves in an external electromagnetic field. Then its Hamiltonian has the form of (23.17) of Part II. Carrying out the replacement of corresponding quantities by operators, i.e. $H \to i\hbar(\partial/\partial t)$, $\mathbf{p} \to -i\hbar\nabla$, we obtain the equation

$$\left(i\hbar\frac{\partial}{\partial t} - e\varphi\right)^2 \psi = c^2\left(-i\hbar\nabla - \frac{e}{c}\mathbf{A}\right)^2 \psi + m^2c^4\psi . \tag{110.1}$$

Equation (110.1) is called the Klein–Gordon–Fock equation. The relativistic invariance of this equation is evident. The Klein–Gordon–Fock equation is a second order wave equation.

Since the relativistic Hamiltonian goes over in the limit into that of classical mechanics, it is natural to assume that for $c \to \infty$ the Klein–Gordon–Fock equation will go over into the Schrödinger equation. We shall show this.

Since the zero point energies in non-relativistic theory and the theory of relativity differ by mc^2, it is convenient to introduce a transformation of the wave function ψ by means of the following relation:

$$\psi(x,t) = \psi'(x,t)\, e^{-imc^2 t/\hbar} \ .$$

Substituting into (110.1) and calculating the derivatives with respect to time, we obtain

$$2i\hbar mc^2 \frac{\partial \psi'}{\partial t} - \hbar^2 \frac{\partial^2 \psi'}{\partial t^2} - 2e\varphi \left[mc^2 \psi' + i\hbar \frac{\partial \psi'}{\partial t} \right] + e^2 \varphi^2 \psi' = c^2 \left(\hat{\mathbf{p}} - \frac{e}{c} \mathbf{A} \right)^2 \psi' \ .$$

(110.2)

We retain only terms proportional to c^2 in this equation. Upon dividing both sides of the equation by $2mc^2$ we arrive at the ordinary Schrödinger equation

$$i\hbar \frac{\partial \psi'}{\partial t} = \frac{\left(\hat{\mathbf{p}} - \frac{e}{c} \mathbf{A} \right)^2}{2m} \psi' + e\varphi \psi' \ .$$

(110.3)

Thus we have shown that the Klein–Gordon–Fock equation goes over into the Schrödinger equation in the non-relativistic limit.

Equation (110.1), like the Schrödinger equation, defines the development of a process in time. The state of the particle is, as before, characterized by the wave function $\psi(x,y,z,t)$. This function depends on the coordinates x, y, z, t and contains no spin variables. Hence it is clear that the Klein–Gordon–Fock equation defines the behaviour of a spin zero particle. In order that this equation may describe particles of spin different from zero it must somehow be modified.

Since the Klein–Gordon–Fock equation is relativistically invariant, the wave function can be multiplied only by a certain constant phase factor in Lorentz transformations. From normalization considerations it follows that this factor must be equal to +1 (but not −1, since the Lorentz transformation is continuous). Under space reflection of the coordinates the wave function ψ can be multiplied by +1 or −1. In other words, under the action of the parity operator the wave function can transform in two ways:

$$\hat{I}\psi(x,y,z,t) = +\psi(-x,-y,-z,t) ,$$

$$\hat{I}\psi(x,y,z,t) = -\psi(-x,-y,-z,t) .$$

Thus the wave function ψ can be either a scalar or a pseudoscalar. For this reason the Klein–Gordon–Fock equation is often said to be a scalar equation.

As an example of the integration of eq. (110.1) we shall consider the case of a free particle. Then the Klein–Gordon–Fock equation can be written in the form

$$-\hbar^2 \frac{\partial^2 \psi}{\partial t^2} = -c^2 \hbar^2 \nabla^2 \psi + m^2 c^4 \psi . \tag{110.4}$$

We see the solution of eq. (110.4) in the form

$$\psi = e^{-iEt/\hbar} \, \psi_1(x,y,z) .$$

Then for the function ψ_1 we find

$$E^2 \psi_1 = -c^2 \hbar^2 \nabla^2 \psi_1 + m^2 c^4 \psi_1 .$$

Rewriting this equation as follows

$$\nabla^2 \psi_1 + \frac{E^2 - m^2 c^4}{c^2 \hbar^2} \psi_1 = 0 , \tag{110.5}$$

we easily find that its solution is a plane wave of the form

$$\psi_1 = a e^{i(\mathbf{p}\cdot\mathbf{r})/\hbar} .$$

Substituting this value for ψ_1 into (110.5), we arrive at the relativistic relation between the energy E and momentum p of a free particle

$$E^2 = c^2 p^2 + m^2 c^4 .$$

This relation is the same as the usual formula for energy in the theory of relativity.

§111. The charge density and probability current for particles of zero spin

We now turn to finding the charge density and probability current for particles described by the scalar wave equation (110.1). The derivation of expressions for these quantities is carried out according to the same scheme as for the Schrödinger equation. Namely, we multiply the Klein–Gordon–Fock equation

$$\left(i\hbar\frac{\partial}{\partial t}-e\varphi\right)^2\psi-c^2\left(-i\hbar\nabla-\frac{e}{c}\mathbf{A}\right)^2\psi-m^2c^4\psi=0 \qquad (111.1)$$

by the adjoint wave function ψ^*.

We multiply the equation conjugate to eq. (111.1) by the wave function ψ. We subtract the second equation from the first. As a result we find

$$-\hbar^2\left[\psi^*\frac{\partial^2\psi}{\partial t^2}-\psi\frac{\partial^2\psi^*}{\partial t^2}\right]-2i\hbar e\varphi\left(\psi^*\frac{\partial}{\partial t}\psi+\psi\frac{\partial}{\partial t}\psi^*\right)-$$

$$\qquad\qquad\qquad\qquad\qquad\qquad\qquad\qquad (111.2)$$

$$-\hbar^2c^2[\psi\nabla^2\psi^*-\psi^*\nabla^2\psi]-i\hbar ec[\psi^*(\nabla\mathbf{A}+\mathbf{A}\nabla)\psi+\psi(\nabla\mathbf{A}+\mathbf{A}\nabla)\psi^*]=0 .$$

The first expression in the bracket is easily transformed into the form

$$\psi^*\frac{\partial^2\psi}{\partial t^2}-\psi\frac{\partial^2\psi^*}{\partial t^2}=\frac{\partial}{\partial t}\left[\psi^*\frac{\partial\psi}{\partial t}-\psi\frac{\partial}{\partial t}\psi^*\right] .$$

The second bracket is simply the derivative with respect to time of the product $\psi^*\psi$. By means of the vector relation

$$\psi\nabla^2\psi^*=\psi\,\nabla\cdot\nabla\,\psi^*=\nabla\cdot(\psi\nabla\psi^*)-\nabla\psi\nabla\psi^*$$

we find that

$$\psi\nabla^2\psi^*-\psi^*\nabla^2\psi=\nabla\cdot[\psi\nabla\psi^*-\psi^*\nabla\psi] .$$

Finally, the last expression in the bracket of formula (111.2) is brought by means of the relation

$$\psi^*(\nabla\mathbf{A})\psi=\psi^*\,\nabla\cdot(\psi\mathbf{A})=\nabla\cdot(\psi^*\psi\mathbf{A})-\psi\mathbf{A}\nabla\psi^*$$

to the form

$$\psi^*(\nabla\mathbf{A}+\mathbf{A}\nabla)\psi+\psi(\nabla\mathbf{A}+\mathbf{A}\nabla)\psi^*=2\nabla\cdot(\psi\psi^*\mathbf{A}) .$$

If one multiplies all terms of eq. (111.2) by the quantity $e/2i\hbar mc^2$ and makes use of the transformations shown, then eq. (111.2) can be written in the form

$$\nabla\cdot\mathbf{j}+\frac{\partial\rho}{\partial t}=0 ,$$

where the charge density ρ is equal to

$$\rho = \frac{e\hbar}{2imc^2} \left[\psi \frac{\partial \psi^*}{\partial t} - \psi^* \frac{\partial \psi}{\partial t} \right] - \frac{e^2}{mc^2} \psi\psi^* \varphi \,. \qquad (111.3)$$

Then we have for the current density the expression

$$\mathbf{j} = \frac{i\hbar e}{2m} [\psi \nabla \psi^* - \psi^* \nabla \psi] - \frac{e^2}{mc^2} \psi\psi^* \mathbf{A} \,. \qquad (111.4)$$

Let us dwell on the meaning of the results obtained. In non-relativistic theory the charge density ρ can be written in the form

$$\rho(x,y,z,t) = eW(x,y,z,t) \,,$$

where $W(x,y,z,t)$ is the probability current, which is in essence a positive quantity. Clearly, relation (111.3) cannot be interpreted in such a way. The expression for ρ can be made negative by a proper choice of the function ψ at the initial instant of time.

Indeed, since the Klein–Gordon–Fock equation is a second order equation with respect to time, arbitrary values of the ψ-function itself and of its derivative with respect to time can be given at the initial instant of time. Choosing different ψ and $\partial\psi/\partial t$ it is possible to arrive at positive as well as negative values of the quantity ρ.

We shall now show that the quantity ρ/e has indeed as its non-relativistic limit the product $\psi^*\psi$. Let the following relation be fulfilled for the derivative of the wave function ψ with respect to time:

$$i\hbar \frac{\partial\psi}{\partial t} = E\psi \,.$$

In this case the expression for the charge density ρ can be written in the form

$$\rho = \frac{eE}{mc^2} \psi^*\psi - \frac{e^2\varphi}{mc^2} \psi^*\psi \,.$$

If we separate the rest energy from the energy E, i.e. if we set $E = mc^2 + E'$, then in this case we easily obtain

$$\frac{\rho}{e} = \psi^*\psi \left[1 + \frac{E' - e\varphi}{mc^2} \right] \,.$$

If the quantity $E' - e\varphi \ll mc^2$, then we have the correct expression for the non-relativistic limit of the quantity ρ/e. We see that in the case of the Klein–Gordon–Fock equation one cannot introduce a positively defined probability current. This fact was the reason why, for a long time, the Klein–Gordon–Fock equation was not applied to real objects.

§112. The concept of the nuclear force field

The Klein–Gordon–Fock equation was subsequently given a new, completely different physical interpretation.

We already know that in addition to electrical interactions other forms of interaction occur in nature. In particular, one such interaction, which does not depend on the electric charge e, is the strong nuclear interaction. It seemed natural to assume that the nuclear interaction could be associated with the presence of a special nucleon charge g in nucleons. One can then try to describe the nuclear interaction by analogy with the interaction of electric charges, introducing the concept of a nuclear force field. This field should be described by a potential similar to the potential φ of the electric field. The attempt was made to give up the interpretation of the Klein–Gordon–Fock equation as the equation for the wave function of one particle. Instead it was proposed to consider the function ψ as the potential of the nuclear field produced by nucleons. Just as photons are quantum particles corresponding to the electromagnetic field, there correspond to the nuclear field π-mesons.

In §67 we have already dwelt on the obvious interpretation of the π-meson exchange and photon exchange as the sources, respectively, of the strong nucleon interaction and the electromagnetic interaction.

Carrying this analogy further, we can go on to write the equation for the nuclear field potential. As such an equation we take the Klein–Gordon–Fock equation in the form

$$\nabla^2 \psi - \frac{1}{c^2} \frac{\partial^2 \psi}{\partial t^2} - \kappa^2 \psi = 0 \,, \qquad \kappa^2 = m^2 c^2 / \hbar^2 \,, \qquad (112.1)$$

where m is the mass of the π-meson.

We shall not consider the quantum theory of nuclear forces, in which mesons are the elementary excitations of a certain field, like photons in the quantum theory of the electromagnetic field.

Since we are interested only in the qualitative aspect of the subject, we shall carry out our reasoning by analogy with the classical theory of the electrostatic field. Assuming that the nuclear field does not depend on time, we write for its potential ψ the equation

$$\nabla^2 \psi - \kappa^2 \psi = 0 \,. \qquad (112.2)$$

This equation is a certain analogue of the equation of the electrostatic field and goes over into the latter for $m \to 0$ (see below). As is well known, in the presence of point charges the equation of the electrostatic field is of the form

$$\nabla^2 \varphi = -4\pi e \delta(\mathbf{r}) \,.$$

Hence in the presence of a nucleon at the point $\mathbf{r} = 0$ it is natural to give eq. (112.2) the form

$$\nabla^2 \psi - \kappa^2 \psi = -4\pi g \delta(\mathbf{r}) \,. \tag{112.3}$$

We shall find the solution of this equation satisfying the condition $\psi \to 0$ for $r \to \infty$.

We seek ψ in the form

$$\psi(r) = \int \psi_{\mathbf{k}} \, e^{i\mathbf{k}\cdot\mathbf{r}} \, d\mathbf{k} \,.$$

Then, using the expansion of the δ-function in a Fourier integral (see Vol. 1, Appendix III) and eq. (112.3), we find for $\psi_{\mathbf{k}}$ the value

$$\psi_{\mathbf{k}} = \frac{g}{2\pi^2} \frac{1}{k^2 + \kappa^2} \,.$$

For the field ψ we obtain an expression which is conveniently written in the form

$$\psi = \frac{g}{2\pi^2} \int \frac{e^{ikr\cos\theta} \, k^2 \, dk}{k^2 + \kappa^2} \sin\theta \, d\theta \, d\varphi \,.$$

The integration over the angles φ and θ gives

$$\psi = \frac{2g}{\pi r} \int_0^\infty \frac{k \sin kr}{k^2 + \kappa^2} \, dk \,. \tag{112.4}$$

In integrating this expression it is convenient to introduce the range of integration over k from $-\infty$ to $+\infty$. In this case we have

$$\psi = \frac{g}{\pi i r} \int_{-\infty}^\infty \frac{k \, e^{ikr}}{k^2 + \kappa^2} \, dk \,.$$

This integral is easily calculated by means of the theory of residues

$$\int_{-\infty}^{\infty} \frac{k\, e^{ikr}}{k^2 + \kappa^2}\, dk = 2\pi i \operatorname{Res}(k=i\kappa) = \pi i\, e^{-\kappa r}\,,$$

whence we obtain the expression for the nuclear field potential

$$\psi = \frac{g}{r} e^{-\kappa r}\,, \tag{112.5}$$

called the Yukawa potential.

Formula (112.5) shows that the potential of nuclear forces decreases exponentially with increasing distance. The effective region in which ψ is different from zero has the size

$$R \approx \kappa^{-1} = \hbar/mc\,.$$

The size of this region is in order of magnitude the same as the range of nuclear forces determined experimentally.

For $m = 0$ the potential ψ goes over into the potential of the electrostatic field

$$\psi = g/r\,.$$

Thus the quantity g indeed plays in the Yukawa potential the same role as the charge e in the electrostatic potential, and can rightfully be called the nucleon charge. It should be stressed that the calculation carried out above can by no means pretend to be a quantitative characteristic of the nuclear force field.

In reality the interaction between nucleons is not of a static character. For a correct treatment of the processes of virtual π-meson exchange it is necessary to quantize the π-meson field ψ defined by the Klein–Gordon–Fock equation. This means that the function ψ and the adjoint function $\psi^\dagger$ must be considered as quantum-mechanical operators in the space of occupation numbers. These operators have matrix elements different from zero for the processes of absorption and emission of π-mesons. The interaction between nucleons should be calculated by methods analogous to those applied in radiation theory.

We have seen in ch. 12 that the mathematical apparatus of radiation theory is based on the application of perturbation theory. The dimensionless interaction constant $e^2/\hbar c$, made up of the charge of the particle and of the universal constants $\hbar$ and c, figures as the small parameter. The strong nuclear interaction can also be characterized by the interaction constant $g^2/\hbar c$. However, and here lies the profound difference from the electromagnetic interaction, the quantity $g^2/\hbar c$ is of the order of ten. Thus the effectiveness of the nuclear interaction exceeds that of the electromagnetic interaction by a factor of more than a thousand. The term 'strong nuclear interaction' is associated with this fact. The large value of the interaction constant $g^2/\hbar c$ makes it impossible to use the apparatus of perturbation theory for the calculation of nuclear interactions.

This fact reflects the change in the physical nature of the interaction in passing from charged particles to nucleons. The smallness of the electromagnetic interaction constant means that the probability of emission of N particles in one act is proportional to $(e^2/\hbar c)^N \ll 1$. In other words, the probability of emission of one (actual or virtual) photon is considerably higher than that of the simultaneous emission of two, three and so on photons.

The situation is different in the case of the strong nuclear interaction. The probability of simultaneous emission of a large number of mesons is of the same order of magnitude as the probability of emission of one meson.

Hence each nucleon should be considered as a particle surrounded by a cloud of virtual π-mesons.

The validity of such a picture is confirmed by the phenomena of multiple production of π-mesons in collisions of high-energy nucleons.

Thus the picture of the π-meson interaction of nucleons turns out to be much more complex than that of the photon interaction of charges. The interaction between two nucleons involves without fail a multitude of π-mesons and its consideration must be based on the solution of the many-body problem. No consequential quantitative theory of the strong nuclear interaction has as yet been developed.

§113. The Dirac equation

In the preceding sections we have considered the relativistically invariant wave equation valid for spin zero particles. We have seen that the quantity ρ/e, which should be interpreted as the probability density, can take on negative as well as positive values.

As can be seen from formula (111.3), this is associated with the fact that the value of ρ/e is determined not only by the initial value of the ψ-function but also by the initial value of the derivative $\partial\psi/\partial t$, defined arbitrarily. It is clear that in order to eliminate this difficulty it is necessary to eliminate the possibility of an arbitrary choice of the derivative $\partial\psi/\partial t$. In other words, it is necessary that the relativistic generalization of the Schrödinger equation contain only the first derivative with respect to time, as does the Schrödinger equation itself. Since, however, all relativistically invariant expressions must involve coordinates and time in the same way, the relativistic generalization of the Schrödinger equation should also involve first derivatives with respect to the coordinates.

The principle of superposition requires that the relativistic wave equation be linear. On the basis of these considerations Dirac formulated the following equation for the description of the motion of a free particle:

$$i\hbar \frac{\partial \psi}{\partial t} = \left(\beta'_x \frac{\partial}{\partial x} + \beta'_y \frac{\partial}{\partial y} + \beta'_z \frac{\partial}{\partial z} + \beta_0 \right) \psi \ . \tag{113.1}$$

Expression (113.1) represents the most general linear form containing only the first derivatives of the function sought. This equation is conveniently rewritten in a somewhat different form by redefining the quantities β'. Namely, we write it in the form

$$i\hbar \frac{\partial \psi}{\partial t} = (\hat{\beta}_x \hat{p}_x + \hat{\beta}_y \hat{p}_y + \hat{\beta}_z \hat{p}_z + \hat{\beta}_0)\psi \ ,$$

where the operators $\hat{p}_x$, $\hat{p}_y$, $\hat{p}_z$ are the ordinary operators of the components of the momentum along the coordinate axes, and the operators $\hat{\beta}_x, \hat{\beta}_y, \hat{\beta}_z, \hat{\beta}_0$ contain no coordinates. We determine the properties of these operators from the following reasoning. Introducing the notation

$$\hat{H} = \hat{\beta}_x \hat{p}_x + \hat{\beta}_y \hat{p}_y + \hat{\beta}_z \hat{p}_z + \hat{\beta}_0 \ ,$$

eq. (113.1) can be written in the form

$$i\hbar \frac{\partial \psi}{\partial t} = \hat{H}\psi \ , \tag{113.2}$$

which is completely, although as yet only formally, similar to the Schrödinger equation.

If it is assumed that the operator $\hat{H}$ indeed represents a Hamiltonian, then there must be the same relation between $\hat{H}$ and the momentum operators as between the energy and momentum in the theory of relativity, i.e.

$$\hat{H}^2 = c^2(\hat{p}_x^2 + \hat{p}_y^2 + \hat{p}_z^2) + m^2 c^4 \ . \tag{113.3}$$

This requirement allows one to define the operators $\hat{\beta}_x, \hat{\beta}_y, \hat{\beta}_z, \hat{\beta}_0$. Indeed, squaring the operator $\hat{H}$ we obtain

$$\hat{H}^2 = \hat{\beta}_x^2 \hat{p}_x^2 + \hat{\beta}_y^2 \hat{p}_y^2 + \hat{\beta}_z^2 \hat{p}_z^2 + \hat{\beta}_0^2 \ +$$

$$+ \ (\hat{\beta}_x \hat{\beta}_y + \hat{\beta}_y \hat{\beta}_x)\hat{p}_x \hat{p}_y + (\hat{\beta}_y \hat{\beta}_z + \hat{\beta}_z \hat{\beta}_y)\hat{p}_z \hat{p}_y + (\hat{\beta}_x \hat{\beta}_z + \hat{\beta}_z \hat{\beta}_x)\hat{p}_x \hat{p}_z \ +$$

$$+ \ (\hat{\beta}_x \hat{\beta}_0 + \hat{\beta}_0 \hat{\beta}_x)\hat{p}_x + (\hat{\beta}_y \hat{\beta}_0 + \hat{\beta}_0 \hat{\beta}_y)\hat{p}_y + (\hat{\beta}_z \hat{\beta}_0 + \hat{\beta}_0 \hat{\beta}_z)\hat{p}_z \ . \qquad (113.4)$$

The operator $\hat{H}^2$ will have the form of (113.3) if the following relations are fulfilled:

$$\hat{\beta}_x^2 = \hat{\beta}_y^2 = \hat{\beta}_z^2 = c^2 \ , \qquad \hat{\beta}_0^2 = m^2 c^4 \ ,$$

$$\hat{\beta}_i \hat{\beta}_k + \hat{\beta}_k \hat{\beta}_i = 0 \quad (i \neq k) \ , \qquad \hat{\beta}_i \hat{\beta}_0 + \hat{\beta}_0 \hat{\beta}_i = 0 \ .$$

Here i and k take on the values x, y, z.

Usually in place of the operators $\hat{\beta}_i$ one introduces operators α_i which differ from the former by constant factors:

$$\hat{\beta}_x = c\alpha_x \ ; \qquad \hat{\beta}_y = c\alpha_y \ ; \qquad \hat{\beta}_z = c\alpha_z \ ; \qquad \hat{\beta}_0 = mc^2 \beta$$

The following relations obviously hold for the operators α and β:

$$\alpha_x^2 = \alpha_y^2 = \alpha_z^2 = \beta^2 = 1 \ , \qquad \alpha_i \beta + \beta \alpha_i = 0 \ ,$$

$$\alpha_i \alpha_k + \alpha_k \alpha_i = 0 \quad (i \neq k) \ . \qquad (113.5)$$

By means of these operators eq. (113.1) can be written in the form

$$i\hbar \frac{\partial \psi}{\partial t} = [c(\alpha_x \hat{p}_x + \alpha_y \hat{p}_y + \alpha_z \hat{p}_z) + mc^2 \beta]\psi \ . \qquad (113.6)$$

This equation is called the Dirac equation.

If a vector operator is introduced by the equality

$$\boldsymbol{\alpha} = \alpha_x \mathbf{i} + \alpha_y \mathbf{j} + \alpha_z \mathbf{k} \ ,$$

then the Dirac equation can be written in a still more compact form

$$i\hbar \frac{\partial \psi}{\partial t} = \hat{H}\psi \ , \qquad \hat{H} = c\,\boldsymbol{\alpha} \cdot \hat{\mathbf{p}} + mc^2 \beta \ . \qquad (113.7)$$

We now seek the explicit form of the operators $\alpha_x, \alpha_y, \alpha_z, \beta$. We note, first of all, that the actions of these operators cannot reduce to the multiplication of the wave function by constant numbers. By means of such operators it would be impossible to satisfy relations (113.5).

Let us try to seek the operators $\alpha_x, \alpha_y, \alpha_z, \beta$ in the form of a set of constant, in general, complex numbers, i.e. in the form of the square matrices

$$\alpha_x = \begin{pmatrix} a_{11} & a_{12} & \cdots & a_{1n} \\ a_{21} & a_{22} & \cdots & a_{2n} \\ \cdot & \cdot & & \cdot \\ \cdot & \cdot & & \cdot \\ \cdot & \cdot & & \cdot \\ a_{n1} & a_{n2} & \cdots & a_{nn} \end{pmatrix} .$$

We first define the number n, which we assume to be the same for the matrices α and β. For this we make the matrices α and β correspond to the determinants

$$\det \alpha_x = \begin{vmatrix} a_{11} & a_{12} & \cdots & a_{1n} \\ a_{21} & a_{22} & \cdots & a_{2n} \\ \cdot & \cdot & & \cdot \\ \cdot & \cdot & & \cdot \\ \cdot & \cdot & & \cdot \\ a_{n1} & a_{n2} & \cdots & a_{nn} \end{vmatrix} .$$

Before proceeding to further investigation of the matrices we note that the following relation must be fulfilled for the determinant of the product of the matrices:

$$\det \alpha_x \beta = \det \alpha_x \, \det \beta . \tag{113.8}$$

From the commutation rules it follows further that

$$\alpha_x \beta = -\beta \alpha_x = -I \beta \alpha_x .$$

Here I is the unit matrix. Making then use of relation (113.8) we find

$$\det \alpha_x \beta = \det \alpha_x \, \det \beta = \det(-I) \det \beta \det \alpha_x .$$

Since the determinants are ordinary numbers, we find that

$$\det(-I) = 1$$

and, consequently,

$$(-1)^n = 1 . \tag{113.9}$$

Thus the number n must be even. If the number n were equal to two, then the matrices sought would be two-by-two matrices. We have already encountered such matrices in §59, where it was shown that there are 4 linearly independent two-by-two numerical matrices: 3 Pauli matrices and a unit matrix. This last commutes with all the Pauli matrices and, consequently, does not satisfy the condition of anti-commutation (113.5). On the other hand, in the case of four-by-four matrices it turns out to be possible to construct matrices with the properties required. Namely, by a simple check it can be seen that the matrices

$$\alpha_x = \begin{pmatrix} 0 & 0 & 0 & 1 \\ 0 & 0 & 1 & 0 \\ 0 & 1 & 0 & 0 \\ 1 & 0 & 0 & 0 \end{pmatrix}, \qquad \alpha_y = \begin{pmatrix} 0 & 0 & 0 & -i \\ 0 & 0 & i & 0 \\ 0 & -i & 0 & 0 \\ i & 0 & 0 & 0 \end{pmatrix}$$

$$\alpha_z = \begin{pmatrix} 0 & 0 & 1 & 0 \\ 0 & 0 & 0 & -1 \\ 1 & 0 & 0 & 0 \\ 0 & -1 & 0 & 0 \end{pmatrix}, \qquad \beta = \begin{pmatrix} 1 & 0 & 0 & 0 \\ 0 & 1 & 0 & 0 \\ 0 & 0 & -1 & 0 \\ 0 & 0 & 0 & -1 \end{pmatrix}$$

$$(113.10)$$

satisfy all the requirements formulated above. Matrices (113.10) can be written in an abbreviated form by making use of the Pauli matrices. Indeed, from definitions (60.14) and (60.15) and (113.10) it is clear that we have the relations

$$\alpha_x = \begin{pmatrix} 0 & \sigma_x \\ \sigma_x & 0 \end{pmatrix}, \qquad \alpha_y = \begin{pmatrix} 0 & \sigma_y \\ \sigma_y & 0 \end{pmatrix},$$

$$\alpha_z = \begin{pmatrix} 0 & \sigma_z \\ \sigma_z & 0 \end{pmatrix}, \qquad \beta = \begin{pmatrix} 1 & 0 \\ 0 & -1 \end{pmatrix}.$$

$$(113.11)$$

The matrices α and β are Hermitian matrices. This can be established by a simple check. If we transpose the matrices and carry out complex conjugation, then the matrices obtained will be the same as the original ones. Hence for these matrices one can write that $\alpha_x^\dagger = \alpha_x$, $\alpha_y^\dagger = \alpha_y$, $\alpha_z^\dagger = \alpha_z$ and $\beta^\dagger = \beta$.

If in place of the four-by-four matrices we had introduced matrices of a higher rank, then the formal set-up of the theory would not be violated. However, as will be clear from what follows, when the four-by-four matrices are introduced the general Dirac equation describes the properties of spin one-half particles.

Taking for α_x, α_y, α_z and β the matrix expression (113.10) in the form of four-by-four matrices, we have to assign four components to the wave function ψ. Indeed, only in this case do the 4 equations, into which the general expression (113.7) resolves when four-by-four matrices are substituted into it, contain four unknown functions. The four-component function ψ (called the Dirac bispinor) can be written in the form of the matrix

$$\psi = \begin{pmatrix} \psi_1 \\ \psi_2 \\ \psi_3 \\ \psi_4 \end{pmatrix}.$$

We write down these equations in explicit form, making use of the rule of multiplication of matrices (see (45.6)):

$$i\hbar \frac{\partial \psi_1}{\partial t} = c(\hat{p}_x - i\hat{p}_y)\psi_4 + c\hat{p}_z\psi_3 + mc^2\psi_1 ,$$

$$i\hbar \frac{\partial \psi_2}{\partial t} = c(\hat{p}_x + i\hat{p}_y)\psi_3 - c\hat{p}_z\psi_4 + mc^2\psi_2 ,$$

$$i\hbar \frac{\partial \psi_3}{\partial t} = c(\hat{p}_x - i\hat{p}_y)\psi_2 + c\hat{p}_z\psi_1 - mc^2\psi_3 ,$$

$$i\hbar \frac{\partial \psi_4}{\partial t} = c(\hat{p}_x + i\hat{p}_y)\psi_1 - c\hat{p}_z\psi_2 - mc^2\psi_4 .$$

The Dirac equation is easily generalized to the case of the motion of a charged particle in an external electromagnetic field. Namely, replacing the momentum operator $\hat{\mathbf{p}}$ by the operator $\hat{\mathbf{p}} - (e/c)\mathbf{A}$ according to the usual scheme and adding the operator $e\varphi$ to the operator $\hat{H}$, where $\mathbf{A}$ and φ are the vector potential and scalar potential of the electromagnetic field, we obtain the Dirac equation

$$i\hbar \frac{\partial \psi}{\partial t} = \left[c\boldsymbol{\alpha} \left(\hat{\mathbf{p}} - \frac{e}{c}\mathbf{A} \right) + e\varphi + mc^2\beta \right] \psi . \tag{113.12}$$

Let us bring the Dirac equation into a more symmetrical form. Multiplying eq. (113.7) by the operator β from the left-hand side we have

$$i\hbar\beta \frac{\partial \psi}{\partial t} = (c\beta\boldsymbol{\alpha}\mathbf{p} + mc^2\beta^2) \psi . \tag{113.13}$$

We now introduce the following system of matrices:

$$\gamma_1 = -i\beta\alpha_x , \qquad \gamma_2 = -i\beta\alpha_y ,$$
$$\gamma_3 = -i\beta\alpha_z , \qquad \gamma_4 = \beta . \tag{113.14}$$

It is easy to verify that the commutation rules for the matrices γ_i are the same as for the matrices α and β, viz.

$$\gamma_i \gamma_k + \gamma_k \gamma_i = 2\delta_{ik} .$$

By means of the matrices γ_i we can rewrite the Dirac equation in the form:

$$\sum_{i=1}^{3} \gamma_i \frac{\partial \psi}{\partial x_i} + \frac{mc}{\hbar} \psi + \frac{\gamma_4}{ic} \frac{\partial \psi}{\partial t} = 0 . \tag{113.15}$$

If we now introduce the coordinate $x_4 = ict$ we can transform the Dirac equation into a highly symmetrical form:

$$\gamma_\mu \frac{\partial \psi}{\partial x_\mu} + \frac{mc}{\hbar} \psi = 0 . \tag{113.16}$$

With the aid of the operators $\hat{p}_\mu$ and A_μ the last equation can be rewritten in the form

$$\left[\gamma_\mu \left(\hat{p}_\mu - \frac{e}{c} A_\mu \right) - imc \right] \psi = 0 . \tag{113.17}$$

§114. The probability density and probability current in Dirac's theory

We shall show, first of all, that the difficulty of interpreting the probability density ρ/e which we have encountered in discussing the Klein–Gordon–Fock equation is absent in the Dirac equation. Following the usual scheme, we write in addition to the Dirac equation

$$i\hbar \frac{\partial \psi}{\partial t} = (-i\hbar c\, \alpha \nabla + mc^2 \beta)\psi , \tag{114.1}$$

the adjoint equation

$$-i\hbar \frac{\partial \psi^\dagger}{\partial t} = i\hbar c \nabla \psi^\dagger \alpha^\dagger + mc^2 \psi^\dagger \beta^\dagger . \tag{114.2}$$

Here we have made use of the rule of conjugation of the product of matrices

$$(ab)^\dagger = b^\dagger a^\dagger .$$

Since the operators α and β are Hermitian, then $\alpha^\dagger = \alpha$, $\beta^\dagger = \beta$, and we obtain

$$-i\hbar \frac{\partial \psi^\dagger}{\partial t} = i\hbar c \nabla \psi^\dagger \alpha + mc^2 \psi^\dagger \beta . \tag{114.3}$$

We multiply eq. (114.1) by $\psi^\dagger$ on the left, and eq. (114.3) by ψ on the right and subtract the second equation from the first. We have

$$i\hbar\left(\psi^\dagger\frac{\partial\psi}{\partial t} + \frac{\partial\psi^\dagger}{\partial t}\,\psi\right) = -i\hbar c\,[\psi^\dagger\boldsymbol{\alpha}\boldsymbol{\nabla}\psi + (\boldsymbol{\nabla}\psi^\dagger\boldsymbol{\alpha})\psi]\ . \tag{114.4}$$

The parenthesis on the right-hand side of eq. (114.4) means that the gradient acts only on the function $\psi^\dagger$. The expression standing in the bracket can easily be transformed by means of the formula

$$\psi^\dagger(\boldsymbol{\alpha}\boldsymbol{\nabla})\psi + (\boldsymbol{\nabla}\psi^\dagger\boldsymbol{\alpha})\psi = \psi^\dagger\boldsymbol{\nabla}\boldsymbol{\alpha}\psi + (\boldsymbol{\nabla}\psi^\dagger\boldsymbol{\alpha})\psi = \boldsymbol{\nabla}(\psi^\dagger\boldsymbol{\alpha}\psi)\ .$$

Equation (114.4) is then written in the form

$$\frac{\partial\psi^\dagger\psi}{\partial t} = -c\,\boldsymbol{\nabla}(\psi^\dagger\boldsymbol{\alpha}\psi)\ . \tag{114.5}$$

Comparing the expression obtained with the general formula (7.5) we see that the essentially positive quantity $\psi^\dagger\psi = \psi_1^*\psi_1 + \psi_2^*\psi_2 + \psi_3^*\psi_3 + \psi_4^*\psi_4$ represents the probability density. The vector defined by the equality $\mathbf{j} = c\psi^\dagger\boldsymbol{\alpha}\psi$ gives the probability current for a particle with the wave function ψ.

Thus, as in the Schrödinger theory, the wave function allows the usual probabilistic interpretation. From the linearity of the Dirac equation and the probabilistic interpretation of the function ψ it follows that the basic propositions of quantum mechanics remain valid: (1) the interpretation of the quantity $|c_m(t)|^2$, where $c_m(t)$ is the coefficient of the expansion

$$\psi = \sum_m c_m\psi_m$$

and ψ_m is the eigenfunction of a certain operator, as the probability of measuring the corresponding eigenvalue, (2) the definition of the mean value

$$\bar{L} = \int\psi^\dagger\hat{L}\psi\,\mathrm{d}V\ .$$

Consequently, the entire structure of quantum mechanics also remains valid.

§115. The solution of the Dirac equation for a free particle

As the simplest example of the solution of the Dirac equation let us consider the motion of a free particle. We shall seek the solution of the Dirac equation for a freely moving particle in the usual way

$$i\hbar\frac{\partial\psi}{\partial t} = (c\boldsymbol{\alpha}\cdot\hat{\mathbf{p}} + mc^2\beta)\psi\ . \tag{115.1}$$

Substituting the wave function $\psi = \psi_0 e^{-iEt/\hbar}$ into (115.1) we obtain the equation for the time-independent wave function ψ_0

$$E\psi_0 = (c\boldsymbol{\alpha}\cdot\hat{\mathbf{p}} + mc^2\beta)\psi_0 . \tag{115.2}$$

We consider further states with a definite momentum and seek the solution of eq. (115.2) in the form of a plane wave

$$\psi_0 = u\, e^{i(\mathbf{p}\cdot\mathbf{r})/\hbar} .$$

Then for the function u we obtain the equation

$$Eu = (c\boldsymbol{\alpha}\cdot\mathbf{p} + mc^2\beta)u . \tag{115.3}$$

We write u in the form

$$u = \begin{pmatrix} u_1 \\ u_2 \\ u_3 \\ u_4 \end{pmatrix} = \begin{pmatrix} w \\ w' \end{pmatrix}, \qquad w = \begin{pmatrix} u_1 \\ u_2 \end{pmatrix}, \qquad w' = \begin{pmatrix} u_3 \\ u_4 \end{pmatrix}. \tag{115.4}$$

Substituting (115.4) into (115.3) and taking into account the representations of matrices α and β (113.11), we find

$$Ew = c\boldsymbol{\sigma}\cdot\mathbf{p}w' + mc^2 w , \tag{115.5}$$

$$Ew' = c\boldsymbol{\sigma}\cdot\mathbf{p}w - mc^2 w' . \tag{115.6}$$

Each of the functions w and w' has two components.

For the system of linear equations obtained to have a solution it is necessary that its determinant reduce to zero

$$\begin{vmatrix} E - mc^2 & -c\boldsymbol{\sigma}\cdot\mathbf{p} \\ -c\boldsymbol{\sigma}\cdot\mathbf{p} & E + mc^2 \end{vmatrix} = 0 . \tag{115.7}$$

Evaluating the determinant we obtain

$$E^2 - m^2 c^4 = +c^2(\boldsymbol{\sigma}\cdot\mathbf{p})^2 .$$

The expression $(\boldsymbol{\sigma}\cdot\mathbf{p})^2$ can easily be transformed by means of the known properties of the Pauli matrices. According to (60.17) we have

$$(\boldsymbol{\sigma}\cdot\mathbf{p})^2 = \mathbf{p}^2 . \tag{115.8}$$

As was to be expected, we arrive at the relation already known between the energy and momentum of a particle

$$E^2 = c^2 p^2 + m^2 c^4 . \tag{115.9}$$

The energy of a particle can take on positive as well as negative values. We have already discussed this problem in the theory of relativity and have seen that within the framework of classical mechanics this fact did not lead to any difficulties, since the energy range of width $2mc^2$ is forbidden. Indeed, in classical mechanics all variables change continuously and a particle has either a positive or a negative energy. Continuous transition from one region into the other is impossible.

In relativistic quantum mechanics there are no grounds for rejecting the negative sign. We shall discuss in detail the meaning of the negative sign of the energy later.

Choosing the plus or minus sign for the energy we can solve the system of eqs. (115.5) and (115.6). By virtue of the homogeneity of the system of equations one of the quantities, either w or w', remains arbitrary.

Let w be an arbitrary quantity. Then

$$w' = \frac{c\,\boldsymbol{\sigma}\cdot\mathbf{p}}{E + mc^2}\, w \, . \tag{115.10}$$

If, on the contrary, the quantity w' is assumed to be arbitrary, then we have

$$w = \frac{c\,\boldsymbol{\sigma}\cdot\mathbf{p}}{E - mc^2}\, w' \, . \tag{115.11}$$

The corresponding wave functions are of the form (for simplicity the direction of the momentum vector is here taken to be the z-axis)

$$u = \begin{pmatrix} A \\ B \\ \dfrac{cp_z A}{E + mc^2} \\ -\dfrac{cp_z B}{E + mc^2} \end{pmatrix}, \qquad v = \begin{pmatrix} \dfrac{cp_z D}{E - mc^2} \\ \dfrac{cp_z F}{E - mc^2} \\ D \\ F \end{pmatrix} \tag{115.12}$$

Here A, B, D and F are arbitrary constants. The character of these expressions will become clearer if we pass to the non-relativistic limit, setting $E \sim mc^2$ or $E \sim -mc^2$ respectively. Then from (115.10) it is seen that in the first case

$$w' = \frac{c\,\boldsymbol{\sigma}\cdot\mathbf{p}}{mc^2}\, w \sim \frac{v}{c}\, w \, .$$

The spinor w' is less than w in the ratio v/c, so that $u \approx \begin{pmatrix} A \\ B \\ 0 \\ 0 \end{pmatrix}$.

For the negative value of energy (115.11) gives

$$w \approx -\frac{c\,\boldsymbol{\sigma}\cdot\mathbf{p}w'}{2mc^2} \approx -\frac{v}{c}w' \qquad \text{and} \qquad v = \begin{pmatrix} 0 \\ 0 \\ D \\ F \end{pmatrix}.$$

Thus in the transition to the non-relativistic approximation two components of the wave function turn out to be small in comparison with two other components. In this case for positive energies w is large in comparison with w', and for negative energies the reverse is true.

The general solution of the Dirac equation for the motion of a free particle can be written in the form of a superposition of wave functions of the type (115.12), i.e. as a Fourier integral of the form

$$\psi(x,y,z,t) = \int u(A,B)\,e^{-iEt/\hbar}\,e^{i(\mathbf{p}\cdot\mathbf{r})/\hbar}\,d\mathbf{p} + \int v(D,F)\,e^{+i|E|t/\hbar}\,e^{i(\mathbf{p}\cdot\mathbf{r})/\hbar}\,d\mathbf{p},$$

where $d\mathbf{p} = dp_x\,dp_y\,dp_z$.

If at the initial instant of time $t = 0$ the following wave function is given

$$\psi(x,y,z,0) = \int \varphi(\mathbf{p})\,e^{i(\mathbf{p}\cdot\mathbf{r})/\hbar}\,d\mathbf{p}\,; \qquad \varphi(\mathbf{p}) = \begin{pmatrix} \varphi_1(\mathbf{p}) \\ \varphi_2(\mathbf{p}) \\ \varphi_3(\mathbf{p}) \\ \varphi_4(\mathbf{p}) \end{pmatrix},$$

then a given set of quantities $\varphi_1, \varphi_2, \varphi_3$ and φ_4 can be defined unambiguously in terms of the four arbitrary coefficients involved in u and v.

Thus a set of two waves, one of which corresponds to the positive energy and the other to the negative energy, forms the total solution of the Dirac equation. It is clear that if the particular solution corresponding to the negative energy were rejected and only the solution with the positive energy were retained, then the system of functions found in this case would be incomplete.

The initial conditions contain four given quantities, whereas u involves only two indeterminate constants A and B. Thus irrespective of other considerations the necessity of taking into account solutions with negative energy follows from the general foundations of quantum mechanics.

In the next section we shall come back to the discussion of the fundamental conclusions which have been drawn from the existence of solutions of the Dirac equation which correspond to a negative energy of the particle.

§116. The concept of the positron

We now turn to the discussion of the formula

$$E = \pm(\mathbf{p}^2 c^2 + m^2 c^4)^{\frac{1}{2}} .\tag{116.1}$$

As has already been pointed out, from the point of view of classical mechanics the negative energy of a free particle has no physical meaning.

In quantum mechanics the situation is different. Namely, discontinuous transitions are possible from states with a positive energy into states with a negative energy. In other words, these two classes of states are no longer separated by an impenetrable barrier. We have already seen that the exclusion of states with a negative energy contradicts the general propositions of quantum mechanics, since the wave functions of states with a positive energy do not form a complete system of functions.

On the other hand, it is impossible to assume the existence of particles with a negative energy. Such particles would possess properties which differ fundamentally from those of all particles observed in nature. As an example we can point to the following: a particle with a negative energy $-|E_1|$ could make a transition into a state with a lower negative energy $-|E_2|$, $|E_2| > |E_1|$. Then the difference $|E_2| - |E_1|$ could be converted into useful work. Such a transition could be carried out continuously, since $|E_2|$ is in no way limited, and a particle with a negative energy could serve as an infinitely large source of work.

In order to avoid difficulties associated with the introduction of observable particles with negative energy into the theory, Dirac introduced the concept of the vacuum as that state of space in which all states with negative energy are occupied by electrons and all states with positive energy are free. According to the Pauli principle, there is one electron in each state with a negative energy.

We assume further that under the influence of an external action one of the electrons is removed from a state with a negative energy. The vacant state with negative energy manifests itself as 'something' with a positive energy, since for the destruction of such a state, i.e. for its occupation it is necessary to add to it an electron with a negative energy. Thus the vacant state with a negative energy should be treated as a particle having a positive energy.

It should be noted that Dirac at first incorrectly assumed that this state corresponded to the proton. Subsequently it was shown theoretically that the particle corresponding to a vacant state with a negative energy must have a mass equal to the mass of the electron and, consequently, it could not be the proton.

Let us consider in more detail the considerations of Dirac about the occupied background of negative energies. Let $N_\alpha^{(-)}(p)$ and $N_\alpha^{(+)}(p)$ denote the numbers of electrons which are respectively in states with a negative and a positive energy and have momentum p and a definite spin orientation. The index α can take on two values according to the direction of spin. These numbers, in accordance with the Pauli principle, can take on only the values 0 or 1. In the vacuum state (the index v) we have

$$N_{\alpha v}^{(-)}(p) = 1 \,, \qquad N_{\alpha v}^{(+)}(p) = 0$$

for all momentum values.

Indeed, all states with negative energy are then occupied, and all states with positive energy are vacant. The energy E_v and charge q_v in the vacuum are defined by the relations

$$E_v = - \sum_{\alpha,p} E(p) N_{\alpha v}^{(-)}(p) \,, \tag{116.2}$$

$$q_v = - |e| \sum_{\alpha,p} N_{\alpha v}^{(-)}(p) \,. \tag{116.3}$$

Here e is the charge of the electron.

Since the momentum and energy of free particles are in no way limited, the values of E_v and q_v are infinitely large. However, according to Dirac, these quantities are in principle not observable. Only those quantities which characterize the departure from the vacuum state are observable.

Further, we write the total energy E of the system and the charge q of the system in the case where there are in space electrons in states with positive energy while there are vacancies in states with negative energy

$$E = \sum_{\alpha,p} [N_\alpha^{(+)}(p) - N_\alpha^{(-)}(p)] E(p) \,, \tag{116.4}$$

$$q = - \sum_{\alpha,p} |e| [N_\alpha^{(+)}(p) + N_\alpha^{(-)}(p)] \,. \tag{116.5}$$

In correspondence with the above only the following differences are observable:

$$E - E_\mathrm{v} = \sum_{\alpha,p} [N_\alpha^{(+)}(p)+(N_{\alpha v}^{(-)}(p)-N_\alpha^{(-)}(p))]\,E(p)\,, \qquad (116.6)$$

$$q - q_\mathrm{v} = -|e| \sum_{\alpha,p} [N_\alpha^{(+)}(p)+N_\alpha^{(-)}(p)-N_{\alpha v}^{(-)}(p)]\,. \qquad (116.7)$$

From formulae (116.6) and (116.7) we see that if a certain state with a negative energy is vacant, i.e. $N_\alpha^{(-)}(p) = 0$, then it corresponds to a positive contribution to the observed values of energy and charge. Indeed, formulae (116.6) and (116.7) involve the expressions $N_{\alpha v}^{(-)}(p) - N_\alpha^{(-)}(p)$. If a state with a negative energy is vacant $N_\alpha^{(-)}(p) = 0$, then $N_{\alpha v}^{(-)}(p) - N_\alpha^{(-)}(p) = 1$. In this case a positive contribution to the energy and charge of the system arises, equal respectively to E_p and $|e|$. Thus we see that the absence of an electron with momentum p in the continuous background of occupied negative states is equivalent to the appearance of an observable particle with a positive charge, positive energy and momentum $-p$. Such a particle, with charge $(+|e|)$ and a mass equal to the mass of the electron, was called a positron. It was discovered by Anderson in cosmic rays a few years after the appearance of Dirac's theory.

Proceeding from Dirac's concepts it turns out to be possible to account for a number of known physical effects. For example, it is obvious that an electromagnetic field can produce an electron–positron pair if the energy of the photon $h\omega$ is greater than $2mc^2$. This energy is necessary for bringing an electron from a state with a negative energy into a state with a positive energy.

The laws of conservation of energy and momentum restrict the possibility of the reaction of electron–positron pair production by a photon. In fact this reaction can take place only in the presence of a third body – for example a nucleus, which takes a part of the momentum. In addition to electron–positron pair production the inverse reaction, positron annihilation, is possible. In the annihilation an electron with a positive energy makes a transition into a vacant state with negative energy. The difference between the energies is emitted in the form of a γ-quantum.

Dirac's theory made it possible to predict not only these phenomena but also to calculate the cross section for both processes. The excellent agreement of the results of the calculations with experimental data was a strong confirmation of the validity of Dirac's ideas. However, the last decade has been marked by very important theoretical and experimental achievements, which will in part be elucidated in what follows. These successes allow one, on the one hand, to show the reality of the existence of the vacuum in the Dirac sense, and, on the other hand, to extend the regions of applicability of relativistic quantum mechanics. As has already been pointed out, the vacuum represents a system of charged particles occupying all possible states. When an external electric charge is brought into the vacuum, or when an electro-magnetic field arises, the vacuum begins to interact with the external fields. For example, Lamb discovered in 1953 that the levels $^2S_{\frac{1}{2}}$ and $^2P_{\frac{1}{2}}$ of the hydrogen atom have somewhat different energies (the Lamb shift). This effect can be accounted for only by an interaction with the vacuum (see §119 and §128).

Thus the concept of the vacuum developed by Dirac is confirmed by a number of diverse experiments.

The symmetry of the theory with respect to electrons and positrons finds its expression in the fact that there exists a unitary operator $\hat{C}$, called the charge conjugation operator, which transforms a particle into its antiparticle. In other words, the action of the operator $\hat{C}$ exchanges an electron and a positron (of the same spin and energy).

If ψ_e and ψ_p denote respectively the wave functions of an electron and of a positron, then by definition we can write for them (see (113.17))

$$\left[\gamma_\mu\left(\hat{p}_\mu - \frac{e}{c}A_\mu\right) - imc\right]\psi_e = 0 , \tag{116.8}$$

$$\left[\gamma_\mu\left(\hat{p}_\mu + \frac{e}{c}A_\mu\right) - imc\right]\psi_p = 0 . \tag{116.9}$$

Then the function ψ_e^* which is complex-conjugate to ψ_e satisfies the equation

$$\left[\gamma_\mu^*\left(\hat{p}_\mu^* - \frac{e}{c}A_\mu^*\right) + imc\right]\psi_e^* = 0 ,$$

or, since

$$\hat{p}_i^* = -\hat{p}_i , \qquad A_i^* = A_i \quad (i=1,2,3) , \qquad \hat{p}_4^* = \hat{p}_4 , \qquad A_4^* = -A_4 ,$$

we have

$$\left[-\gamma_4^*\left(\hat{p}_4 + \frac{e}{c}A_4\right) + \gamma_i^*\left(\hat{p}_i + \frac{e}{c}A_i\right) - imc\right]\psi_e^* = 0 . \tag{116.10}$$

From the comparison of (116.9) and (116.10) it is natural to assume that

$$\psi_p = \hat{C}\psi_e^* , \qquad \psi_e^* = \hat{C}^{-1}\psi_p . \tag{116.11}$$

Substituting (116.11) into (116.9) we have

$$\left[\gamma_i \left(\hat{p}_i + \frac{e}{c} A_i \right) + \gamma_4 \left(\hat{p}_4 + \frac{e}{c} A_4 \right) - imc \right] \hat{C} \psi_e^* = 0 .$$

Multiplying this equation from the left by $\hat{C}^{-1}$, we have

$$\hat{C}^{-1} \left[\gamma_i \left(\hat{p}_i + \frac{e}{c} A_i \right) + \gamma_4 \left(\hat{p}_4 + \frac{e}{c} A_4 \right) - imc \right] \hat{C} \psi_e^* = 0 ,$$

or

$$\left[\left(\hat{p}_i + \frac{e}{c} A_i \right) \hat{C}^{-1} \gamma_i \hat{C} + \left(\hat{p}_4 + \frac{e}{c} A_4 \right) \hat{C}^{-1} \gamma_4 \hat{C} - imc \right] \psi_e^* = 0 . \quad (116.12)$$

For (116.12) to be identical with (116.10) it is necessary that the following equalities be fulfilled:

$$\hat{C}^{-1} \gamma_i \hat{C} = \gamma_i^* , \qquad \hat{C}^{-1} \gamma_4 \hat{C} = -\gamma_4^* = -\gamma_4 .$$

If γ_i and γ_4 are defined by formulae (113.14), then

$$\hat{C} = \gamma_2 = \begin{vmatrix} 0 & 0 & 0 & 1 \\ 0 & 0 & -1 & 0 \\ 0 & 1 & 0 & 0 \\ -1 & 0 & 0 & 0 \end{vmatrix} .$$

From definition (116.11) it is seen directly that the operator $\hat{C}$ commutes with the Hamiltonian. Thus one can introduce two wave functions ψ_e and ψ_p which are completely equivalent and interrelated by the following relations conserved in time:

$$\psi_p = \hat{C} \psi_e^* = \gamma_2 \psi_e^* , \qquad \psi_e = \hat{C}^{-1} \psi_p^* = \gamma_2^{-1} \psi_p^* .$$

The two wave functions describe particles of the same (positive) energy, mass and spin, but with different signs of the charge and of the magnetic moment. The introduction of charge-conjugate wave functions for equivalent particles removes to a certain extent the logical difficulties associated with the simplified interpretation of the vacuum as a background filled by particles of negative energy.

In following chapters we shall describe in more detail the modern theory of fields and of elementary particles.

§117. The spin of particles described by the Dirac equation

Although we have so far used the concept of the spin of a particle extensively, the spin operator was introduced in a purely formal way, as a necessary tool for the description of experimental data. We shall now show that the existence of spin follows directly from the Dirac equation. For this we shall consider the conservation laws resulting from the Dirac equation.

Since in Dirac's theory all the general propositions of quantum mechanics are preserved, to find the conservation laws it is only necessary to find the commutator with the Hamiltonian. The difference from Schrödinger's theory lies in the fact that the Hamiltonian now has the form of (113.7).

If the Hamiltonian does not depend on time (and for this it is necessary that the external field potentials be time-independent), then the energy conservation law holds. In this respect there is no difference between the Schrödinger and the Dirac theories.

For a particle moving in the vacuum the total angular momentum must also be conserved. Hence there must exist a total angular momentum operator commuting with the Hamiltonian.

An interesting result is obtained when the operator for the orbital angular momentum operator $\hat{\mathbf{L}} = \hat{\mathbf{r}} \times \hat{\mathbf{p}}$ is commuted with the Hamiltonian.

For our purposes we restrict ourselves to the case of a free particle. Choosing an arbitrarily oriented z-axis, we have

$$\hat{H}\hat{L}_z - \hat{L}_z\hat{H} = (c\boldsymbol{\alpha}\cdot\mathbf{p} + mc^2\beta)\hat{L}_z - \hat{L}_z(c\boldsymbol{\alpha}\cdot\hat{\mathbf{p}} + mc^2\beta) \ .$$

Since the operator

$$\hat{L}_z = \frac{\hbar}{i}\left(y\frac{\partial}{\partial x} - x\frac{\partial}{\partial y}\right)$$

commutes with the operators $\hat{\beta}$ and $\hat{\alpha}_z\hat{p}_z$, we then obtain

$$\hat{H}\hat{L}_z - \hat{L}_z\hat{H} = c\alpha_x(\hat{p}_x\hat{L}_z - \hat{L}_z\hat{p}_x) + c\alpha_y(\hat{p}_y\hat{L}_z - \hat{L}_z\hat{p}_y) \ . \qquad (117.1)$$

Making use of the property of commutation of the momentum components with the angular momentum components, we find

$$\hat{H}\hat{L}_z - \hat{L}_z\hat{H} = i\hbar c(\alpha_y\hat{p}_x - \alpha_x\hat{p}_y) \ . \qquad (117.2)$$

We obtain analogous results for the other momentum components.

Thus the orbital angular momentum is not a constant of the motion and is not conserved. To find the quantity playing the role of the total angular momentum we introduce the operator $\hat{\mathbf{J}} = \hat{\mathbf{L}} + \hat{\mathbf{s}}$, where $\hat{\mathbf{s}}$ is an unknown operator. We require that the operator $\hat{\mathbf{J}}$ commute with the operator $\hat{H}$:

$$\hat{H}\hat{J}_i - \hat{J}_i\hat{H} = 0 \quad \text{or} \quad (\hat{H}\hat{L}_i - \hat{L}_i\hat{H}) + (\hat{H}\hat{s}_i - \hat{s}_i\hat{H}) = 0 .$$

Substituting $i = z$ and using the value of the commutator (117.2), we have

$$\hat{H}\hat{s}_z - \hat{s}_z\hat{H} = i\hbar c(\alpha_x\hat{p}_y - \alpha_y\hat{p}_x) . \tag{117.3}$$

We try to satisfy this equation, setting

$$\hat{s}_z = A\alpha_x\alpha_y , \tag{117.4}$$

where A is an unknown constant.

We further calculate the commutator $\hat{H}\hat{s}_z - \hat{s}_z\hat{H}$. Making use of (113.5), we obtain

$$A\hat{H}\alpha_x\alpha_y - A\alpha_x\alpha_y\hat{H} =$$
$$= Ac(\alpha_x\hat{p}_x + \alpha_y\hat{p}_y)\,\alpha_x\alpha_y - Ac\alpha_x\alpha_y(\alpha_x\hat{p}_x + \alpha_y\hat{p}_y) =$$
$$= 2Ac(\alpha_y\hat{p}_x - \alpha_x\hat{p}_y) .$$

Comparing this expression with formula (117.3), we find the value $A = -\frac{1}{2}i\hbar$. Thus the operator s_z is equal to

$$\hat{s}_z = -\frac{i\hbar}{2}\alpha_x\alpha_y = -\frac{i\hbar}{2}\begin{pmatrix} 0 & \sigma_x \\ \sigma_x & 0 \end{pmatrix}\begin{pmatrix} 0 & \sigma_y \\ \sigma_y & 0 \end{pmatrix} = -\frac{i\hbar}{2}\begin{pmatrix} \sigma_x\sigma_y & 0 \\ 0 & \sigma_x\sigma_y \end{pmatrix} =$$

$$= -\frac{i\hbar}{2}\begin{pmatrix} i\sigma_z & 0 \\ 0 & i\sigma_z \end{pmatrix} = \frac{\hbar}{2}\begin{pmatrix} \sigma_z & 0 \\ 0 & \sigma_z \end{pmatrix} = \frac{\hbar}{2}\begin{bmatrix} 1 & 0 & 0 & 0 \\ 0 & -1 & 0 & 0 \\ 0 & 0 & 1 & 0 \\ 0 & 0 & 0 & -1 \end{bmatrix} . \tag{117.5}$$

The two other vector components $\hat{s}_x$ and $\hat{s}_y$ are obtained from analogous calculations

$$\hat{s}_x = -\frac{1}{2}i\hbar\alpha_y\alpha_z ; \qquad \hat{s}_y = -\frac{1}{2}i\hbar\alpha_z\alpha_x .$$

We now find the operator $\hat{s}^2 = \hat{s}_x^2 + \hat{s}_y^2 + \hat{s}_z^2$. Making use of the properties of the operators $\alpha_x, \alpha_y, \alpha_z$, we find

$$\hat{s}^2 = \tfrac{3}{4}\hbar^2\hat{I} = \hbar^2\tfrac{1}{2}(1+\tfrac{1}{2})\hat{I} . \tag{117.6}$$

We now turn to the discussion of the results obtained. It is evident that the quantity **J**, conserved in time, should be considered as the total angular momentum of the particle. In its turn the total angular momentum is the sum of the orbital and the intrinsic spin angular momenta of the particle. The operators $\hat{s}_z$ and $\hat{s}^2$ in formulae (117.5) and (117.6) are brought to diagonal form. The spin component along the z-axis can then take on the two values $\pm\frac{1}{2}\hbar$. The eigenvalues of the operator $\hat{s}^2$ have the form of $\hbar^2 s(s+1)$, where $s = \frac{1}{2}$. Hence it is obvious that the particle has spin $\frac{1}{2}\hbar$.

§118. The transition from the Dirac equation to the Pauli equation. The magnetic moment of a particle

Let us now see how the Dirac equation transforms if the transition to the non-relativistic approximation is carried out in it. We shall consider the general case where the particle moves in an external electromagnetic field, so that the Dirac equation has the form of (113.12). Just as in the limiting transition in the scalar relativistic equation, we first of all separate the rest energy, i.e. we carry out a transformation of the form

$$\psi = \psi' e^{-imc^2 t/\hbar} .$$

Then for the function ψ' we obtain the equation

$$i\hbar \frac{\partial \psi'}{\partial t} = \left[c\boldsymbol{\alpha} \cdot \left(\hat{\mathbf{p}} - \frac{e}{c} \mathbf{A} \right) + mc^2(\beta - 1) + e\varphi \right] \psi' . \tag{118.1}$$

If the wave function is written in the form $\psi' = \begin{pmatrix} w \\ w' \end{pmatrix}$, then just as for a free particle we obtain the equations for w and w':

$$i\hbar \frac{\partial w}{\partial t} = c\boldsymbol{\sigma} \cdot \left(\hat{\mathbf{p}} - \frac{e}{c} \mathbf{A} \right) w' + e\varphi w ,$$

$$i\hbar \frac{\partial w'}{\partial t} = c\boldsymbol{\sigma} \cdot \left(\hat{\mathbf{p}} - \frac{e}{c} \mathbf{A} \right) w - 2mc^2 w' + e\varphi w' . \tag{118.2}$$

As always, the limiting transition to the non-relativistic approximation corresponds to a formal expansion in powers of c. We assume at first that in the general case of the motion of a particle in a field, as well as for a free particle, $w' \sim c^{-1} w$. Then in the second of eqs. (118.2) we can disregard the terms $i\hbar(\partial w'/\partial t)$ and $e\varphi w'$, since they are small in comparison with the quantities

$$c\boldsymbol{\sigma} \cdot \left(\hat{\mathbf{p}} - \frac{e}{c} \mathbf{A} \right) w \quad \text{and} \quad mc^2 w' ,$$

which are proportional to c. We then obtain for the spinor w' the expression

$$w' = \frac{1}{2mc} \boldsymbol{\sigma} \cdot \left(\hat{\mathbf{p}} - \frac{e}{c} \mathbf{A} \right) w , \tag{118.3}$$

which is in agreement with our assumption.

Substituting (118.3) into the first of eqs. (118.2) we find

$$i\hbar \frac{\partial w}{\partial t} = \frac{\left[\boldsymbol{\sigma} \cdot \left(\mathbf{p} - \frac{e}{c}\mathbf{A}\right)\right]^2}{2m} w + e\varphi w .$$ (118.4)

We evaluate the square of the operator in the explicit form

$$\left[\boldsymbol{\sigma} \cdot \left(\hat{\mathbf{p}} - \frac{e}{c}\mathbf{A}\right)\right]^2 = \left[\sigma_x\left(\hat{p}_x - \frac{e}{c}A_x\right) + \sigma_y\left(\hat{p}_y - \frac{e}{c}A_y\right) + \sigma_z\left(\hat{p}_z - \frac{e}{c}A_z\right)\right]^2 .$$

In multiplying it should be recalled that the operators $\hat{\mathbf{p}}$ and $\mathbf{A}$ do not commute with each other. Carrying out the multiplication we find

$$\left[\boldsymbol{\sigma} \cdot \left(\hat{\mathbf{p}} - \frac{e}{c}\mathbf{A}\right)\right]^2 =$$

$$\sigma_x^2\left(\hat{p}_x - \frac{e}{c}A_x\right)^2 + \sigma_y^2\left(\hat{p}_y - \frac{e}{c}A_y\right)^2 + \sigma_z^2\left(\hat{p}_z - \frac{e}{c}A_z\right)^2$$

$$+ \sigma_x\sigma_y\left(\hat{p}_x - \frac{e}{c}A_x\right)\left(\hat{p}_y - \frac{e}{c}A_y\right) + \sigma_y\sigma_x\left(\hat{p}_y - \frac{e}{c}A_y\right)\left(\hat{p}_x - \frac{e}{c}A_x\right)$$

$$+ \sigma_x\sigma_z\left(\hat{p}_x - \frac{e}{c}A_x\right)\left(\hat{p}_z - \frac{e}{c}A_z\right) + \sigma_z\sigma_x\left(\hat{p}_z - \frac{e}{c}A_z\right)\left(\hat{p}_x - \frac{e}{c}A_x\right)$$

$$+ \sigma_y\sigma_z\left(\hat{p}_y - \frac{e}{c}A_y\right)\left(\hat{p}_z - \frac{e}{c}A_z\right) + \sigma_z\sigma_y\left(\hat{p}_z - \frac{e}{c}A_z\right)\left(\hat{p}_y - \frac{e}{c}A_y\right) .$$ (118.5)

According to (60.16) we have for the Pauli matrices $\sigma_x^2 = \sigma_y^2 = \sigma_z^2 = 1$. We see that the sum of the first three terms is brought to the form

$$\sigma_x^2\left(\hat{p}_x - \frac{e}{c}A_x\right)^2 + \sigma_y^2\left(\hat{p}_y - \frac{e}{c}A_y\right)^2 + \sigma_z^2\left(\hat{p}_z - \frac{e}{c}A_z\right)^2 = \left(\hat{\mathbf{p}} - \frac{e}{c}\mathbf{A}\right)^2 .$$

We carry out the further transformation only with the terms

$$\sigma_x\sigma_y\left(\hat{p}_x - \frac{e}{c}A_x\right)\left(\hat{p}_y - \frac{e}{c}A_y\right) + \sigma_y\sigma_x\left(\hat{p}_y - \frac{e}{c}A_y\right)\left(\hat{p}_x - \frac{e}{c}A_x\right) ,$$ (118.6)

since the remaining expressions transform in a way analogous to (118.6). The matrices σ_x and σ_y anticommute and, consequently, expression (118.6) can be rewritten in the form

$$\frac{e}{c}\sigma_x\sigma_y\left[-\hat{p}_xA_y - A_x\hat{p}_y + \hat{p}_yA_x + A_y\hat{p}_x\right] .$$ (118.7)

Making use of the commutation properties of the operators $\hat{p}_x$ and $\hat{p}_y$ with the operators depending on coordinates (26.10), we have

$$\frac{e}{c} \sigma_x \sigma_y \left[-i\hbar \frac{\partial A_x}{\partial y} + i\hbar \frac{\partial A_y}{\partial x} \right] = \frac{ie\hbar}{c} \sigma_x \sigma_y \left[\frac{\partial A_y}{\partial x} - \frac{\partial A_x}{\partial y} \right] =$$

$$= \frac{ie\hbar}{c} \sigma_x \sigma_y (\nabla_z \times \mathbf{A}) = \frac{ie\hbar}{c} \sigma_x \sigma_y \mathscr{H}_z .$$

Since, according to (60.16), $\sigma_x \sigma_y = i\sigma_z$, then we finally have

$$\frac{ie\hbar}{c} \sigma_x \sigma_y \mathscr{H}_z = - \frac{e\hbar}{c} \sigma_z \mathscr{H}_z .$$

Carrying out analogous transformations with the remaining terms of (118.5) we obtain

$$\left[\boldsymbol{\sigma} \cdot \left(\hat{\mathbf{p}} - \frac{e}{c} \mathbf{A} \right) \right]^2 = \left(\hat{\mathbf{p}} - \frac{e}{c} \mathbf{A} \right)^2 - \frac{e\hbar}{c} \boldsymbol{\sigma} \cdot \mathscr{H} . \tag{118.8}$$

Substituting (118.8) into (118.4) we find

$$i\hbar \frac{\partial w}{\partial t} = \left[\frac{\left(\hat{\mathbf{p}} - \frac{e}{c} \mathbf{A} \right)^2}{2m} + e\varphi - \frac{e\hbar}{2mc} \boldsymbol{\sigma} \cdot \mathscr{H} \right] w . \tag{118.9}$$

We see that in the transition to the non-relativistic approximation the Dirac equation automatically goes over into the Pauli equation. Hence from Dirac's theory it is seen that there results, not only the existence of the spin of particles (equal to $\frac{1}{2}\hbar$), but also the existence of the intrinsic magnetic moment of particles

$$\mu = \frac{e\hbar}{2mc} . \tag{118.10}$$

We can now define more precisely the problem as to what are the particles having spin $\frac{1}{2}\hbar$ to which the Dirac equation can be applied. If m is understood to be the mass of the electron, then good agreement is obtained between the calculated and measured values of the magnetic moment.

Thus the Dirac equation describes the behaviour of electrons with a high degree of accuracy. The Dirac equation also makes it possible, apparently, to describe well the properties of the neutrino, a particle with rest mass $m = 0$ (see §123).

However, attempts to apply the Dirac equation to heavy particles of spin $\frac{1}{2}$, the proton and the neutron, have not led to very satisfactory results. On the other hand, it has also been possible to obtain some general and very important conclusions from the Dirac equation for heavy particles.

It turns out that the behaviour of fast protons and neutrons described qualitatively also fits the framework of the Dirac equation. Of particular importance is the fact that the basic idea of Dirac's theory, the existence of antiparticles, has received direct confirmation for mesons as well as for nucleons.

The antiproton $\bar{p}$, a particle with negative elementary charge and a mass equal to that of the proton, was discovered in 1955 in the reaction $p + p \rightarrow p + (\bar{p}+p) + p$ using an accelerator. Somewhat later the reaction $p + \bar{p} \rightarrow n + \bar{n}$ with the production of antineutrons was observed. The antineutron differs from the neutron in the sign of its magnetic moment and in its parity. When antiparticles are annihilated other particles are produced. For example, when protons and antiprotons are annihilated π- and K-mesons are produced.

In spite of all these facts, quantitative calculations and, in particular, calculations of the magnetic moment are in disagreement with experimental data. If m in formula (118.10) is assumed to be the mass of the proton, then a value differing from the experimental value by a factor of 2.7 is obtained for its magnetic moment.

This disagreement of theory with experiment is apparently associated with the fact that heavy particles, protons and neutrons, interact strongly with the meson field. Herein lies their difference from electrons, which interact relatively weakly with the electromagnetic field*.

§119. The hydrogen atom in Dirac's theory

Although the motion of the electron in the hydrogen atom corresponds to non-relativistic velocities, finding the relativistic corrections to the hydrogen energy levels was of great interest, since Schrödinger's theory could not account for the appearance of fine structure in the hydrogen spectrum.

In §38 it was found that the energy levels of the hydrogen atoms depend only on the principal quantum number. However, experiment shows that the principal quantum number characterizes the energy levels only approximately. In reality the excited levels are split into close sub-levels. As a result a splitting of the spectral lines, clearly observable in an ordinary spectrometer and

---

* For more detailed considerations on the possibility of applying the Dirac equation to nucleons see A.I.Akhiezer and V.B.Berestetskii, *Quantum electrodynamics* (Interscience Publishers, New York, 1965).

particularly accurately measured by means of modern radiospectroscopic methods, was observed in the hydrogen spectrum. It turns out that this splitting of levels is associated with the spin—orbit interaction and that it follows from Dirac's theory.

The Dirac equation for stationary state motion in the Coulomb field is of the form

$$[c\boldsymbol{\alpha}\cdot\hat{\mathbf{p}}+mc^2\beta]\,\psi = \left(E+\frac{Ze^2}{r}\right)\psi\;.$$

The Dirac equation, as well as the Schrödinger equation, allows exact solution for the Coulomb field. However, in contrast to the Schrödinger equation, the Dirac equation does not lead to distinct laws of conservation of total angular momentum (see §117). Calculations show that only in the non-relativistic approximation can one speak of constant values of the orbital and spin angular momenta. In this case it turns out that the Hamiltonian assumes the form*

$$\hat{H} = \frac{\hat{\mathbf{p}}^2}{2m} + U + \frac{1}{2mc^2}\frac{1}{r}\frac{\partial U}{\partial r}\,(\hat{\mathbf{s}}\cdot\hat{\mathbf{L}}) + \mathrm{O}\left(\frac{1}{c^2}\right), \tag{119.1}$$

where the first two terms are the same as the Hamiltonian of the Schrödinger equation, and the third term represents the spin—orbit interaction energy. Terms of the order of $1/c^2$, which are not written out, contain relativistic corrections to the kinetic and potential energies which do not have an obvious interpretation.

Solving the Dirac equation leads to the following expression for the energy of the electron:

$$E = mc^2 - \frac{Z^2e^4m}{2\hbar^2n^2} - \left(\frac{Ze^2}{\hbar c}\right)^4\frac{mc^2}{2n^4}\left(\frac{n}{(j+\frac{1}{2})} - \frac{3}{4}\right), \tag{119.2}$$

where $j = l + \frac{1}{2}$ is the eigenvalue of the total angular momentum operator; the other quantities have the same meaning as in formula (38.17). The energy levels depend now not only on n but also on j. For convenience of comparison with non-relativistic results the formula (119.2) was obtained from the accurate formula by expansion in powers of $Ze^2/\hbar c$.

Accidental degeneracy (see §38) is removed, and energy levels with one and the same value of n but with different j have different values. However, this splitting of levels is very small in comparison with the spacing between neighbouring levels with different n.

---

* See L.Schiff, *Quantum mechanics* (McGraw-Hill Book Company, New York, 1949).

The degeneracy of states with the same value of j is conserved. For example, for $n = 2$ there are the following three states: $2S_{\frac{1}{2}}$, $2P_{\frac{1}{2}}$ and $2P_{\frac{3}{2}}$. The first two states are degenerate, since they correspond to $n = 2$ and $j = \frac{1}{2}$.

Up to relatively recent times it was assumed that Dirac's theory gave the fine structure of hydrogen levels with a very high degree of accuracy. The distribution of terms, the selection rules and the intensities of lines given by the theory were exactly the same as those found experimentally. It was only in 1953 that Lamb, using radiospectroscopic methods for the measurement, discovered that the $2S_{\frac{1}{2}}$ and $2P_{\frac{1}{2}}$ levels have slightly different energies.

This disagreement between formula (119.2) obtained from the Dirac theory and experiment is associated with a fundamental property of matter, the reality of the vacuum, and in the end not only does it not contradict Dirac's theory but is one of its most brilliant confirmations. New mathematical methods by means of which the Lamb shift was found from the Dirac theory will be described in ch. 14.

§120. The invariance of the Dirac equation with respect to reflection, rotation and Lorentz transformation of coordinates

In §113 we have considered some properties of the Dirac equation. Let us now show that this equation satisfies the conditions of invariance with respect to reflection, rotation and Lorentz transformations. The rotation of the spatial system of coordinates and the Lorentz transformation are linear and orthogonal transformations. We can write them in the form

$$x'_\mu = a_{\mu\nu} x_\nu , \qquad x_\nu = a_{\mu\nu} x'_\mu , \qquad a_{\mu\nu} a_{\mu\rho} = \delta_{\nu\rho} . \qquad (120.1)$$

We find the transformation of the wave function

$$\psi' = S\psi , \qquad (120.2)$$

which leaves the Dirac equation invariant under linear transformations (120.1). The transformed wave function satisfies the Dirac equation

$$\gamma_\mu \frac{\partial \psi'}{\partial x'_\mu} + \frac{mc}{\hbar} \psi' = 0 . \qquad (120.3)$$

The derivatives $\partial/\partial x'_\mu$ can be transformed by means of the relation

$$\frac{\partial}{\partial x'_\mu} = \frac{\partial}{\partial x_\nu} \frac{\partial x_\nu}{\partial x'_\mu} = a_{\mu\nu} \frac{\partial}{\partial x_\nu} . \qquad (120.4)$$

Making use of (120.4) we transform eq. (120.3) into the form

$$a_{\mu\nu}\gamma_{\mu}S\frac{\partial\psi}{\partial x_{\nu}}+\frac{mc}{\hbar}S\psi = 0 \ . \tag{120.5}$$

If there exists a matrix S^{-1} for which the conditions

$$S^{-1}a_{\mu\nu}\gamma_{\mu}S = \gamma_{\nu} \quad \text{or} \quad S^{-1}\gamma_{\mu}S = \sum_{\nu} a_{\mu\nu}\gamma_{\nu} , \qquad S^{-1}S = 1 \tag{120.6}$$

are fulfilled, then, multiplying eq. (120.5) by the matrix S^{-1}, we arrive at eq. (113.16).

Let us now find the explicit form of the matrix of the linear transformation S for rotation of the spatial system of coordinates and Lorentz transformations. In the case of the rotation of the system of coordinates in the $(x_1 x_2)$-plane the coefficients $a_{\mu\nu}$ are defined by the relations

$$x_1' = x_1 \cos\varphi + x_2 \sin\varphi \ ,$$
$$x_2' = -x_1 \sin\varphi + x_2 \cos\varphi \ . \tag{120.7}$$

We shall now show that if the matrix S is chosen in the form

$$S = e^{\frac{1}{2}\varphi\gamma_1\gamma_2} , \tag{120.8}$$

then relations (120.6) are fulfilled. For this we expand the exponential in a series

$$S = 1 + \frac{\varphi}{2}\gamma_1\gamma_2 + \frac{\varphi^2}{2!4}(\gamma_1\gamma_2)^2 + \frac{\varphi^3}{3!8}(\gamma_1\gamma_2)^3 + \frac{\varphi^4}{4!16}(\gamma_1\gamma_2)^4 + \dots .$$

Further, making use of the expressions

$$(\gamma_1\gamma_2)^2 = \gamma_1\gamma_2\gamma_1\gamma_2 = -\gamma_1\gamma_1\gamma_2\gamma_2 = -1 \ ,$$
$$(\gamma_1\gamma_2)^3 = (\gamma_1\gamma_2)^2\gamma_1\gamma_2 = -\gamma_1\gamma_2 \ ,$$
$$(\gamma_1\gamma_2)^4 = (\gamma_1\gamma_2)^2(\gamma_1\gamma_2)^2 = 1 \ ,$$

we find

$$S = \left(1 - \frac{\varphi^2}{2!4} + \frac{\varphi^4}{4!16} - \dots\right) + \gamma_1\gamma_2\left(\frac{\varphi}{2} - \frac{\varphi^3}{3!8} + \frac{\varphi^5}{5!32} - \dots\right). \tag{120.9}$$

It is easily seen that the matrix S is equal to

$$S = \cos\tfrac{1}{2}\varphi + \gamma_1\gamma_2 \sin\tfrac{1}{2}\varphi \ . \tag{120.10}$$

Let us now check relations (120.6). The equality $S^{-1}S = 1$ is obvious. Let us find, for example, the expression

$$S^{-1}\gamma_1 S = (\cos\tfrac{1}{2}\varphi - \gamma_1\gamma_2 \sin\tfrac{1}{2}\varphi)\gamma_1(\cos\tfrac{1}{2}\varphi + \gamma_1\gamma_2 \sin\tfrac{1}{2}\varphi) .$$

Using the properties of the γ-matrices and elementary trigonometric formulae, we find

$$S^{-1}\gamma_1 S = \gamma_1 \cos\varphi + \gamma_2 \sin\varphi = a_{11}\gamma_1 + a_{12}\gamma_2 = a_{1\nu}\gamma_\nu ,$$

which is in complete agreement with (120.6).

We now turn to Lorentz transformations. According to §10 of Part II, the Lorentz transformation can be treated as a rotation through an imaginary angle $\varphi = i\chi$ in the $(x_1 x_4)$-plane:

$$x_1' = x_1 \cosh\chi - x_0 \sinh\chi , \qquad \tanh\chi = v/c ;$$

$$x_0' = -x_1 \sinh\chi + x_0 \cosh\chi , \qquad \sinh\chi = \frac{v}{c(1-v^2/c^2)^{\frac{1}{2}}} .$$

The matrix S can be found in analogy to (120.8), replacing the angle φ by $i\chi$. Then S assumes the form

$$S = e^{\frac{1}{2}i\chi\gamma_1\gamma_4} = \cosh\tfrac{1}{2}\chi + i\gamma_1\gamma_4 \sinh\tfrac{1}{2}\chi . \tag{120.11}$$

Besides the rotation transformation and the Lorentz transformation it is necessary to consider the transformation of inversion in the origin. Under the inversion of coordinates the spatial coordinates change according to the formulae

$$x_1 \to -x_1' , \qquad x_2 \to -x_2' , \qquad x_3 \to -x_3' , \qquad x_4 = x_4' . \tag{120.12}$$

We have to require that the equation

$$\sum_{i=1}^{3} \gamma_i \frac{\partial\psi'}{\partial x_i'} + \gamma_4 \frac{\partial\psi'}{\partial x_4'} + \frac{mc}{\hbar}\psi' = 0$$

remain invariant under the replacement (120.12), and that the wave function undergo the transformation $\psi' = \hat{I}\psi$. It is easily seen that the requirement of invariance will be fulfilled if the operator $\hat{I}$ is of the form

$$\hat{I} = a\beta , \tag{120.13}$$

where a is a certain number. Indeed, making use of (120.12) and (120.13) we obtain

$$-\sum_{i=1}^{3} \gamma_i \hat{I} \frac{\partial \psi}{\partial x_i} + \hat{I}\gamma_4 \frac{\partial \psi}{\partial x_4} + \frac{mc}{\hbar} \hat{I}\psi = 0 .$$

Multiplying this equation from the left by β and dividing it by a, we obtain eq. (113.16).

A double inversion transformation brings the system back into its initial state i.e. it corresponds to rotation over the angle 2π. In the last case ψ may change sign. Hence we find the condition imposed upon the quantity

$$a^2 = \pm 1 . \tag{120.14}$$

For what follows we shall need the laws of transformation of the function

$$\bar{\psi} = \psi^\dagger \gamma_4 . \tag{120.15}$$

They can be introduced if it is noted that the function $\psi^\dagger$ satisfies the equation

$$\sum_{i=1}^{3} \frac{\partial \psi^\dagger}{\partial \psi_i} \gamma_i + \frac{mc}{\hbar} \psi^\dagger - \frac{\partial \psi^\dagger}{ic\,\partial t} \gamma_4 = 0 . \tag{120.16}$$

The requirements of invariance of this equation with respect to the rotation of the spatial system of coordinates and the Lorentz transformation lead to the condition

$$\bar{\psi}' = \bar{\psi}S^{-1} . \tag{120.17}$$

Under the inversion transformation we find $\bar{\psi}' = a^* \bar{\psi}\gamma_4$.

§121. The laws of transformation of bilinear combinations made up of wave functions

Later, in discussing one of the basic problems of modern physics, the problem of the interactions between elementary particle, we shall have to make use of certain properties of bilinear combinations made up of the wave function ψ and the function $\bar{\psi}$ conjugate to it.

As we shall see in what follows, it is necessary for a relativistically invariant formulation of the laws of interaction of nuclear particles to know the laws of transformation of the bilinear combinations of the quantities mentioned under the Lorentz transformation, spatial rotation and inversion. A simple calculation shows that from the components of the wave function and the

γ-matrices one can construct certain bilinear combinations which possess the following transformation properties:

$$\overline{\psi}\psi \qquad \text{one component (scalar)},$$
$$\overline{\psi}\gamma_i\psi \qquad \text{four components (4-vector)},$$
$$\overline{\psi}\gamma_i\gamma_5\psi \qquad \text{four components (pseudovector)},$$
$$\overline{\psi}\gamma_i\gamma_k\psi \qquad \text{six components } i \neq k \text{ (4-tensor of the second rank)},$$
$$\overline{\psi}\gamma_5\psi \qquad \text{one component (pseudoscalar)}.$$

Here the following notation is introduced:

$$\gamma_5 = \gamma_1\gamma_2\gamma_3\gamma_4 = -\begin{pmatrix} 0 & 0 & 1 & 0 \\ 0 & 0 & 0 & 1 \\ 1 & 0 & 0 & 0 \\ 0 & 1 & 0 & 0 \end{pmatrix}. \qquad (121.1)$$

The quantity γ_5 has the following properties:

$$\gamma_5^2 = 1, \qquad \gamma_5\gamma_\mu + \gamma_\mu\gamma_5 = 0, \qquad \mu = 1, 2, 3, 4.$$

The validity of these relations is easily seen by direct check. We turn to the proof of these transformation properties. When a Lorentz transformation and a spatial rotation are made, one can write by virtue of (120.2) and (120.17)

$$\overline{\psi}'\psi' = \overline{\psi}S^{-1}S\psi = \overline{\psi}\psi.$$

Because $\overline{\psi}'\psi' = a^*\overline{\psi}\gamma_4 a\gamma_4\psi = \overline{\psi}\psi$ the quantity $\overline{\psi}\psi$ also remains invariant under reflection of the system of coordinates. Thus we see that the quantity $\psi\psi$ is invariant with respect to an orthogonal transformation.

Further, we shall show that the four quantities $\overline{\psi}\gamma_i\psi$ transform as the components of a four-dimensional vector. When the rotation of the system of coordinates and a Lorentz transformation are carried out we can write

$$\overline{\psi}'\gamma_i\psi' = \overline{\psi}S^{-1}\gamma_i S\psi.$$

In correspondence with formula (120.6) we find

$$\overline{\psi}'\gamma_i\psi' = a_{i\nu}\overline{\psi}\gamma_\nu\psi. \qquad (121.2)$$

When inversion of coordinates is carried out we obtain

$$\overline{\psi}'\gamma_i\psi' = a^*a\overline{\psi}\gamma_4\gamma_i\gamma_4\psi = -\overline{\psi}\gamma_i\psi \qquad (i \neq 4). \qquad (121.3)$$

Thus the quantity $\overline{\psi}\gamma_i\psi$ changes sign under the inversion. Formulae (121.2) and (121.3) show that the four components indeed form a four-dimensional vector.

We shall now show that the quantity $\bar{\psi}\gamma_5\psi$ represents a pseudoscalar. Under inversion of coordinates we have

$$\bar{\psi}'\gamma_5\psi' = a^*a\bar{\psi}\gamma_4\gamma_1\gamma_2\gamma_3\gamma_4\gamma_4\psi \ .$$

Using the form of the matrix γ_5 and the commutation property of the matrices $\gamma_i\gamma_k + \gamma_k\gamma_i = 2\delta_{ik}$, we easily find

$$\bar{\psi}'\gamma_5\psi' = -\bar{\psi}\gamma_5\psi \ ,$$

which proves the statement which we have already formulated. The quantities $\bar{\psi}\gamma_i\gamma_5\psi$ do not change sign under reflection, and transform as the components of a vector under the rotation and Lorentz transformations. Consequently, we can state that these quantities are the components of a four-dimensional axial vector or pseudovector.

We can convince ourselves of the tensor character of the quantities $\bar{\psi}\gamma_i\gamma_k\psi$ in an analogous way:

$$\bar{\psi}'\gamma_i\gamma_k\psi' = \bar{\psi}S^{-1}\gamma_i\gamma_k S\psi = \bar{\psi}S^{-1}\gamma_i SS^{-1}\gamma_k S\psi = a_{il}a_{km}\bar{\psi}\gamma_l\gamma_m\psi \ ,$$

which is the same as the definition of a tensor.

§ 122. The concept of weak interactions. Parity non-conservation

We have already seen that in addition to the electromagnetic interaction there is also another form of interaction; the strong interaction between nucleons.

It turns out that in addition to the strong interaction there is one more form of interaction which is also of non-electromagnetic character and is called the weak interaction (see below, § 130).

Weak interactions, which cannot bind nucleons in the nucleus, play an important role in the physics of elementary and nuclear particles. They are responsible for the radioactive decay of nuclei with the emission of light particles, electrons and neutrinos. In other words, the weak interaction between elementary particles leads to β-decay.

The theory of weak interactions has recently achieved considerably successes. However, a consideration of the relevant problems is possible only within the framework of the quantum field theory and hence we shall confine ourselves only to some comments. First of all we note that the Dirac equation (113.7) can be considered as the equation for a certain electron–positron field ψ. We have already mentioned such a field approach in §112 where we considered the Klein–Gordon–Fock equation. In the field description particles are considered as the excitation quanta of the corresponding field (for example, photons are the excitation quanta of the electromagnetic field (see §101 and §102)). Then the function ψ should be considered as an operator in the space of occupation numbers (see formula (99.26) of the theory of second quantization). Of course, passing to the 'field' description we give up the one-particle interpretation of the Dirac equation. The operator ψ has non-zero matrix elements corresponding to the absorption of an electron and the production of a positron, whereas the operator $\psi^\dagger$ has non-zero matrix elements corresponding to the production of an electron and the absorption of a positron. Such considerations are general and apply also to other particles (μ-mesons, neutrinos, nucleons and so on).

Let us now consider any process, for example the decay of a μ-meson, with the emission of a neutrino and an antineutrino

$$\mu \to e + \nu + \bar{\nu}\,.$$

We recall that by definition the neutrino is understood to be the particle emitted in the positron decay of the proton

$$p \to e^+ + n + \nu\,,$$

and the antineutrino the particle emitted in the β-decay of the neutron

$$n \to p + e^- + \bar{\nu}\,.$$

Experimental data available at present show that these particles are not identical.

The process of decay of the μ-meson involves four particles with spin $\frac{1}{2}$, four fermions.

For the description of the μ-meson, the electron, and the neutrino we introduce respectively the operators ψ_μ, ψ_e and ψ_ν each of which satisfies the corresponding Dirac equation. The basic problem consists now in choosing the interaction leading to the decay. For this it is necessary to formulate the interaction Hamiltonian

$$\hat{\mathbf{H}}' = \int \hat{\mathbf{H}}'\, dV\,, \tag{122.1}$$

where $\hat{\mathbf{H}}'$ is the density of the interaction Hamiltonian.

Since the ψ'_s are operators in the space of occupation numbers, the density of the interaction Hamiltonian must contain these operators, as in non-relativistic physics (§99).

From the structure of expression (122.1) it is seen that the density of the interaction Hamiltonian $\hat{H}'$ (we shall sometimes omit the word 'density') must be a relativistic scalar (invariant with respect to the rotation and Lorentz transformations). Until the mid-fifties there were no doubts as to the existence of symmetry with respect to 'the right' and to 'the left', i.e. it was assumed that the parity conservation law holds for all interactions. Hence it was assumed that the density $\hat{H}'$ must also be invariant with respect to the inversion transformation. The requirement of relativistic invariance strongly restricts the class of possible expressions for $\hat{H}'$. Namely, since in the theory of relativity any interaction has the character of a short-range action, the values of the characteristics of all particles (the operators ψ) must be taken at one point of space and at one instant of time.

For the process of β-decay involving four fermions

$$A + B \rightarrow C + D . \tag{122.2}$$

Fermi proposed the simplest law of interaction in the form

$$\hat{H}' \sim (\bar{\psi}_C \Gamma \psi_A)(\bar{\psi}_D \Gamma \psi_B) + \text{Herm.conj.} , \tag{122.3}$$

where Herm.conj. denotes the Hermitian conjugate expression, and the value of all operators ψ_i is taken at one point. The quantity Γ can have the following forms:

$$\Gamma_1 = 1 \qquad \text{scalar covariant ,}$$
$$\Gamma_2 = \gamma_\mu \qquad \text{vector covariant ,}$$
$$\Gamma_3 = \sigma_{\mu\nu} \qquad \text{tensor covariant ,}$$
$$\Gamma_4 = \gamma_\mu \gamma_5 \qquad \text{pseudovector covariant ,}$$
$$\Gamma_5 = \gamma_5 \qquad \text{pseudoscalar covariant ,}$$

where $\sigma_{\mu\nu} = -\frac{1}{2} i (\gamma_\mu \gamma_\nu - \gamma_\nu \gamma_\mu)$, $\mu, \nu = 1, 2, 3, 4$, and the summation in (122.3) is carried out from one to four over repeated vector indices.

The Hamiltonian density (122.3) does not involve the derivatives of the operators ψ and $\bar{\psi}$. This form of the interaction Hamiltonian is called 'coupling without derivatives'. We shall come back to the problem of the absence of derivatives in the law of interaction below.

We have, in §121, established the transformation properties of bilinear combinations of the type $(\bar{\psi}_C \Gamma \psi_A)$. Since the operator $\hat{H}'$ contains products in which the quantities Γ are involved twice, it is a scalar for all Γ. Thus, for example, for the vector covariant of the interaction we have

$$\hat{H}' = g_2 (\bar{\psi}_C \gamma_\mu \psi_A)(\bar{\psi}_D \gamma_\mu \psi_B) + \text{Herm.conj.} ,$$

where the constant g_2 is called the coupling constant or the interaction constant of the vector covariant. The vector covariant is constructed as the scalar product of two four-dimensional vectors (the summation from 1 to 4 is carried our over μ). The addition of the Hermitian conjugate terms makes the operator Hermitian.

In the general case the Hamiltonian density represents the sum of all five types of interaction. The expression written satisfies the requirements of the theory of relativity and, besides the characteristics of the particles, the operators ψ, contains only the interaction constant and the matrices involved in the Dirac equation.

We now come back to the process of decay of the μ-meson with the emission of a neutrino and an antineutrino. Since the operator ψ_ν describes the emission of an antineutrino as well as the absorption of a neutrino, the process of decay of the μ-meson is equivalent to a process with the absorption of a neutrino

$$\mu + \nu \to e + \nu .$$

Correspondingly $\hat{H}'$ is of the form

$$\hat{H}' = \sum_{k=1}^{5} g_k (\overline{\psi}_e \Gamma_k \psi_\nu)(\overline{\psi}_\nu \Gamma_k \psi_\mu) + \text{Herm.conj.} . \qquad (122.4)$$

The use of the Hamiltonian (122.3) led to some success in the construction of a theory of β-decay. As will be clear from what follows, the Hamiltonian (122.4) became a basis for working out the modern theory of β-decay.

The further considerable development of the theory was associated with the discovery of parity non-conservation in weak interactions. The assumption of parity non-conservation in weak interactions was made by Lee and Yang* on the basis of available data on two types of K-meson decay.

K-mesons represent a group of elementary particles (a positive, a negative and two neutral ones) having zero spin and a mass of about 966 electron masses. All K-mesons are unstable and decay with a lifetime of 1.2×10^{-8} sec for the charged mesons and 10^{-10} sec and 6×10^{-8} sec for the two neutral mesons. It turns out that in addition to the decay into a μ-meson and a neutrino, K-mesons can decay according to the schemes

---

* T.D.Lee and C.N.Yang, *New properties of symmetry of elementary particles*, Phys. Rev. 102 (1956) 290; 104 (1956) 254.

$$K^+ \to \pi^+ + \pi^0 , \qquad\qquad (\theta\text{-decay}) ,$$

$$K^+ \to \begin{cases} \pi^+ + \pi^- + \pi^+ , \\ \\ \pi^+ + \pi^0 + \pi^0 \end{cases} \qquad (\tau\text{-decay}) .$$

The possibility of decay of the K-meson into two or three π-mesons directly contradicts the parity conservation law. Indeed, the analysis of the properties of π-mesons and of their angular distribution shows that the parity of a system of two mesons differs from that of a system of three mesons.

Still more definite indications of parity non-conservation were obtained subsequently in studying the β-decay of polarized nuclei of ^{60}Co. The nuclei of ^{60}Co have a spin σ different from zero. This fact imposes certain requirements upon the angular distribution of the β-electrons emitted by them. Namely, it follows from the parity conservation law that the distribution of the electrons must possess symmetry with respect to the direction of the vector σ. The number of electrons emerging at angles θ and $180° - \theta$ with respect to the direction of σ must be the same. Indeed, if the number of electrons entering solid angle $d\Omega$ is written in the form

$$dI = F(\theta) d\Omega ,$$

where F is a certain function of the angle θ between the vectors p and σ, then this relation should not be violated under the inversion transformation.

The vector σ, being an axial vector, does not change, whereas the polar vector p changes sign under the inversion transformation. Hence the angle θ transforms under inversion: $\theta \to 180° - \theta$. Thus the parity conservation law requires the invariance of the distribution function

$$F(\theta) = F(180° - \theta) .$$

Direct measurements showed that the angular distribution of the β-electrons emitted by polarized nuclei of ^{60}Co does not possess the symmetry mentioned. On the contrary, the electrons emerge preferentially in the direction opposite to the orientation of the spin of the nucleus. Thus the β-decay of polarized nuclei demonstrates directly the violation of the parity conservation law.

The parity conservation law, as we have seen in §33, is associated with symmetry properties of space. Violation of parity would mean that space possesses no symmetry and that the notions of 'right' and 'left' in it are of absolute character. Such an interpretation would lead to extremely grave difficulties in interpretating all the laws of physics. It appeared to be completely incomprehensible how space, could be asymmetric while remaining homogeneous and isotropic.

A way out of this difficulty was proposed by Landau. According to Landau's hypothesis, the particles themselves are asymmetric, not space. Landau* proposed the principle of combined parity, according to which all physical laws must remain invariant under combined inversion, space inversion and the simultaneous replacement of particles by antiparticles (so-called charge conjugation). As examples of the latter we can mention the replacement of electrons by positrons, protons by antiprotons and so on.

Parity non-conservation in weak interactions leads to the fact that the Hamiltonian $\hat{\mathbf{H}}'$ must no longer necessarily be a scalar with respect to reflection. Consequently, in the general case the Hamiltonian (122.3) must be supplemented, by introducing into it terms which change sign under the reflection of coordinates

$$\hat{\mathbf{H}}' = \sum_{k=1}^{5} [\, g_k (\bar{\psi}_C \Gamma_k \psi_A)(\bar{\psi}_D \Gamma_k \psi_B) + g_k' (\bar{\psi}_C \Gamma_k \psi_A)(\bar{\psi}_D \Gamma_k \gamma_5 \psi_B)\,] + \text{Herm.conj.} \tag{122.5}$$

The second component of each term of the sum is a pseudoscalar. The constants g_k', generally speaking, are not the same as the constants g_k. One would think that the increase in the number of constants makes the interpretation of available experimental data and its comparison with conclusions from theory difficult. However, as a matter of fact, parity non-conservation opened new possibilities and led to the formulation of the universal law of the four-fermion interaction.

§123. Two-component neutrino theory. The universal four-fermion interaction

The discovery of parity non-conservation in weak interactions made it possible to formulate the theory of the longitudinal or two-component neutrino**. The theory of the two-component neutrino is based on the assumption that the mass of the neutrino is not simply small but exactly equal to zero. Since the neutrino has spin one-half, it is described by the Dirac

---

* L.D.Landau, Soviet Physics JETP 5 (1957) 336.

** L.D.Landau, Societ Physics JETP 5 (1957) 336; Nuclear Physics 3 (1957) 127; A.Salam, Nuovo Cimento 5 (1957) 299.

equation, which for $m = 0$ for states with given momentum $\mathbf{p}$ is of the form (see (115.3))

$$Eu = (\boldsymbol{\alpha} \cdot \mathbf{p})u , \qquad E = \pm|\mathbf{p}| . \tag{123.1}$$

(In this section and further on we use the system of units in which $\hbar = 1$ and $c = 1$.) We can pass from eq. (123.1) for the four-component function u to the equation for two-component functions. Setting $u = \frac{1}{2}\sqrt{2} \binom{w}{w'}$ and taking into account that $\boldsymbol{\alpha} = \binom{0 \ \ \boldsymbol{\sigma}}{\boldsymbol{\sigma} \ \ 0}$, we rewrite eq. (123.1) in the form

$$Ew = (\boldsymbol{\sigma} \cdot \mathbf{p})w' , \qquad Ew' = (\boldsymbol{\sigma} \cdot \mathbf{p})w . \tag{123.2}$$

Adding up and subtracting eq. (123.2) we obtain

$$E\varphi_+ = (\boldsymbol{\sigma} \cdot \mathbf{p})\varphi_+ , \qquad E\varphi_- = -(\boldsymbol{\sigma} \cdot \mathbf{p})\varphi_- , \tag{123.3}$$

where

$$\varphi_+ = \frac{1}{\sqrt{2}}(w+w') , \qquad \varphi_- = \frac{1}{\sqrt{2}}(w-w') .$$

We see that the two-component functions φ_+ and φ_- satisfy equations of the first order. Of course, if parity were conserved we could not make use of the superposition of the functions w and w', since these functions transform differently under the inversion of the system of coordinates. Indeed, since $\boldsymbol{\sigma}$ is an axial vector and $\mathbf{p}$ is a polar vector, the product $(\boldsymbol{\sigma} \cdot \mathbf{p})$ is a pseudoscalar. Then from (123.2) we see that if w transforms under the reflection as a polar spinor, w' transforms as a pseudospinor, and vice versa.

Choosing the direction of the vector $\mathbf{p}$ to be the z-axis, we obtain from (123.3)

$$\begin{aligned} \sigma_z\varphi_+ = \varphi_+ \quad &\text{for} \quad E = |\mathbf{p}| , \\ \sigma_z\varphi_+ = -\varphi_+ \quad &\text{for} \quad E = -|\mathbf{p}| \end{aligned} \tag{123.4}$$

and

$$\begin{aligned} \sigma_z\varphi_- = -\varphi_- \quad &\text{for} \quad E = |\mathbf{p}| , \\ \sigma_z\varphi_- = \varphi_- \quad &\text{for} \quad E = -|\mathbf{p}| . \end{aligned} \tag{123.5}$$

We see that the functions φ_+ and φ_- describe states whose polarization (the spin component along the z-axis) is unambiguously related to the sign of the energy. Thus the function φ_- describes the state polarized against the direction of the momentum for $E = |\mathbf{p}|$ and along the direction of the momentum for $E = -|\mathbf{p}|$, whereas the function φ_+ describes the state polarized along the direction of the momentum for $E = |\mathbf{p}|$ and against the direction of the momentum for $E = -|\mathbf{p}|$.

In the theory of the two-component neutrino it is assumed that the neutrino ($E=|\mathbf{p}|$) and antineutrino ($E=-|\mathbf{p}|$) are described by the function φ_-, i.e. that the neutrino is always polarized against the direction of the momentum, and the antineutrino always along the direction of the momentum. Of course, it might equally well be assumed that the neutrino and antineutrino are described by the function φ_+, but this would lead to conclusions which are in disagreement with experimental data. If the energy of the antineutrino is also assumed to be positive $E = |\mathbf{p}|$, then the antineutrino will be described by the function φ_+ (see the first equation of (123.4) and the second equation of (123.5)).

Under the inversion of the system of coordinates the axial vector $\boldsymbol{\sigma}$ does not change whereas the polar vector $\mathbf{p}$ reverses its direction. Consequently, the neutrino then goes over into the antineutrino and vice versa, in accordance with the ideas put forward by Landau (combined parity conservation).

The entire conclusion presented is based on the assumption that the mass of the neutrino is exactly equal to zero. This also follows immediately from the following obvious considerations. If the mass of the neutrino were not equal to zero, then it would move with a velocity less than the velocity of light. There would then exist an inertial system of coordinates, moving with respect to the laboratory system of coordinates, with a velocity larger than the velocity of the neutrino, in which the direction of the momentum of the neutrino would be reversed. Since the direction of the spin does not change under such a transformation, we would have in one inertial system of coordinates a neutrino, and in the other an antineutrino, i.e. we would arrive at a contradiction, since the neutrino and antineutrino by assumption are not identical.

The theory of the two-component neutrino can easily be formulated within the framework of the usual mathematical apparatus, i.e. by means of four-component functions. Namely, it is easily seen (see (113.16)) that if the bispinor ψ is the solution of the Dirac equation with a rest mass equal to zero

$$\gamma_\mu \frac{\partial \psi}{\partial x_\mu} = 0 \, ,$$

then the functions ψ_+ and ψ_- will also be solutions of this equation

$$\psi_+ = \frac{1}{\sqrt{2}} (1-\gamma_5)\psi \, , \qquad \psi_- = \frac{1}{\sqrt{2}}(1+\gamma_5)\psi \qquad (123.6)$$

or respectively for states with a definite momentum

$$u_+ = \frac{1}{\sqrt{2}}(1-\gamma_5)u , \qquad u_- = \frac{1}{\sqrt{2}}(1+\gamma_5)u , \qquad (123.7)$$

where u satisfies eq. (123.1).

It is easily seen that the functions u_+ and u_- are expressed in terms of φ_+ and φ_-. Indeed, setting $u = \frac{1}{2}\sqrt{2}\binom{w}{w'}$ and taking into account that $\gamma_5 = -\binom{0\ 1}{1\ 0}$, where the four-by-four matrix is written in terms of two-by-two matrices, we obtain

$$u_+ = \frac{1}{2}\binom{w+w'}{w+w'} = \frac{1}{\sqrt{2}}\binom{\varphi_+}{\varphi_+} ,$$

$$u_- = \frac{1}{2}\binom{w-w'}{-(w-w')} = \frac{1}{\sqrt{2}}\binom{\varphi_-}{-\varphi_-} . \qquad (123.8)$$

The function ψ_-

$$\psi_- = u_- \, e^{i(\mathbf{p}\cdot\mathbf{r}-Et)} , \qquad (123.9)$$

describes the neutrino for $E = |\mathbf{p}|$. It is the eigenfunction of the spin component operator $\hat{s}_z$,

$$\hat{s}_z = \begin{pmatrix} \sigma_z & 0 \\ 0 & \sigma_z \end{pmatrix} ,$$

corresponding to the eigenvalue -1,

$$\hat{s}_z u_- = -u_- \quad \text{for} \quad E = |\mathbf{p}| \qquad \text{(neutrino)} .$$

The antineutrino $(E=|\mathbf{p}|)$ is described by the function u_+ and ψ_+ respectively. Under the action of the operator $\hat{s}_z$ we have

$$\hat{s}_z u_+ = u_+ \quad \text{for} \quad E = |\mathbf{p}| \qquad \text{(antineutrino)} .$$

We also note that the functions ψ_+ and ψ_- are eigenfunctions of the operator γ_5. Indeed, since $\gamma_5^2 = 1$, it follows from (123.6) that

$$\gamma_5\psi_+ = -\psi_+ , \qquad \gamma_5\psi_- = \psi_- . \qquad (123.10)$$

The operator γ_5 is called the helicity operator. To the eigenvalue $\gamma_5 = +1$ there corresponds left-handed helicity, while to the eigenvalue $\gamma_5 = -1$ there corresponds right-handed helicity. The above results can be given an obvious interpretation in terms of the helicity operator: there is a strong correlation between the direction of the momentum vector and the direction of the spin vector of a particle. For the neutrino the spin $\boldsymbol{\sigma}$ is antiparallel to $\mathbf{p}$ $(\gamma_5=1)$,

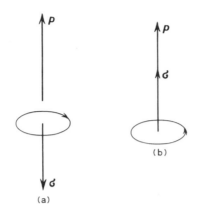

Fig. V.29

whereas for the antineutrino it is parallel ($\gamma_5 = -1$). If the spin is represented in an obvious way as the rotation of the particle, then the neutrino rotates as a left-handed helix about the axis **p** (fig. V.29). Under the space inversion the direction of **p** reverses, whereas the vector σ remains unchanged. There is no neutrino having a spin of 'irregular' orientation. Hence under the reflection of space coordinates it is necessary to allow for the transformation of a neutrino into an antineutrino, in correspondence with the principle of combined parity.

Gell-Mann and Feynman* put forward the hypothesis that the property of helicity is of a general character and is a characteristic of all fermions and not only of the neutrino. According to this hypothesis the transformation (123.6) should hold for all four fermions involved in the process of weak interaction (122.2). This means that in the general expression (122.5) the operators ψ_C, ψ_D, ψ_A and ψ_B should be replaced respectively by the operators

$$\chi_C = \frac{1}{\sqrt{2}}(1+\gamma_5)\psi_C , \qquad \chi_A = \frac{1}{\sqrt{2}}(1+\gamma_5)\psi_A ,$$

$$\chi_D = \frac{1}{\sqrt{2}}(1+\gamma_5)\psi_D , \qquad \chi_B = \frac{1}{\sqrt{2}}(1+\gamma_5)\psi_B .$$

Let us elucidate this assumption. We note, first of all, that the operators χ

* R. Feynman and M. Gell-Mann, Phys. Rev. 109 (1958) 193.

are actually two-component operators. They are expressed analogously to (123.8) in terms of the spinors w and w' involved in

$$\chi = \frac{1}{2} \begin{pmatrix} w-w' \\ -(w\ w') \end{pmatrix}.$$

Since $\chi = \frac{1}{2}\sqrt{2}(1+\gamma_5)\psi$, then, acting on both sides of the equation with the operator

$$\frac{1}{\sqrt{2m}} \left[\gamma_\mu \left(\frac{\partial}{\partial x_\mu} - ieA_\mu \right) - m \right]$$

and taking into account the Dirac equation

$$\left[\gamma_\mu \left(\frac{\partial}{\partial x_\mu} - ieA_\mu \right) + m \right] \psi = 0 \, ,$$

we obtain

$$-\frac{1}{\sqrt{2m}} \left[\gamma_\mu \left(\frac{\partial}{\partial x_\mu} - ieA_\mu \right) - m \right] \chi = \psi \, . \tag{123.11}$$

Substituting ψ in the form of (123.11) into the Dirac equation we obtain the equation of second order satisfies by the operator χ

$$\left[\left(\frac{\partial}{\partial x_\mu} - ieA_\mu \right)^2 + e\sigma_{\mu\nu}F_{\mu\nu} - m^2 \right] \chi = 0 \, , \tag{123.12}$$

where

$$\sigma_{\mu\nu} = -\frac{1}{2}i(\gamma_\mu\gamma_\nu - \gamma_\nu\gamma_\mu)$$

and

$$F_{\mu\nu} = \frac{\partial}{\partial x_\mu} A_\nu - \frac{\partial}{\partial x_\nu} A_\mu \, .$$

Feynman and Gell-Mann assumed that the operator χ is more fundamental than the operator ψ and thus that the interaction Hamiltonian (122.3) should not contain the derivatives of the operator χ. Hence, in view of (123.11) the interaction Hamiltonian should involve the operator χ and not ψ.

Such an assumption leads immediately to the fact that of all the co-variants of interaction only the vector covariant and the axial-vector covariant (with the same constants) turn out to be possible, whereas all other covariants give zero.

Let us show, for example, that the pseudoscalar covariant reduces to zero.

$$\hat{H}' = \tfrac{1}{4}g_5 \overline{((1+\gamma_5)\psi_C}\gamma_5(1+\gamma_5)\psi_A)\overline{((1+\gamma_5)\psi_D}\gamma_5(1+\gamma_5)\psi_B) +$$

$$+ \tfrac{1}{4}g_5' \overline{((1+\gamma_5)\psi_C}\gamma_5(1+\gamma_5)\psi_A)\overline{((1+\gamma_5)\psi_D}(1+\gamma_5)\psi_B) + \text{Herm.conj.} \ .$$

But

$$\gamma_5(1+\gamma_5) = 1 + \gamma_5 \ .$$

Consequently the two terms are identical. Further,

$$\overline{(1+\gamma_5)\psi} = ((1+\gamma_5)\psi)^\dagger \gamma_4 = \psi^\dagger (1+\gamma_5)\gamma_4 = \overline{\psi}(1-\gamma_5) \ .$$

Hence the following terms appear in the parentheses:

$$(1-\gamma_5)(1+\gamma_5) = 1 - \gamma_5^2 = 0 \ .$$

The scalar and tensor covariants also give zero.

Taking into account that

$$\gamma_5(1+\gamma_5) = (1+\gamma_5) \ , \qquad (1+\gamma_5)^2 = 2(1+\gamma_5) \ ,$$

we have for the vector and axial-vector covariants

$$\hat{H}' = \tfrac{1}{4}(g_2+g_2')[\overline{\psi}_C(1-\gamma_5)\gamma_\nu(1+\gamma_5)\psi_A][\overline{\psi}_D(1-\gamma_5)\gamma_\nu(1+\gamma_5)\psi_B] +$$

$$+ \tfrac{1}{4}(g_4+g_4')[\overline{\psi}_C(1-\gamma_5)\gamma_\nu\gamma_5(1+\gamma_5)\psi_A] \times$$

$$\times [\overline{\psi}_D(1-\gamma_5)\gamma_\nu\gamma_5(1+\gamma_5)\psi_B] + \text{Herm.conj.} =$$

$$= \tfrac{1}{4}(g_2+g_2'+g_4+g_4') \times$$

$$\times [\overline{\psi}_C\gamma_\nu 2(1+\gamma_5)\psi_A][\overline{\psi}_D\gamma_\nu 2(1+\gamma_5)\psi_B] + \text{Herm.conj.} \equiv$$

$$\equiv f(\overline{\psi}_C\gamma_\nu(1+\gamma_5)\psi_A)(\overline{\psi}_D\gamma_\nu(1+\gamma_5)\psi_B) + \text{Herm.conj.} \ . \tag{123.13}$$

On the basis of a careful analysis of experimental data Sudarshan and Marshak arrived at exactly the same form of the Hamiltonian for the four-fermion interaction.

The interaction Hamiltonian (123.13) gives the universal law of the four-fermion interaction with only one coupling constant f. In analyzing concrete processes by means of the Hamiltonian (123.13) it is also necessary to take into account the so-called leptonic charge conservation law. Leptons are light particles taking part in weak interaction processes, namely: electrons e^-, μ^--mesons and the neutrino ν. The particles e^+, μ^+ and $\overline{\nu}$ are called antileptons. Leptons are assigned the leptonic charge $+1$, and antileptons -1. For other particles, for example nucleons, the leptonic charge is assumed to be equal to zero. The total leptonic charge (the algebraic sum of the leptonic charges) must be conserved in the reaction (see ch. 15).

Let us consider, for example, the decay of the μ^--meson

$$\mu^- \to e^- + \nu + \overline{\nu}. \tag{123.14}$$

A decay with the emission of two neutrinos (or antineutrinos) is evidently forbidden by the leptonic charge conservation law.

The interaction Hamiltonian (123.14), in correspondence with (123.13), is of the form

$$\hat{H}' = f(\overline{\psi}_e \gamma_\mu (1+\gamma_5)\psi_\nu)(\overline{\psi}_\nu \gamma_\mu (1+\gamma_5)\psi_{\mu^-}) + \text{Herm.conj.} \,. \tag{123.15}$$

For the process of β-decay of the neutron we have, correspondingly

$$\hat{H}' = f(\overline{\psi}_e \gamma_\mu (1+\gamma_5)\psi_\nu)(\overline{\psi}_p \gamma_\mu (1+\gamma_5)\psi_n) + \text{Herm.conj.} \,. \tag{123.16}$$

Knowing the interaction Hamiltonian it is easy to determine the probability of the corresponding process by ordinary methods of perturbation theory. The universal law of the four-fermion interaction, proposed by Gell-Mann and Feynman and by Sudarshan and Marshak is quantitatively confirmed by a vast amount of experimental data.

Some Problems
of Quantum Electrodynamics

§124. The Green's function of the Dirac equation

The theory of the interaction of non-relativistic charged particles with the electromagnetic field, presented in ch. 12, is easily generalized to the case of relativistic particles. However, calculations of higher approximations of perturbation theory (expansion in powers of $e^2/\hbar c$) led to diverging expressions whose physical meaning was not clear. Thus, for example, the intrinsic energy of the electron turned out to be infinite, as in classical electrodynamics. Corrections to the scattering cross sections calculated in the second order approximation of perturbation theory also turned out to be infinitely large, and so on. All this pointed to a limited region of applicability of the mathematical apparatus of quantum electrodynamics. At the same time good agreement with experimental data on cross sections for different processes calculated in the first non-vanishing approximation of perturbation theory indicated the validity of the general ideas and methods of the theory.

An increase in the accuracy of experimental methods of investigation led recently to the establishment of new facts which had no explanation in quantum electrodynamics. Namely, in 1947, in addition to the discovery by Lamb of the shift of the $2^2S_{\frac{1}{2}}$ and $2^2P_{\frac{1}{2}}$ levels of the hydrogen atom which, according to the Dirac theory, should coincide, Rabi established that the value of the magnetic moment of the electron differs somewhat from a Bohr magneton. The discovery of these phenomena led to a further intense development of quantum electrodynamics. Very important roles in the development of the theory were played by the studies of Bethe, Feynman, Dyson, Schwinger, Tomonaga and others*. In particular, Feynman proposed a new method of calculation which made it possible to simplify considerably all the calculations and also to give them an obvious physical meaning**. Within the framework of this book we can present only the most general outlines of Feynman's method. A detailed exposition of Feynman's method as well as numerous examples of its application to concrete problems can be found in the articles and monographs cited below.

The method of Green's function is the basis of the mathematical apparatus of Feynman's theory. We now turn directly to the exposition of the method proposed by Feynman. First of all we write the Dirac equation in a form which is more compact and convenient for these calculations. For this we introduce the operator $\hat{\nabla}$

$$\hat{\nabla} \equiv \sum_{\mu=1}^{4} \gamma_\mu \frac{\partial}{\partial x_\mu} \equiv \gamma_\mu \frac{\partial}{\partial x_\mu} ,$$

where $x_4 = ix_0 = it$. In this notation the Dirac equation has the form[†]

$$(\hat{\nabla} + m)\psi = 0 . \tag{124.1}$$

Analogously to what we did in non-relativistic theory for the Schrödinger equation (see §29) we introduce the Green's function $K(2,1)$ of the Dirac equation (124.1). The Green's function $K(2,1)$ by definition satisfies the equation

$$(\hat{\nabla}_2 + m)K(2,1) = i^{-1}\delta^4(2,1) . \tag{124.2}$$

Here and in what follows the numbers 1 and 2 denote the set of four coor-

---

* A detailed bibliography is given in the book of S.Schweber, H.Bethe and F.de Hoffman, *Mesons and fields* (Row, Peterson and Company, Evanston, Illinois and White Plains, New York, 1956). For a detailed exposition of quantum electrodynamics see also A.I.Akhiezer and V.B.Berestetskii, *Quantum electrodynamics* (Interscience Publ., New York, 1965).

** R.P.Feynman, Phys. Rev. 76 (1949) 749. See also the monographs cited above.

† We use a notation somewhat different from that introduced by Feynman. A similar notation is adopted, for example, in the book of A.I.Akhiezer and V.B.Berestetskii. We also note (as on p. 510) that in this chapter we assume that $\hbar = 1$ and $c = 1$.

dinates x_μ, and $\hat{\nabla}_2$ is the operator acting on the variables $x_{2\mu}$. The symbol $\delta^4(2,1)$ denotes a 4-dimensional δ-function equal to

$$\delta^4(2,1) = \delta^4(x_2 - x_1) = \delta(\mathbf{r}_2 - \mathbf{r}_1)\delta(t_2 - t_1) . \tag{124.3}$$

We seek the solution of eq. (124.1) in the momentum representation. In other words, we expand the function $K(2,1)$ in the Fourier integral

$$K(2,1) = \int_{-\infty}^{\infty} S(p) \exp\left[ip_\mu(x_{2\mu} - x_{1\mu})\right] \mathrm{d}^4p , \tag{124.4}$$

where

$$\mathrm{d}^4p = \mathrm{d}^3p\mathrm{d}p_0 = \mathrm{d}p_x\mathrm{d}p_y\mathrm{d}p_z\mathrm{d}p_0 = \mathrm{d}p_1\mathrm{d}p_2\mathrm{d}p_3\mathrm{d}p_0 .$$

p_μ is the 4-dimensional momentum vector, and $p_4 = ip_0$. The summation from 1 to 4 is carried out over the index μ. In order not to overload the formulae with indices, we shall in what follows omit the index μ if this cannot lead to misunderstanding. Also expanding $\delta^4(2,1)$ in a Fourier integral according to the formula (see Appendix III in Vol. 1)

$$\delta^4(2,1) = \frac{1}{(2\pi)^4} \int_{-\infty}^{\infty} e^{ip(x_2 - x_1)} \mathrm{d}^4p \tag{124.5}$$

and substituting expressions (124.4) and (124.5) into (124.2), we find

$$(\hat{\nabla}_2 + m) \int S(p)^{ip(x_2 - x_1)} \mathrm{d}^4p = \frac{1}{(2\pi)^4 i} \int e^{ip(x_2 - x_1)} \mathrm{d}^4p . \tag{124.6}$$

The action of the operator $(\hat{\nabla}_2 + m)$ gives

$$(\hat{\nabla}_2 + m) e^{ip(x_2 - x_1)} = (i\hat{p} + m) e^{ip(x_2 - x_1)} , \tag{124.7}$$

where $\hat{p} = p_\mu\gamma_\mu$. (We stress that in this chapter the mark $\hat{}$ has a meaning different from that in the preceding chapters of the book.)

Equating the Fourier components in (124.6) and taking into account (124.7), we can write the formal solution for $S(p)$

$$S(p) = \frac{1}{(2\pi)^4}\frac{1}{i}\frac{1}{i\hat{p} + m} = \frac{-1}{(2\pi)^4}\frac{i\hat{p} - m}{(i\hat{p}+m)(i\hat{p}-m)} = \frac{i}{(2\pi)^4}\frac{i\hat{p} - m}{\hat{p}^2 + m^2} , \tag{124.8}$$

where

$$\hat{p}^2 = \sum_{\mu,\nu} p_\mu p_\nu \gamma_\mu \gamma_\nu = \frac{1}{2} \sum_{\mu,\nu} (\gamma_\mu \gamma_\nu + \gamma_\nu \gamma_\mu) p_\mu p_\nu = \sum_\mu p_\mu p_\mu = p^2 .$$

We have made use of the anticommutativity of the matrices γ. For $K(2,1)$ we have correspondingly

$$K(2,1) = \frac{i}{(2\pi)^4} \int \frac{i\hat{p} - m}{p^2 + m^2} \, e^{ip(x_2 - x_1)} \, d^4p . \tag{124.9}$$

This expression is conveniently written in the form

$$K(2,1) = i(\hat{\nabla}_2 - m) I(2,1) , \tag{124.10}$$

where $I(2,1)$ is an integral which depends only on ordinary variables but not on the Dirac matrices and is equal to

$$I(2,1) = \frac{1}{(2\pi)^4} \int \frac{e^{ip(x_2 - x_1)}}{p^2 + m^2} \, d^4p . \tag{124.11}$$

We carry out the integration over the variable p_0:

$$I(2,1) = -\frac{1}{(2\pi)^4} \int e^{i\mathbf{p}\cdot(\mathbf{r}_2 - \mathbf{r}_1)} \, d^3p \int \frac{e^{-ip_0(t_2 - t_1)}}{p_0^2 - E_p^2} \, dp_0 ,$$

where $E_p = +(\mathbf{p}^2 + m^2)^{\frac{1}{2}}$. If p_0 is considered as a certain complex variable, then the integration in the plane of this complex variable is carried out over the entire real axis. However, the integrand has poles on this axis at the points $p_0 = E_p$ and $p_0 = -E_p$. Consequently, for the integral (124.11) to have a definite meaning it is necessary to define the rule of circumventing these poles. Feynman proposed the following rule: the left pole is circumvented from below, and the right pole from above. To carry this out one has to add to the mass m an infinitesimal negative imaginary part which in the final result should be made to tend to zero: $m \rightarrow m - i\delta, \delta > 0$.

As a matter of fact, E_p then also receives an infinitesimal negative imaginary part, and correspondingly the poles of the integrand are situated as shown in fig. V.30. We can now carry out the integration, closing the contour of integration with an infinitely large semicircle and calculating residues at corresponding poles. Since an exponential function stands within the integral sign, the contour of integration is closed below at $t_2 > t_1$ and above at $t_2 < t_1$. Correspondingly for $t_2 > t_1$ the residue is taken at the point $p_0 = E_p$, and for $t_2 < t_1$ at the point $p_0 = -E_p$. Thus we obtain

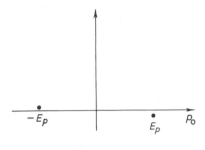

Fig. V.30

$$I(2,1) = \frac{i}{16\pi^3} \int \frac{1}{E_p} \exp\left[i\mathbf{p}\cdot(\mathbf{r}_2 - \mathbf{r}_1) - iE_p(t_2 - t_1)\right] d^3p \qquad (t_2 > t_1),$$

$$I(2,1) = \frac{i}{16\pi^3} \int \frac{1}{E_p} \exp\left[i\mathbf{p}\cdot(\mathbf{r}_2 - \mathbf{r}_1) + iE_p(t_2 - t_1)\right] d^3p \qquad (t_2 < t_1),$$

$$(124.12)$$

We see that since $E_p > 0$, then for $t_2 > t_1$ only states with a positive energy give a contribution, whereas for $t_2 < t_1$ correspondingly only states with a negative energy contribute. We note that the result obtained differs substantially from that obtained in non-relativistic theory. Indeed, in non-relativistic theory Green's function was assumed (see (29.3)) to be equal to zero for $t_2 < t_1$. We would also obtain an analogous expression in the relativistic case if both poles were circumvented from above, which corresponds to the replacement $p_0 \to p_0 + i\delta$ in the integrand.

A calculation analogous to that which we have just carried out shows that for such a replacement, there would correspond to a time $t_2 > t_1$ a summation over positive as well as negative energies, and for $t_2 < t_1$ we would obtain $I = 0$. However, from what follows it will be seen that the use of the Green's function proposed by Feynman and defined by formulae (124.12) is much more convenient.

By means of the Green's function introduced it is possible to construct the solution of the Dirac equation, i.e. to obtain a formula analogous to the non-relativistic relation (29.2). For this purpose it is simplest to make use of Gauss' theorem in 4-space ($d^4x = d^3x dt$)

$$i \int \frac{\partial F_\mu(x)}{\partial x_\mu} d^4x = \int_S F_\mu(x) n_\mu d\sigma(x) , \qquad (124.13)$$

where F is an arbitrary 4-vector, S is the surface bounding the given 4-dimensional volume, and $n(x)$ is the external normal to this surface at point x. Setting

$$F_\mu(x') = K(x-x')\gamma_\mu\psi(x') ,$$

we obtain

$$\frac{\partial F_\mu}{\partial x'_\mu} = \frac{\partial K(x-x')}{\partial x'_\mu}\gamma_\mu\psi(x') + K(x-x')\gamma_\mu\frac{\partial\psi(x')}{\partial x'_\mu} =$$

$$= -\frac{\partial K(x-x')}{\partial x_\mu}\gamma_\mu\psi(x') + K(x-x')\gamma_\mu\frac{\partial\psi(x')}{\partial x'_\mu} .$$

From (124.10) it follows that

$$\frac{\partial K(x-x')}{\partial x_\mu}\gamma_\mu = \gamma_\mu\frac{\partial K(x-x')}{\partial x_\mu} = \hat{\nabla}_x K(x-x') .$$

Making use of relations (124.1) and (124.2) we obtain

$$\frac{\partial F_\mu(x')}{\partial x'_\mu} = -(\hat{\nabla}_x+m)K(x-x')\psi(x') = i\delta^4(x-x')\psi(x') .$$

Substituting this expression into the 4-dimensional integral written above and using (124.13) we obtain

$$\psi(x) = -\int_S K(x-x')\gamma_\mu\psi(x')n_\mu(x')d\sigma(x') .$$

Denoting point x in terms of point 2, and point x' in terms of point 1, we can rewrite the relation obtained in the form

$$\psi(2) = \int_{t_{1'}} K(2,1)\gamma_4\psi(1)d^3x_1 - \int_{t'_1} K(2,1')\gamma_4\psi(1')d^3x_{1'} . \qquad (124.14)$$

Here two infinite space-like planes $t = t_1$ and $t = t_{1'}$, where $t_1 < t_2 < t_{1'}$, are chosen as the surface of integration. The integration over time-like surfaces can be dropped, since they are as spatially distant from point 2 as one wishes, and the function $K(2,1)$, as can be shown, decreases exponentially to zero in space-like directions as spatial distances increase indefinitely[*].

The function $K(2,1)$ contains a summation only over states with a positive energy, and the function $K(2,1')$, for $t_2 < t_{1'}$, only over states with a negative

* See R.P.Feynman, Phys. Rev. 76 (1949) 749.

energy. Hence the first integral in (124.14) differs from zero for the components $\psi(1)$ corresponding to particles with a positive energy, and the second integral is correspondingly not equal to zero for the components $\psi(1')$ corresponding to particles with a negative energy.

We see that the wave function of a particle at point 2 of the 4-dimensional space is defined by the Green's function and by the values of $\psi(1)$ and $\psi(1')$. Analogously, setting $F_\mu = \bar\psi(x')\gamma_\mu K(x'-x)$ it is easy to find the expressions for the function $\bar\psi(2)$:

$$\bar\psi(2) = \int_{t_1'>t_2} \bar\psi(1')\gamma_4 K(1',2)\,\mathrm{d}^3x_{1'} - \int_{t_1<t_2} \bar\psi(1)\gamma_4 K(1,2)\mathrm{d}^3x_1 , \qquad (124.15)$$

where the function $\bar\psi(x) = \psi^\dagger(x)\gamma_4$ and satisfies the equation

$$\frac{\partial\bar\psi}{\partial x_\mu}\,\gamma_\mu - m\bar\psi = 0$$

or

$$\bar\psi(\hat\nabla - m) = 0 .$$

In this form of notation the operator $\hat\nabla$ acts on functions standing on its left.

The components of the wave function which correspond to negative energies E are interpreted in Feynman's theory as the amplitudes of probability of finding the particle in the positron state, i.e. in a state with positive energy $+E$ and charge $+e$. Thus the function $\psi(2)$ is given if the amplitude of the electron state $\psi(1)$, at instant of time $t_1 < t_2$, and the amplitude of the positron state $\psi(1')$ at instant of time $t_{1'} > t_2$, are known.

The phase factor involved in the function $K(2,1)$ for $t_2 < t_1$ depends on time according to the law

$$\exp[iE_p(t_2-t_1)] = \exp[-iE_p|t_2-t_1|]$$

(see (124.12)). In other words, the time factor of the function $K(2,1')$ depends on $(t_2-t_{1'})$ in the same way as the phase factor of the wave function to particles with a positive energy. In accordance with this, positron states can be considered as states of a particle which has a positive energy but moves in the opposite direction along the time axis. To this there corresponds the fact that the state of a positron must be given at an instant of time $t_{1'} > t_2$ (the second integral in formula (124.14)).

We now assume that there is an external electromagnetic field. In our notation the Dirac equation in this case is written in the form

$$(\hat{\nabla}-ie\hat{A}+m)\psi = 0 , \tag{124.16}$$

where $\hat{A} = A_\mu \gamma_\mu$, and $A_4 = i\varphi$ (φ is the scalar potential).

Green's function is as usual defined by the equation

$$(\hat{\nabla}_x - ie\hat{A} + m)K^A(x\ x') = -i\delta(x-x') .$$

The function $K^A(2,1)$, as well as the function $K(2,1)$, contains in its expansion only components corresponding to positive energies for $t_2 > t_1$ and to negative energies for $t_2 < t_1$. Relations (124.14) and (124.15) remain valid, provided that $K(2,1)$ in them is replaced by $K^A(2,1)$.

Just as in the non-relativistic case, an integral equation satisfied by the function $K^A(2,1)$ can be formulated. Namely

$$K^A(2,1) = K(2,1) - e \int K(2,3)\hat{A}(3)K^A(3,1)d^4x_3 . \tag{124.17}$$

The derivation of integral equation (124.17) does not differ from that of integral equation (29.17). As we have shown in §58, an equation of such a type is conveniently solved by a method of successive approximations (see (58.5)).

§125. Green's function for a system of two particles

The expression found above for the wave function of one particle must be generalized to the case of a system of interacting particles. The simplest example of such a system is one consisting of two particles interconnected by an interaction of electromagnetic character. We note, first of all, that the Green's function of a system of two non-interacting particles is equal to the product of the Green's functions of each of the particles:

$$K(3,4;1,2) = K_a(3,1)K_b(4,2) . \tag{125.1}$$

Here $K_a(3,1)$ is the Green's function of the free particle a moving from point 1 to point 3. The quantity $K_b(4,2)$ for the particle b has an analogous meaning.

In the case of two interacting particles the Green's function given by formula (125.1) can be considered as the zero order approximation with respect to the interaction. Let us now find the Green's function $K^{(1)}(3,4;1,2)$ in the first approximation with respect to the interaction. That is, we consider two charged particles which are described by the Dirac equation. In order to write the interaction operator it is convenient, following Feynman, to consider at first the non-relativistic approximation and then to carry out the corresponding generalization to the case of relativistic particles. In the non-relativistic approximation the interaction between particles is described by the Coulomb law, and the function $K^{(1)}(3,4;1,2)$ by analogy with formula (58.3) is defined by the relation

$$K^{(1)}(3,4;1,2) = -ie^2 \int K_a(3,5)K_b(4,6)\frac{1}{r_{56}}\delta(t_{56})K_a(5,1)K_b(6,2)\,dx_5^4 dx_6^4,$$

(125.2)

where $r_{56} = |\mathbf{r}_5 - \mathbf{r}_6|$. The meaning of the expression for $K^{(1)}(3,4;1,2)$ is easily understood if it is compared with a diagram (fig. V.31) which is interpreted as follows. Particle a moves from point 1 to point 3, passing through the intermediate point 5. The line 2–6–4 describes the motion of particle b. To the line 1–5 of the diagram there corresponds the function of motion $K_a(5,1)$, and to the line 2–6 the function $K_b(6,2)$. The interaction between the particles takes place at points 5 and 6. The dotted line corresponds to the expression $(e^2/r_{56})\delta(t_{56})$, where $r_{56} = |\mathbf{r}_5 - \mathbf{r}_6|$ is the spatial distance between points 5 and 6, and $t_{56} = t_5 - t_6$, where t_5 and t_6 are the instants of time at which particles a and b arrive at points 5 and 6. The δ-function of the time argument means that in the non-relativistic approximation one has to disregard the time lag and to consider particles at points 5 and 6 at one and the same instant of time $t = t_5 = t_6$. The lines 5–3 and 6–4 correspond to the motion of free particles after the interaction (the functions $K_a(3,5)$ and $K_b(4,6)$ in (125.2)).

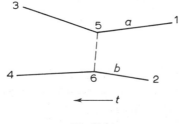

Fig. V.31

The generalization of expression (125.2) to the relativistic case involves first of all taking into account the interaction lag. At first sight it may seem that $\delta(t_{56})$ should be replaced by $\delta(t_{56} - r_{56})$, where r_{56} defines the time lag (in our notation the velocity of propagation of interaction $c = 1$). However, such a replacement would be incorrect. Indeed, the electromagnetic interaction represents the exchange of photons which have a positive energy.

However, the expansion of the δ-function in a Fourier integral contains positive as well as negative frequencies. Hence for the transition from the Coulomb interaction to the relativistic generalization taking into account the time lag, the δ-function should be replaced by the function δ_+ defined by the relation

$$\delta_+(x) = \lim_{\epsilon \to 0} \int_0^\infty e^{-i\omega(x-i\epsilon)} \frac{d\omega}{\pi} \equiv \lim_{\epsilon \to 0} \frac{1}{x - i\epsilon} \frac{1}{\pi i} . \tag{125.3}$$

The analogue of the δ-function defined in such a way contains the expansion only in terms of positive frequencies. Since t_{56} takes on positive as well as negative values, one takes the symmetrized combination

$$\frac{1}{2r_{56}} [\delta_+(t_{56} - r_{56}) + \delta_+(-t_{56} - r_{56})] = \delta_+(t_{56}^2 - r_{56}^2) = \delta_+(-x_{56}^2) .$$

This equality is immediately found by means of formula (125.3).

Furthermore, it is necessary to take into account that if the particles are moving, then in addition to the Coulomb interaction there is an electromagnetic interaction (see (25.27) of Part II). This leads to the fact that the interaction is defined by the expression $(1 - \mathbf{v}_5 \cdot \mathbf{v}_6) e^2 \delta_+(-x_{56}^2)$.

We shall obtain the interaction operator if we replace the velocity vectors $\mathbf{v}_5$ and $\mathbf{v}_6$ by the operators $\boldsymbol{\alpha}_a$ and $\boldsymbol{\alpha}_b$; each of the operators acts respectively on the variables of particle a and particle b. (Indeed, the velocity operator can easily be found by the formula $\mathbf{v} = [\hat{H}, \hat{\mathbf{r}}]$. But $\hat{H}$ (see (113.7)) is equal to $\hat{H} = i^{-1} \boldsymbol{\alpha} \nabla + \beta m$, and, commuting, we obtain $[\hat{H}, \mathbf{r}] = \boldsymbol{\alpha}$.) Then, by virtue of (113.14), one can write for the interaction operator in the relativistic case

$$(1 - \boldsymbol{\alpha}_a \cdot \boldsymbol{\alpha}_b) e^2 \delta_+(-x_{56}^2) = e^2 \beta_a \beta_b \gamma_{a\mu} \gamma_{b\mu} \delta_+(-x_{56}^2) .$$

To obtain the final expression for the function $K^{(1)}(3,4;1,2)$ we have to establish the connection between relativistic and non-relativistic Green's functions. For this we compare formulae (29.3) and (124.14). We see that the following correspondence holds:

$$K_{\text{non-rel}} \to K_{\text{rel}} \beta .$$

Thus for particles by the Dirac equation we have

$$K^{(1)}(3,4;1,2) \beta_a \beta_b = -ie^2 \int K_a(3,5) K_b(4,6) \gamma_{a\mu} \gamma_{b\mu} \delta_+(-x_{56}^2) \times$$

$$\times K_a(5,1) K_b(6,2) \beta_a \beta_b d^4 x_5 d^4 x_6 . \tag{125.4}$$

Multiplying (125.4) from the left by $\beta_a \beta_b$, we find finally

$$K^{(1)}(3,4;1,2) = -ie^2 \int K_a(3,5)K_b(4,6)\gamma_{a\mu}\gamma_{b\mu}\delta_+(-x_{56}^2) \times$$

$$\times K_a(5,1)K_b(6,2)d^4x_5 d^4x_6 =$$

(125.5)

$$= e^2 \int K_a(3,5)K_b(4,6)\gamma_{a\mu}D(-x_{56}^2)\gamma_{b\mu}K_a(5,1)K_b(6,2)d^4x_5 d^4x_6 \ .$$

The function $D(-x_{56}^2) = -i\delta_+(-x_{56}^2)$ is usually called the propagation function of the virtual photon. Thus we see that, taking into account relativistic effects, the diagram in fig. V.31 can be interpreted as follows. The functions K correspond to solid (electron) lines, the function D corresponds to the dotted line, and to the vertices there correspond the matrices $e\gamma_{a\mu}$ and $e\gamma_{b\mu}$.

All calculations in Feynman's theory are substantially simplified if they are carried out in the momentum representation. The form of the function K in the p-representation is given by formula (124.8). There remains to be determined the Fourier component of the function δ_+. We shall show that the following relation holds:

$$\delta_+(-x^2) = \frac{1}{4\pi^3} \int \frac{e^{ikx}}{k^2 - i\epsilon} d^4k \ ,$$

(125.6)

where ϵ is an infinitesimal quantity. It defines the rule for circumventing poles. We can convince ourselves of the validity of relation (125.6) by calculating the integral on the right directly:

$$\int \frac{e^{ikx}}{k^2 - i\epsilon} d^4k = \int e^{i\mathbf{k}\cdot\mathbf{r}} d^3k \int_{-\infty}^{\infty} \frac{e^{-ik_0x_0}}{k^2 - k_0^2 - i\epsilon} dk_0 \ .$$

We assume, for example, that $x_0 > 0$. Closing the contour of integration below and finding the residues, we obtain

$$2\pi i \int \frac{e^{i\mathbf{k}\cdot\mathbf{r}}}{2|\mathbf{k}|} e^{-ik|x_0|} d^3k \ .$$

Writing d^3k in the form $k^2 dk\, d\Omega$, integrating over angles and taking into account (125.3) we obtain the relation sought.

§126. Feynman diagrams

We shall now consider the rules for calculation of the probabilities of transition from one state into another by means of the mathematical apparatus presented in the preceding sections. For simplicity we shall first consider one particle (for example, an electron) which makes a transition from one state into another under the action of an external electromagnetic field. Let the electron at the initial instant of time $t = t_1$ be in the state $\psi(\mathbf{r}_1,t_1) = \psi(1)$, and at the instant of time $t = t_2$ let it be in the state $\psi(\mathbf{r}_2,t_2) = \psi(2)$ corresponding to a positive energy. The probability of transition into a particular state $\psi_n(\mathbf{r}_2,t_2)$ is, as always, defined by the square of the modulus of the corresponding amplitude of the expansion of the function $\psi(\mathbf{r}_2,t_2)$ in terms of the function $\psi_n(\mathbf{r}_2,t_2)$

$$M = \int \psi_n^\dagger(\mathbf{r}_2,t_2)\psi(\mathbf{r}_2,t_2)\,\mathrm{d}^3x_2 \ . \tag{126.1}$$

Expressing the function $\psi(\mathbf{r}_2,t_2)$ in terms of Green's function according to formula (124.14), we obtain

$$M = \int \psi_n^\dagger(\mathbf{r}_2,t_2)K^A(\mathbf{r}_2,t_2;\mathbf{r}_1,t_1)\beta\psi(\mathbf{r}_1,t_1)\,\mathrm{d}^3x_1\,\mathrm{d}^3x_2 \ .$$

In place of the Green's function K^A we can write its expansion in a series of successive approximations. We then obtain an expression for the transition amplitude M in the form of a perturbation theory series. Thus, for example, the transition amplitude in the first approximation of perturbation theory is equal to

$$M^{(1)} = -e \int \psi_n^\dagger(2)K(2,3)\hat{A}(3)K(3,1)\beta\psi(1)\,\mathrm{d}^3x_1\,\mathrm{d}^3x_2\,\mathrm{d}^4x_3 \ . \tag{126.2}$$

This expression can be written in a more compact form, using the relations

$$\psi(3) = \int K(3,1)\beta\psi(1)\,\mathrm{d}^3x_1 \ ,$$
$$\bar{\psi}_n(3) = \int \bar{\psi}_n(2)\beta K(2,3)\,\mathrm{d}^3x_2 \ .$$

Then for the transition amplitude we have

$$M^{(1)} = -e \int \bar{\psi}_n(3)\hat{A}(3)\psi(3)\,\mathrm{d}^4x_3 \ . \tag{126.3}$$

It is easy to obtain, in an analogous way, the second approximation of the transition amplitude

$$M^{(2)} = (-e)^2 \int \bar{\psi}_n(3)\hat{A}(3)K(3,4)\hat{A}(4)\psi(4)\,\mathrm{d}^4x_3\,\mathrm{d}^4x_4 \ . \tag{126.4}$$

If the initial and final states are described by plane waves, then formulae (126.3) and (126.4) are conveniently rewritten in the momentum representation. Setting

$$\bar{\psi}_n(3) = \bar{u}(\mathbf{p}_2)\mathrm{e}^{-ip_2x_3} \ ,$$
$$\psi(3) = u(\mathbf{p}_1)\mathrm{e}^{ip_1x_3}$$

and using the Fourier representation of the operator $\hat{A}$

$$\hat{A}(3) = \int \hat{a}(k)\,e^{ikx_3}\,d^4k \,, \tag{126.5}$$

we obtain for the transition amplitude of the first order (126.3)

$$\begin{aligned}
M^{(1)} &= -e \int d^4x_3 \int e^{-ip_2x_3+ikx_3+ip_1x_3}\bar{u}(\mathbf{p}_2)\hat{a}(k)u(\mathbf{p}_1)\,d^4k = \\
&= -e(2\pi)^4 \int \bar{u}(\mathbf{p}_2)\hat{a}(k)\delta^4(k+p_1-p_2)u(\mathbf{p}_1)\,d^4k = \\
&= -e(2\pi)^4\bar{u}(\mathbf{p}_2)\hat{a}(p_2-p_1)u(\mathbf{p}_1) \,. \tag{126.6}
\end{aligned}$$

For the transition amplitude of the second order we have, correspondingly,

$$\begin{aligned}
M^{(2)} &= (-e)^2 \int \bar{u}(\mathbf{p}_2)\hat{a}(k_1)S(p)\hat{a}(k)u(\mathbf{p}_1)d^4pd^4kd^4k_1 \times \\
&\quad \times \int e^{-ip_2x_3+ik_1x_3+ip(x_3-x_4)+ikx_4+ip_1x_4}\,d^4x_3 d^4x_4 = \\
&= e^2(2\pi)^8 \int \bar{u}(\mathbf{p}_2)\hat{a}(k_1)\delta^4(k_1+p-p_2)S(p)\hat{a}(k)\times \\
&\quad \times \delta^4(p_1+k-p)u(\mathbf{p}_1)\,d^4pd^4kd^4k_1 = \\
&= e^2(2\pi)^8 \int \bar{u}(\mathbf{p}_2)\hat{a}(p_2-p_1-k) \times \\
&\quad \times \frac{1}{i(2\pi)^4\,[i(\hat{p}_1+\hat{k})+m]}\hat{a}(k)u(\mathbf{p}_1)\,d^4k \,. \tag{126.7}
\end{aligned}$$

Formulae (126.6) and (126.7) can be associated with pictorial representations, called Feynman diagrams. As will be shown below, to each line and to each crossing of lines (called a vertex) in the Feynman diagram there corresponds a definite factor in the transition amplitude. In the case of complex processes such diagrams make it possible to simplify the construction of expressions for the transition amplitudes. In a Feynman diagram we represent the states of electrons and positrons by solid lines, and the states of the electromagnetic field by dotted lines. The arrows on the lines show the order of writing the terms of the transition amplitude. To an increase in time there corresponds the motion of the particle from the right to the left.

Let us consider the simplest Feynman diagram (fig. V.32) corresponding to the following process: an electron with momentum $\mathbf{p}_1$ was scattered by an external electromagnetic field and made a transition into a new state with momentum $\mathbf{p}_2$. The probability amplitude of this transition is given by formula (126.6). In fig. V.32 the free electron with momentum $\mathbf{p}_1$ is represented by the solid line AB. This straight line corresponds to the first factor in the transition amplitude $M^{(1)}$ (the factors are numbered from the right to the left), the bispinor $u(\mathbf{p}_1)$. At point B the electron is scattered by the electromagnetic field represented by the dotted line. The crossing of the solid and dotted lines (the vertex) in the Feynman diagram corresponds to the operator

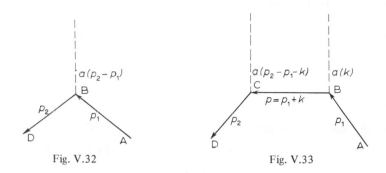

Fig. V.32   Fig. V.33

$-e\hat{a}(k)$ in the transition amplitude $M^{(1)}$ multiplied by the δ-function of the momenta of all three particles. The electron with momentum $\mathbf{p}_2$ is represented by the straight line BD. In the amplitude $M^{(1)}$ the bispinor $\bar{u}(\mathbf{p}_2)$ corresponds to it.

We see that the order of the process with respect to the charge e is defined by the number of vertices in the Feynman diagram. This is particularly clearly seen from the consideration of the Feynman diagram for a process of the second order (fig. V.33). This diagram corresponds to the process of electron scattering in the second approximation of perturbation theory. The line AB (called the external line) represents the motion of the free electron. To it there corresponds the bispinor $u(\mathbf{p}_1)$ in the transition amplitude $M^{(2)}$. The scattering of the electron takes place at vertex B. In the transition amplitude $M^{(2)}$ there corresponds to vertex B the factor $-e\hat{a}(k)$ and the momentum δ-function $\delta^4(p_1+k-p)$. The line BC joining the two vertices is called the internal line. To it there corresponds in $M^{(2)}$ the factor S, the Fourier component of the Green's function defined by formula (124.8). The external field acts at vertex C. To the vertex C there corresponds in $M^{(2)}$ the operator $-e\hat{a}(k_1)$ and the δ-function $\delta^4(k_1+p-p_2)$. The external line CD represents the motion of the electron with momentum $\mathbf{p}_2$. To the line CD in $M^{(2)}$ there corresponds the bispinor $\bar{u}(\mathbf{p}_2)$. Since the law of conservation of 4-momentum in the transition from state $\mathbf{p}_1$ into state $\mathbf{p}_2$ is fulfilled for arbitrary values of the wave vector k, the integration is carried out over the vector k (or k_1). The value of the numerical factor in the expression for the amplitude is determined by the number of δ-functions involved in it, which is equal to the number of vertices. Each vertex brings into $M^{(2)}$ a factor of $(2\pi)^4$.

The Feynman diagram for processes involving positrons can be constructed in exactly the same way. For example, the diagram in fig. V.33 also describes positron scattering in the second approximation of perturbation theory. Since in Feynman's theory the positron is considered as an electron moving backward in time, this diagram defines the transition amplitude of the positron from a state with momentum $-\mathbf{p}_2$ into a state with momentum $-\mathbf{p}_1$. For the quantity $M^{(2)}$ in this case we have

$$M^{(2)} = \frac{1}{i} e^2 (2\pi)^4 \int \bar{v}(\mathbf{p}_2) \hat{a}(p_2 - p_1 - k) \frac{1}{i(\hat{p}_1 + \hat{k}) + m} \hat{a}(k) v(\mathbf{p}_1) \, d^4 k \ . \tag{126.8}$$

Here v is the Dirac bispinor corresponding to a state with a negative energy.

The relations derived make it possible to consider processes associated with the emission and absorption of free electrons in addition to scattering processes in an external electromagnetic field. For this the operator $\hat{A}$ in the general formulae (126.3) and (126.4) must correspond to the pole of one emitted or one absorbed photon. In accordance with formulae (102.3) and (102.5), there corresponds to the pole of the emitted photon the vector potential

$$A_\mu = (2\pi/\omega)^{\frac{1}{2}} e_\mu e^{-ikx} \ , \tag{126.9}$$

and to the absorbed photon

$$A_\mu = (2\pi/\omega)^{\frac{1}{2}} e_\mu e^{ikx} \ . \tag{126.10}$$

Here e_μ is the polarization vector, and k denotes a four-dimensional wave vector. Since only one photon is absorbed or emitted, the matrix elements of the operators $\hat{a}$ and $\hat{a}^\dagger$ are equal to one. The diagrams shown in fig. V.34 describe processes involving two free photons. Thus, for example, these diagrams describe the process of Compton scattering, i.e. the scattering of a photon by a free electron. In this process the photon before scattering had wave vector $\mathbf{k}_1$, and after scattering $\mathbf{k}_2$. In the diagram shown in fig. V.34a

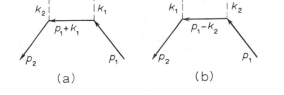

Fig. V.34

this corresponds to the fact that the photon is first absorbed and then emitted by the electron. The diagram shown in fig. V.34b also corresponds to the same process, where first the photon $\mathbf{k}_2$ is emitted and then the photon $\mathbf{k}_1$ is absorbed. Of course, the words 'first' and 'then' refer only to the order of writing the factors in the transition amplitude and have no other, physical meaning.

The Feynman diagrams shown in fig. V.34 make it possible to write immediately the transition amplitude without having to carry out each time special calculations of the type carried out in deriving formulae (126.6) and (126.7). The total transition amplitude is defined by the sum of the amplitudes corresponding to diagrams V.34a and V.34b.

It has the form

$$
M = \frac{e^2}{i}(2\pi)^4 \left[\bar{u}(\mathbf{p}_2) \left(\frac{2\pi}{\omega_2}\right)^{\frac{1}{2}} \hat{e}_2 \frac{1}{i(\hat{p}_1 + \hat{k}_1) + m} \left(\frac{2\pi}{\omega_1}\right)^{\frac{1}{2}} \hat{e}_1 u(\mathbf{p}_1) + \right.
$$
$$
\left. + \bar{u}(\mathbf{p}_2) \left(\frac{2\pi}{\omega_1}\right)^{\frac{1}{2}} \hat{e}_1 \frac{1}{i(\hat{p}_1 - \hat{k}_2) + m} \left(\frac{2\pi}{\omega_2}\right)^{\frac{1}{2}} \hat{e}_2 u(\mathbf{p}_1) \right] \delta^4(p_1 + k_1 - p_2 - k_2) ,
$$

(126.11)

where $\hat{e}_1 = e_{1\mu}\gamma_\mu$, $\hat{e}_2 = e_{2\mu}\gamma_\mu$, and e_1 and e_2 are the polarization vectors of the photon before and after scattering.

The appearance of the δ-function is easily understood if it is taken into account that in the given case expressions of the type (126.9) and (126.10), containing no integration over k, are substituted for the external field operator of the form of (126.5) in the expression of the type (126.4). Thus in an expression of the type (126.7) there will be no integration over k and k_1. After the integration over p there will remain one δ-function, expressing the law of conservation of energy and momentum in the Compton process.

Diagrams of such a type describe, for example, the process of annihilation of an electron with momentum $\mathbf{p}_1$ and of a positron with momentum $-\mathbf{p}_2$. Two photons with momenta $\mathbf{k}_1$ and $\mathbf{k}_2$ are produced in the annihilation (fig. V.35). The amplitude of the two-photon annihilation of the pair, according to the same rules, is written in the form

$$
M = \frac{1}{i} e^2 (2\pi)^5 \frac{1}{(\omega_1 \omega_2)^{\frac{1}{2}}} \times
$$
$$
\times \left[\bar{v}(\mathbf{p}_2) \left(\hat{e}_1 \frac{1}{i(\hat{p}_1 - \hat{k}_2) + m} \hat{e}_2 + \hat{e}_2 \frac{1}{i(\hat{p}_1 - \hat{k}_1) + m} \hat{e}_1 \right) u(\mathbf{p}_1) \right] \times
$$
$$
\times \delta^4(p_1 - p_2 - k_1 - k_2) .
$$

(126.12)

As another example let us consider the electron bremsstrahlung, i.e. the

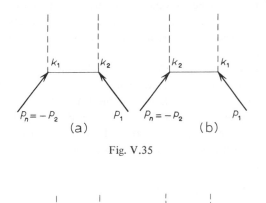

Fig. V.35

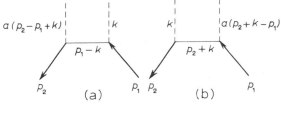

Fig. V.36

radiation arising when a fast electron traverses the field of a nucleus. The diagram corresponding to this process is shown in fig. V.36. An electron with momentum $\mathbf{p}_1$ scattered by an external field a emits a photon of momentum $\mathbf{k}$ and polarization $\mathbf{e}$ and makes a transition into a state with momentum $\mathbf{p}_2$. In this case two processes, shown in the diagrams of fig. V.36a and b, are possible. The total transition amplitude, in accordance with the rules presented, is of the form

$$M = \frac{e^2}{i} \frac{(2\pi)^{\frac{9}{2}}}{\omega^{\frac{1}{2}}} \bar{u}(\mathbf{p}_2) \left[\hat{a}(p_2 - p_1 + k) \frac{1}{i(\hat{p}_1 - \hat{k}) + m} \hat{e} + \right.$$

$$\left. + \hat{e} \frac{1}{i(\hat{p}_2 + \hat{k}) + m} \hat{a}(p_2 + k - p_1) \right] u(\mathbf{p}_1) . \tag{126.13}$$

Let us now consider processes associated with the interaction of two particles, for example the scattering of an electron by a μ-meson. The transition amplitude is again calculated by expanding the wave function of the system in terms of the products of the wave functions of the free particles. The wave function of a system of two particles is defined by the Green's function (125.5). These calculations lead to graphs which are plotted according to the same principles as for one particle. In the first approximation of perturbation theory there corresponds to the scattering process the Feynman diagram shown in fig. V.37.

Here the solid lines AB and CD correspond to the motion of free particles with momenta p_1 and p_2. The electromagnetic interaction between particles amounts to photon exchange. A virtual photon, which is emitted by the second particle (vertex D), is absorbed at vertex B. Lines BE and DF correspond to the motion of the particles after the interaction with momenta p_3 and p_4. The transition amplitude is constructed according to the usual rules and has the form

$$M = e^2 (2\pi)^8 \bar{u}_a(\mathbf{p}_3) \bar{u}_b(\mathbf{p}_4) \gamma_{a\mu} D_f(p_1 - p_3) \times$$

$$\times \gamma_{b\mu} u_a(\mathbf{p}_1) u_b(\mathbf{p}_2) \delta^4(p_3 + p_4 - p_1 - p_2) . \qquad (126.14)$$

Here D_f is the Fourier component of the propagation function of the virtual photon D, which, according to (125.5) and (125.6), is given by the formula

$$D_f(k) = - \frac{i}{4\pi^3} \frac{1}{k^2} . \qquad (126.15)$$

Let us dwell briefly on corrections which arise in higher approximations of perturbation theory. The diagrams corresponding to these corrections must,

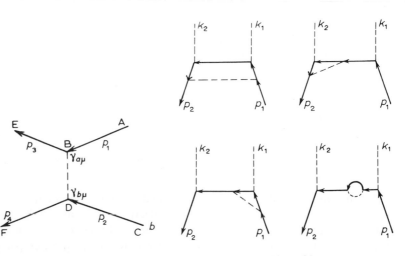

Fig. V.37.   Fig. V.38

naturally, contain a larger number of vertices in comparison with the corresponding basic diagram (to each vertex there corresponds the smallness parameter e). Namely, it is the number of vertices that determines the order of smallness of the correction considered.

Let us consider, for example, diagrams corresponding to the next approximation in the theory of the Compton effect. As is easily understood, to the· diagram in fig. V.34a there correspond the corrections in fig. V.38, and analogously for the diagram in fig. V.34b. All these diagrams differ from the initial diagram (fig. V.34) by the presence of an internal photon line. This line corresponds, as we have already seen, to the emission and absorption of a virtual quantum. Hence these corrections are usually called radiative corrections. The calculation of these corrections involves particular difficulties and requires the so-called renormalization method. We shall not dwell on these problems*.

§127. The Compton effect

To illustrate the technique of calculation of cross sections in Feynman's theory we shall consider the theory of the Compton effect in more detail. The transition amplitude for this process has already been obtained by means of a Feynman diagram (fig. V.34) in the form of (126.11). Separating the δ-function, we write the transition amplitude in the form

$$M = M_{21}\delta^4(p_1+k_1-p_2-k_2),\qquad(127.1)$$

where

$$M_{21} = \frac{ie^2(2\pi)^5}{(\omega_1\omega_2)^{\frac{1}{2}}}\,\bar{u}(\mathbf{p}_2)\left[\hat{e}_2\frac{i(\hat{p}_1+\hat{k}_1)-m}{(\hat{p}_1+\hat{k}_1)^2+m^2}\hat{e}_1+\hat{e}_1\frac{i(\hat{p}_1-\hat{k}_2)-m}{(\hat{p}_1-\hat{k}_2)^2+m^2}\hat{e}_2\right]u(\mathbf{p}_1).$$

The probability of the Compton effect is given by the formula

$$P'_{21} = |M_{21}|^2(\delta^4(p_1+k_1-p_2-k_2))^2.\qquad(127.2)$$

In order to eliminate the square of the δ-function it is convenient to make use of the definition of the 4-dimensional δ-function

$$\delta^4(p) = \frac{1}{(2\pi)^4}\int e^{ipx}\,\mathrm{d}^4x\ .$$

---

* See, for example, A.I.Akhiezer and V.B.Berestetskii, *Quantum electrodynamics* (Interscience Publ., New York, 1965).

At the point $p = 0$, which, as is seen from (127.2), is the only one playing an important role, the 4-dimensional integral is equal to $VT/(2\pi)^4$, where V is the normalization volume, and T is the duration of the process. Choosing V to be unity, we have for the transition probability P_{21} per unit time

$$P_{21} = \frac{1}{T} P'_{21} = \frac{1}{(2\pi)^4} |M_{21}|^2 \delta^4 (p_1 + k_1 - p_2 - k_2) . \qquad (127.3)$$

The final state of the system is defined by the momentum of the electron p_2 and of the scattered photon k_2. The number of final states in the interval of momenta dp_2 and dk_2 is given by the usual relation $dp_2 dk_2 / (2\pi)^6$. The probability of transition into the interval of final states $dp_2 dk_2$ is written in the form

$$dW_{21} = \frac{1}{(2\pi)^4} |M_{21}|^2 \delta^4 (p_1 + k_1 - p_2 - k_2) \frac{dp_2 dk_2}{(2\pi)^6} . \qquad (127.4)$$

The 4-dimensional δ-function expresses the law of conservation of energy and momentum

$$p_1 + k_1 = p_2 + k_2 . \qquad (127.5)$$

From this relation it is easy to define the frequency of the scattered photon as a function of the scattering angle ϑ, i.e. the angle between the vectors k_1 and k_2

$$p_1^2 + k_1^2 + 2p_1 k_1 = p_2^2 + k_2^2 + 2p_2 k_2 ,$$

but

$$p_1^2 = \mathbf{p}_1^2 - E_1^2 = -m^2 = p_2^2 ,$$

$$k_1^2 = \mathbf{k}_1^2 - \omega_1^2 = k_2^2 = 0$$

and, consequently, $p_1 k_1 = p_2 k_2$. Making use of (127.5) we have

$$p_1 k_1 = p_1 k_2 + k_1 k_2 .$$

We assume for simplicity that the electron was initially at rest: $\mathbf{p}_1 = 0$, $E_1 = m$. After a simple calculation (see (17.11) of Part II) we find

$$\omega_2 = \frac{\omega_1}{1 + m^{-1} \omega_1 (1 - \cos \vartheta)} . \qquad (127.6)$$

Integrating (127.4) over the three momentum components we obtain

$$dW_{21} = \frac{1}{(2\pi)^{10}} |M_{21}|^2 \delta(m+\omega_1-E_2-\omega_2)\omega_2^2 \, d\omega_2 d\Omega . \quad (127.7)$$

We integrate this expression over frequency ω_2. It should then be recalled that the energy E_2 is also a function of ω_2,

$$E_2 = [m^2+p_2^2]^{\frac{1}{2}} = [m^2+(k_1-k_2)^2]^{\frac{1}{2}} = [m^2+\omega_1^2+\omega_2^2-2\omega_1\omega_2 \cos \vartheta]^{\frac{1}{2}} .$$

We introduce the new variable $y = E_2 + \omega_2$

$$dW_{21} = \frac{1}{(2\pi)^{10}} |M_{21}|^2 \delta(m+\omega_1-y)\omega_2^2 \frac{1}{\partial y/\partial \omega_2} \, dy d\Omega . \quad (127.8)$$

Integrating with respect to y we have

$$dW_{21} = \frac{1}{(2\pi)^{10}} |M_{21}|^2 \omega_2^2 \frac{1}{\partial y/\partial \omega_2}\bigg|_{\omega_2+E_2=m+\omega_1} d\Omega , \quad (127.9)$$

but

$$\frac{\partial y}{\partial \omega_2} = \frac{\partial(E_2+\omega_2)}{\partial \omega_2} = 1 + \frac{1}{E_2} (\omega_2-\omega_1 \cos \vartheta) . \quad (127.10)$$

Since the value of ω_2 satisfying the energy conservation law is used, then substituting (127.6) into (127.10), we find for the transition probability (127.9)

$$dW_{21} = \frac{1}{(2\pi)^{10}} \frac{\omega_2^3 E_2}{m\omega_1} |M_{21}|^2 \, d\Omega . \quad (127.11)$$

We shall obtain the cross section for the process by dividing the transition probability by the incident photon flux density. For a normalization such that there is 1 photon in volume $V = 1$ the flux density is numerically equal to the velocity of light c. In the system of units that we have chosen $c = 1$ and the cross section turns out to be numerically equal to dW_{21}. We also set (see (127.1))

$$M_{21} = ie^2 \frac{(2\pi)^5}{2m(\omega_1\omega_2)^{\frac{1}{2}}} (\bar{u}(p_2)\hat{Q}u(p_1)) ,$$

where

$$Q = \left[\hat{e}_2 \frac{i(\hat{p}_1+\hat{k}_1) - m}{(\hat{p}_1+\hat{k}_1)^2 + m^2} \hat{e}_1 + \hat{e}_1 \frac{i(\hat{p}_1-\hat{k}_2) - m}{(\hat{p}_1-\hat{k}_2)^2 + m^2} \hat{e}_2 \right] 2m .$$

The expression for the operator Q can be somewhat simplified if it is taken into account that

$$(\hat{p}_1 + \hat{k}_1)^2 + m^2 = p_1^2 + k_1^2 + 2p_1 k_1 + m^2 = 2p_1 k_1 = -2\omega_1 m \ .$$

Analogously for the denominator of the second fraction one can write

$$(\hat{p}_1 - \hat{k}_2)^2 + m^2 = p_1^2 + k_2^2 - 2p_1 k_2 + m^2 = -2p_1 k_2 = 2\omega_2 m \ .$$

Then for $\hat{Q}$ we have

$$\hat{Q} = \hat{e}_1 \frac{i(\hat{p}_1 - \hat{k}_2) - m}{\omega_2} \hat{e}_2 - \hat{e}_2 \frac{i(\hat{p}_1 + \hat{k}_1) - m}{\omega_1} \hat{e}_1 \ . \qquad (127.12)$$

Correspondingly the differential cross section is equal to

$$d\sigma = \tfrac{1}{4} r_0^2 \frac{\omega_2^2 E_2}{m\omega_1^2} |\bar{u}(\mathbf{p}_2) \hat{Q} u(\mathbf{p}_1)|^2 \, d\Omega \ , \qquad (127.13)$$

where $r_0 = e^2/m$ is the classical radius of the electron.

The expression obtained describes a process in which the electron and photon in the initial and final states have definite polarizations. If the electrons in the initial state are not polarized and we are not interested in the polarization in the final state, then the cross section must be averaged over the spin states of the electron in the initial state and summed over the final spin states. Consequently we have to determine the quantity

$$\overline{d\sigma} = \frac{1}{2} \sum_{\sigma_2} \sum_{\sigma_1} \tfrac{1}{4} r_0^2 \frac{\omega_2^2 E_2}{m\omega_1^2} |\bar{u}_{\sigma_2}(\mathbf{p}_2) \hat{Q} u_{\sigma_1}(\mathbf{p}_1)|^2 \, d\Omega =$$

$$= \tfrac{1}{8} r_0^2 \frac{\omega_2^2 E_2}{m\omega_1^2} \sum_{\sigma_1, \sigma_2} (u_{\sigma_2}^\dagger \beta \hat{Q} u_{\sigma_1})(u_{\sigma_1}^\dagger \hat{Q}^\dagger \beta u_{\sigma_2}) \ , \qquad (127.14)$$

where u_{σ_1} and u_{σ_2} are states with a definite polarizations. For example, $(\hat{s} \cdot \mathbf{p}_2) u_{\sigma_2} = \sigma_2 |\mathbf{p}_2| u_{\sigma_2}$, i.e. σ_2 is the spin component along the direction of motion, $\sigma_2 = \pm\tfrac{1}{2}$. The operator $\hat{s}$ is defined in §117.

The sum over σ_1 involved in (127.14) is conveniently rewritten in the form

$$\sum_{\sigma_1} (u_{\sigma_2}^\dagger \beta \hat{Q} u_{\sigma_1})(u_{\sigma_1}^\dagger \hat{Q}^\dagger \beta u_{\sigma_2}) = \sum_{\sigma_1} (\sigma_2 |\beta \hat{Q}| \sigma_1)(\sigma_1 |\hat{Q}^\dagger \beta| \sigma_2) \ . \qquad (127.15)$$

The summation is carried out over two spin states σ_1 with a positive energy. If the summation were also carried out over states with a negative energy, i.e. over all possible states (for given momentum), then expression (127.15) would be considerably simplified and would represent the matrix element $\langle \sigma_2 | \beta \hat{Q} \hat{Q}^\dagger \beta | \sigma_2 \rangle$ (in correspondence with the rule of multiplication of matrices (45.6)).

In order to extend the summation to all four intermediate states, the following method is used in calculations of such a type: the auxiliary operator $\hat{R}$, called the projection operator, is introduced

$$\hat{R} = \frac{\hat{H} + |E|}{2|E|} = \frac{\boldsymbol{\alpha} \cdot \hat{\mathbf{p}} + \beta m + |E|}{2|E|} . \tag{127.16}$$

The action of this operator on the functions u and v which correspond to positive and negative energies is defined by the equations

$$\frac{\hat{H} + |E|}{2|E|} u = u ; \qquad \frac{\hat{H} + |E|}{2|E|} v = 0 . \tag{127.17}$$

Replacing the function u_{σ_1} in (127.15) by the function

$$\frac{\boldsymbol{\alpha} \cdot \hat{\mathbf{p}}_1 + \beta m + |E_1|}{2|E_1|} u_{\sigma_1}$$

we can formally extend the summation also to states with a negative energy, since in the presence of the projection operator, in correspondence with (127.17), they give no contribution to the result.

In place of the matrices α_k and β we introduce the matrices γ_μ

$$\alpha_k = i\gamma_4 \gamma_k \quad (k=1,2,3) , \qquad \beta = \gamma_4 .$$

Substituting the matrices γ_μ into (127.16) we obtain

$$\hat{R} = -\frac{1}{2E} (i\gamma_\mu p_\mu - m)\gamma_4 = -\frac{1}{2E} (i\hat{p} - m)\gamma_4 . \tag{127.18}$$

The sum (127.15) can be rewritten in the form

$$\sum_{\sigma_1, E_1} (u_{\sigma_2}^\dagger \gamma_4 \hat{Q} \hat{R}_1 u_{\sigma_1})(u_{\sigma_1}^\dagger \hat{Q}^\dagger \gamma_4 u_{\sigma_2}) = (u_{\sigma_2}^\dagger \gamma_4 \hat{Q} \hat{R}_1 \hat{Q}^\dagger \gamma_4 u_{\sigma_2}) . \tag{127.19}$$

Further, we calculate the sum over the components of the spin in the final state of the electron σ_2. Here it is also convenient to pass to the summation over all four states, introducing the projection operator $\hat{R}_2$

$$S = \sum_{\sigma_2, E_2} (u_{\sigma_2}^\dagger \gamma_4 \hat{Q} \hat{R}_1 \hat{Q}^\dagger \gamma_4 \hat{R}_2 u_{\sigma_2}) . \tag{127.20}$$

The summation is carried out over states with a positive energy as well as over states with a negative energy. We see that expression (127.20) represents the sum of diagonal matrix elements, i.e.

$$S = \text{Tr}\,(\gamma_4 \hat{Q} \hat{R}_1 \hat{Q}^\dagger \gamma_4 \hat{R}_2) = \text{Tr}\,(\hat{Q} R_1 \hat{Q}^\dagger \gamma_4 \hat{R}_2 \gamma_4)\,, \qquad (127.21)$$

since a cyclic permutation of matrices can be carried out under the sign Tr, as can easily be checked directly. Substituting this expression into (127.14) and taking into account that $\gamma_4^2 = 1$, $\hat{\mathbf{p}}_1 = 0$, $E_1 = m$, we obtain

$$d\sigma = \tfrac{1}{32} r_0^2 \frac{\omega_2^2}{m^2 \omega_1^2} \text{Tr}\,[\hat{Q}(i\hat{p}_1 - m)\gamma_4 \hat{Q}^\dagger \gamma_4 (i\hat{p}_2 - m)]\,. \qquad (127.22)$$

For any operator of the form of $\hat{A}$, the fourth component of which is imaginary while three components are real, the following equality holds:

$$\gamma_4 \hat{A}^\dagger \gamma_4 = -\hat{A}\,. \qquad (127.23)$$

For the product of operators we have correspondingly

$$\gamma_4 \hat{A}^\dagger \hat{B}^\dagger \hat{C}^\dagger \gamma_4 = \gamma_4 \hat{A}^\dagger \gamma_4 \gamma_4 \hat{B}^\dagger \gamma_4 \gamma_4 \hat{C}^\dagger \gamma_4 = (-\hat{A})(-\hat{B})(-\hat{C})\,. \qquad (127.24)$$

Then the expression $\gamma_4 \hat{Q}^\dagger \gamma_4$ is rewritten in the form

$$\gamma_4 \{\omega_1^{-1} \hat{e}_1^\dagger [i(\hat{p}_1 + \hat{k}_1)^\dagger + m]\hat{e}_2^\dagger - \omega_2^{-1} \hat{e}_2^\dagger [i(\hat{p}_1 - \hat{k}_2)^\dagger + m]\hat{e}_1^\dagger\} \gamma_4 =$$
$$= \{\omega_1^{-1} \hat{e}_1 [-i(\hat{p}_1 + \hat{k}_1) + m]\hat{e}_2 + \omega_2^{-1} \hat{e}_2 [i(\hat{p}_1 - \hat{k}_2) - m]\hat{e}_1\}\,. \qquad (127.25)$$

We now sum the cross section over the final states of the photon and average over the initial states. In order to calculate the cross section for the case where the incident photon is polarized along axis 1 and the scattered photon along axis 2, one has to substitute the values $\hat{e}_1 = \gamma_1$ and $\hat{e}_2 = \gamma_2$ in the expression (127.12) for $\hat{Q}$. Since, however, the incident photons are not polarized and we are not interested in the polarization of the scattered photons, we have to substitute γ_ν for $\hat{e}_1$ and γ_μ for $\hat{e}_2$ and to sum over all values of the indices ν and μ. We then have

$$d\sigma = \frac{r_0^2 \omega_2^2}{32 m^2 \omega_1^2} \times$$

$$\qquad (127.26)$$

$$\times \text{Tr}\,\{[\omega_2^{-1} \gamma_\nu (i(\hat{p}_1 - \hat{k}_2) - m)\gamma_\mu - \omega_1^{-1} \gamma_\mu (i(\hat{p}_1 + \hat{k}_1) - m)\gamma_\nu](i\hat{p}_1 - m) \times$$
$$\times [\omega_2^{-1} \gamma_\mu (i(\hat{p}_1 - \hat{k}_2) - m)\gamma_\nu - \omega_1^{-1} \gamma_\nu (i(\hat{p}_1 + \hat{k}_1) - m)\gamma_\mu](i\hat{p}_2 - m)\},$$

The summation is carried out over the twice repeated indices μ and ν.

Although a free photon can be polarized in two directions perpendicular to the direction of motion, the summation over μ and ν can actually be carried out over all four values. This is associated with the fact that there are no real photons polarized in the direction of motion and along the time axis, and taking them into account formally does not change the final result*.

Relation (127.26) is conveniently rewritten in the form

$$d\sigma = \frac{r_0^2 \omega_2^2}{32m^2 \omega_1^2} (\mathrm{Tr}\, F_1 + \mathrm{Tr}\, F_2) , \qquad (127.27)$$

where

$$F_1 = [\omega_2^{-1}\gamma_\nu(i\hat{q}_2-m)\gamma_\mu - \omega_1^{-1}\gamma_\mu(i\hat{q}_1-m)\gamma_\nu](i\hat{p}_1-m) \times$$
$$\times [\gamma_\mu \omega_2^{-1}(i\hat{q}_2-m)\gamma_\nu](i\hat{p}_2-m) , \qquad (127.28)$$

$$q_1 = p_1 + k_1 , \qquad q_2 = p_1 - k_2 .$$

The expression for F_2 is obtained from F_1 by the replacement

$$q_1 \to q_2 , \quad q_2 \to q_1 ; \quad \omega_1 \to -\omega_2 , \quad \omega_2 \to -\omega_1 .$$

For further calculations it is convenient to use Feynman's formulae, which are easily checked by direct calculation:

$$\begin{aligned}
\gamma_\nu \gamma_\nu &= 4 , \\
\gamma_\nu \hat{A} \gamma_\nu &= -2\hat{A} , \\
\gamma_\nu \hat{A}_1 \hat{A}_2 \gamma_\nu &= 4(A_1 A_2) , \\
\gamma_\nu \hat{A}_1 \hat{A}_2 \hat{A}_3 \gamma_\nu &= -2\hat{A}_3 \hat{A}_2 \hat{A}_1 .
\end{aligned} \qquad (127.29)$$

Making use of these expressions it is easy to carry out the summation over μ and ν in (127.27).

Thus setting $F_1 = F_1' + F_1''$ we find

$$\mathrm{Tr}\, F_1' = \mathrm{Tr}\, [\gamma_\nu \omega_2^{-1}(i\hat{q}_2-m)\gamma_\mu(i\hat{p}_1-m)\gamma_\mu \omega_2^{-1}(i\hat{q}_2-m)\gamma_\nu(i\hat{p}_2-m)] =$$
$$= \mathrm{Tr}\, [\omega_2^{-1}(i\hat{q}_2-m)\gamma_\mu(i\hat{p}_1-m)\gamma_\mu \omega_2^{-1}(i\hat{q}_2-m)\gamma_\nu(i\hat{p}_2-m)\gamma_\nu] =$$
$$= 4\omega_2^{-2}\, \mathrm{Tr}\, [(i\hat{q}_2-m)(i\hat{p}_1+2m)(i\hat{q}_2-m)(i\hat{p}_2+2m)] , \qquad (127.30)$$

$$\mathrm{Tr}\, F_1'' = -\,\mathrm{Tr}\, [\gamma_\mu \omega_1^{-1}(i\hat{q}_1-m)\gamma_\nu(i\hat{p}_1-m)\gamma_\mu \omega_2^{-1}(i\hat{q}_2-m)\gamma_\nu(i\hat{p}_2-m)] =$$
$$= -\,4\omega_1^{-1}\omega_2^{-1}\, \mathrm{Tr}\, \{[2i(q\,q_2)\hat{p}_1 + m\hat{q}_2\hat{p}_1 - im\hat{q}_2(i\hat{q}_1-m) -$$
$$-\, im\hat{p}_1(i\hat{q}_1-m) + m^2(i\hat{q}_1+2m)](i\hat{p}_2-m)\} . \qquad (127.31)$$

In calculating the traces it is necessary to use the following rules:

---

* See, for example, A.I.Akhiezer and V.B.Berestetskii, *Quantum electrodynamics* (Interscience Publ., New York, 1965).

(1) the trace of the product of an odd number of vectors $\hat{A}$ is equal to zero;
(2) the trace of a scalar quantity is equal to its quadrupoled value;

(3) $\qquad \mathrm{Tr}\,\hat{A}_1\hat{A}_2 = 4(A_1A_2)$;

(4) $\quad \mathrm{Tr}\,\hat{A}_1\hat{A}_2\hat{A}_3\hat{A}_4 = 4[(A_1A_2)(A_3A_4)+(A_1A_4)(A_2A_3)-(A_1A_3)(A_2A_4)$.

$$(127.32)$$

The first two rules are trivial. Rules (3) and (4) are easily proved, making use of the identity

$$\hat{A}_1\hat{A}_2 + \hat{A}_2\hat{A}_1 = 2(A_1A_2). \qquad (127.33)$$

If we take the trace of the left-hand and right-hand sides and carry out the cyclic permutation of the vectors $\hat{A}_1$ and $\hat{A}_2$ under the sign of trace, then we immediately obtain rule (3). We find the trace of $\hat{A}_1\hat{A}_2\hat{A}_3\hat{A}_4$. Making use of the identity (127.33) we have

$$\mathrm{Tr}\,\hat{A}_1\hat{A}_2\hat{A}_3\hat{A}_4 = -\,\mathrm{Tr}\,\hat{A}_1\hat{A}_2\hat{A}_4\hat{A}_3 + 2\,\mathrm{Tr}\,\hat{A}_1\hat{A}_2(A_3A_4) =$$

$$= \mathrm{Tr}\,\hat{A}_1\hat{A}_4\hat{A}_2\hat{A}_3 - 2\,\mathrm{Tr}\,(A_2A_4)\hat{A}_1\hat{A}_3 + 8(A_1A_2)(A_3A_4) =$$

$$= -\mathrm{Tr}\,\hat{A}_4\hat{A}_1\hat{A}_2\hat{A}_3 + 2\,\mathrm{Tr}\,\hat{A}_2\hat{A}_3(A_1A_4) - 8(A_2A_4)(A_1A_3) + 8(A_1A_2)(A_3A_4) =$$

$$= -\mathrm{Tr}\,\hat{A}_1\hat{A}_2\hat{A}_3\hat{A}_4 + 8(A_2A_3)(A_1A_4) - 8(A_2A_4)(A_1A_3) + 8(A_1A_2)(A_3A_4). $$

From this equality we obtain rule (4).

By means of the above rules we find

$$\mathrm{Tr}\,F_1' = \frac{16}{\omega_2^2}\,[2(q_2p_1)(q_2p_2)-(q_2^2+m^2)(p_1p_2)+4m^2(q_2p_1)-$$

$$-4m^2q_2^2+4m^2(q_2p_2)+4m^4]\,,$$

$$\mathrm{Tr}\,F_1'' = \frac{16}{\omega_1\omega_2}\,[2(q_1q_2)(p_1p_2)+m^2(q_2p_1)+m^2(q_2q_1)+m^2(q_2p_2) +$$

$$+m^2(q_1p_1)+m^2(p_1p_2)+m^2(q_1p_2)+2m^4]\,. \qquad (127.34)$$

Carrying out the replacement mentioned, we obtain from these expressions $\mathrm{Tr}\,F_2$.

Performing the necessary transformations, after several long but simple calculations, we arrive at the well-known Klein–Nishina formula

$$d\sigma = \tfrac{1}{2}r_0^2\left(\frac{\omega_2}{\omega_1}\right)^2\left(\frac{\omega_1}{\omega_2}+\frac{\omega_2}{\omega_1}-\sin^2\vartheta\right)d\Omega\,, \qquad (127.35)$$

which plays an important role in applications.

For low photon energies $\omega_1 \ll m$, $\omega_2 = \omega_1$ (see (127.6)) and formula (127.35) in the limit reduces to the classical Thomson formula

$$d\sigma = \tfrac{1}{2}r_0^2(1+\cos^2 \vartheta)d\Omega ,$$

obtained in §36 of Part I.

§128. The shift of the terms of the hydrogen atom under the action of the vacuum field (the Lamb shift)

The importance of Feynman's method of calculation does not, of course, reduce to a simplification and standardization of calculations.

As we have already pointed out in §124, the Feynman formalism made it possible to obtain in an obvious form the solution of a number of important problems of quantum electrodynamics. They include, in particular, the Lamb shift of atomic terms already mentioned.

The phenomenon of the Lamb shift yields a very obvious illustration of the validity of those concepts which were assumed as the basis of the quantum theory of radiation and the theory of the positron. In the quantum theory of radiation it was assumed that in a vacuum there is an electromagnetic field. This is the field which corresponds to the zero-point oscillations of field oscillators. The set of electromagnetic field oscillators in states with zero energy is often said to represent the 'electromagnetic vacuum'. In the electromagnetic vacuum, corresponding to the field state with the lowest energy, there is a field strength different from zero. More precisely, the mean (over time) values of the squares of field strengths $(\mathcal{E})^2$ and $(\mathcal{H})^2$ are different from zero.

The existence of the vacuum field had no effect on the phenomena of emission, absorption and scattering which were considered in ch. 12. All these phenomena were associated with the transitions of field oscillators from non-excited (zero) states into excited states and vice versa. Hence in the course of a number of years the properties of the electromagnetic vacuum were not related to directly observed phenomena.

In the theory of positrons it is assumed that in addition to the electromagnetic vacuum there is an electron—positron vacuum, or a background of occupied states with negative energies, which was considered in detail in §116. It turns out that the existence of the electromagnetic and electron—positron vacua is shown directly not only in processes occurring at large energies (for example, in the Compton effect or in the process of pair production) but also in features of the behaviour of particles at small energies, in particular in the phenomenon of the Lamb shift. The phenomenon of the Lamb shift can be studied rigorously by means of the Feynman formalism. It turns out, however, that this effect can also be discussed without using a relatively complicated mathematical apparatus, on the basis of simple and direct considerations*.

For this we shall first of all discuss the problem as to what magnitude the mean square value of the field strength can have at an arbitrary point of vacuum.

To calculate the mean square value of the field strength in vacuum we consider the normalization volume V_0. The zero-point oscillation frequency ω has energy $\frac{1}{2}\hbar\omega$. One can write the obvious equality

$$\tfrac{1}{2}\hbar\omega = \frac{1}{8\pi} \overline{\int (\boldsymbol{\mathcal{E}}_{0\omega}^2 + \boldsymbol{\mathcal{H}}_{0\omega}^2)\,\mathrm{d}V} = \frac{1}{4\pi} \overline{\int \boldsymbol{\mathcal{E}}_{0\omega}^2\,\mathrm{d}V} = \frac{\boldsymbol{\mathcal{E}}_{0\omega}^2 V_0}{8\pi}, \quad (128.1)$$

where $\boldsymbol{\mathcal{E}}_{0\omega}$ and $\boldsymbol{\mathcal{H}}_{0\omega}$ are the field strength amplitudes of the field in vacuum corresponding to zero-point oscillations with frequency ω; the bar denotes averaging over the oscillation period. Equality (128.1) makes it possible to find the mean square value of the amplitude of the zero-point oscillations of the field with frequency ω:

$$\boldsymbol{\mathcal{E}}_{0\omega}^2 = \frac{4\pi\hbar\omega}{V_0}. \tag{128.2}$$

Let us consider the electron in the hydrogen atom. This electron is acted upon by the Coulomb field of the nucleus and by the fluctuations of the zero point field of the vacuum. Hence a random motion under the action of the vacuum field will be superposed on the orbital motion of the electron.

Let $U(\mathbf{r})$ denote the potential energy of the electron at point $\mathbf{r}$. We now assume that the coordinate of the electron can be written as $\mathbf{r} = \mathbf{r}_0 + \mathbf{r}'$, where $\mathbf{r}_0$ is the ordinary value of the coordinate of the electron smoothly varying in its orbital motion, and $\mathbf{r}'$ is its small displacement under the action of a random force, the fluctuating field. Then the change in the mean potential energy of the electron undergoing random displacements can be written in the form

$$\langle \Delta U \rangle = \langle U(\mathbf{r}_0 + \mathbf{r}') - U(\mathbf{r}_0) \rangle \approx \left\langle x_i' \frac{\partial U}{\partial x_i} + \tfrac{1}{2}(x_i' x_k') \frac{\partial^2 U}{\partial x_i \partial x_k} \right\rangle =$$

$$= \tfrac{1}{2} \nabla^2 U \langle (x_i')^2 \rangle = \tfrac{1}{6}(\nabla^2 U)\langle (\mathbf{r}')^2 \rangle . \tag{128.3}$$

---

* See T. Welton, Phys. Rev. 74 (1948) 1157.

Here the bracket $\langle \rangle$ denotes the mean over all possible values of the random quantity $\mathbf{r}'$. In averaging we have takne into account that $\langle x_i' \rangle = 0$ and that by virtue of the spatial isotropy of random displacements

$$\langle x_i' x_k' \rangle = \tfrac{1}{3} \langle (\mathbf{r}')^2 \rangle .$$

The value of the potential energy in the factor $\nabla^2 U$ is evidently taken at the value $\mathbf{r} = \mathbf{r}_0$.

The potential energy of the electron in the atom without the perturbation caused by the vacuum field does not depend on the state of the vacuum field and the sign in it is dropped.

In the Coulomb field of the proton one can write for $\nabla^2 U(\mathbf{r}_0)$

$$\nabla^2 U(\mathbf{r}_0) = 4\pi e^2 \delta(\mathbf{r}_0) ,$$

so that

$$\langle U \rangle = U(\mathbf{r}_0) + \tfrac{2}{3}\pi e^2 \delta(\mathbf{r}_0)\langle \mathbf{r}'^2 \rangle . \qquad (128.4)$$

To obtain the shift of the atomic term we have to take the mean value of (128.4) over the state of the electron in the atom. We then have

$$\Delta E_{\text{Lamb}} = \overline{\langle U \rangle - U(\mathbf{r}_0)} = \overline{\langle \Delta U \rangle} = \tfrac{2}{3}\pi e^2 \int \delta(\mathbf{r}_0) |\psi_n(\mathbf{r}_0)|^2 \langle \mathbf{r}'^2 \rangle \, \mathrm{d}V_0 , \qquad (128.5)$$

where ψ_n is the wave function of the electron in the atom. Making use of the properties of the δ-function, we find

$$\Delta E_{\text{Lamb}} = \overline{\langle \Delta U \rangle} = \tfrac{2}{3}\pi e^2 |\psi_n(0)|^2 \langle \mathbf{r}'^2 \rangle . \qquad (128.6)$$

The calculation of $\langle \mathbf{r}'^2 \rangle$, the mean square displacement of the electron under the action of the zero-point oscillations of the field, can be carried out relatively simply if only relatively low oscillation frequencies of the field are taken into account.

We shall assume that the displacement of the electron under the action of the field proceeds independently of the orbital motion. Neglecting relativistic effects, one can write the equations of motion in the form

$$m \frac{\mathrm{d}^2 \mathbf{r}_\omega'}{\mathrm{d}t^2} = e\boldsymbol{\mathcal{E}}_{0\omega} \sin(\mathbf{k}\cdot\mathbf{r} - \omega t) ,$$

whence

$$\mathbf{r}_\omega' = -\frac{e\boldsymbol{\mathcal{E}}_{0\omega}}{m\omega^2} \sin(\mathbf{k}\cdot\mathbf{r} - \omega t)$$

and, correspondingly,

$$\overline{\langle (\mathbf{r}'_\omega)^2 \rangle} = \frac{e^2}{2m^2\omega^4}\, \mathcal{E}^2_{0\omega} = \frac{2\pi e^2 \hbar}{m^2 \omega^3 V_0},$$ (128.7)

where the bar denotes average over time. Here $\mathbf{r}'_\omega$ denotes the displacement under the action of the zero-point oscillations of the field with frequency ω.

Since zero-point oscillations with different frequencies are independent, their contribution to the total mean square displacement of the electron is found by simple summation. Consequently, we can write for the total mean square displacement the expression

$$\langle \mathbf{r}'^2 \rangle = \int \overline{\langle (\mathbf{r}'_\omega)^2 \rangle}\, \frac{\omega^2\, d\omega V_0}{\pi^2 c^3} = \frac{2e^2\hbar}{\pi c^3 m^2} \int_{\omega_{min}}^{\omega_{max}} \frac{d\omega}{\omega},$$ (128.7′)

where the integration is carried out over all possible frequencies of zero-point oscillations.

If there were no electron–positron vacuum, then the frequencies of the zero-point oscillations of the field could take on values as large as one wished and the formula obtained would not make sense.

It turns out, however, that at frequencies larger than the minimum frequency of pair production $\omega_1 = 2mc^2/\hbar$ an interaction arises between the zero-point oscillations of the field and the occupied background of negative energies (electromagnetic and electron–positron vacuums). This interaction can be pictured in an obvious way as the interaction between the fluctuations of the 'current' associated with the random displacement of the electron with positive energy and the fluctuations of 'currents' associated with random displacements of the electrons of the background of occupied states, caused by the action of the zero-point oscillations of the electromagnetic vacuum. Since by virtue of the Pauli principle all electrons tend to avoid each other (see §67), it turns out that the fluctuations of the electrons of the background are in the opposite phase with respect to the fluctuations of the electron with positive energy. As a result they are mutually cancelled and the mean square displacement of the electron turns out to be considerably smaller than that given by formula (128.7). Simplifying the actual state of affairs, it can be said that at frequencies larger than ω_1 the mean square displacement of the electron reduces to zero.

Hence as the upper limit of integration over the frequencies of the field in (128.7′) one has to choose the quantity $\omega_{max} = \omega_1$. In defining the minimum frequency ω_{min} it is necessary to take into account that the electron considered is not free but is bound in an atom.

The frequency $\omega_{\min}$ is in order of magnitude equal to the Rydberg frequency of the electron in the hydrogen atom, i.e.

$$\omega_{\min} = \omega_0 = 2\pi R = \frac{me^4}{2\hbar^3},$$

where R is the Rydberg constant equal to $|E_0|/2\pi\hbar$, and E_0 is the energy of the ground state of the atom. Making use of these expressions for $\omega_{\min}$ and $\omega_{\max}$ we find

$$\langle \mathbf{r}'^2 \rangle = \frac{2e^2\hbar}{\pi c^3 m^2} \ln \frac{2mc^2}{\hbar\omega_0} = \frac{2e^2}{\pi\hbar c} \left(\frac{\hbar}{mc}\right)^2 \ln \frac{2mc^2}{\hbar\omega_0} \tag{128.8}$$

and for the shift of the term we find

$$\Delta E_{\text{Lamb}} = \frac{4e^4}{3\hbar c} \left(\frac{\hbar}{mc}\right)^2 |\psi_n(0)|^2 \ln \frac{2mc^2}{\hbar\omega_0}. \tag{128.9}$$

Formula (128.9) shows that the shift of the levels of the electron in the hydrogen atom under the action of the vacuum on the electron occurs only in s-states. Indeed, only in s-states is the quantity $|\psi_n(0)|^2$ different from zero. This shift is always positive: the level in an s-state must lie above that defined by formula (119.2).

The calculation of ΔE_{Lamb} is of an absolute character, and the numerical value of the shift (with certain additional corrections which are not taken into account in the simplified derivation presented above) is equal to 1057.19 Mc. The experimental value of this quantity turned out to be equal to 1057.77 ± 0.1 Mc. The perfect agreement between the calculated and measured values of the Lamb shift is an obvious confirmation of the general concepts of the reality of the 'vacuum'.

Analogous calculations, on which we shall not dwell, made it possible to find the correction to the magnetic moment of the electron already mentioned (see the article of Welton cited earlier). Of particular importance was the solution in quantum electrodynamics of a number of problems of principle. One succeeded in constructing a quantitative theory which made it possible to calculate with any degree of accuracy the probabilities of all possible processes associated with the interaction of electrons with each other and with the electromagnetic field.

The difficulties of principle of the theory associated with diverging expressions, for example the difficulty frequently mentioned that the intrinsic mass (or energy) of the electron goes to infinity, could, to a certain degree, be removed. In expressions for the intrinsic energy of a particle and in similar relations one succeeded in separating finite observable quantities, whereas diverging expressions describe only quantities which are in principle not observable.

This procedure, called renormalization, cannot be presented here, and we refer the reader to the specialist literature, for example to the monographs of S.Schweber, H.Bethe and F.de Hoffman or A.I.Akhiczer and V.B.Berestetskii, which have frequently been cited before.

15

Fundamentals of the
Theory of Elementary Particles

§129. The classification and properties of elementary particles

At the present time a large number (of the order of two hundred) of elementary particles have been discovered which can transform into each other but which do not consist, in the usual sense of the word, of smaller entities. They can be divided into two large groups: stable particles and short-lived particles (resonances). The first term is a relative one, since the group of stable particles contains, for example, both the electron which lives infinitely long, and the π^0-meson, whose lifetime is of the order of 10^{-16} sec. Stability is understood only in the sense that the lifetime of these particles is much longer than the characteristic time of $10^{-24}-10^{-23}$ sec for a light signal to traverse a distance of 10^{-13} cm which is typical of the 'size' of elementary particles.

The stable particles are divided into the following four classes:

1. The class of photons, which comprises the quanta of classical fields. This class includes the photon itself, a quantum of the electromagnetic field, and sometimes the graviton, a quantum of the gravitational field.

2. The class of leptons (light particles) contains the electron, the muon, which was earlier called the μ-meson, two neutrinos (the electron neutrino and the muon neutrino) and the four corresponding antiparticles. The elec-

tron neutrino is produced in the decay of the neutron, and the muon neutrino in the decay of the muon:

$$n \to p + e^- + \bar{\nu}_e , \qquad \mu^- \to e^- + \bar{\nu}_e + \nu_\mu$$

(the bar distinguishes antiparticles from the corresponding particles).

3. The class of mesons (particles of medium mass) comprises 3 π-mesons (π^- is the antiparticle of π^+, while the particle and antiparticle for π^0 are identical), 4 K-mesons (two particles and two antiparticles) and 1 η-meson (which is its own antiparticle).

4. The class of baryons (heavy particles) contains two nucleons (the proton and the neutron), 1 Λ-hyperon, 3 Σ-hyperons, 2 Ξ-hyperons, 1 Ω-hyperon and the corresponding antiparticles.

The term 'resonance' arose in connection with the fact that data on short-lived particles were initially obtained from scattering experiments. Characteristic resonance maxima were observed in total cross sections at definite values of the energy of the scattered particles. Thus, for example, in experiments with π-meson scattering the existence of the ρ-meson was discovered.

The group of short-lived particles comprises only baryon resonances and meson resonances – in all more than 150 particles. The most important among them are the mesons ω (1), φ (1), ρ (3), K^* (4) and the baryon resonances Δ_{1236} (4), Σ_{1385} (3), Ξ_{1530} (2) along with their antiparticles. Furthermore, the f-meson, which was theoretically predicted by Pomeranchuk and subsequently discovered experimentally, is of interest.

The set of all mesons and baryons and their resonances forms a large group of particles which are at present called hadrons. Also the term hadenons, referring to leptons and to the photon, has begun to be used in Russian particle physics literature.

Let us briefly enumerate the basic properties of elementary particles.

1. Each particle possesses a rest mass, which is measured in MeV. The range of mass values of different particles is rather wide: from 0 (photon and neutrino) up to 3000 MeV and more (Δ_{3230}). The initial classification of particles (their division into the four classes mentioned above) was based only on the values of their masses. But it turned out that this classification, just as with the mass of atoms in the periodic system of elements, is rather loose. This is particularly clearly seen in examples of resonances.

2. A very important characteristic of a particle is the value of its spin σ. The photon has spin 1 (with certain reservations, since its rest mass is equal to zero). Leptons have spin $\frac{1}{2}$, stable mesons spin 0, baryons (except for Ω) spin $\frac{1}{2}$, and the Ω-hyperon spin $\frac{3}{2}$. Among resonances there are particles with spin values from 0 up to $\frac{19}{2}$. The spin of the meson resonances listed above

(except for f) is equal to 1, the spin of the baryon resonances is equal to $\frac{3}{2}$, and that of the f-meson is 2. According to the Pauli–Lüders theorem, proved on the basis of most general principles which do not depend on actual dynamics, the spin of a particle unambiguously defines the type of statistics: particles with half-integer spin (leptons, baryons and baryon resonances) obey Fermi–Dirac statistics, while those with integer spin (photon, mesons and meson resonances) obey Bose–Einstein statistics. Furthermore, the spin of the particle determines the transformation properties of its wave function with respect to Lorentz transformations. Spin zero particles are described by a (pseudo-)scalar wave function, spin one-half particles by a spinor wave function, spin one-half particles by a (pseudo-)vector wave function and so on.

3. The parity P of a particle determines the transformation properties of its wave function with respect to the space inversion transformation. All stable mesons have odd parity and, having zero spin, are described by a pseudoscalar wave function. Meson resonances of spin 1 ($\omega, \varphi, \rho, K^*$) also have odd parity. Since under space inversion the vector components change sign, the wave function of these particles is vectorial. The parity of the f-meson is even. Baryons and their resonances can be assigned only relative parity. If it is assumed by definition that the parities of the proton, neutron and Λ-hyperon are even, then the parities of all the baryons enumerated above and of their resonances will also be even, whereas those of the antibaryons will be odd.

4. Each particle is characterized by the value of its electric charge, which (the charge of the electron being assumed to be equal to unity) can take on only integer values. At present a large number of neutral particles and particles with a charge equal in absolute value to unity are known. The charge of six particles (Δ-resonances) is equal to +2.

5. In order to characterize elementary particles, the lepton number L and baryon number B are introduced. By definition, for leptons $L = +1$, $B = 0$; for antileptons $L = -1$, $B = 0$; for baryons $L = 0$, $B = +1$, for antibaryons $L = 0$, $B = -1$; for mesons and photons $L = B = 0$. In all reactions involving elementary particles these quantum numbers are conserved, and their importance is due just to this fact.

6. All hadrons are divided into small families whose members are denoted by one and the same symbol (for example π). These families are called isomultiplets. The particles constituting an isomultiplet have about the same mass but different charges. Each isomultiplet is assigned a definite value of isospin T, which defines the number of members of the multiplet $N = 2T + 1$. Thus the isospin of the nucleon and of the K-meson is equal to $\frac{1}{2}$, the isospin of the Σ-hyperon and π-meson is equal to 1, that of the Δ-resonance is equal to $\frac{3}{2}$ and so on.

Table 3

Stable elementary particles

Class	Particle	Mass (MeV)	Spin and parity	Lifetime (sec)	Basic decays	S	T	T_3
Photon $B, L=0$	γ	0	1^-	∞	—	—	—	—
Leptons $B=0, L=1$	ν_e	0	$\frac{1}{2}$	∞	—	—	—	—
	ν_μ	0	$\frac{1}{2}$	∞	—	—	—	—
	e^-	0.511	$\frac{1}{2}$	∞	—	—	—	—
	μ^-	105.66	$\frac{1}{2}$	2.2×10^{-6}	$e^- \bar{\nu}_e \nu_\mu$	—	—	—
Mesons $B=0, L=0$	π^+	139.58	0^-	2.6×10^{-8}	$\mu^+ \nu_\mu$	0	1	$+1$
	π^0	134.98		0.89×10^{-16}	$\gamma\gamma$			0
	π^-	139.58		2.6×10^{-8}	$\mu^- \bar{\nu}_\mu$			-1
	K^+	493.8	0^-	1.24×10^{-8}	$\mu^+ \nu_\mu; \pi^+ \pi^0; \pi^+ \pi^- \pi^+$	$+1$	$\frac{1}{2}$	$+\frac{1}{2}$
	$K^0 \left\{ \begin{matrix} K^0_S \\ \\ K^0_L \end{matrix} \right.$	497.9		0.87×10^{-10} 5.73×10^{-8}	$\pi^+ \pi^-; \pi^0 \pi^0$ $\pi e \nu_e; \pi \mu \nu_\mu;$ $\pi^0 \pi^0 \pi^0; \pi^+ \pi^- \pi^0$			$-\frac{1}{2}$
	η	548	0^-	$\sim 10^{-17}$	$\gamma\gamma; \pi^+ \pi^- \pi^0;$ $\pi^0 \pi^0 \pi^0; \pi^0 \gamma\gamma$	0	0	0
Baryons $B=1, L=0$	p	938.25	$\frac{1}{2}^+$	∞	—	0	$\frac{1}{2}$	$+\frac{1}{2}$
	n	939.55		$\sim 10^3$	$pe^- \bar{\nu}_e$			$-\frac{1}{2}$
	Λ	1115.6	$\frac{1}{2}^+$	2.54×10^{-10}	$p\pi^-; n\pi^0$	-1	0	0
	Σ^+	1189.5	$\frac{1}{2}^+$	0.8×10^{-10}	$p\pi^0; n\pi^+$	-1	1	$+1$
	Σ^0	1192.6		$<1 \times 10^{-14}$	$\Lambda\gamma$			0
	Σ^-	1197.4		1.65×10^{-10}	$n\pi^-$			-1
	Ξ^0	1314.7	$\frac{1}{2}^+?$	3×10^{-10}	$\Lambda\pi^0$	-2	$\frac{1}{2}$	$+\frac{1}{2}$
	Ξ^-	1321.2		1.74×10^{-10}	$\Lambda\pi^-$			$-\frac{1}{2}$
	Ω	1674	$\frac{3}{2}^+?$	$\sim 1 \times 10^{-10}$	$\Xi\pi; \Lambda\bar{K}^+$	-3	0	0

Table 4

The most important resonances

Class	Particle	Mass (MeV)	Spin and parity	Width [a] (MeV)	Basic decays	S	T
Meson resonances $B=0, L=0$	ω	783	1^-	12	$\pi^+\pi^-\pi^0; \pi^0\gamma$	0	0
	φ	1019	1^-	4	$K^+K^-; K_L K_S; \pi^+\pi^-\pi^0$	0	0
	f	1250	2^+	110	$\pi\pi$	0	0
	$\rho^\pm$	774	$\Big\} 1^-$	$\Big\} 128$	$\Big\} \pi\pi$	$\Big\} 0$	$\Big\} 1$
	ρ^0	780					
	K^*	892	1^-	50	$K\pi$	+1	$\frac{1}{2}$
Baryon resonances $B=1, L=0$	Δ_{1236}	1236	$\frac{3}{2}^+$	120	$N\pi$	0	$\frac{3}{2}$
	Σ_{1385}	1382	$\frac{3}{2}^+$	40	$\Lambda\pi; \Sigma\pi$	−1	1
	Ξ_{1530}	1530	$\frac{3}{2}^+$	7	$\Xi\pi$	−2	$\frac{1}{2}$

[a] The lifetime τ is connected with the width Γ by the uncertainty relation $\Delta E \Delta \sim \hbar$, i.e. $\tau = \hbar/\Gamma$.

7. The different particles constituting an isomultiplet differ from each other by the isospin projection T_3 onto the third axis of the fictitious isospace. Depending on the value of isospin T, its projection T_3 can take on integer or half integer values. The concept of isospin was initially introduced only for the nucleon and pion. In this case T_3 is related in the following way to the value of the electric charge of the particle:

$$Q = T_3 + \tfrac{1}{2}B .$$ (129.1)

8. After the discovery of the K-mesons and hyperons (the so-called strange particles) it was necessary to modify the above formula:

$$Q = T_3 + \tfrac{1}{2}(B+S)$$ (129.2)

(the Gell-Mann—Nishijima relation). The new quantum number S was called strangeness. In a wide range of phenomena, for example in hadron production reactions, the strangeness is conserved, which immediately made it possible to explain certain incomprehensible features of these processes (say, the fact that strange particles are always produced in pairs). Recently instead of strangeness the physicist prefers to use another quantum number, the hypercharge Y, which is closely related to S:

$$Y = B + S .$$ (129.3)

We shall not dwell on some other quantum numbers, which are introduced to characterize elementary particles (time parity and charge parity, G-parity, muonic charge and so on).

A summary of properties is given in tables 3 and 4.

§130. The types of interactions of elementary particles

Elementary particles can take part in very different types of interaction: a particle may annihilate with its antiparticle, fast particles in collisions are scattered and new particles are produced, many particles are unstable and disintegrate and so on. At present four types of elementary particle interaction, sharply differing from each other in strength and other properties, are known.

1. The electromagnetic interaction: the interaction of charged particles with photons, and by means of them also with each other. Because of virtual processes neutral particles may also take part in the electromagnetic interaction. Examples of reactions caused by the electromagnetic interaction are the transformations

$$e^- + e^+ \to 2\gamma \, , \qquad \gamma + e^- \to \gamma + e^- \quad \text{(Compton effect)} \, ,$$

$$\pi^0 \to 2\gamma \, , \qquad \Sigma^0 \to \Lambda + \gamma \, ,$$

and so on. The electromagnetic interaction is of infinite range; times of $10^{-16}-10^{-14}$ sec are characteristic for reactions caused by it. The strength of the electromagnetic interaction is determined by the charge of the particle or by the dimensionless coupling constant which, in this particular case, is the fine structure constant $\alpha = e^2/\hbar c = 1/137$. The smallness of this constant allows the electromagnetic interaction to be considered as a perturbation, which explains the successes of quantum electrodynamics (see ch. 14).

2. The strong interaction: the interaction of hadrons which is responsible for their scattering, for production reactions and for resonance decays. Typical examples are

$$\pi + N \to \pi + N \, , \qquad \pi^- + p \to \Sigma^- + K^+ \, ,$$

$$\Delta \to N + \pi \, , \qquad \rho \to 2\pi \, ,$$

and so on. The strong interaction is of short range, the radius of action being of the order of 10^{-13} cm; times of $10^{-24}-10^{-23}$ sec are characteristic of it. The intensity of the strong interaction is characterized by a parameter g, which is an analogue of the electric charge. For the interaction of π-mesons with nucleons the dimensionless coupling constant is equal to $g^2/\hbar c \approx 14$, so that the strong interaction is three orders of magnitude greater than the electromagnetic interaction. Therefore perturbation theory is inadequate for its analysis. Up to now there is no complete theory of strong interactions.

3. The weak interaction: this is responsible for the slow decays of elementary particles, for example

$$p \to n + e^- + \bar{\nu}_e \, , \qquad \mu^- \to e^- + \bar{\nu}_e + \nu_\mu \, ,$$

$$K^+ \to \pi^+ + \pi^0 \ (\theta\text{-decay}) \, , \qquad K^+ \to \pi^+ + \pi^- + \pi^+ \ (\tau\text{-decay}) \, .$$

It can in principle also give rise to other reactions (for example, the scattering of neutrinos by electrons), but such processes have extremely small cross sections and have not been observed experimentally. The weak interaction is of an even shorter range than the strong interaction (its range may be of the order of 10^{-17} cm), and its characteristic times are $10^{-10}-10^{-6}$ sec. To characterize the intensity of the weak interaction one cannot introduce a natural dimensionless constant as in the preceding cases. This is associated with the fact that the analogue of electric charge, the so-called Fermi constant G, has a dimensionality different from that of e. The dimensionless constant of the weak interaction must involve a certain mass. If the mass of the π-meson is taken, then $G^2(\hbar c)^{-2}(\hbar/\mu c) \approx 5 \times 10^{-14}$, i.e. the weak interaction is less strong than the electromagnetic interaction by about 11 orders of magnitude. Nevertheless, strictly speaking, perturbation theory cannot be used in this case, because the weak interaction is non-renormalizable (see ch. 14).

4. The gravitational interaction: this is characterized by an extremely small dimensionless coupling constant $\kappa M/\hbar c = 2 \times 10^{-39}$ (κ is the gravitational constant, and M is the mass of the nucleon), which allows one to disregard it at the present state of development of the theory of elementary particles.

The interactions enumerated above differ not only in strength but also in their conservation laws, which is, apparently, even more essential. In all the interactions the energy, momentum, angular momentum, electric charge and baryon and lepton numbers are conserved. However, there are no universal conservation laws for isospin T, its projection T_3, strangeness S (or hypercharge Y) and parity P.

1. The strong interaction is the most symmetric; in the reactions to which it gives rise all the quantum numbers mentioned above are conserved. In particular, the isospin conservation law is an expression of charge independence: all the terms of an isomultiplet behave in the same way with respect to strong interactions.

2. 'Switching on' the electromagnetic interaction violates the equivalence of the particles contained in an isomultiplet, since they have different charges. Apparently, it is just this interaction which is responsible for the existence of the small differences between the masses of these particles. Since switching on the electromagnetic interaction specifies a definite direction in isospace, the total isospin T will no longer be conserved, but its projection T_3 is still conserved. Also the laws of conservation of strangeness S and of parity P hold.

3. The weak interaction is the least symmetric. None of the four quantum numbers mentioned above is conserved in the corresponding reactions. In particular, the parity conservation law is no longer valid (see ch. 13). Moreover, as shown by experiments with neutral K-mesons, the weak interaction is apparently also not invariant under time reversal.

§131. Symmetry groups in quantum mechanics

We have already stressed that no consistent and complete theory of strong interactions has been developed up to now. In particular, dynamical equations describing the behaviour of particles under the strong interaction have not been formulated. Therefore the study of the general symmetry properties of strong interactions assumes a special role. It allows one to obtain a satisfactory classification of hadrons and to derive a number of quantitative relations.

In § 129 it was pointed out that all hadrons are divided into small families, isomultiplets, which can be assigned definite values of the isospin T. The members of a given multiplet differ in the isospin projection T_3, which determines the value of electric charge, and when the electromagnetic interaction is 'switched off' they have strictly the same mass. In strong interactions the quantum numbers T and T_3 are conserved.

We have frequently encountered analogous situations already. For example, when a non-relativistic particle moves in a central field its possible states are also grouped into definite sets which are characterized by different values of the angular momentum J. The wave functions belonging to the same set differ in the angular momentum component J_3 and corresponds to one and the same energy, i.e. form a degenerate energy level. In the motion of the particle the quantum numbers J and J_3 are conserved.

The invariance of theory with respect to a definite class of transformations in real or in a certain fictitious space (in our quantum-mechanical example with respect to space rotations) is characteristic of all similar cases. The set of transformations is closed, i.e. successive application of the allowed transformations again leads to an allowed transformation. Furthermore, there is an identity or unit transformation, and to each transformation there corresponds an inverse transformation. Such invariance (symmetry) transformations are said to form a group. Its elements can be denoted by g, and successive application of two transformations g_1 and g_2 will be written in the form of a product $g_2 g_1$ (in just this order). It is easily seen that all space rotations form a group – a three-dimensional rotation group, denoted by O(3). Lorentz transformations are another example of transformations forming a group.

The concept of the invariance of a theory with respect to a given group of transformations involves two aspects: definite transformation properties of the wave functions ψ and definite transformation properties of the Hamiltonian $\hat{H}$.

With respect to a group of transformations g the entire Hilbert space of the wave functions is broken into invariant sub-spaces, i.e. there are sets of wave functions which transform, according to a given law, only into each other:

$$\psi' = U(g)\psi .$$

$$(131.1)$$

It is required that to the product $g_2 g_1$ of the elements of the group there correspond the product of the operators $U(g)$:

$$U(g_2, g_1) = U(g_2)U(g_1) .\qquad(131.2)$$

In this case the set of operators $U(g)$ is said to form a representation of the given group. The dimensionality of the space (the maximum number of linearly independent wave functions) in which these operators act is said to be the dimension of the given representation. If the invariant sub-space does not contain any invariant sub-spaces of lower dimension, then one speaks of an irreducible representation. Otherwise the representation is said to be reducible. In what follows the set of wave functions transformable according to an irreducible representation of a symmetry group will be called a multiplet.

From the theory of angular momentum it is known that in space rotations the spherical functions $Y_J^{J_z}(\theta, \varphi)$ corresponding to a given angular momentum J and to all its possible projections J_z transform only into each other. There corresponds to them an irreducible representation with dimension $2J+1$ of the group O(3), i.e. all spherical functions corresponding to the angular momentum J form a $(2J+1)$-dimensional multiplet of the rotation group.

Let us consider the requirements which, in the invariant theory, are imposed upon the transformation properties of the Hamiltonian. We act on the Schrödinger equation

$$i\hbar \frac{\partial \psi}{\partial t} = \hat{H}\psi \qquad(131.3)$$

with the operator $U(g)$ of the representation according to which the wave function ψ transforms. Assuming that $U(g)$ commutes with the operator $\partial/\partial t$ (we shall not need the analysis of the more general case), we obtain

$$i\hbar \frac{\partial (U\psi)}{\partial t} = U\hat{H}U^{-1} U\psi$$

or

$$i\hbar \frac{\partial \psi'}{\partial t} = U\hat{H}U^{-1} \psi' . \qquad(131.4)$$

The invariance of the theory means the identity of the form of the Schrödinger equation for the initial wave function ψ and for the transformed wave function ψ'. Hence

$$\hat{H} = U\hat{H}U^{-1}$$

or

$$[\hat{H}, U(g)] = 0 . \qquad (131.5)$$

Thus the requirement of invariance of a theory with respect to transformations of a certain group leads to the commutativity of the Hamiltonian with all the representation operators of this group.

We shall consider below only the so-called Lie groups, whose elements are single-valued differentiable functions of a finite number of real parameters. The latter are chosen in such a way that a unit element corresponds to their zero values. Thus for a Lie group:

$$g = g(\alpha_1, ..., \alpha_n) , \qquad g(0,...,0) = I . \qquad (131.6)$$

The number n of all independent real parameters of the Lie group is said to be its dimension. If the transformation g differs infinitely little from the identity transformation, i.e. if it is infinitesimal, then there correspond to it infinitesimal values of the parameters α_k. In this case, taking into account (131.6) one can write

$$g(\alpha_1, ..., \alpha_n) \approx I + \sum_{k=1}^{n} \alpha_k \left. \frac{\partial g(\alpha_1, ..., \alpha_n)}{\partial \alpha_k} \right|_{\alpha_1 = ... = \alpha_n = 0} = I + i \sum_{k=1}^{n} \alpha_k l_k . \qquad (131.7)$$

The quantities

$$l_k = -i \left. \frac{\partial g(\alpha_1, ..., \alpha_n)}{\partial \alpha_k} \right|_{\alpha_1 = ... = \alpha_n = 0} \qquad (131.8)$$

(the factor $-i$ is introduced for convenience), called group generators, represent square matrices whose order is equal to the dimension of the space in which the group transformations act. The rotation group generators are the 3×3 matrices J_x, J_y, J_z presented in (51.17). To see this it is sufficient to take the matrices of finite rotations about the corresponding coordinate axes and to make use of the definition (131.8).

It is obvious that the operators $U(g)$ of the representations of the Lie group depend on the same parameters α_k. Taking into account that to the unit element of the group there corresponds the unit representation operator, one can write

$$U(g) = U(\alpha_1, ..., \alpha_n) , \qquad U(0, ..., 0) = I . \qquad (131.9)$$

The N-dimensional representation operators corresponding to infinitesimal transformations have the form

$$U(g) = I + \sum_{k=1}^{n} \alpha_k \left. \frac{\partial U(\alpha_1,...,\alpha_n)}{\partial \alpha_k} \right|_{\alpha_1=...=\alpha_n=0} \equiv I + i \sum_{k=1}^{n} \alpha_k L_k . \tag{131.10}$$

Here the quantities

$$L_k = -i \left. \frac{\partial U(\alpha_1,...,\alpha_n)}{\partial \alpha_k} \right|_{\alpha_1=...=\alpha_n=0} \tag{131.11}$$

called representation generators, are $N \times N$ matrices. For the rotation group such operators are, to within the factor $\frac{1}{2}\hbar$, the operators $\hat{J}_x, \hat{J}_y, \hat{J}_z$ of the angular momentum components. In particular, spinor representation generators are the same as the Pauli matrices (see (61.13)). It can be shown that the operators $U(g)$ of a given representation, which correspond to finite transformations, are of the form

$$U(g) = \exp\left(i \sum_{k=1}^{n} \alpha_k L_k\right) \tag{131.12}$$

(see, for example, formulae (61.13) and (61.12)).

In §46 we have seen that only in the case of unitary transformations of the wave functions does the physical content of a theory not change, which singles out of the multitude of all representations a very important class of unitary representations for which

$$U^\dagger U = U U^\dagger = I . \tag{131.13}$$

From (131.13) and (131.10) we have

$$I = U^\dagger U \approx \left(I - i \sum_{k=1}^{n} \alpha_k L_k^\dagger\right)\left(I + i \sum_{k=1}^{n} \alpha_k L_k\right) \approx I + i \sum_{k=1}^{n} \alpha_k (L_k - L_k^\dagger) .$$

Hence

$$L_k^\dagger = L_k , \tag{131.14}$$

i.e. unitary representation generators are Hermitian matrices. Hence for infinitesimal transformations

$$U(g) \approx I + i \sum_{k=1}^{n} \alpha_k L_k ; \qquad U^\dagger(g) = U^{-1}(g) \approx I - i \sum_{k=1}^{n} \alpha_k L_k . \tag{131.15}$$

We also note that for the Lie groups the relation (131.5), valid in the invariant theory, is evidently equivalent to the condition

$$[\hat{H}, L_k] = 0 . \tag{131.16}$$

The importance of studying symmetry groups in quantum mechanics is due to the following facts:

1. There exist Hermitian mutually commuting combinations of the representation generators which also commute with them. They are called invariant operators, and their basic property is that on a given multiplet these operators are simply multiples of the unit operators (the Shur lemma). This means that all the wave functions of one multiplet are eigenfunctions of any invariant operator with one and the same eigenvalue. Thus a natural set of quantum numbers arises equal in number to the invariant operators characterizing the multiplet as a whole. In the rotation group there is one invariant operator $\hat{\mathbf{J}}^2 = \hat{J}_x^2 + \hat{J}_y^2 + \hat{J}_z^2$, so that in this case each multiplet is characterized by the value of the angular momentum J.

2. Among the representation generators there may exist several mutually commuting generators. Their number is determined by the properties of the group and is said to be its rank. The basis functions of a multiplet can be chosen in such a way that they are eigenfunctions of these generators. The corresponding eigenvalues are the quantum numbers classifying the wave functions belonging to the given multiplet. In the rotation group there are no mutually commuting generators, i.e. its rank is equal to 1. Therefore wave funcitons with a given angular momentum can be assigned only one more quantum number, for example the value of the component J_z.

3. The generators and invariant operators described form a set of Hermitian operators which commute with each other and with the Hamiltonian (see (131.16)). Hence there correspond to them conserved and at the same time measurable physical quantities. A consequence of the theory of invariance with respect to space rotations is the conservation of angular momentum $\mathbf{J}$ and of its component J_z.

4. From the above it follows that if the wave function of the initial state of a system belongs to a certain multiplet, then as a result of a reaction (scattering or decay) the system will make a transition to a new state whose wave function belongs to the same multiplet. This establishes definite selection rules for the reactions.

5. From (131.16) and the Shur lemma it follows that the eigenvalues of the Hamiltonian (the values of the energy or mass of the elementary particles) are the same for the wave functions of a given multiplet. This accounts for the presence of degeneracy and allows one to establish the multiplicity, which is equal to the dimension of the multiplet. In a theory invariant with respect to the rotation group there occurs a degeneracy in J_z, its multiplicity being equal to $2J+1$.

§132. The isogroup SU(2) and its representations

The existence of hadron multiplets with the properties already described suggests that the strong interaction of elementary particles is invariant with respect to a certain group of transformations. It turned out that this latter is the group SU(2), which we shall henceforth call the isogroup. From the mathematical point of view it is closely related (is almost equivalent) to the rotation group O(3).

By the group SU(2) is meant the set of all unitary and unimodular (determinant equal to 1) 2×2 matrices:

$$g^\dagger g = g g^\dagger = I \; ; \qquad \det g = 1 \; . \tag{132.1}$$

We can make a useful geometrical interpretation of unitary 2×2 matrices. We take the two-dimensional complex space of vectors x, which we shall write in the form of columns, and introduce the Hermitian conjugate vectors $x^\dagger$:

$$x = \begin{pmatrix} x_1 \\ x_2 \end{pmatrix} ; \qquad x^\dagger = (x_1^*, x_2^*) \equiv (x^1, x^2) \; . \tag{132.2}$$

In this space we consider the linear transformation

$$x' = gx \qquad \text{or} \qquad x_i' = g_i^j x_j \tag{132.3}$$

(the superscript numbers the column, while the subscript numbers the row; they run over the values 1, 2, repeated indices implying summation). Second-order unitary matrices correspond to the transformation matrices (132.3), which do not change the quadratic form

$$x^\dagger x = x^i x_i = x_1^* x_1 + x_2^* x_2 \; . \tag{132.4}$$

According to §131 the infinitesimal matrix g can be written as

$$g \approx I + i\epsilon_\alpha \tau_\alpha \; . \tag{132.5}$$

Here τ_α are the generators of the group SU(2), and ϵ_α are its parameters whose number n (the dimension of the group) is to be defined. The requirement of unitarity of the matrix g leads to the hermiticity of its generators τ_α. Noting that, to terms of the second order of smallness in ϵ_α, for the infinitesimal matrix (132.5)

$$1 = \det g \approx I + i\epsilon_\alpha \, \mathrm{Tr}\, \tau_\alpha$$

we arrive at the conclusion that the trace of the generators τ_α is equal to zero. Thus they are 2 × 2 matrices with the properties

$$\tau_\alpha^\dagger = \tau_\alpha \,, \qquad \mathrm{Tr}\,\tau_\alpha = 0 \,. \tag{132.6}$$

These restrictions impose 5 (4 plus 1) conditions upon the 8 real parameters of the complex 2×2 matrix. Hence there are 3 independent matrices with the properties (132.6), which just determines the dimension of the group SU(2). One can take as its generators the Pauli matrices, which are Hermitian and have trace zero:

$$\tau_1 = \begin{pmatrix} 0 & 1 \\ 1 & 0 \end{pmatrix}, \qquad \tau_2 = \begin{pmatrix} 0 & -i \\ i & 0 \end{pmatrix}, \qquad \tau_3 = \begin{pmatrix} 1 & 0 \\ 0 & -1 \end{pmatrix}. \tag{132.7}$$

Among these there are no mutually commuting ones, so that the rank of SU(2) is equal to 1. The sum of the squares of the Pauli matrices commutes with each of them:

$$[\tau^2, \tau_\alpha] = 0 \,, \quad \text{where} \quad \tau^2 = \tau_1^2 + \tau_2^2 + \tau_3^2 \,. \tag{132.8}$$

Let us now go on to the construction of the representations of the group SU(2).

1. The trivial representation is the simplest one. Its multiplets are one-dimensional, i.e. contain one wave function φ each, which does not change under the transformation (132.3):

$$\varphi' = \varphi \,. \tag{132.9}$$

The dimension of this representation is 1, and its generators are the numbers 0. Such a representation is called a scalar representation, and φ is said to be a scalar.

2. Let us consider the two-dimensional space of vectors φ_i, which transform according to the same law as the vectors x:

$$\varphi_i' = g_i^j \varphi_j \quad \text{or} \quad \varphi_i' \approx (I + i\epsilon_\alpha \tau_\alpha)_i^j \varphi_j \,. \tag{132.10}$$

As a result, two-dimensional multiplets arise whose members transform according to the representation of the same dimension. From (132.10) it follows that its generators are the matrices τ_α themselves:

$$(L_\alpha)_i^j = (\tau_\alpha)_i^j \,. \tag{132.11}$$

This representation is called the spinor representation, and the quantities φ_i are said to be spinors (see the formalism presented in ch. 8).

3. Let us take two quantities φ^i which transform in the same way as the components of the vector $x^\dagger$:

$$(\varphi^i)' = (g^\dagger)_j^i \varphi^j \quad \text{or} \quad (\varphi^i)' \approx (I - i\epsilon_\alpha \tau_\alpha)_j^i \varphi^j \,. \tag{132.12}$$

As a result we obtain a two-dimensional representation, which is said to be conjugate to the spinor representation. Its generators are the matrices

$$(L_\alpha)^j_i = -(\tau_\alpha)^j_i .$$
(132.13)

In fact this representation is equivalent to the spinor representation, i.e. from the quantities φ^i it is possible to form linear combinations which transform according to the law (132.10). But its introduction is very convenient from the formal point of view.

4. All other representations can be constructed from the spinor representation and its conjugate. Let us consider the set of $2(p+q)$ quantities $\varphi^{j_1 \cdots j_q}_{i_1 \cdots i_p}$ which transform as the product of p spinors and q conjugate spinors, i.e. according to the law

$$(\varphi^{j_1 \cdots j_q}_{i_1 \cdots i_p})' = g^{i'_1}_{i_1} \cdots g^{i'_p}_{i_p} (g^\dagger)^{j_1}_{j'_1} \cdots (g^\dagger)^{j_q}_{j'_q} \varphi^{j'_1 \cdots j'_q}_{i'_1 \cdots i'_p} .$$
(132.14)

For the infinitesimal transformation we have

$$(\varphi^{j_1 \cdots j_q}_{i_1 \cdots i_p})' = (I + i\epsilon_{\alpha_1} \tau_{\alpha_1})^{i'_1}_{i_1} \cdots (I - i\epsilon_{\beta_q} \tau_{\beta_q})^{j_q}_{j'_q} \varphi^{j_1 \cdots j'_q}_{i'_1 \cdots i'_p}$$
(132.15)

so that the generators of this representation have the form

$$(L_\alpha)^{i'_1 \cdots i'_p j_1 \cdots j_q}_{i_1 \cdots i_p j'_1 \cdots j'_q} = \sum_{s=1}^{p} \delta^{i'_1}_{i_1} \cdots \delta^{i'_{s-1}}_{i_{s-1}} (\tau_\alpha)^{i'_s}_{i_s} \delta^{i'_{s+1}}_{i_{s+1}} \cdots \delta^{i'_p}_{i_p} \delta^{j_1}_{j'_1} \cdots \delta^{j_q}_{j'_q} -$$

$$- \sum_{s=1}^{q} \delta^{i'_1}_{i_1} \cdots \delta^{i'_p}_{i_p} \delta^{j_1}_{j'_1} \cdots \delta^{j_{s-1}}_{j'_{s-1}} (\tau_\alpha)^{j_s}_{j'_s} \delta^{j_{s+1}}_{j'_{s+1}} \cdots \delta^{j_q}_{j'_q} .$$
(132.16)

The representation obtained in such a way is said to be the direct product of p spinor representations and q conjugate representations. Symbolically

$$\underbrace{(1,0) \otimes \cdots \otimes (1,0)}_{p \text{ times}} \underbrace{\otimes (0,1) \otimes \cdots \otimes (0,1)}_{q \text{ times}}$$

(the notation is obvious).

If the function $\varphi^{j_1 \cdots j_q}_{i_1 \cdots i_p}$ is symmetric with respect to a certain pair of subscripts (or superscripts), then this property is conserved under the transformation (132.13). Furthermore, the invariance of the quadratic form (132.4) leads to the conservation of the trace with respect to any pair in superscripts and subscripts, i.e. quantities of the type $\varphi^{\cdots i \cdots}_{\cdots i \cdots}$. Hence the set of all functions $\varphi^{j_1 \cdots j_q}_{i_1 \cdots i_p}$ is divided into individual multiplets whose members transform only into each other, i.e. the representation constructed is reducible. In the given case the condition of irreducibility amounts to the fact that $\varphi^{j_1 \cdots j_q}_{i_1 \cdots i_p}$ must be symmetric separately with respect to all the subscripts and superscripts, the trace with respect to any pair in the subscripts and superscripts being bound to reduce to zero. It is easy to count the number of independent quantities $\varphi^{j_1 \cdots j_q}_{i_1 \cdots i_p}$ of such a type, i.e. to determine the dimension of the corresponding irreducible representation. The total number of subscripts (superscripts) which are equal to 1 can vary from 0 up to p (up to q), the other indices being equal to 2. Under the condition of symmetry the order of succession of the indices does not matter, so that the number of different completely symmetric quantities $\varphi^{j_1 \cdots j_q}_{i_1 \cdots i_p}$ is equal to $(p+1)(q+1)$. The fact that the traces are equal to zero imposes upon them pq conditions, so that in all we have

$$N = (p+1)(q+1) - pq = p + q + 1$$

independent components. The irreducible representation described will be denoted by the symbol (p,q), and its dimension by $N(p,q)$. We have

$$N(p,q) = (p+q) + 1 . \tag{132.17}$$

We stress once more that the superscripts and subscripts are equivalent, so that the irreducible representation is in essence defined by the total number of indices, which is equal to $p+q$.

Let us consider the example (important for what follows) of the direct product of the spinor representation by the conjugate representation, i.e. the representation $(1,0) \otimes (0,1)$. Functions transforming according to this representation form a square 2×2 matrix. In correspondence with (132.13)

$$(\varphi^j_i)' = g^{i'}_i (g^\dagger)^j_{j'} \varphi^{j'}_{i'} . \tag{132.18}$$

For infinitesimal transformations (132.15) gives

$$(\varphi^j_i)' \approx (I + i\epsilon_\alpha \tau_\alpha)^{i'}_i (I - i\epsilon_\beta \tau_\beta)^j_{j'} \varphi^{j'}_{i'}$$

$$\approx \{\delta^{i'}_i \delta^j_{j'} + i\epsilon_\alpha [(\tau_\alpha)^{i'}_i \delta^j_{j'} - (\tau_\alpha)^j_{j'} \delta^{i'}_i]\} \varphi^{j'}_{i'} \tag{132.19}$$

so that the generators of this representation have the form

$$(L_\alpha)_{ij}^{i'j} = (\tau_\alpha)_i^{i'} \delta_j^j - \delta_i^{i'} (\tau_\alpha)_j^j . \tag{132.20}$$

To decompose it into irreducible representations it suffices to separate from the matrix φ_i^j the non-zero trace, i.e. to rewrite it in the form

$$\varphi_i^j \equiv (\varphi_i^j - \tfrac{1}{2}\delta_i^j \varphi_k^k) + \tfrac{1}{2}\delta_i^j \varphi_k^k . \tag{132.21}$$

The last term is invariant with respect to transformations from the group SU(2), i.e. is a scalar. The trace of the matrix standing in parentheses is equal to zero. It contains three independent components transforming according to the three-dimensional irreducible representation which is called the vector representation. Symbolically the expansion (132.21) is written in the form

$$(1,0) \otimes (0,1) = (1,1) \oplus (0,0) . \tag{132.22}$$

From (132.15) and (132.8) it follows that for the group SU(2) the quantities $\hat{L}^2$ are invariant operators:

$$[\hat{L}^2, \hat{L}_\alpha] = 0 \quad \text{where} \quad \hat{L}^2 = \hat{L}_1^2 + \hat{L}_2^2 + \hat{L}_3^2 . \tag{132.23}$$

Making use of (132.16) and (132.7) it can be shown that any matrix $\varphi_{i_1 \dots i_p}^{j_1 \dots j_q}$ of the multiplet (p,q) is an eigenfunction of the operator $\hat{L}^2$, where

$$\hat{L}^2 \varphi_{i_1 \dots i_p}^{j_1 \dots j_q} = (p+q)(p+q+2)\varphi_{i_1 \dots i_p}^{j_1 \dots j_q} . \tag{132.24}$$

The eigenvalues of the operator $\hat{L}^2$ depend only on the type of multiplet (more precisely, only on its dimension) and are its characteristics.

To classify the basis elements of a multiplet, which can be chosen to be combinations of the matrices $\varphi_{i_1 \dots i_p}^{j_1 \dots j_q}$ with only one non-zero element, use can be made of the eigenvalues of the diagonal operator $\hat{L}_3$. Since the rank of the group SU(2) is equal to 1, this quantum number is sufficient. From (132.16) and (132.7) we conclude that if among p subscripts there are p_2 and among q superscripts q_2 indices equal to 2, then

$$\hat{L}_3 \varphi_{i_1 \dots i_p}^{j_1 \dots j_q} = [(p-q) - 2(p_2 - q_2)]\varphi_{i_1 \dots i_p}^{j_1 \dots j_q} . \tag{132.25}$$

The neighbouring eigenvalues of the generator $\hat{L}_3$, which correspond to the basis elements of a given multiplet, differ by 2. From (132.25) it is seen that the minimum of these is equal to $-(p+q)$ and the maximum to $+(p+q)$, so that there are in all $p+q+1$ independent terms of the multiplet, and we again arrive at formula (132.17) for the dimension of the irreducible representation.

§133. Isomultiplets of elementary particles

We now suppose that the strong interaction of elementary particles is invariant with respect to the group SU(2). This means, first of all, that the wave functions of hadrons transform according to certain of its irreducible representations, i.e. they are the products of the ordinary wave function, depending on spatial coordinates and on the spin projection, and the iso-matrix:

$$\Psi = \psi(x,y,z,s_z) \cdot \phi^{j_1 \cdots j_q}_{i_1 \cdots i_p} , \tag{133.1}$$

In this case an individual hadron can conveniently be compared with a wave function in which the matrix $\phi^{j_1 \cdots j_q}_{i_1 \cdots i_p}$ is one of the basis elements. Thus all hadrons turn out to be distributed in isomultiplets, which are characterized by the eigenvalues of the generator L^2. Individual hadrons within an isomultiplet are classified by the eigenvalues of the generator L_3. In consequence of the assumed invariance of the strong interaction with respect to the isogroup, these two quantum numbers will be conserved in all reactions due to it. Furthermore, quantum-mechanical degeneracy must occur, i.e. the masses of the particles belonging to one isomultiplet will be strictly the same. When the electromagnetic interaction, which is not considered to be invariant with respect to the isogroup, is switched on, this degeneracy is removed and the isomultiplets split into individual particles with somewhat different masses. An analogue of this is, for example, the Zeeman effect (§74), in which the application of an external magnetic field violating the invariance with respect to the rotation group removes the earlier degeneracy of levels with respect to the angular momentum component J_z.

We note that for hadrons involved in the same isomultiplet all quantum numbers (spin σ, parity P, baryonic number B and strangeness S or hyper-charge Y) differing from the eigenvalues of L_3 must be the same. This follows from the fact that the corresponding operators commute with the generators L_α (i.e. are invariant operators), and from the Shur lemma.

Instead of the generators L_α it is convenient to introduce the operators

$$\hat{T}_\alpha = \tfrac{1}{2} L_\alpha \tag{133.2}$$

which are called isospin component operators. Then as an invariant operator it is natural to take the operator of the square of the isospin

$$\hat{T}^2 = \hat{T}_1^2 + \hat{T}_2^2 + \hat{T}_3^2 = \tfrac{1}{4} L^2 . \tag{133.3}$$

Formulae (132.24) and (132.25) now assume the form

$$\hat{T}^2 \varphi_{i_1 \dots i_p}^{j_1 \dots j_q} = \tfrac{1}{4}(p+q)(p+q+2)\varphi_{i_1 \dots i_p}^{j_1 \dots j_q} \equiv T(T+1)\varphi_{i_1 \dots i_p}^{j_1 \dots j_q} \, , \qquad (133.4)$$

$$\hat{T}_3 \varphi_{i_1 \dots i_p}^{j_1 \dots j_q} = [\tfrac{1}{2}(p-q)-(p_2-q_2)]\varphi_{i_1 \dots i_p}^{j_1 \dots j_q} \equiv T_3 \varphi_{i_1 \dots i_p}^{j_1 \dots j_q} \, . \qquad (133.5)$$

The quantum number

$$T = \tfrac{1}{2}(p+q) \qquad (133.6)$$

classifies the hadron multiplets and is called the isospin, and the quantity

$$T_3 = \tfrac{1}{2}(p-q) - (p_2-q_2) \qquad (133.7)$$

classifies the basis elements of the isomultiplet, i.e. the individual hadrons belonging to it. It can take on values from $-T$ up to $+T$, and is said to be the third component of isospin. Formula (132.17) for the dimension of the irreducible representation, i.e. for the number of different particles involved in the isomultiplet, takes the form

$$N(p,q) \equiv N(T) = 2T + 1 \, . \qquad (133.8)$$

For the wave functions which are the eigenfunctions of the operators $\hat{T}^2$ and $\hat{T}_3$ with eigenvalues T and T_3 we shall henceforth sometimes use Dirac's notation $|T,T_3\rangle$. Finally, in correspondence with the empirical Gell-Mann—Nishijima relation (§129) we shall assume by definition that the charge operator is

$$\hat{Q} = \hat{T}_3 + \tfrac{1}{2}(B+S)\hat{I} = \hat{T}_3 + \tfrac{1}{2}Y\hat{I} \qquad (133.9)$$

where $\hat{I}$ is the unit operator in the space of functions $\varphi_{i_1 \dots i_p}^{j_1 \dots j_q}$.

The above relations point to a close relation of the group SU(2) to the rotation group O(3), the isospin T being a complete analogue of the angular momentum J. This relationship allows one to make use of the entire mathematical apparatus of angular momentum theory presented in §§51 and 52. In particular, the formalism of the Clebsch—Gordan coefficients (§52) is applicable for the actual decomposition of the direct product of representations into irreducible representations.

Let us now consider actual isomultiplets of hadrons. The proton and neutron have the same masses and are identical with respect to strong interactions. They form an isodoublet, i.e. their wave functions are of the form

$$\Psi_p = \psi_p \begin{pmatrix} 1 \\ 0 \end{pmatrix} = \begin{pmatrix} \psi_p \\ 0 \end{pmatrix} \equiv \begin{pmatrix} p \\ 0 \end{pmatrix} \equiv N_1 \; ; \quad \Psi_n = \psi_n \begin{pmatrix} 0 \\ 1 \end{pmatrix} = \begin{pmatrix} 0 \\ n \end{pmatrix} \equiv N_2 \; . \qquad (133.10)$$

The wave function of a nucleon is written as

$$N_i = \begin{pmatrix} p \\ n \end{pmatrix} . \qquad (133.11)$$

Under the normalization condition

$$\sum_{\sigma_z = \pm \frac{1}{2}} \int dV(|p|^2 + |n|^2) = 1 \qquad (133.12)$$

the two terms of this expression corresponding to the proton and neutron are interpreted as the probabilities of observing the nucleon in the proton state and in the neutron state. The spinor representation has dimension 2, so that the nucleon isodoublet is to be assigned isospin $\frac{1}{2}$. The generators of the representation are the matrices τ_α, so that the isospin components T_3 are the eigenvalues of the operator $\frac{1}{2}\tau_3$. To find them use can be made of formula (133.7), but in the given case they are easily determined directly:

$$\hat{T}_3 \begin{pmatrix} p \\ 0 \end{pmatrix} = \frac{1}{2} \begin{pmatrix} 1 & 0 \\ 0 & -1 \end{pmatrix} \begin{pmatrix} p \\ 0 \end{pmatrix} = \frac{1}{2} \begin{pmatrix} p \\ 0 \end{pmatrix},$$

$$\hat{T}_3 \begin{pmatrix} 0 \\ n \end{pmatrix} = \frac{1}{2} \begin{pmatrix} 1 & 0 \\ 0 & -1 \end{pmatrix} \begin{pmatrix} 0 \\ n \end{pmatrix} = -\frac{1}{2} \begin{pmatrix} 0 \\ n \end{pmatrix}, \qquad (133.13)$$

so that for the proton $T_3 = +\frac{1}{2}$, and for the neutron $T_3 = -\frac{1}{2}$. In correspondence with (133.8), taking into account that for the nucleon $B = 1$, $S = 0$, we obtain

$$\hat{Q}_N = \hat{T}_3 + \frac{1}{2}\hat{I} = \frac{1}{2}(\hat{\tau}_3 + \hat{I}) = \begin{pmatrix} 1 & 0 \\ 0 & 0 \end{pmatrix}, \qquad (133.14)$$

so that

$$\hat{Q}_N \begin{pmatrix} p \\ 0 \end{pmatrix} = \begin{pmatrix} 1 & 0 \\ 0 & 0 \end{pmatrix} \begin{pmatrix} p \\ 0 \end{pmatrix} = \begin{pmatrix} p \\ 0 \end{pmatrix},$$

$$\hat{Q}_N \begin{pmatrix} 0 \\ n \end{pmatrix} = \begin{pmatrix} 1 & 0 \\ 0 & 0 \end{pmatrix} \begin{pmatrix} 0 \\ n \end{pmatrix} = \begin{pmatrix} 0 \\ 0 \end{pmatrix}. \qquad (133.15)$$

Thus the charges of the proton and neutron are equal respectively to +1 and 0, as is to be expected.

The Ξ-hyperon, the Ξ_{1530} resonance, the K-meson and the K*-meson resonance are also isodoublets. The relations written above are valid for them, with the only difference that for the Ξ-hyperon and its resonance the hypercharge is $Y = -1$ ($B=1$, $S=-2$), so that for these particles the charge operator is equal to

$$\hat{Q}_\Xi = \hat{T}_3 - \tfrac{1}{2}\hat{I} = \tfrac{1}{2}(\hat{\tau}_3 - \hat{I}) = \begin{pmatrix} 0 & 0 \\ 0 & -1 \end{pmatrix}. \tag{133.16}$$

Thus there are the following hadron isodoublets:

$$N_i = \begin{pmatrix} p \\ n \end{pmatrix}; \qquad \Xi_i = \begin{pmatrix} \Xi^0 \\ \Xi^- \end{pmatrix}, \qquad (\Xi_{1530})_i = \begin{pmatrix} \Xi^0_{1530} \\ \Xi^-_{1530} \end{pmatrix},$$

$$K_i = \begin{pmatrix} K^+ \\ K^0 \end{pmatrix}, \qquad K_i^* = \begin{pmatrix} K^{*+} \\ K^{*0} \end{pmatrix}. \tag{133.17}$$

It is natural to assume that the wave functions of the antiproton and antineutron transform according to the representation conjugate to the nucleon representation, i.e.

$$\Psi_{\bar{p}} = \psi_{\bar{p}}(1,0) = (\psi_{\bar{p}},0) \equiv (\bar{p},0) ,$$

$$\Psi_{\bar{n}} = \psi_{\bar{n}}(0,1) = (0,\bar{n}) . \tag{133.18}$$

Hence the wave function of the antinucleon is written in the form

$$\bar{N}^i = (\bar{p},\bar{n}) . \tag{133.19}$$

The isospin of the antinucleon is also equal to $\tfrac{1}{2}$, but, because all the generators have now changed sign, $\hat{T}_3 = -\tfrac{1}{2}\hat{\tau}_3$ and hence for the antiproton $T_3 = -\tfrac{1}{2}$, and for the antineutron $T_3 = +\tfrac{1}{2}$ (it should be noted that, according to the rules of matrix multiplication by a wave function having the form of a row, the operators which are square matrices act from the right and not from the left). It is obvious that the antinucleon charge operator $\hat{Q}_{\bar{N}}$ is equal to $-\hat{Q}_{\bar{N}}$ so that the antiproton has $Q = -1$ and the antineutron $Q = 0$. All these statements may also be applied in an obvious way to the isodoublets of other antiparticles.

There exist three π-mesons: π^+, π^0 and π^-, with about the same masses, identical with respect to the strong interaction. It is natural to assume that they form an isotriplet, i.e. their wave functions transform according to the vector representation, forming a second-order matrix with trace zero:

$$\pi_i^j = \begin{pmatrix} \pi_1^1 & \pi_1^2 \\ \pi_2^1 & \pi_2^2 \end{pmatrix}. \tag{133.20}$$

The condition $\operatorname{Tr} \pi_i^j = 0$ requires that

$$\pi_1^1 = -\pi_2^2 . \tag{133.21}$$

To the isotriplet there corresponds the isospin $T = 1$. Making use of formula (133.7) we find that to the components π_1^1, π_1^2 and π_2^1 there correspond respectively $T_3 = 0$, $T_3 = +1$ and $T_3 = -1$. Since for π-mesons $B = S = 0$, then

$$\hat{Q}_\pi = \hat{T}_3 \tag{133.22}$$

i.e. their charges are the same as the values of the component T_3. Hence

$$\pi_1^1 = -\pi_2^2 \sim \pi^0 , \qquad \pi_1^2 \sim \pi^+ , \qquad \pi_2^1 \sim \pi^- . \tag{133.23}$$

By requiring that the isomatrices involved in the wave function of each π-meson be normalized to one, i.e. that they form the orthonormalized basis of the isotriplet, and taking into account (133.20) and (133.23), we finally obtain for the total wave function of the π-meson

$$\pi_i^j = \begin{pmatrix} \pi^0/\sqrt{2} & \pi^+ \\ \pi^- & -\pi^0/\sqrt{2} \end{pmatrix}. \tag{133.24}$$

Taking into account that the components π_i^j transform in terms of each other according to the law (132.18) and making use of the properties of unitary unimodular matrices, it is easily shown that the wave functions

$$\pi^+ = \pi_1^2 , \qquad \pi^0 = (\pi_1^1 - \pi_2^2)/\sqrt{2} , \qquad \pi^- = \pi_2^1 \tag{133.25}$$

transform as the ordinary components of a three-dimensional vector. This accounts for the name of the representation (132.1). Hence the wave function of the π-meson can be written in the form

$$\boldsymbol{\pi} = \begin{pmatrix} \pi^+ \\ \pi^0 \\ \pi^- \end{pmatrix}. \tag{133.26}$$

In such a formalism the generators of the vector representation and the isospin component operators $\hat{T}_3$ will be Hermitian 3×3 matrices with trace zero. Making use of (51.17) we can write immediately

$$\hat{T}_1 = \frac{1}{\sqrt{2}} \begin{pmatrix} 0 & 1 & 0 \\ 1 & 0 & 1 \\ 0 & 1 & 0 \end{pmatrix}; \qquad \hat{T}_2 = \frac{1}{2} \begin{pmatrix} 0 & -i & 0 \\ i & 0 & -i \\ 0 & i & 0 \end{pmatrix}; \qquad \hat{T}_3 = \begin{pmatrix} 1 & 0 & 0 \\ 0 & 0 & 0 \\ 0 & 0 & -1 \end{pmatrix}. \quad (133.27)$$

By direct check we see indeed that

$$\hat{T}_3 \begin{pmatrix} \pi^+ \\ 0 \\ 0 \end{pmatrix} = \begin{pmatrix} \pi^+ \\ 0 \\ 0 \end{pmatrix}; \qquad \hat{T}_3 \begin{pmatrix} 0 \\ \pi^0 \\ 0 \end{pmatrix} = \begin{pmatrix} 0 \\ 0 \\ 0 \end{pmatrix}; \qquad \hat{T}_3 \begin{pmatrix} 0 \\ 0 \\ \pi^- \end{pmatrix} = -\begin{pmatrix} 0 \\ 0 \\ \pi^- \end{pmatrix}. \quad (133.28)$$

The results obtained apply automatically also to other hadron isotriplets: Σ, ρ, Σ_{1385}. Since in these cases $Y = 0$, the charge operator also has the form (133.22).

Since the representation conjugate to (1,1) coincides with itself, and their generators are also the same, the structure of antiparticle isotriplets is identical with that of the matrix (133.24). For example,

$$\bar{\Sigma}_i^j = \begin{pmatrix} \overline{\Sigma}^0/\sqrt{2} & \overline{\Sigma}^- \\ \overline{\Sigma}^+ & -\overline{\Sigma}^0/\sqrt{2} \end{pmatrix} \qquad (133.29)$$

(we note that the charge of $\overline{\Sigma}^-$ is positive, and that of $\overline{\Sigma}^+$ is negative). For mesons there are no quantum numbers, except of the component T_3, which would distinguish between particles and antiparticles (for the Σ-hyperon such quantum numbers are the baryonic number and strangeness). Hence the antiparticle with respect to π^+ is π^- and vice versa, while in the case of the π^0-meson the particle is identical with the antiparticle. This accounts for the fact that for π-mesons the particles and antiparticles are contained in the same isomultiplet, whereas, say, Σ and $\overline{\Sigma}$ form two different isomultiplets. An analogous situation also occurs in the case of ρ-mesons

$$\rho_i^j = \begin{pmatrix} \rho^0/\sqrt{2} & \rho^+ \\ \rho^- & -\rho^0/\sqrt{2} \end{pmatrix} \qquad \text{where} \quad \bar{\rho}^+ = \rho^-; \quad \bar{\rho}^- = \rho^+; \quad \bar{\rho}^0 = \rho^0.$$
$$(133.30)$$

The four nucleon resonances Δ^{++}, Δ^+, Δ^0 and Δ^- form an isoquartet i.e. their wave functions transform according to the irreducible representation of dimension 4, forming the matrix Δ_{ijk} symmetric in any pair of indices. To the isoquartet there corresponds the isospin $T = \frac{3}{2}$. From (133.7) it follows that for the components $\Delta_{111}, \Delta_{112} = \Delta_{121} = \Delta_{211}, \Delta_{212} = \Delta_{221} = \Delta_{122}$ and Δ_{222} the isospin component T_3 is equal respectively to $+\frac{3}{2}$, $+\frac{1}{2}$, $-\frac{1}{2}$ and $-\frac{3}{2}$. Since for the Δ-resonances $Y = 1$ ($B=1, S=0$), then

$$\hat{Q}_\Delta = \hat{T}_3 + \tfrac{1}{2}\hat{I} \tag{133.31}$$

so that

$$\Delta_{111} \sim \Delta^{++} , \qquad \Delta_{112} = \Delta_{121} = \Delta_{211} \sim \Delta^+ ,$$
$$\Delta_{122} = \Delta_{212} = \Delta_{221} \sim \Delta^0 , \qquad \Delta_{222} \sim \Delta^- . \tag{133.32}$$

From the normalization condition it follows that in the first and last relations the factor of proportionality is equal to 1, while in the second and third cases it is equal to $1/\sqrt{3}$. The isoquartet $\bar{\Delta}$ is filled in an obvious way.

There are hadrons, for example Λ- and Ω-hyperons and η-, ω- and φ-mesons, which are isosinglets, i.e. their wave functions transform according to the isoscalar representation. Since its generators are equal to zero, $T = T_3 = 0$ for an isosinglet and $Q = \tfrac{1}{2}Y$. The antiparticles $\bar{\eta}, \bar{\omega}, \bar{\varphi}$ are the same as the corresponding particles.

§134. The wave functions of a system of nucleons and π-mesons

Let us now consider the three simplest composite systems: NN, N$\bar{\text{N}}$ and πN, which are of considerable interest.

1. We decompose the wave function of the system nucleon–antinucleon

$$\varphi_i^j = \bar{N}^j N_i \tag{134.1}$$

into the irreducible parts (§132):

$$\varphi_i^j = (\bar{N}^j N_i - \tfrac{1}{2}\delta_i^j \bar{N}^k N_k) + \tfrac{1}{2}\delta_i^j \bar{N}^k N_k \equiv \chi_i^j + \delta_i^j \chi . \tag{134.2}$$

The last term is an isoscalar, while the expression standing in parentheses is an isovector. Recalling the analysis of the matrix (133.20) we obtain

$$|1,+1\rangle = a\chi_1^2 = a\bar{n}p , \qquad |1,0\rangle = b\chi_1^1 = b\tfrac{1}{2}(\bar{p}p - \bar{n}n) ,$$
$$|1,-1\rangle = c\chi_2^1 = c\bar{p}n , \qquad |0,0\rangle = d\chi = d\tfrac{1}{2}(\bar{p}p + \bar{n}n) \tag{134.3}$$

where a, b, c and d are normalization coefficients. Assuming that the spatial-spin parts of all the wave functions $\bar{N}^j$ and N_i are normalized to unity, from the condition of orthonormalization of the basis states (134.3) we have $a^2 = c^2 = 1, b^2 = d^2 = 2$. Thus, finally,

$$|1,+1\rangle = \bar{n}p \,, \qquad |1,0\rangle = \frac{1}{\sqrt{2}}(\bar{p}p - \bar{n}n) \,,$$

$$|1,-1\rangle = \bar{p}n \,, \qquad |0,0\rangle = \frac{1}{\sqrt{2}}(\bar{p}p + \bar{n}n) \,.$$

(134.4)

If the system $N\bar{N}$ is in the $^1S_0$-state (spins antiparallel), then its total spin is equal to 0, and the parity is odd (the relative parities of the particle and antiparticle are opposite). Thus one can construct from the nucleon and antinucleon a pseudoscalar isotriplet and a pseudoscalar isosinglet, i.e. a π-meson and an η-meson. This result underlies the composite model of the π-meson, proposed in 1949 by Fermi and Yang. The isotriplet state $^3S_1$ of the $N\bar{N}$ pair can be compared to the ρ-meson.

2. Considering the wave function of a system of two nucleons

$$\varphi_{ij} = N_i' N_j''$$

(134.5)

we can separate it into parts symmetric and antisymmetric with respect to the indices:

$$\varphi_{ij} = N_i' N_j'' = \tfrac{1}{2}(N_i' N_j'' + N_j' N_i'') + \tfrac{1}{2}(N_i' N_j'' - N_j' N_i'') \equiv \chi_{[ij]} + \chi_{\{ij\}} \,.$$

(134.6)

The second term contains one non-zero independent component ($i=1, j=2$), i.e. an isoscalar, while the first term is an isovector. Making use of formula (133.7), we obtain

$$|1,+1\rangle = a\chi_{11} = ap'p'' \,, \qquad |1,0\rangle = b\chi_{[12]} = b\tfrac{1}{2}(p'n'' + n'p'') \,,$$

$$|1,-1\rangle = c\chi_{22} = cn'n'' \,, \qquad |0,0\rangle = d\chi_{\{12\}} = d\tfrac{1}{2}(p'n'' - n'p'') \,.$$

(134.7)

The normalization coefficients are defined as before. We have

$$|1,+1\rangle = p'p'' \,, \qquad |1,0\rangle = \frac{1}{\sqrt{2}}(p'n'' + n'p'') \,,$$

$$|1,-1\rangle = n'n'' \,, \qquad |0,0\rangle = \frac{1}{\sqrt{2}}(p'n'' - n'p'')$$

(134.8)

(see relations (66.4) and (66.5)). The first three functions, corresponding to isospin 1, are symmetric, whereas the last function, corresponding to the isoscalar system pn, is antisymmetric in the isovariables.

In the given formalism the proton and neutron are considered as two states of one particle, the nucleon. Hence the total wave function of a system of two nucleons considered as identical particles must possess definite symmetry properties with respect to their exchange. Since the type of symmetry does not depend on what pair of particles is exchanged, we exchange two protons. They obey the Pauli statistics, i.e. when their coordinates and spins are exchanged the wave function changes sign. On the other hand, from (134.8) it is seen that when the isovariables of two protons are exchanged the wave function does not change. Thus nucleons obey the generalized Pauli principle according to which the total wave function of a system of nucleons is antisymmetric with respect to the exchange of any of their pairs. Hence it follows, in particular, that the wave function of a system NN with $T = 1$ describes a state whose angular momentum differs from that of the isoscalar ($T{=}0$) state.

3. From the wave function of a system of a π-meson and a nucleon

$$\varphi_{ij}^k = N_i \pi_j^k \tag{134.9}$$

we separate the parts symmetric and antisymmetric with respect to subscripts:

$$\varphi_{ij}^k = \tfrac{1}{2}(N_i \pi_j^k + N_j \pi_i^k) + \tfrac{1}{2}(N_i \pi_j^k - N_j \pi_i^k) \equiv \varphi_{[ij]}^k + \varphi_{\{ij\}}^k . \tag{134.10}$$

The second term contains two non-zero independent components $\varphi_{[12]}$, i.e. it is an isospinor ($T{=}\tfrac{1}{2}$). But the first matrix is still reducible, since its traces are not equal to zero. Separating them, we get

$$\varphi_{[ij]}^k = \tfrac{1}{2}(N_i \pi_j^k + N_j \pi_i^k - \tfrac{1}{3}\delta_i^k N_m \pi_j^m - \tfrac{1}{3}\delta_j^k N_m \pi_i^m) +$$
$$+ \tfrac{1}{6}(\delta_i^k N_m \pi_j^m + \delta_j^k N_m \pi_i^m) \equiv \chi_{ij}^k + (\delta_i^k \chi_j + \delta_j^k \chi_i) . \tag{134.11}$$

The second term represents an isospinor. It is easily seen that its components are the same as the two components $\varphi_{\{12\}}^k$. From (133.6) we conclude that the first term corresponds to isospin $T = \tfrac{3}{2}$, i.e. that it contains four independent components. Making use of (133.7), we have

$$|\tfrac{3}{2},+\tfrac{3}{2}\rangle \sim \chi_{11}^2 , \qquad |\tfrac{3}{2},+\tfrac{1}{2}\rangle \sim \chi_{11}^1 , \qquad |\tfrac{3}{2},-\tfrac{1}{2}\rangle \sim \chi_{22}^2 ,$$
$$|\tfrac{3}{2},-\tfrac{3}{2}\rangle \sim \chi_{22}^1 , \qquad |\tfrac{1}{2},+\tfrac{1}{2}\rangle \sim \chi_1 , \qquad |\tfrac{1}{2},-\tfrac{1}{2}\rangle \sim \chi_2 . \tag{134.12}$$

Writing the components explicitly and defining the factors of proportionality from the normalization conditions, we finally obtain

$$|\tfrac{3}{2},+\tfrac{3}{2}\rangle = p\pi^+ , \qquad |\tfrac{3}{2},+\tfrac{1}{2}\rangle = \sqrt{\tfrac{2}{3}}p\pi^0 - \sqrt{\tfrac{1}{3}}n\pi^+ , \qquad |\tfrac{3}{2},-\tfrac{1}{2}\rangle = \sqrt{\tfrac{1}{3}}p\pi^- + \sqrt{\tfrac{2}{3}}n\pi^0 ,$$
$$|\tfrac{3}{2},-\tfrac{3}{2}\rangle = n\pi^- , \qquad |\tfrac{1}{2},+\tfrac{1}{2}\rangle = \sqrt{\tfrac{1}{3}}p\pi^0 + \sqrt{\tfrac{2}{3}}n\pi^+ , \qquad |\tfrac{1}{2},-\tfrac{1}{2}\rangle = \sqrt{\tfrac{2}{3}}p\pi^- - \sqrt{\tfrac{1}{3}}n\pi^0 .$$
$$\tag{134.13}$$

One can also arrive at this result by means of the formalism developed in §§ 51 and 52. According to the rules of addition of angular momenta the isospin of a system consisting of a π-meson ($T=1$) and a nucleon ($T=\frac{1}{2}$) can take on the values $\frac{3}{2}$ and $\frac{1}{2}$. To construct the wave functions of the system corresponding to definite values of T and T_3 use can be made of the general formula (52.3), which in our notation is written as

$$|T,T_3\rangle = \sum_{t_3=\pm\frac{1}{2}} C^T_{T_3-t_3,t_3}|1,T_3-t_3\rangle|\tfrac{1}{2},t_3\rangle. \tag{134.14}$$

Taking the Clebsch–Gordan coefficients from the table 1, §52, we have

$$|\tfrac{3}{2},+\tfrac{3}{2}\rangle = C^{\frac{3}{2}}_{1,\frac{1}{2}}|1,+1\rangle|\tfrac{1}{2},+\tfrac{1}{2}\rangle = p\pi^+ ,$$

$$|\tfrac{3}{2},+\tfrac{1}{2}\rangle = C^{\frac{3}{2}}_{1,-\frac{1}{2}}|1,+1\rangle|\tfrac{1}{2},-\tfrac{1}{2}\rangle + C^{\frac{3}{2}}_{0,\frac{1}{2}}|1,0\rangle|\tfrac{1}{2},+\tfrac{1}{2}\rangle = \sqrt{\tfrac{2}{3}}p\pi^0 - \sqrt{\tfrac{1}{3}}n\pi^+ ,$$

$$|\tfrac{3}{2},-\tfrac{1}{2}\rangle = C^{\frac{3}{2}}_{0,-\frac{1}{2}}|1,0\rangle|\tfrac{1}{2},-\tfrac{1}{2}\rangle + C^{\frac{3}{2}}_{-1,\frac{1}{2}}|1,-1\rangle|\tfrac{1}{2},+\tfrac{1}{2}\rangle = \sqrt{\tfrac{1}{3}}p\pi^- + \sqrt{\tfrac{2}{3}}n\pi^0 ,$$

$$|\tfrac{3}{2},-\tfrac{3}{2}\rangle = C^{\frac{3}{2}}_{-1,-\frac{1}{2}}|1,-1\rangle|\tfrac{1}{2},-\tfrac{1}{2}\rangle = n\pi^- ,$$

$$|\tfrac{1}{2},+\tfrac{1}{2}\rangle = C^{\frac{1}{2}}_{1,-\frac{1}{2}}|1,+1\rangle|\tfrac{1}{2},-\tfrac{1}{2}\rangle + C^{\frac{1}{2}}_{0,\frac{1}{2}}|1,0\rangle|\tfrac{1}{2},+\tfrac{1}{2}\rangle = \sqrt{\tfrac{1}{3}}p\pi^0 + \sqrt{\tfrac{2}{3}}n\pi^+ ,$$

$$|\tfrac{1}{2},-\tfrac{1}{2}\rangle = C^{\frac{1}{2}}_{0,-\frac{1}{2}}|1,0\rangle|\tfrac{1}{2},-\tfrac{1}{2}\rangle + C^{\frac{1}{2}}_{-1,\frac{1}{2}}|1,-1\rangle|\tfrac{1}{2},+\tfrac{1}{2}\rangle = \sqrt{\tfrac{2}{3}}p\pi^- - \sqrt{\tfrac{1}{3}}n\pi^0 .$$

Thus we again arrive at relations (134.13).

One can easily reverse them and express the wave function of a system consisting of a π-meson and a nucleon in terms of the functions $|T,T_3\rangle$:

$$p\pi^+ = |\tfrac{3}{2},+\tfrac{3}{2}\rangle ,$$

$$p\pi^- = \sqrt{\tfrac{1}{3}}|\tfrac{3}{2},-\tfrac{1}{2}\rangle + \sqrt{\tfrac{2}{3}}|\tfrac{1}{2},-\tfrac{1}{2}\rangle ,$$

$$p\pi^0 = \sqrt{\tfrac{2}{3}}|\tfrac{3}{2},+\tfrac{1}{2}\rangle + \sqrt{\tfrac{1}{3}}|\tfrac{1}{2},+\tfrac{1}{2}\rangle , \tag{134.15}$$

$$n\pi^+ = -\sqrt{\tfrac{1}{3}}|\tfrac{3}{2},+\tfrac{1}{2}\rangle + \sqrt{\tfrac{2}{3}}|\tfrac{1}{2},+\tfrac{1}{2}\rangle ,$$

$$n\pi^- = |\tfrac{3}{2},-\tfrac{3}{2}\rangle ,$$

$$n\pi^0 = \sqrt{\tfrac{2}{3}}|\tfrac{3}{2},-\tfrac{1}{2}\rangle - \sqrt{\tfrac{1}{3}}|\tfrac{1}{2},-\tfrac{1}{2}\rangle .$$

§135. Isotopically invariant interaction

In §67 we pointed out that nuclear forces, i.e. the forces acting between two nucleons of the atomic nucleus, possess the property of charge independence. This property is most simply described within the framework of the isospin formalism. The hypothesis of charge independence of nuclear forces says that the interaction of two protons, two neutrons and of a proton with a neutron in the same spatial-spin states is identical. From the generalized Pauli principle it follows that if the two nucleons are in the isotriplet state (symmetric with respect to isospin variables), then their coordinate wave function is antisymmetric. For the proton and neutron in the isosinglet state, the spin-coordinate part of the wave function is symmetric. This means that the interaction of two nucleons in states $|1,+1\rangle$, $|1,0\rangle$ and $|1,-1\rangle$, other things being equal, will be the same, whereas the interaction in state $|0,0\rangle$ is, in general, essentially different.

As an example let us consider the deuteron, which consists of a proton and a neutron. The coordinate wave function of its ground state ($l=0$) is symmetric, and the spin wave function is also symmetric; the total angular momentum of the deuteron is equal to 1, i.e. it is in the triplet spin state $^3S_1$ with a small admixture of state $^3D_1$ (see §76). Hence the isospin part of the wave function must be antisymmetric and form an isosinglet. It is known from experiment that the binding energy of the deuteron is equal to 2.23 MeV. On the other hand, if the deuteron were in the state $^1S_0$, then its isospin wave function would be symmetric, i.e. it would be an isovector. In this case there are no bound states.

Thus, under the assumption of charge independence, nuclear forces are determined by the total isospin T, rather than by its component T_3. Consequently, in the phenomenological theory of nuclear forces, in which the interaction between nucleons is described by a certain potential, the Hamiltonian may contain only the invariant operator $\hat{T}^2$, the square of the total isospin of the system of nucleons, which can be written in the form

$$\hat{T}^2 = (\hat{T}'+\hat{T}'')^2 = \tfrac{1}{4}(\tau'+\tau'')^2 = \tfrac{1}{4}\tau'^2 + \tfrac{1}{4}\tau''^2 + \tfrac{1}{2}(\tau'\cdot\tau''). \quad (135.1)$$

Here $\tau' = \{\tau'_1,\tau'_2,\tau'_3\}$ and $\tau'' = \{\tau''_1,\tau''_2,\tau''_3\}$ are Pauli matrices acting on the isospin indices of the first and second nucleons respectively. Since the operators τ'^2 and τ''^2 are multiples of the unit operator, the dependence of the interaction Hamiltonian on the isospin variables is defined only by the scalar product $(\tau'\cdot\tau'')$:

$$\hat{H}_{\text{int}} = U_1 + U_2(\tau'\cdot\tau''). \quad (135.2)$$

Here U_1 and U_2 depend only on the coordinates and on the ordinary spin; they are written down in §76.

It is easily verified that the Hamiltonian (135.2) commutes with the isospin component operators $\hat{T}_\alpha$, and thus also with the operator $\hat{T}^2$:

$$[\hat{H}_{\text{int}},\hat{T}_\alpha] = [\hat{H}_{\text{int}},\hat{T}^2] = 0 . \quad (135.3)$$

Consequently, if the Coulomb interaction of protons and the small difference in the masses of the proton and neutron are disregarded, then in the system of interacting nucleons there holds not only the law of conservation of the isospin component, expressing the trivial fact of charge conservation, but also the law of conservation of the total isospin, which may serve as a formulation of charge independence.

Let us now turn to the interaction of nucleons with π-mesons. As we know from §67, it is just this interaction which is responsible for the existence of nuclear forces, i.e. for the interaction of nucleons, which we have considered above purely phenomenologically. A logical discussion of the corresponding problems is possible only within the framework of the quantum field theory. Therefore we shall confine ourselves to some comments. Analogous to what we did in §122 in describing the weak interaction, we shall consider the nucleon function N_i and the π-meson function π_j^k as operators in the space of occupation numbers. The operator N_i corresponds to the destruction of a nucleon and to the creation of an antinucleon, whereas the operator $(N^\dagger)_i$ creates a nucleon and destroys an antinucleon. The operator π_j^k creates and destroys π-mesons.

The basic processes of the interaction considered are those of virtual production and absorption of π-mesons by nucleons (see §67). Hence the interaction Hamiltonian density must have the general structure $N^\dagger N \pi$. Since the π-meson operator is a pseudoscalar, relativistic invariance and parity conservation require that it be multiplied by the pseudoscalar combination of $N^\dagger$ and N. Recalling the results of §121, we arrive at the expression $\bar{N}\gamma_5 N \pi$. In consequence of the charge independence of the πN interaction, the Hamiltonian density must be an isoscalar. One can form from the matrices $\bar{N}^i$, N_j and π_k^l only one combination of this type: $\bar{N}^i N_j \pi_i^j$. Thus, finally,

$$\hat{H}_{\text{int}} = \sqrt{2}g\bar{N}^i\gamma_5 N_j \pi_i^j \tag{135.4}$$

where g is the strong πN interaction constant (an analogue of the electric charge). The factor $\sqrt{2}$ is introduced for historical reasons.

Making use of the explicit form of the nucleon matrices and the π-meson matrices $\bar{N}^i$, N_j and π_k^l (see (133.17), (133.18) and (133.24)), from (135.4) we obtain

$$\hat{H}_{\text{int}} = g[\sqrt{2}\bar{p}\gamma_5 n\pi^+ + \sqrt{2}\bar{n}\gamma_5 p\pi^- + (\bar{p}\gamma_5 p - \bar{n}\gamma_5 n)\pi^0] . \tag{135.5}$$

Consequently, on the assumption of charge independence the following relation holds between the constants of the πN interaction:

$$g_{pn\pi^+} : g_{pn\pi^-} : g_{pp\pi^0} : g_{nn\pi^0} = 1 : 1 : \frac{1}{\sqrt{2}} : \left(-\frac{1}{\sqrt{2}}\right). \qquad (135.6)$$

Taking into account the explicit form of the Pauli matrices and passing to the π-meson vector function $\boldsymbol{\pi}$ (133.26), it is easily verified that the Hamiltonian density (135.4) can be rewritten as

$$\hat{H}_{\text{int}} = g\overline{N}\gamma_5 \boldsymbol{\tau} N \boldsymbol{\pi} \qquad (135.7)$$

where it is assumed that the isovectors $\boldsymbol{\tau}$ and $\boldsymbol{\pi}$ are multiplied in the scalar way. In older studies only this form of notation was used, which accounts for the appearance of the factor $\sqrt{2}$ in expression (135.4). The interaction of other baryons with mesons is described in an analogous way. For example, for the system $\pi, \Sigma, \overline{\Sigma}$ it is easily found that

$$\hat{H}_{\text{int}} = \sqrt{2}g'\overline{\Sigma}_i^j \gamma_5 \Sigma_j^k \pi_k^i . \qquad (135.8)$$

In conclusion we stress once more that the electromagnetic interaction of nucleons violates the invariance with respect to transformations of the isogroup SU(2), and the results formulated above are no longer valid. For this interaction the Hamiltonian density can be written on the basis of the same considerations as those which were used in obtaining (135.4). Taking into account that the operator of creation and destruction of photons is the vector potential A_μ, we shall have

$$\hat{H}_{\text{int}} = e\overline{N}\gamma_\mu \hat{Q} N A_\mu \qquad (135.9)$$

where $\hat{Q}$ is the nucleon charge operator given by formula (133.14). Hence

$$\hat{H}_{\text{int}} = \tfrac{1}{2}e\overline{N}\gamma_\mu (I + \tau_3) N A_\mu \qquad (135.10)$$

i.e. the Hamiltonian density contains an isovector part (the term with τ_3) in addition to the isoscalar part. The presence of the former violates the conservation of the total isospin T, although its component T_3 is conserved. But the intensity of the electromagnetic interaction is much smaller than that of the strong interaction, so that electromagnetic corrections can frequently be neglected, being considered at most as a small perturbation.

Some quantitative consequences of the hypothesis of charge independence of the strong interaction, which is equivalent to total isospin conservation, are presented in the next section.

§136. The scattering of nucleons and π-mesons

We apply the isospin formalism to the analysis of the processes of scattering of nucleons by nucleons and of π-mesons by nucleons. Generalization to the case of other hadrons presents no difficulty.

Let several reactions

$$a_i + b_i \to c_i + d_i \tag{136.1}$$

be considered, all particles of the type a, b, c and d belonging to one and the same isomultiplet. For the wave functions of the initial and final states we shall make use of the Dirac notation $|a_i b_i\rangle$ and $|c_i d_i\rangle$. The scattering amplitude $f^{(i)}$ is proportional to the matrix element

$$M^{(i)} = \langle c_i d_i | a_i b_i \rangle \tag{136.2}$$

the square of the modulus of which defines the differential and, after integrating over angles, total cross sections for the process.

We first assume that the state of the particles before scattering has definite values of the isospin T and of its component T_3, i.e. that its wave function is $|T,T_3\rangle$. If the part of the matrix element corresponding to Coulomb scattering is separated, then from charge independence it follows that in the reaction the isospin does not change:

$$\langle T',T_3 | T,T_3 \rangle = 0 \quad \text{for} \quad T' \neq T, \tag{136.3}$$

i.e. the wave function of the final state is also a function of the type $|T,T_3\rangle$. Furthermore, the matrix element corresponding to the scattering due to the strong interaction cannot depend on the values of the component T_3, but is defined by the isospin T (and by other quantum numbers), by momenta, by spins and so on. We denote

$$\langle T,T_3 | T,T_3 \rangle \equiv M^{(T)} . \tag{136.4}$$

The validity of the isospin formalism as applied to the class of problems considered lies in the fact that the matrix element of any real process of the set (136.1) can be expressed in terms of a small number of (in most cases two) matrix elements $M^{(T)}$ corresponding to the scattering in a definite isospin state. For this it suffices to expand the wave functions $|a_i b_i\rangle$ and $|c_i d_i\rangle$ in terms of the wave functions $|T,T_3\rangle$, substitute these expansions into (136.2) and make use of formulae (136.3)–(136.4). Thus one can establish a number of relations between the cross sections for different processes corresponding to the same initial and the same final spatial-spin states of the particles involved in the scattering.

1. As the first example let us consider the scattering of protons by protons and of neutrons by protons. First of all, from (134.8) we express the wave functions of the initial and final states, i.e. of the systems pp or np, in terms of the basis functions of the isotriplet and isosinglet:

$$|p'p''\rangle = |1,+1\rangle , \qquad |n'p''\rangle = \frac{1}{\sqrt{2}}(|1,0\rangle - |0,0\rangle) , \qquad |p'n''\rangle = \frac{1}{\sqrt{2}}(|1,0\rangle + |0,0\rangle) .$$
$$(136.5)$$

Then for the process of scattering $p' + p'' \rightarrow p' + p''$ we find that

$$M^{(\text{pp})} = \langle p'p''|p'p''\rangle = \langle 1,+1|1,+1\rangle = M^{(1)} . \qquad (136.6)$$

In neutron–proton scattering the following two processes are possible:
ordinary elastic scattering

$$n' + p'' \rightarrow n' + p''$$

and charge-exchange scattering

$$n' + p'' \rightarrow p' + n'' .$$

For the first of these

$$M^{\text{elas}} = \langle n'p''|n'p''\rangle = \tfrac{1}{2}[(\langle 1,0| - \langle 0,0|)(|1,0\rangle - |0,0\rangle)] = \tfrac{1}{2}[M^{(1)} + M^{(0)}] \qquad (136.7)$$

and for the second

$$M^{\text{ch.ex}} = \langle p'n''|n'p''\rangle = \tfrac{1}{2}[(\langle 1,0| + \langle 0,0|)(|1,0\rangle - |0,0\rangle)] = \tfrac{1}{2}[M^{(1)} - M^{(0)}] . \qquad (136.8)$$

The elastic scattering cross sections are proportional to the squares of the moduli of the matrix elements, hence

$$\frac{d\sigma^{(\text{pp})}}{d\Omega} \sim |M^{(1)}|^2 , \qquad \frac{d\sigma^{\text{elas}}}{d\Omega} \sim \tfrac{1}{4}|M^{(1)} + M^{(0)}|^2 ,$$

$$\frac{d\sigma^{\text{ch.ex}}}{d\Omega} \sim \tfrac{1}{4}|M^{(1)} - M^{(0)}|^2 . \qquad (136.9)$$

Summing the last two expressions, we obtain the total neutron–proton scattering cross section, which is determined experimentally by the total number of protons and neutrons scattered at a given angle. Finally

$$\frac{d\sigma^{(\text{pp})}}{d\Omega} \sim |M^{(1)}|^2 , \qquad \frac{d\sigma^{(\text{np})}}{d\Omega} \sim \tfrac{1}{2}|M^{(1)}|^2 + \tfrac{1}{2}|M^{(0)}|^2 . \qquad (136.10)$$

Since the angular dependence of the matrix elements $M^{(1)}$ and $M^{(0)}$ can be essentially different, then, in spite of charge independence, the behaviour of the proton–proton and neutron–proton scattering cross sections as functions of the angular variable can be different. Experiment shows that this is indeed so. In the energy range of 300–500 MeV in the centre-of-mass system the first cross section is almost independent of the scattering angle, whereas the second has a minimum at $\theta = \frac{1}{2}\pi$, increasing sharply in the backward direction and to a lesser degree in the forward direction.

2. The example of the reaction of π-meson production with formation of a deuteron in nucleon–nucleon collisions is somewhat more interesting:

$$p + p \rightarrow d + \pi^+ ,$$

$$n + p \rightarrow d + \pi^0 .$$

Since the deuteron isospin is equal to zero (see the beginning of §135), then

$$|\pi^+ d\rangle = |1,+1\rangle , \qquad |\pi^0 d\rangle = |1,0\rangle . \tag{136.11}$$

Therefore

$$M^{(pp)} = \langle \pi^+ d|pp\rangle = \langle 1,+1|1,+1\rangle = M^{(1)} ,$$

$$M^{(np)} = \langle \pi^0 d|np\rangle = \frac{1}{\sqrt{2}} [\langle 1,0|(|1,0\rangle - |0,0\rangle)] = \frac{1}{\sqrt{2}} M^{(1)} . \tag{136.12}$$

From this follows the relation between the cross sections:

$$\frac{d\sigma^{(pp)}/d\Omega}{d\sigma^{(np)}/d\Omega} = 2 \tag{136.13}$$

which has been confirmed experimentally.

3. The scattering of charged π-mesons by protons is an even more interesting case:

$$\pi^+ + p \rightarrow \pi^+ + p ,$$

$$\pi^- + p \rightarrow \pi^- + p ,$$

$$\pi^- + p \rightarrow \pi^0 + n .$$

Denoting the matrix elements and cross sections referring to these processes by the symbol of the π-meson in the final state and making use of formulae (134.15), we obtain

$$M^{(+)} = \langle \pi^+ p | \pi^+ p \rangle = \langle \tfrac{3}{2}, +\tfrac{3}{2} | \tfrac{3}{2}, +\tfrac{3}{2} \rangle = M^{(\tfrac{3}{2})} \,,$$

$$M^{(-)} = \langle \pi^- p | \pi^- p \rangle = [(\sqrt{\tfrac{1}{3}} \langle \tfrac{3}{2}, -\tfrac{1}{2} | + \sqrt{\tfrac{2}{3}} \langle \tfrac{1}{2}, -\tfrac{1}{2} |) \times$$

$$\times \, (\sqrt{\tfrac{1}{3}} | \tfrac{3}{2}, -\tfrac{1}{2} \rangle + \sqrt{\tfrac{2}{3}} | \tfrac{1}{2}, -\tfrac{1}{2} \rangle)] = \tfrac{1}{3} M^{(\tfrac{3}{2})} + \tfrac{2}{3} M^{(\tfrac{1}{2})} \,,$$

$$M^{(0)} = \langle \pi^0 n | \pi^- p \rangle = [(\sqrt{\tfrac{2}{3}} \langle \tfrac{3}{2}, -\tfrac{1}{2} | - \sqrt{\tfrac{1}{3}} \langle \tfrac{1}{2}, -\tfrac{1}{2} |) \times$$

$$\times \, (\sqrt{\tfrac{1}{3}} | \tfrac{3}{2}, -\tfrac{1}{2} \rangle + \sqrt{\tfrac{2}{3}} | \tfrac{1}{2}, -\tfrac{1}{2} \rangle)] = \tfrac{\sqrt{2}}{3} M^{(\tfrac{3}{2})} - \tfrac{\sqrt{2}}{3} M^{(\tfrac{1}{2})} \,,$$

hence

$$f^{(+)} = f^{(\tfrac{3}{2})} \,, \qquad f^{(-)} = \tfrac{1}{3} f^{(\tfrac{3}{2})} + \tfrac{2}{3} f^{(\tfrac{1}{2})} \,, \qquad f^{(0)} = \tfrac{\sqrt{2}}{3} f^{(\tfrac{3}{2})} - \tfrac{\sqrt{2}}{3} f^{(\tfrac{1}{2})} \,. \qquad (136.14)$$

Eliminating the amplitudes $f^{(\tfrac{3}{2})}$ and $f^{(\tfrac{1}{2})}$, we have the relation

$$f^{(+)} - f^{(-)} = \sqrt{2} f^{(0)} \qquad (136.15)$$

from which the so-called triangle relations follow:

$$|\sqrt{\sigma^{(+)}} - \sqrt{\sigma^{(-)}}| \leqslant \sqrt{2\sigma^{(0)}} \leqslant \sqrt{\sigma^{(+)}} + \sqrt{\sigma^{(-)}} \,. \qquad (136.16)$$

Under some additional assumptions regarding the properties of the amplitudes, more interesting relations can be obtained between the cross sections. Thus in the case $f^{(\tfrac{1}{2})} \approx 0$

$$\sigma^{(+)} : \sigma^{(-)} : \sigma^{(0)} = 9 : 1 : 2 \,. \qquad (136.17)$$

If it is assumed that $f^{(\tfrac{1}{2})} \approx f^{(\tfrac{3}{2})}$, then

$$\sigma^{(+)} : \sigma^{(-)} : \sigma^{(0)} = 1 : 1 : 0 \,. \qquad (136.18)$$

Finally, if $f^{(\tfrac{3}{2})} \approx 0$, then

$$\sigma^{(+)} : \sigma^{(-)} : \sigma^{(0)} = 0 : 2 : 1 \,. \qquad (136.19)$$

Experiment shows that for an energy of 120 MeV of the incident π-meson the total cross sections are in the ratio $93:11:22 \approx 9:1:2$, i.e. in this energy range the scattering in the state with isospin $T = \tfrac{3}{2}$ is dominating. For energies above 200 MeV the amplitude $f^{(\tfrac{1}{2})}$ also begins to give a considerable contribution.

There is also another, simpler method of obtaining the relations between the cross sections, which does not require knowledge of the Clebsch–Gordan coefficients and is especially useful in those cases where their calculation is for any reason difficult. It is called the method of invariant amplitudes, and will be demonstrated by the example of the scattering of charged π-mesons by nucleons.

Under the assumption of charge independence the isospin T does not change in scattering. This means that the total amplitude of the scattering of one isomultiplet by another must be an isoscalar. In our case it is constructed of the wave functions N_i and π_i^j of the initial state and the wave functions $\bar{N}^i$ and $\bar{\pi}_i^j$ of the final state which transform according to the conjugate representations (we denote by a bar the functions describing the final state; at the same time the bar is the symbol of the conjugate representation). The matrices of these wave functions are of the form

$$N_i = \begin{pmatrix} p \\ n \end{pmatrix}, \qquad \pi_i^j = \begin{pmatrix} \pi^0/\sqrt{2} & \pi^+ \\ \pi^- & -\pi^0/\sqrt{2} \end{pmatrix},$$

$$\bar{N}^i = (\bar{p}, \bar{n}), \qquad \bar{\pi}_i^j = \begin{pmatrix} \bar{\pi}^0/\sqrt{2} & \bar{\pi}^- \\ \bar{\pi}^+ & -\bar{\pi}^0/\sqrt{2} \end{pmatrix}. \tag{136.20}$$

One can form from them the two independent isoscalars

$$\bar{N}^i N_i \bar{\pi}_j^k \pi_k^j \quad \text{and} \quad \bar{N}^i \bar{\pi}_i^j \pi_j^k N_k$$

so that the amplitude is written in the form of a linear combination

$$f = f_1(\bar{N}^i N_i \bar{\pi}_j^k \pi_k^j) + f_2(\bar{N}^i \bar{\pi}_i^j \pi_j^k N_k). \tag{136.21}$$

From (136.20) we have for the isospin part of the wave functions of the particles of the initial and final states

for the proton $p \to N_1 = 1$, $\bar{p} \to \bar{N}^1 = 1$

for the neutron $n \to N_2 = 1$, $\bar{n} \to \bar{N}^2 = 1$

for the π^+-meson $\pi^+ \to \pi_1^2 = 1$, $\bar{\pi}^+ \to \bar{\pi}_2^1 = 1$

for the π^--meson $\pi^- \to \pi_2^1 = 1$, $\bar{\pi}^- \to \bar{\pi}_1^2 = 1$

for the π^0-meson $\pi^0 \to \pi_1^1 = \dfrac{1}{\sqrt{2}}, \pi_2^2 = -\dfrac{1}{\sqrt{2}}, \bar{\pi}^0 \to \bar{\pi}_1^1 = \dfrac{1}{\sqrt{2}}, \bar{\pi}_2^2 = -\dfrac{1}{\sqrt{2}}$

(all other components are equal to zero).

Substituting these wave functions into the amplitude (136.21), we obtain

$$f^{(+)} = f_1, \qquad f^{(-)} = f_1 + f_2, \qquad f^{(0)} = -f_2/\sqrt{2}, \tag{136.22}$$

where f_1 and f_2 are functions of spatial-spin variables, unknown but the same for all pion–nucleon processes. From (136.22) there follows, in particular, relation (136.15), and therefore also the triangle inequalities. It is known from experiment that in the region of high energies and small angles the cross section of the charge-exchange process $\pi^- + p \to \pi^0 + n$ is small compared to elastic cross sections. Hence assuming $f_2 \approx 0$ we obtain the approximate equality of the differential cross sections in the forward direction for the elastic scattering of π^+ and π^--mesons by protons:

$$\frac{d\sigma^{(+)}}{d\Omega}\bigg|_{\theta \to 0} \approx \frac{d\sigma^{(-)}}{d\Omega}\bigg|_{\theta \to 0} . \qquad (136.23)$$

The amplitudes f_i can be expressed in terms of the amplitudes $f^{(T)}$ and vice versa. For this it suffices to compare relations (136.22) and (136.14), hence

$$f_1 = f^{(\frac{3}{2})} ; \qquad f_2 = -\tfrac{2}{3} f^{(\frac{3}{2})} + \tfrac{2}{3} f^{(\frac{1}{2})} . \qquad (136.24)$$

In concluding this section we indicate a simple method of determining the number of independent invariant amplitudes. According to the rule of addition of angular momenta we find possible values of the isospin T of the initial and final states. By virtue of charge independence only transitions with conservation of T are possible, hence the number of independent amplitudes is defined by the number of values of the isospin which occur in both the initial and final states. In our example the isospin of the initial and final states are equal to $\frac{3}{2}$ and $\frac{1}{2}$, for which there are two amplitudes: $f^{(\frac{3}{2})}$ and $f^{(\frac{1}{2})}$, or f_1 and f_2.

§137. The unitary group SU(3) and its representations

In the preceding sections we studied some consequences of the isospin invariance of the strong interaction based on the group SU(2) and one most suitable for the description of the symmetry properties of nucleons and π-mesons. However, after the discovery of strange particles the framework of this group turned out to be too narrow, because its rank is equal to 1 and it gives only one conserved additive quantum number (the isospin component T_3) by means of which the terms of a given isomultiplet are classified. By an additive quantum number is meant a quantity whose value for a certain system is equal to the sum of its values for the subsystems. In this sense, T is not an additive characteristic. At the same time there is at least one more characteristic of this type, the hypercharge Y (or strangeness), so that it is natural to try to group several isomultiplets with different hypercharges into one supermultiplet. For this a group of rank 2 is necessary, the mutually commuting generators of which give two simultaneously measurable conserved quantum numbers characterizing the terms of the supermultiplet, so that they can be identified with T_3 and Y. On the other hand, from the requirement that the new theory contain the results of the old it follows that the isogroup must be a part (subgroup) of a new larger symmetry group. These requirements can most simply and naturally be satisfied if one postulates the approximate invariance of the strong interaction with respect to the group SU(3) which we shall henceforth call unitary (in the narrow sense of the word)*. The corresponding mathematical apparatus is very close to the formalism presented in §132 the results of which we shall frequently refer to.

The group SU(3) is understood to be the set of all unitary and unimodular matrices of third order which correspond to linear transformations in a three-dimensional complex space conserving the quadratic form $x^\dagger x = x^i x_i = x_1^* x_1 + x_2^* x_2 + x_3^* x_3$ (the indices i, j and so on now run over the values 1, 2, 3). The generators of this group are Hermitian square 3×3 matrices λ_α with trace zero:

$$\lambda_\alpha^\dagger = \lambda_\alpha , \qquad \mathrm{Tr}\, \lambda_\alpha = 0 . \tag{137.1}$$

Nine conditions of hermiticity and 1 condition of equality to zero of the trace are imposed upon 18 real parameters of the complex 3×3 matrices, so that there are 8 independent matrices with the properties (137.1), which defines the dimension of the group SU(3). Among the matrices λ_α there are two mutually commuting ones (the rank of the group SU(3) is equal to 2) which can simultaneously be diagonalized. We choose the representation in which λ_3 and λ_8 are diagonal:

$$\lambda_1 = \begin{pmatrix} 0 & 1 & 0 \\ 1 & 0 & 0 \\ 0 & 0 & 0 \end{pmatrix}, \quad \lambda_2 = \begin{pmatrix} 0 & -i & 0 \\ i & 0 & 0 \\ 0 & 0 & 0 \end{pmatrix}, \quad \lambda_3 = \begin{pmatrix} 1 & 0 & 0 \\ 0 & -1 & 0 \\ 0 & 0 & 0 \end{pmatrix}, \quad \lambda_4 = \begin{pmatrix} 0 & 0 & 1 \\ 0 & 0 & 0 \\ 1 & 0 & 0 \end{pmatrix},$$

$$\lambda_5 = \begin{pmatrix} 0 & 0 & -i \\ 0 & 0 & 0 \\ i & 0 & 0 \end{pmatrix}, \quad \lambda_6 = \begin{pmatrix} 0 & 0 & 0 \\ 0 & 0 & 1 \\ 0 & 1 & 0 \end{pmatrix}, \quad \lambda_7 = \begin{pmatrix} 0 & 0 & 0 \\ 0 & 0 & -i \\ 0 & i & 0 \end{pmatrix}, \quad \lambda_8 = \frac{1}{\sqrt{3}}\begin{pmatrix} 1 & 0 & 0 \\ 0 & 1 & 0 \\ 0 & 0 & -2 \end{pmatrix}. \tag{137.2}$$

Hence it follows that if all parameters ω_α of the group SU(3), except for the first three, are set equal to zero, then we obtain the isogroup SU(2).

The representations of the group SU(3) are constructed in a completely analogous way to §132. However, the mutually conjugated representations

---

* It should be noted that in 1961–1964 the situation was not completely clear. since it was impossible to make an unambiguous choice between SU(3) and the so-called group G_2; some preference even was given to the latter. The problem was finally solved in 1964 when the Ω-hyperon was discovered (see §138).

will now not be equivalent, so that the given irreducible representation (p,q) is defined by two numbers, p and q (rather than one number $p+q$). The definitions and relations (132.13)–(132.16) remain valid as before if τ_α is replaced in them by λ_α and if it is assumed that the Roman indices run over values from 1 to 3, and Greek indices from 1 to 8.

Let us find the number of independent components of the matrix $\varphi^{j_1 \ldots j_q}_{i_1 \ldots i_p}$ which is symmetric separately in all subscripts and superscripts and which has zero trace over any pair of superscript and subscript, i.e. let us determine the dimension $N(p,q)$ of the irreducible representation (p,q). By virtue of symmetry only components differing in one of the numbers p_1, p_2, p_3 (the numbers of ones, twos and threes among the subscripts) are different. Since $p_1 + p_2 + p_3 = p$, then for a given p_1 the number p_2 can vary from 0 up to $p - p_1$, which gives $p - p_1 + 1$ different components. Now varying p_1 from 0 to p, we obtain the total number of different components for fixed superscripts:

$$N(p) = \sum_{p_1=0}^{p} (p - p_1 + 1) = \tfrac{1}{2}(p+1)(p+2) .$$

Analogously, the number of different components for fixed subscripts is equal to $N(q) = \tfrac{1}{2}(q+1)(q+2)$, there are in all $N(p)N(q)$ components. But they are still not independent because they are related by the conditions that the traces are equal to zero. By virtue of symmetry it is sufficient that the trace with respect to one of the pairs of superscript and subscript reduce to zero; this trace will be a matrix with $p-1$ subscripts and $q-1$ superscripts, having thus $N(p-1)N(q-1)$ components. Thus $N(p,q) = N(p)N(q) - N(p-1)N(q-1)$ and, finally,

$$N(p,q) = \tfrac{1}{2}(p+1)(q+1)(p+q+2) . \tag{137.3}$$

Let us enumerate the most important representations of the group SU(3):

1. φ $(0,0) \equiv 1$ unitary scalar or singlet ($N=1$),

2. φ_i and φ^i $(1,0) \equiv 3$ and $(0,1) \equiv \bar{3}$ unitary spinors or triplets ($N=3$),

3. φ_{ij} and φ^{ij} $(2,0) \equiv 6$ and $(0,2) \equiv \bar{6}$ sextets ($N=6$),

4. φ^j_i $(1,1) \equiv 8$ unitary vector or octet ($N=8$),

5. φ_{ijk} and φ^{ijk} $(3,0) \equiv 10$ and $(0,3) \equiv \overline{10}$ decuplets ($N=10$),

6. φ^k_{ij} and φ^{jk}_i $(2,1) \equiv 15$ and $(1,2) \equiv \overline{15}$ 15-plets ($N=15$),

7. φ^k_{ij} $(2,2) \equiv 27$ ($N=27$),

and so on. A very convenient notation is given here, which immediately indicates the dimension of the irreducible representation. Attention should be drawn, for example, to the representations $(2,0)$ and $(1,1)$; the corresponding wave functions have the same number of indices (namely 2), but the dimensions of the multiplets are different (6 and 8 respectively). Such a situation cannot be encountered in the group $SU(2)$ by virtue of the equivalence of its mutually conjugate representations.

Let us now write down the decompositions of some direct products of representations into irreducible representations:

$$(1,0)\otimes(0,1) = (0,0)\oplus(1,1) \tag{137.4a}$$

or $\qquad\qquad 3\otimes\bar{3} = 1\oplus8 \ ,$

$$(1,0)\otimes(1,0)\otimes(1,0) = (0,0)\oplus(1,1)\oplus(1,1)\oplus(3,0) \tag{137.4b}$$

or $\qquad\qquad 3\otimes3\otimes3 = 1\oplus8\oplus8\oplus10 \ ,$

$$(1,0)\otimes(1,0)\otimes(0,1) = (1,0)\oplus(1,0)\oplus(0,2)\oplus(2,1) \tag{137.4c}$$

or $\qquad\qquad 3\otimes3\otimes\bar{3} = 3\oplus3\oplus\bar{6}\oplus15 \ ,$

$$(1,1)\otimes(1,1) = (0,0)\oplus(1,1)\oplus(1,1)\oplus(3,0)\oplus(0,3)\oplus(2,2) \tag{137.4d}$$

or $\qquad\qquad 8\otimes8 = 1\oplus8\oplus8\oplus10\oplus\overline{10}\oplus27 \ ,$

$$(1,1)\otimes(3,0) = (1,1)\oplus(3,0)\oplus(2,2)\oplus(3,1) \tag{137.4e}$$

or $\qquad\qquad 8\otimes10 = 8\oplus10\oplus27\oplus35 \ .$

Let us prove, for example, decompositions (137.4a) and (137.4d). In the first case the function $\varphi_i\chi^j$ contains the non-zero trace $\varphi_i\chi^i$ which is a scalar (representation $(0,0)$); the function $\varphi_i\chi^j - \frac{1}{3}\delta_i^j\varphi_k\chi^k$, which remains after the separation of the non-zero trace, transforms according to the irreducible representation $(1,1)$. The proof of formula (137.4d) is somewhat more complex. First of all, if we symmetrize the function $\varphi_i^j\chi_k^l$ with respect to the subscripts and superscripts and separate non-zero traces, then we obtain representation $(2,2)$. Further, two different traces with respect to only one of the pairs of indices (we recall that $\varphi_i^i = \chi_i^i = 0$) after the separation of the non-zero traces with respect to the remaining pair of indices, i.e. the two functions

$$\varphi_i^j\chi_k^i - \frac{1}{3}\delta_k^j\varphi_i^l\chi_l^i \quad \text{and} \quad \varphi_i^j\chi_j^k - \frac{1}{3}\delta_i^k\varphi^j\chi_j^l$$

transform according to the two representations $(1,1)$ and $(1,1)'$. The total trace $\varphi_i^j\chi_j^i$ is a unitary scalar, so that we have the representation $(0,0)$. Finally, on separating symmetric parts there arise two functions of the type

$$\psi_{[ik]}^{\{jl\}} \equiv \varphi_i^j\chi_k^l + \varphi_k^j\chi_i^l - \varphi_i^l\chi_k^j - \varphi_k^l\chi_i^j \quad \text{and} \quad \psi_{\{ik\}}^{[jl]} \equiv \varphi_i^j\chi_k^l + \varphi_i^l\chi_k^j - \varphi_k^j\chi_i^l - \varphi_k^l\chi_i^j$$

containing 10 independent components each. It can be shown that the lower pair $\{ik\}$ is equivalent to one upper index and vice versa, hence we obtain the representations (3,0) and (0,3). We can convince ourselves of the validity of this decomposition by comparing the dimensions on the left-hand and right-hand sides of formula (137.4d): $8 \times 8 = 1+8+8+10+10+27 = 64$. Other formulae (137.4) are proved in an analogous way.

The group SU(3) has two invariant operators whose eigenvalues serve for the classification of irreducible representations. But we have already unambiguously characterized the representations also by two numbers p and q. Hence we shall not write down and analyse the invariant operators; we shall not need them in what follows.

For the classification of the basis elements of a multiplet, which can be chosen to be definite combinations of the matrices $\varphi_{i_1 \cdots i_p}^{j_1 \cdots j_q}$ with only one non-zero element, we shall make use of the eigenvalues of the diagonal generators Λ_3, Λ_8. From the explicit expressions of the type (132.15) for the generators Λ_α it follows that they act on the functions $\varphi_{i_1 \cdots i_p}^{j_1 \cdots j_q}$ with one non-zero element in the following way:

$$\Lambda_\alpha \varphi_{i_1 \cdots i_p}^{j_1 \cdots j_q} = \sum_{s=1}^{p} (\lambda_\alpha)_{i_s}^{i'_s} \varphi_{i_1 \cdots i_{s-1} i'_s i_{s+1} \cdots i_p}^{j_1 \cdots j_q} - \sum_{s=1}^{q} (\lambda_\alpha)_{j'_s}^{j_s} \varphi_{i_1 \cdots i_p}^{j_1 \cdots j_{s-1} j'_s j_{s+1} \cdots j_q} . \tag{137.5}$$

Let there be among p subscripts, p_3 indices 3 and $p-p_3$ indices 1 and 2, and among q superscripts, q_3 indices 3 and $q-q_3$ indices 1 and 2. Since the matrix λ_8 gives $1/\sqrt{3}$ in application to each superscript and subscript 1, 2, and $-2/\sqrt{3}$ in application to any index 3, then according to (137.5) the eigenvalues of the generators Λ_8 are equal to

$$\Lambda_8 = \sqrt{3}[\tfrac{1}{3}(p-q)-p_3+q_3] . \tag{137.6}$$

Analogously, for the eigenvalues of the generator Λ_3 we have

$$\Lambda_3 = (p_1-p_2-q_1+q_2) . \tag{137.7}$$

If no threes were present among the indices, i.e. if the group SU(2) were considered, then the relations $p_1+p_2 = p$ and $q_1+q_2 = q$ would be fulfilled. In this case formula (137.7) would go over into (132.25).

Two quantum numbers Λ_3, Λ_8 are not sufficient, however, for the unambiguous classification of the basis elements of a given multiplet. They are complemented, say, by the eigenvalue of the operator $\Lambda^2 = \Lambda_1^2 + \Lambda_2^2 + \Lambda_3^2$. From (137.2) it follows that

$$\lambda^2 \equiv \lambda_1^2 + \lambda_2^2 + \lambda_3^2 = 3 \begin{pmatrix} 1 & 0 & 0 \\ 0 & 1 & 0 \\ 0 & 0 & 0 \end{pmatrix} \qquad (137.8)$$

and hence the matrix λ^2 does not commute with all the matrices λ_α. By virtue of the fact that the expressions for the generators are completely analogous to those given in §132 (see (132.15)) this statement is valid also for Λ^2. For this reason the situation with the eigenvalues of the operator Λ^2 is somewhat more complex, since there may correspond several eigenvalues of the operator Λ^2 to the component $\varphi_{i_1 \ldots i_p}^{j_1 \ldots j_q}$ referring to the eigenvalues $\Lambda_3 = \Lambda_8 = 0$, i.e. a peculiar degeneracy occurs. It is easily shown that

$$\Lambda^2 \varphi_{i_1 \ldots i_p}^{j_1 \ldots j_q} = (p_1 + p_2 + q_1 + q_2)(p_1 + p_2 + q_1 + q_2 + 2), \varphi_{i_1 \ldots i_p}^{j_1 \ldots j_q}, \qquad (137.9)$$

but this statement will be valid generally only if $\Lambda_3 \neq 0$ or $\Lambda_8 \neq 0$. In the case $\Lambda_3 = \Lambda_8 = 0$ the factor in the right-hand side of (137.9) gives, however, the maximum eigenvalue of the operator Λ^2.

§138. The eightfold way formalism and unitary multiplets

Suppose that in nature there were a superstrong interaction of elementary particles invariant with respect to the group SU(3). Then the wave functions of hadrons would transform according to some of its irreducible representations, i.e. they are the products of ordinary spatial-spin wave functions with the unitary matrices $\varphi_{i_1 \ldots i_p}^{j_1 \ldots j_q}$. As a result all hadrons would be distributed in unitary multiplets, characterized by a pair of numbers p and q, spin, parity, baryonic number and other quantities not associated with the group SU(3) (see §133). Individual hadrons within a multiplet are classified by the eigenvalues of the generators Λ_3 and Λ_8 and of the operator Λ^2, which will be identified below with T_3, Y and T^2.

If it is assumed that there is only the superstrong interaction, then these quantum numbers must be conserved, the transitions occurring as a result of the reactions being possible only within one unitary multiplet, as in the case of isospin invariance. Furthermore, quantum-mechanical degeneracy must result, and the masses of the particles contained in one unitary multiplet must all be strictly the same. On the other hand, experiment shows that the masses of the particles having different hypercharge (for the same spin, parity and baryonic number) differ sharply: for example, the mass difference between the nucleon and Ξ-hyperon amounts to about 30% of the mass of the latter. This means that the ordinary strong interaction must already essentially violate the SU(3) invariance, and to a much larger degree than the electromagnetic interaction violates the isospin symmetry. But in the strong interaction Y, T and T_3 are still conserved, and this allows one to obtain the transformation properties of the Hamiltonian of the interaction violating the symmetry with respect to unitary transformations from the group SU(3). This in its turn makes it possible to obtain definite relations between the masses of the particles contained in a given unitary multiplet.

Instead of the generators Λ_1, Λ_2, Λ_3 and Λ_8 it is convenient to introduce the operators

$$\hat{T}_1 = \tfrac{1}{2}\Lambda_1 , \qquad \hat{T}_2 = \tfrac{1}{2}\Lambda_2 , \qquad \hat{T}_3 = \tfrac{1}{2}\Lambda_3 \tag{138.1}$$

and

$$\hat{Y} = \frac{1}{\sqrt{3}} \Lambda_8 . \tag{138.2}$$

It is natural to identify the operators $\hat{T}_1$, $\hat{T}_2$, $\hat{T}_3$ with the operators of the isospin components, and the operator

$$\hat{T}^2 \equiv \hat{T}_1^2 + \hat{T}_2^2 + \hat{T}_3^2 = \tfrac{1}{4}\Lambda^2 \tag{138.3}$$

with the operator of the isospin squared. From (137.7) and (137.9) it follows that the eigenvalues of $\hat{T}_3$ and $\hat{T}^2$ are equal to (see however the remark at the end of §137)

$$T_3 = \tfrac{1}{2}(p_1 - p_2 - q_1 + q_2) \tag{138.4}$$

and

$$T(T+1) = \tfrac{1}{4}(p_1 + p_2 + q_1 + q_2)(p_1 + p_2 + q_1 + q_2 + 2) . \tag{138.5}$$

The operator $\hat{Y} = \Lambda_8/\sqrt{3}$ can be identified with the hypercharge operator, so that, according to (137.6),

$$Y = \tfrac{1}{3}(p-q) - p_3 + q_3 . \tag{138.6}$$

Such an identification is not unambiguous and corresponds to the so-called eightfold formalism or eightfold way proposed in 1961 by Gell-Mann and independently by Ne'eman. Its suitability is justified a posteriori, since within the framework of the eightfold way the real elementary particles are described in the best of all possible ways. Another choice of the operator $\hat{Y}$ is considered in §141.

From (138.6) and the requirement that the hypercharge has an integer value it follows that it is necessary to consider only those representations of the group SU(3) for which the difference $p-q$ is a multiple of 3

$$p - q = 3n \qquad (138.7)$$

i.e. the representations

$$(0,0) = 1 , \quad (1,1) = 8 , \quad (3,0) = 10 , \quad (0,3) = \overline{10} , \quad (2,2) = 27 , \quad (138.8)$$

and so on. In view of the importance of the representation $(1,1) = 8$ in the approach considered, it is called the eightfold formalism. Finally, we note that in correspondence with the Gell-Mann—Nishijima relation (§129, 133) we assume by definition that the charge operator is

$$\hat{Q} = \hat{T}_3 + \tfrac{1}{2}\hat{Y} = \tfrac{1}{2}\Lambda_3 + \frac{1}{2\sqrt{3}}\,\Lambda_8 . \qquad (138.9)$$

Let us now consider concrete unitary multiplets of hadrons. We turn first of all to the octet with the wave function φ_i^j, and investigate its content in isospin T, its component T_3, hypercharge Y and electric charge Q. Making use of relations (138.3)–(138.6) and (138.9), one can draw up table 5. In the last three columns are shown the stable particles, pseudoscalar and vector mesons and baryons with spin $\tfrac{1}{2}$, which have the corresponding quan-

Table 5

Component	T	T_3	Y	Q	Meson 0^-	Meson 1^-	Baryon
φ_1^1	$1,0^*$	0	0	0	π^0,η	$\rho^0;\omega,\varphi$	Σ^0,Λ
φ_2^2	$1,0^*$	0	0	0	π^0,η	$\rho^0;\omega,\varphi$	Σ^0,Λ
φ_1^2	1	$+1$	0	$+1$	π^+	ρ^+	Σ^+
φ_2^1	1	-1	0	-1	π^-	ρ^-	Σ^-
φ_1^3	$\tfrac{1}{2}$	$+\tfrac{1}{2}$	$+1$	$+1$	K^+	K^{*+}	p
φ_2^3	$\tfrac{1}{2}$	$-\tfrac{1}{2}$	$+1$	0	K^0	K^{*0}	n
φ_3^1	$\tfrac{1}{2}$	$-\tfrac{1}{2}$	-1	-1	$\overline{K}^+$	$\overline{K}^{*+}$	Ξ^-
φ_3^2	$\tfrac{1}{2}$	$+\tfrac{1}{2}$	-1	0	$\overline{K}^0$	$\overline{K}^{*0}$	Ξ^0
φ_3^3	0	0	0	0	η	ω,φ	Λ

tum numbers. It is remarkable that all stable pseudoscalar mesons, whose number is just 8, and all 8 stable baryons (except for Ω^- whose spin is equal to $\frac{3}{2}$) are involved here. Hence it is natural to assume that these two groups of particles just form unitary octets. The matrices of their wave functions have the following form:

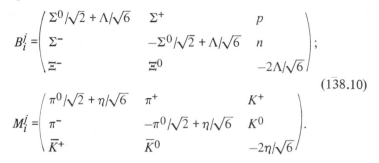

$$B_i^j = \begin{pmatrix} \Sigma^0/\sqrt{2} + \Lambda/\sqrt{6} & \Sigma^+ & p \\ \Sigma^- & -\Sigma^0/\sqrt{2} + \Lambda/\sqrt{6} & n \\ \Xi^- & \Xi^0 & -2\Lambda/\sqrt{6} \end{pmatrix};$$

$$M_i^j = \begin{pmatrix} \pi^0/\sqrt{2} + \eta/\sqrt{6} & \pi^+ & K^+ \\ \pi^- & -\pi^0/\sqrt{2} + \eta/\sqrt{6} & K^0 \\ \bar{K}^+ & \bar{K}^0 & -2\eta/\sqrt{6} \end{pmatrix}.$$

(138.10)

The choice of just such coefficients is dictated by the requirement that the trace of the matrix be equal to zero and by the normalization considerations (see § 133). The meson octet contains particles together with antiparticles. This is accounted for by the fact that there now remain no quantum numbers which are not involved in the group SU(3) by means of which one could distinguish, say, between K^+ and $\bar{K}^+$ (see § 133). For baryons, however, such a number exists (the baryonic number B), so that the corresponding antiparticles form an independent octet:

$$\bar{B}_i^j = \begin{pmatrix} \bar{\Sigma}^0/\sqrt{2} + \bar{\Lambda}/\sqrt{6} & \bar{\Sigma}^- & \bar{\Xi}^- \\ \bar{\Sigma}^+ & -\bar{\Sigma}^0/\sqrt{2} + \bar{\Lambda}/\sqrt{6} & \bar{\Xi}^0 \\ \bar{p} & \bar{\eta} & -2\bar{\Lambda}/\sqrt{6} \end{pmatrix}. \quad (138.11)$$

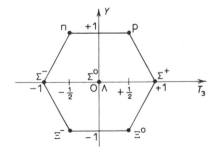

Fig. V.39

Unitary multiplets are conveniently presented in diagrams called weight diagrams. In this case an orthogonal reference frame is chosen, the hypercharge being plotted on an axis and the isospin component on the other axis. Thus, for example, for the baryon octet we have fig. V.39. The situation with vector mesons is somewhat more complex, because at a place analogous to that occupied by the η-meson or Λ-hyperon there are two pretenders, ω and φ. We shall discuss it at the end of this section.

Let us now draw up the table of quantum numbers of the particles which can be contained in the decuplet whose wave function is $\varphi_{[ijk]}$, table 6.

Table 6

Component	T	T_3	Y	Q	Baryon $\frac{3}{2}^+$
φ_{111}	$\frac{3}{2}$	$+\frac{3}{2}$	$+1$	$+2$	Δ^{++}_{1236}
φ_{112}	$\frac{3}{2}$	$+\frac{1}{2}$	$+1$	$+1$	Δ^{+}_{1236}
φ_{122}	$\frac{3}{2}$	$-\frac{1}{2}$	$+1$	0	Δ^{0}_{1236}
φ_{222}	$\frac{3}{2}$	$-\frac{3}{2}$	$+1$	-1	Δ^{-}_{1236}
φ_{113}	1	$+1$	0	$+1$	Σ^{+}_{1385}
φ_{123}	1	0	0	0	Σ^{0}_{1385}
φ_{223}	1	-1	0	-1	Σ^{-}_{1385}
φ_{133}	$\frac{1}{2}$	$+\frac{1}{2}$	-1	0	Ξ^{0}_{1530}
φ_{233}	$\frac{1}{2}$	$-\frac{1}{2}$	-1	-1	Ξ^{-}_{1530}
φ_{333}	0	0	-2	-1	?

At the time when an analogous table was drawn up (1961–1962) the place at which the question mark is standing was unoccupied, because no particle with hypercharge -2 was known. Thus the eightfold way predicted the existence of a new hyperon with spin $\frac{3}{2}^+$ and with the quantum numbers indicated in the table. Moreover, the mass of this particle (about 1680 MeV) was also known approximately (see §140). At the beginning of 1964 such a particle was indeed discovered: this is the rather stable Ω^--hyperon (see the table of elementary particles given in §129). This fact eliminated any doubt as to the validity of the eightfold way and, in general, of unitary symmetry, which is now as classical as isospin symmetry.

The weight diagram for the decuplet is given in fig. V.40.

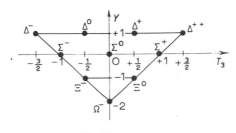

Fig. V.40.

Formula (138.7) also allows for the existence of unitary singlets. Since in this case all generators reduce to zero, then for a unitary scalar particle $T = T_3 = Y = Q = 0$. However, among the multitude of resonance states there is up to now none whose wave function could be quite trustworthily considered as a unitary scalar.

That is why unitary singlets play a decisive role in resolving the difficulty mentioned above with vector mesons. It can be assumed that real ω- and φ-mesons represent different superpositions of the unitary singlet state φ' and the octet state ω' analogous to the η-meson. In the case of strict unitary symmetry such a mixing of components of multiplets of different nature is forbidden, but as a result of the violation of the symmetry by the real strong interaction it is no longer impossible. Thus vector mesons form a nonet and not an octet. Apropos of this we shall confine ourselves only to the remarks of general character made above, referring the reader to the literature.

Let us now enumerate, without comment, some unitary multiplets in which resonances are distributed. It turns out that there exists a family of 9 mesons 2^+ (it contains, in particular, the f^0-meson mentioned in § 129), which also form a nonet. The nonet of mesons 1^+, octet of baryon resonances $\frac{3}{2}^-$ and octets of baryon resonances $\frac{5}{2}^+$ and $\frac{7}{2}^+$ are somewhat more doubtful. At present insufficient data are available for a final solution of the problem.

§ 139. Some consequences of strict unitary symmetry

In this section some physical consequences of the hypothesis of strict unitary symmetry for hadrons are described briefly. It should be stressed immediately that they cannot pretend to be in good agreement with experimental data, since the real strong interaction already violates to a considerable degree the SU(3) invariance of the theory. A more realistic scheme is outlined in the next section.

Let us consider first of all the interaction of stable baryons of spin $\frac{1}{2}^+$ (octet B_i^j) with pseudoscalar mesons (octet M_i^j). In quantum field theory the baryon and meson functions are considered to be operators in the space of occupation numbers. Analogously to what was done in §135, the interaction Hamiltonian density is to be constructed in the form of an invariant combination of the functions $\bar{B}_i^j$, B_k^l and M_m^n. We first of all form unitary scalars from these functions. Noting that the two octets

$$\bar{B}_i^j B_j^k - \tfrac{1}{3}\delta_i^k \bar{B}_m^j B_j^m \quad \text{and} \quad \bar{B}_j^k B_i^j - \tfrac{1}{3}\delta_i^k \bar{B}_j^m B_m^j$$

can be constructed from $\bar{B}_i^j$ and B_k^l, we contract these matrices with the meson function M_k^i. Taking into account that $\delta_i^k M_k^i = M_i^i = 0$ we arrive at the two unitary scalars $\bar{B}_i^j B_j^k M_k^i$ and $\bar{B}_j^k B_i^j M_k^i$ which exhaust the possible invariants. Usually their sum and difference are chosen, so that the baryon–meson interaction is described by the following Hamiltonian density:

$$\hat{H}_{\text{int}} = \frac{1}{\sqrt{2}} g^{(\text{F})} [\bar{B}_i^j \gamma_5 B_j^k - \bar{B}_j^k \gamma_5 B_i^j] M_k^i +$$

$$+ \frac{1}{\sqrt{2}} g^{(\text{D})} [\bar{B}_i^j \gamma_5 B_j^k + \bar{B}_j^k \gamma_5 B_i^j] M_k^i \qquad (139.1)$$

(the matrix γ_5 is introduced on the basis of the same considerations as in §135) containing two independent coupling constants $g^{(\text{F})}$ and $g^{(\text{D})}$. For definite reasons the first term of the Hamiltonian (139.1) is called F-coupling, and the second D-coupling.

Making use of (138.10) and (138.11) we write the components of the baryon and meson matrices in terms of the wave functions of isomultiplets:

$$B_{33} = -\frac{2}{\sqrt{6}} \Lambda, \qquad B_b^a = \Sigma_b^a + \frac{1}{\sqrt{6}} \Lambda \delta_b^a, \qquad B_a^3 = N_a, \qquad B_3^a = \Xi^a,$$

$$\bar{B}_{33} = -\frac{2}{\sqrt{6}} \bar{\Lambda}, \qquad \bar{B}_b^a = \bar{\Sigma}_b^a + \frac{1}{\sqrt{6}} \bar{\Lambda} \delta_b^a, \qquad \bar{B}_a^3 = \bar{\Xi}_a, \qquad \bar{B}_3^a = \bar{N}^a,$$

$$M_3^3 = -\frac{2}{\sqrt{6}} \eta, \qquad M_b^a = \pi_b^a + \frac{1}{\sqrt{6}} \eta \delta_b^a, \qquad M_a^3 = K_a, \qquad M_3^a = \bar{K}^a,$$

(a, $b = 1$, 2 are isospin indices). Substituting them into (139.1), we obtain

$$\hat{H}_{int} = -\sqrt{2}g^{(F)}\pi_a^c\bar{\Sigma}_c^b\gamma_5\Sigma_b^a + \frac{1}{\sqrt{2}}(g^{(D)}+g^{(F)})\,\pi_a^c\bar{N}^a\gamma_5 N_c +$$

$$+\frac{1}{\sqrt{2}}(g^{(D)}-g^{(F)})\,\pi_a^c\bar{\Xi}_c\gamma_5\Xi^a + \frac{1}{\sqrt{3}}g^{(D)}\pi_a^c[\bar{\Lambda}\gamma_5\Sigma_c^a + \bar{\Sigma}_c^a\gamma_5\Lambda] +$$

$$+\sqrt{\tfrac{2}{3}}g^{(D)}\eta\bar{\Sigma}_b^a\gamma_5\Sigma_a^b + \frac{1}{2\sqrt{3}}(3g^{(F)}-g^{(D)})\,\eta\bar{N}^a\gamma_5 N_a -$$

$$-\frac{1}{2\sqrt{3}}(3g^{(F)}+g^{(D)})\,\eta\bar{\Xi}_a\gamma_5\Xi^a - \sqrt{\tfrac{2}{3}}g^{(D)}\eta\bar{\Lambda}\gamma_5\Lambda +$$

$$+\frac{1}{\sqrt{2}}(g^{(D)}+g^{(F)})\,K_a\bar{\Sigma}_b^a\gamma_5\Xi^b + \frac{1}{\sqrt{2}}(g^{(D)}+g^{(F)})\,\bar{K}^a\bar{\Xi}_b\gamma_5\Sigma_a^b +$$

$$+\frac{1}{\sqrt{2}}(g^{(D)}-g^{(F)})\,K_a\bar{N}^b\gamma_5\Sigma_b^a + \frac{1}{\sqrt{2}}(g^{(D)}-g^{(F)})\,\bar{K}^a\bar{\Sigma}_a^b\gamma_5 N_b +$$

$$+\frac{1}{2\sqrt{3}}(3g^{(F)}-g^{(D)})\,K_a\bar{\Lambda}\gamma_5\Xi^a + \frac{1}{2\sqrt{3}}(3g^{(F)}-g^{(D)})\,\bar{K}^a\bar{\Xi}_a\gamma_5\Lambda -$$

$$-\frac{1}{2\sqrt{3}}(3g^{(F)}+g^{(D)})\,K_a\bar{N}^a\gamma_5\Lambda - \frac{1}{2\sqrt{3}}(3g^{(F)}+g^{(D)})\,\bar{K}^a\bar{\Lambda}\gamma_5 N_a\,. \quad (139.2)$$

Thus 12 constants of the couplings $\Sigma\Sigma\pi$, $NN\pi$, $\Sigma\Lambda\pi$, $\Xi\Xi\pi$, $\Sigma\Sigma\eta$, $NN\eta$, $\Xi\Xi\eta$, $\Lambda\Lambda\eta$, $\Sigma\Xi K$, $N\Sigma K$, $\Lambda\Xi K$ and $N\Lambda K$ are expressed in terms of only two parameters $g^{(F)}$ and $g^{(D)}$. Each term contained in (139.2) is invariant with respect to isospin transformations, and can be written in explicit form (see §135), as a result of which 64 terms arise, each corresponding to the interaction of actual baryons with a meson.

Let us now consider how the relations between the cross sections for different processes are obtained in unitary invariant theory. Let there be several reactions

$$a_i + b_i \rightarrow c_i + d_i\,,$$

the particles a_i, b_i, c_i and d_i belonging respectively to the unitary multiplets $(p^{(a)},q^{(a)})$, $(p^{(b)},q^{(b)})$, $(p^{(c)},q^{(c)})$ and $(p^{(d)},q^{(d)})$. In accordance with the general scheme described in §136, it is necessary in order to obtain the relations between the amplitudes of the reactions mentioned to proceed in the following way.

1. We decompose the direct products of the representations

$$(p^{(a)},q^{(a)})\otimes(p^{(b)},q^{(b)}) \quad \text{and} \quad (p^{(c)},q^{(c)})\otimes(p^{(d)},q^{(d)})$$

into irreducible representations.

2. We write down the independent amplitudes corresponding to the transition from a certain multiplet of the first decomposition to the same multiplet of the second decomposition.

3. Making use of the table of Clebsch—Gordan coefficients of the group SU(3) we expand the wave functions $|a_i b_i\rangle$ and $|c_i d_i\rangle$ of the initial and final states in terms of the basis functions of the multiplets involved in the corresponding expansions. These functions are defined by the type of representation and by the eigenvalues of the isospin operator, isospin component operator and hypercharge operator, so that they should be written in the form $|p,q;T,T_3,Y\rangle$.

4. We substitute the expansions of wave functions into the matrix element of the transition $\langle c_i d_i | a_i b_i \rangle$, as a result of which it turns out to be expressed in terms of a relatively small number of matrix elements of transitions between multiplets of the same type as in item 2.

As an example let us consider the scattering of pseudoscalar mesons by stable baryons of spin $\frac{1}{2}$ (it can easily be calculated that there are in all 27 such processes). In this case all the particles of both initial and final states belong to octets. Making use of the decomposition (137.4d) we conclude that there are 8 independent amplitudes corresponding to the transitions

$$1 \to 1, \quad 10 \to 10, \quad \overline{10} \to \overline{10}, \quad 27 \to 27,$$

$$8' \to 8', \quad 8 \to 8, \quad 8 \to 8', \quad 8' \to 8.$$

However, making use of the invariance of theory with respect to time reversal, it can be shown that the last two amplitudes are expressed in terms of each other, so that there are 7 independent amplitudes in terms of which the 27 amplitudes of real processes are expressed.

In the case of the scattering of pseudoscalar mesons by $\frac{1}{2}^+$ baryons with the production of a pseudoscalar meson and a baryon resonance $\frac{3}{2}^+$, for example

$$\pi^+ + p \to \eta + \Delta^{++}$$

we have respectively for the initial and final states the decomposition (137.4d) and (137.4e), so that the amplitudes of the real processes are expressed in terms of only 4 independent transition amplitudes

$$8 \to 8, \quad 8' \to 8, \quad 10 \to 10, \quad 27 \to 27.$$

However, the tables of Clebsch—Gordan coefficients of the group SU(3) are very cumbersome, and in each actual case it is much more convenient to make use of the method of invariant amplitudes described at the end of

§136. Let us again consider the scattering of pseudoscalar mesons by $\frac{1}{2}^+$ baryons. The total scattering amplitude, which in our case must be constructed from the wave functions B_i^j and M_k^l of the initial state and the wave functions $\bar{B}_i^j$ and $\bar{M}_k^l$ of the final state, is a unitary invariant of the most general form. It can be written in the form of the following combination of nine scalar terms:

$$f = f_1(\bar{B}B)(\bar{M}M) + f_2(\bar{B}M)(\bar{M}B) + f_3(\bar{B}\bar{M})(BM) + f_4(\bar{B}BM\bar{M}) + f_5(\bar{B}B\bar{M}M) +$$

$$+ f_6(\bar{B}M\bar{M}B) + f_7(\bar{B}\bar{M}MB) + f_8(\bar{B}MB\bar{M}) + f_9(\bar{B}\bar{M}BM) . \qquad (139.3)$$

For brevity, the traces of the products of matrices of the type B and M are here denoted by parentheses, so that, for example,

$$(\bar{B}B)(\bar{M}M) \equiv \bar{B}_i^j B_j^i \bar{M}_k^l M_l^k ,$$

$$(\bar{B}\bar{M}BM) \equiv \bar{B}_i^j \bar{M}_j^k B_k^l M_l^i$$

and so on. The quantities f_α involved in the amplitude (139.3) are functions of spatial-spin variables, which are unknown but the same for all meson–baryon processes.

We already know that the processes considered are described by eight independent amplitudes (without taking into account the invariance with respect to time reversal). Hence a relation must exist between the nine invariants involved in (139.3). Indeed, it does:

$$(\bar{B}B)(\bar{M}M) + (\bar{B}M)(\bar{M}B) + (\bar{B}\bar{M})(BM) = \qquad (139.4)$$

$$= (\bar{B}B\bar{M}M) + (\bar{B}BM\bar{M}) + (\bar{B}\bar{M}MB) + (\bar{B}M\bar{M}B) + (\bar{B}MB\bar{M}) + (\bar{B}\bar{M}BM)$$

as can easily be seen by direct calculation. In general, the establishment of relations of the type (139.4) in practice turns out to be difficult, but this is not important, because if all the 9 terms are considered to be formally independent, the final results will automatically involve the functions f_α in the form of just eight independent combinations.

As is known, under the operation of time reversal every wave function goes over into its complex conjugate, the initial and final states being exchanged*; in our case

$$B_i^j, \bar{B}_{i'}^{j'}, M_k^l, \bar{M}_{k'}^{l'} \rightarrow B_{i'}^{j'}, \bar{B}_i^j, M_{k'}^{l'}, \bar{M}_k^l . \qquad (139.5)$$

When such an exchange is made the first seven terms of (139.3) do not change, while the last two go over into each other, so that from the invariance

* See, for example, L.D.Landau and E.M.Lifshitz, *Quantum mechanics* (Pergamon Press, Oxford, 1965).

with respect to time reversal it follows that $f_8 = f_9$; taking into account relation (139.4) they can be set equal to zero and one can operate with only the first seven invariant amplitudes.

Now, by means of a procedure quite analogous to that described at the end of §136 the amplitudes of all the 27 real processes can be expressed in terms of 7 independent functions f_α, for example

$$f(\pi^- p \to K^+ \Sigma^-) = f_3$$
$$f(K^0 p \to K^+ \Xi^0) = f_3$$
$$f(K^- p \to K^- p) = f_1 + f_2 + f_4 + f_6 \qquad (139.6)$$
$$f(\pi^- p \to \pi^- p) = f_1 + f_6$$
$$f(K^- p \to \pi^- \Sigma^+) = f_2 + f_4$$

and so on. On eliminating the functions f_α we shall have the relations between the amplitudes of different processes. Thus from (139.6) it follows that

$$f(\pi^- p \to K^+ \Sigma^-) = f(\overline{K}^0 p \to K^+ \Xi^0) \qquad (139.7)$$

and

$$f(K^- p \to K^- p) - f(\pi^- p \to \pi^- p) = f(K^- p \to \pi^- \Sigma^+) \qquad (139.8)$$

and hence for the cross sections we obtain

$$\sigma(\pi^- p \to K^+ \Sigma^-) = \sigma(\overline{K}^0 p \to K^+ \Xi^0) \qquad (139.9)$$

and

$$\sqrt{\sigma(\pi^- p \to \pi^- p)} - \sqrt{\sigma(K^- p \to K^- p)} \leqslant \sqrt{\sigma(K^- p \to \pi^- \Sigma^+)}. \qquad (139.10)$$

The cross section on the left-hand side of equality (139.9) has a large value, while the cross section on the right-hand side is small, so that in analysing these processes it is necessary to take into account the violation of unitary symmetry by the strong interactions. On the other hand, inequality (139.10) is fulfilled in the entire energy range. Moreover, for large energies, when the cross section $\sigma(K^- p \to \pi^- \Sigma^+)$ is very small, (139.10) goes over into the equality

$$\sigma(\pi^- p \to \pi^- p) \approx \sigma(K^- p \to K^- p) \qquad (139.10a)$$

which is in good agreement with experimental data.

It should be noted that the interpretation of the theoretical predictions of strict unitary symmetry and their comparison with experiment is a rather complex problem, since it is necessary to state beforehand the energy range in which the violation of SU(3) invariance can be disregarded. Furthermore, the cross sections for different processes involve kinematical factors which are different, because they contain the masses of the particles taking part in the reactions. On the other hand, the relations written above assume equality of the masses of the particles contained in the same multiplet. Hence from the relations between amplitudes one actually obtains relations not between cross sections but between their ratios to kinematical factors, so that one has to take from experiment certain 'corrected' values of the cross sections.

§ 140. Some aspects of violated unitary symmetry

We have above frequently stressed that the real strong interaction violates strict unitary symmetry but, if electromagnetism is disregarded, the isospin T and hypercharge Y are still conserved. Therefore the interaction Hamiltonian density cannot be a unitary scalar, but must be an isoscalar (T=0) and must correspond to zero hypercharge (Y=0). Thus among the components of unitary multiplets one has to find those for which $T = Y = 0$. From formula (138.5) for the maximum value of the isospin, it is seen that all the indices of these components must be threes, so that the equalities $p_3 = p$ and $q_3 = q$ are fulfilled. Then on the basis of expression (138.6) for hypercharge we arrive immediately at the condition $p - q = 0$. Thus zero isospin and zero hypercharge are possessed only by the components of symmetric multiplets of the type (p,p), i.e. of the multiplets $(1,1) = 8$, $(2,2) = 27$ and so on, each index of which is equal to 3. Hence the strong interaction Hamiltonian density must have the following general structure

$$\hat{H}_{int} = g_0 \hat{H}_0 + g_1 \hat{H}_3^3 + g_2 \hat{H}_{33}^{33} + \dots . \tag{140.1}$$

As soon as the terms violating unitary symmetry are introduced the earlier quantum-mechanical degeneracy will be removed, i.e. the masses of the particles belonging to the same multiplet must split. For the derivation of mass formulae we introduce the operator $\hat{M}$ whose eigenvalues are equal to the masses of isomultiplets with definite hypercharge (in consequence of isospin invariance they do not depend on the component T_3):

$$\hat{M}|p,q;T,Y\rangle = m|p,q;T,Y\rangle . \tag{140.2}$$

Under the assumption of strict unitary symmetry the mass operator will be an invariant of the group SU(3), and its eigenvalues in a given multiplet will be the same. It is natural to assume that, when unitary symmetry is violated,

$\hat{M}$ acquires a structure analogous to (140.1), it being assumed that $g_2 \ll g_1$; hence

$$\hat{M} = \hat{M}_0 + \hat{M}_3^3 . \tag{140.3}$$

It is convenient to introduce the matrix M of the mass operator

$$M = \langle p,q;T',Y' | \hat{M} | p,q;T,Y \rangle \tag{140.4}$$

which from (140.2) is diagonal; its elements define the masses of the terms of the multiplet. This matrix represents a certain bilinear combination of the wave functions $\varphi^{j_1 \cdots j_q}_{i_1 \cdots i_p}$ of a given multiplet (p,q) and the wave function $\bar{\varphi}^{i'_1 \cdots i'_p}_{j'_1 \cdots j'_q}$ of the conjugate multiplet (q,p). From (140.3) it follows that it must contain an invariant part and a term corresponding to the 3–3 component of the octet. From the wave functions mentioned a scalar can be constructed:

$$m_0 \bar{\varphi}^{i_1 \cdots i_p}_{j_1 \cdots j_q} \varphi^{j_1 \cdots j_q}_{i_1 \cdots i_p}$$

and two octets

$$a_1 \bar{\varphi}^{i_1 \cdots i_p - 1j}_{j_1 \cdots j_q} \varphi^{j_1 \cdots j_q}_{i_1 \cdots i_p - 1i} - \tfrac{1}{3} a_1 \delta^j_i \bar{\varphi}^{i_1 \cdots i_p}_{j_1 \cdots j_q} \varphi^{j_1 \cdots j_q}_{i_1 \cdots i_p}$$

and

$$a_2 \bar{\varphi}^{i_1 \cdots i_p}_{j_1 \cdots j_q - 1i} \varphi^{j_1 \cdots j_q - 1j}_{i_1 \cdots i_p} - \tfrac{1}{3} a_2 \delta^j_i \bar{\varphi}^{i_1 \cdots i_p}_{j_1 \cdots j_q} \varphi^{j_1 \cdots j_q}_{i_1 \cdots i_p} ,$$

hence

$$M = a_0 \bar{\varphi}^{i_1 \cdots i_p}_{j_1 \cdots j_q} \varphi^{j_1 \cdots j_q}_{i_1 \cdots i_p} + a_1 \bar{\varphi}^{i_1 \cdots i_p - 1,3}_{j_1 \cdots j_q} \varphi^{j_1 \cdots j_q}_{i_1 \cdots i_p - 1,3} + a_2 \bar{\varphi}^{i_1 \cdots i_p}_{j_1 \cdots j_q - 1,3} \varphi^{j_1 \cdots j_q - 1,3}_{i_1 \cdots i_p} \tag{140.5}$$

where $a_0 \equiv m_0 - \tfrac{1}{3}(a_1 + a_2)$, and m_0, a_1 and a_2 are certain parameters. Thus in the general case the mass formula contains at the most 3 parameters (for multiplets of the type $(3n,0)$ and $(0,3n)$ only two, since in these cases there are only superscripts or subscripts), of which m_0 corresponds to the mass of the members of the multiplet under the assumption of strict unitary symmetry. We note that, according to Feynman, for boson multiplets it is necessary to consider the matrix M^2 instead of the matrix M, since bosons, in contrast to fermions, obey an equation of second order.

For the octet of baryons of spin $\tfrac{1}{2}^+$ formula (140.5) goes over into

$$M = a_0 \bar{B}^i_j B^j_i + a_1 \bar{B}^3_j B^j_3 + a_2 \bar{B}^i_3 B^3_i . \tag{140.6}$$

Making use of (138.10) and calculating the matrix elements corresponding to each baryon, we obtain

$$m_N = a_0 + a_2 , \qquad m_\Xi = a_0 + a_1 ,$$
$$m_\Sigma = a_0 , \qquad m_\Lambda = a_0 + \tfrac{2}{3}(a_1 + a_2) .$$

From this follows the Gell-Mann mass formula

$$3m_\Lambda + m_\Sigma = 2(m_N + m_\Xi) \tag{140.7}$$

which is in good agreement with experiment: the right-hand side adds up to 4518 MeV, while the left-hand side adds up to 4535 MeV, i.e. the accuracy is to within about 0.4%. Taking into account that in the pseudoscalar meson octet there is a K-meson in place of the nucleon, and in place of the Ξ-hyperon there is a $\bar{K}$-meson, the masses of the K and $\bar{K}$ being equal, we have from an analogue of (140.6)

$$3m_\eta^2 + m_\pi^2 = 4m_K^2 . \tag{140.7'}$$

This relation agrees with experiment to within 5%. In the case of vector mesons the situation is made more complex by the presence of $\omega - \varphi$ mixing (see §138), and we shall not discuss it.

For the decuplet of particles of spin $\tfrac{3}{2}^+$ formula (140.5) goes over into

$$M = b_0 \bar{B}^{ijk} B_{ijk} + b_1 \bar{B}^{3ij} B_{3ij} . \tag{140.8}$$

Hence, making use of the results of §138, we obtain

$$m_\Delta = b_0 , \qquad m_{\Sigma^*} = b_0 + \tfrac{1}{3}b_1,$$
$$m_{\Xi^*} = b_0 + \tfrac{2}{3}b_1, \qquad m_\Omega = b_0 + b_1.$$

From these formulae the interval rule follows:

$$m_{\Sigma^*} - m_\Delta = m_{\Xi^*} - m_{\Sigma^*} = m_\Omega - m_{\Xi^*} . \tag{140.9}$$

For the first equality we have 147 and 145 MeV (accuracy of the order of 1%), and from the second equality the mass of the Ω-hyperon can be predicted to be 1676 MeV, which was brilliantly confirmed by experiment: $m_\Omega^{\exp} = 1675$ MeV.

Okubo has derived a general mass formula valid for all unitary multiplets:

$$m^{2-|B|} = a(p,q) + b(p,q)BY + c(p,q)[T(T+1) - \tfrac{1}{4}Y^2] . \tag{140.10}$$

In the case of violated SU(3) symmetry the amplitudes of the scattering of one unitary multiplet by another will no longer be invariant; in addition to the scalar part they will contain the 3−3 components of the octet. As an example we write down the first term of formula (139.3), which now assumes the form

$$f_1 = f_{1,0}(\bar{B}B)(\bar{M}M) + f_{1,1}\bar{B}_i^3 B_3^i(\bar{M}M) + f_{1,2}\bar{B}_3^i B_i^3(\bar{M}M) +$$
$$+ f_{1,3}(\bar{B}B)\bar{M}_i^3 M_3^i + f_{1,4}(\bar{B}B)M_3^i \bar{M}_i^3 . \qquad (140.11)$$

In view of the enormous number of independent arbitrary parameters arising in such a scheme, its heuristic value decreases sharply, since the physical information that it provides becomes very small.

However, if one adopts the reasonable and experimentally confirmed hypothesis of Okun and Pomeranchuk, that asymptotically at very high energies the cross sections for charge exchange inelastic processes are negligibly small in comparison with the cross sections for ordinary elastic scattering, then in this energy range only the first term will remain in (139.3). Then under the assumption of strict unitary symmetry we obtain the asymptotic equality of the amplitudes of the elastic scattering of the π-, η-, K- and $\bar{\text{K}}$-mesons by baryons:

$$f_\pi^\infty = f_\eta^\infty = f_K^\infty = f_{\bar{K}}^\infty . \qquad (140.12)$$

In the case of violated symmetry it is necessary to make use of formula (140.11); hence follows the relation between the amplitudes similar to the mass relation:

$$f_\pi^\infty + 3f_\eta^\infty = 2(f_K^\infty + f_{\bar{K}}^\infty) . \qquad (140.13)$$

In concluding this section we note that one can by the same procedure investigate the isospin symmetry violated by the electromagnetic interaction using the Hamiltonian density (135.10). We leave it to the reader to obtain on his own, as a useful excercise, the following mass formulae for the isotriplet Σ and isoquartet Δ:

$$m_{\Sigma^0} = \tfrac{1}{2}(m_{\Sigma^+} + m_{\Sigma^-}) \qquad (140.14)$$

and

$$m_{\Delta^{++}} - m_{\Delta^-} = 3(m_{\Delta^+} - m_{\Delta^0}) . \qquad (140.15)$$

§141. Composite models in the unitary symmetry scheme. Quarks

At the beginning of §134 we pointed out that the π-meson (and generally speaking, also all other non-strange particles) can be conceived of as a particle consisting of a nucleon and an antinucleon. It is natural to try to formulate an analogous 'minimum' model in which all hadrons are constructed from a small number of some particles said, in a certain sense, to be fundamental. For this it would be necessary to add to the nucleon, which is a carrier of a baryonic number and an isospin (and this means also an electric charge), at least one particle which possesses strangeness. The most economic model of such a type was proposed in 1956 by Sakata, who chose as fundamental particles p, n, Λ and $\bar{p}$, $\bar{n}$, $\bar{\Lambda}$ and assumed that there is an attraction between any fundamental baryon and antibaryon, and a repulsion between two baryons or antibaryons. The wave functions of the hadrons known at that time were constructed as follows:

$$\pi^+ = p\bar{n}\,, \qquad \pi^- = \bar{p}n\,, \qquad \pi^0 = \frac{1}{\sqrt{2}}\,(p\bar{p}-n\bar{n})\,;$$

$$K^+ = p\bar{\Lambda}\,, \qquad K^0 = n\bar{\Lambda}\,, \qquad K^- = \bar{p}\Lambda\,, \qquad \bar{K}^0 = \bar{n}\Lambda\,;$$

$$\Sigma^+ = p\bar{n}\Lambda = \pi^+\Lambda\,, \qquad \Sigma^- = \bar{p}n\Lambda = \pi^-\Lambda\,, \qquad \Sigma^0 = \frac{1}{\sqrt{2}}\,(p\bar{p}-n\bar{n})\Lambda = \pi^0\Lambda\,;$$

$$\Xi^- = \bar{p}\Lambda\Lambda = K^-\Lambda\,, \qquad \Xi^0 = \bar{n}\Lambda\Lambda = \bar{K}^0\Lambda\,. \qquad\qquad (141.1)$$

This model was developed by Markov, Okun and others, and made it possible to obtain a large number of interesting physical results.

It turns out that the Sakata model fits the scheme of unitary symmetry very well, if the mass difference between the nucleon and Λ-hyperon is neglected. For this it is sufficient to assume that these particles form the triplet $(1,0)$, and the corresponding antiparticles the conjugate triplet $(0,1)$:

$$S_i = \begin{pmatrix} p \\ n \\ \Lambda \end{pmatrix}, \qquad S^i = (\bar{p},\bar{n},\bar{\Lambda})\,. \qquad\qquad (141.2)$$

The isospin component operator (138.1) for p, n and Λ gives the correct values $+\frac{1}{2}$, $-\frac{1}{2}$ and 0, but the hypercharge operator (138.2) must be modified in such a way that instead of leading to fractional values it leads to $Y = +1$, $+1, 0$, respectively. We assume by definition

$$\hat{Y} = \frac{1}{\sqrt{3}}\Lambda_8 + \tfrac{2}{3}B\hat{I}\,, \qquad\qquad (141.3)$$

so that instead of (138.6) we shall now have

$$Y = \tfrac{1}{3}(p-q) - p_3 + q_3 + \tfrac{2}{3}B\,. \qquad\qquad (141.4)$$

On the basis of the requirement that the hypercharge be an integer, and taking into account that for mesons $B = 0$, we arrive at the old relation $p - q = 3n$, i.e. in the Sakata model these particles must also fill unitary singlets, octuplets and so on. But for baryons $B = 1$, and instead of (138.7) we obtain

$$p - q = 3n - 2 \tag{141.5}$$

i.e. baryons must fill unitary triplets $(1,0) = 3$, sextets $(0,2) = 6$, 15-plets $(2,1) = 15$ and so on. Making use of formulae (138.5) and (141.4) we immediately find the values of isospin and hypercharge for the components of these multiplets (see table 7).

Table 7

Components	Y	T	Components	Y	T
φ_a	1	$\frac{1}{2}$	φ^3_{ab}	2	1
φ_3	0	0	φ^3_{3a}	1	$\frac{1}{2}$
φ^{33}	2	0	$\varphi^c_{ab} + \frac{1}{3}[\delta^c_a\varphi^3_{3b} + \delta^c_b\varphi^3_{3a}]$	1	$\frac{3}{2}$
φ^{3a}	1	$\frac{1}{2}$	φ^3_{33}	0	0
φ^{ab}	0	1	$\varphi^b_{3a} + \frac{1}{2}\delta^b_a\varphi^3_{33}$	0	1
$(a,b=1,2)$			φ^a_{33}	-1	$\frac{1}{2}$

The distribution of hadrons over the multiplets mentioned corresponds to their wave functions (141.1). Indeed, mesons are made up of a 'sakaton' S_i and the 'antisakaton' $\bar{S}^i$, and their wave functions transform according to the direct product $3 \otimes \bar{3}$ which in its decomposition (137.4a) contains just the representations 1 and 8. Baryons are made up of two sakatons and an anti-sakaton, and the decomposition (137.4c) of the direct product $3 \otimes 3 \otimes \bar{3}$ includes just the necessary representations $3, \bar{6}$ and 15.

It is seen from the table that the Σ-hyperon must be placed at least in a sextet which includes a nucleon-like particle and a particle with $Y = +2$, $T = 0$ which must have spin $\frac{1}{2}^+$. These particles have up to now not been discovered, although there are no prohibitions imposed upon their existence. The Ξ-hyperon must be contained in a 15-plet in which a large number of unoccupied places remains, and the Ω-hyperon cannot be included in any of the lower multiplets. Thus the classification of hadrons based on the Sakata model is much less satisfactory than in the eightfold way formalism.

Furthermore, it leads to a number of conclusions contradicting experiment: for example, in the Sakata model the observed process

$$\bar{p} + p \rightarrow K_L^0 + K_S^0$$

is forbidden.

Wishing to preserve all the advantages of composite models on the one hand, and those of the eightfold way formalism on the other hand, Gell-Mann and independently Zweig in 1964 proposed to renounce the modification of the hypercharge operator (138.2) and at the same time to assume that there is a unitary triplet

$$q_i = \begin{pmatrix} q_1 \\ q_2 \\ q_3 \end{pmatrix}, \qquad \bar{q}^i = (\bar{q}_1, \bar{q}_2, \bar{q}_3) \tag{141.6}$$

of particles possessing very unusual properties. From (138.4)–(138.6) and (138.9) it follows that they have the quantum numbers given in table 8 (the baryon number is by definition equal to $\frac{1}{3}$) (all quantum numbers of the antiparticles, except for T, have the opposite sign), i.e. the electric charge, baryonic number and hypercharge of these particles are fractional numbers. Their name, quarks (something incomprehensible and mystical from one of the novels of the Irish writer J.Joyce), is due to just these properties.

Table 8

Particle	Q	T	T_3	S	B	Y
q_1	$+\frac{2}{3}$	$\frac{1}{2}$	$+\frac{1}{2}$	0	$\frac{1}{3}$	$+\frac{1}{3}$
q_2	$-\frac{1}{3}$	$\frac{1}{2}$	$-\frac{1}{2}$	0	$\frac{1}{3}$	$+\frac{1}{3}$
q_3	$-\frac{1}{3}$	0	0	-1	$\frac{1}{3}$	$-\frac{2}{3}$

Mesons are made up of a quark and an antiquark, and from (137.4a) are distributed over unitary singlets and octets. If the pair $q\bar{q}$ is in a $^1S_0$-state, then its total spin is equal to zero and its parity is odd, and we obtain pseudoscalar mesons with the wave functions

$$\pi^+ = q_1\bar{q}_2 , \qquad \pi^- = \bar{q}_1 q_2 , \qquad \pi^0 = \frac{1}{\sqrt{2}}(q_1\bar{q}_1 - q_2\bar{q}_2) ;$$

$$K^+ = q_1\bar{q}_3 , \qquad K^0 = q_2\bar{q}_3 , \qquad K^- = \bar{q}_1 q_3 , \qquad \bar{K}^0 = \bar{q}_2 q_3 ;$$

$$\eta = \frac{1}{\sqrt{6}}(q_1\bar{q}_1 + q_2\bar{q}_2 - 2q_3\bar{q}_3) ; \tag{141.7}$$

$$X^0 = \frac{1}{\sqrt{3}}(q_1\bar{q}_1 + q_2\bar{q}_2 + q_3\bar{q}_3)$$

where X^0 is a resonance in the $\pi\pi\eta$ system with a mass of 960 MeV, with $T = Y = 0$, and which is a unitary singlet. If the pair $q\bar{q}$ is in a $^3S_1$-state, then its spin is equal to 1^-, and we arrive at the vector mesons ρ, K^*, ω' and φ' whose wave functions are constructed analogously to (141.7). Since for the quark $B = \frac{1}{3}$, a system of three such particles will have a baryonic number equal to 1 and, hence it is natural to identify it with a baryon (we recall that in the Sakata model baryons are made up of two particles and an antiparticle). The wave function of the system qqq transforms according to the representation 3⊗3⊗3 and therefore it follows from the decomposition (137.4b) that in the quark model, as well as in the eightfold way formalism, baryons fill unitary singlets, octets and decuplets. If the spins of two quarks are parallel, then there are 9 $\frac{1}{2}^+$ states, of which one is a unitary singlet, while the 8 remaining states belong to a unitary octet. If the spins of all three quarks are parallel, then we obtain 10 $\frac{3}{2}^+$ states forming a decuplet.

We shall not consider the dynamic consequences of the quark model and some of its inherent difficulties, but refer the reader to the corresponding literature*. We shall dwell only briefly on the problem of the reality of the existence of these unusual particles. If quarks indeed exist, then their world must be almost independent of the ordinary world: from the fact that their charge is fractional it follows that the lightest of the quarks must be absolutely stable; they can be produced only in the form of quark–antiquark pairs, for example as a result of the bombardment of ordinary matter by cosmic rays. Therefore in the course of time the total number of quarks contained in the Earth's crust and in the waters of the oceans must increase progressively. However, numerous attempts to find these 'relic' quarks by means of precision apparatus have not yet been successful. Nor has anyone succeeded in discovering them in experiments with accelerators; only a lower

* See, for example, the reviews of E.M.Levin and L.L.Frankfurt in Usp. Fiz. Nauk 94 (1968) 243, and a rather popular review article of Ya.B.Zeldovich in Soviet Phys. Usp. 8 (1965) 489.

limit for their mass has been established: $m_q > 5$ GeV (5×10^3 MeV). This points to the fact that if hadrons are actually made up of quarks, then their binding energy must be colossal. In this situation an ever increasing number of physicists (including Gell-Mann himself) are beginning to be inclined to the idea that even if quarks do exist they cannot be in a free state, but are similar to quasi-particles, for example to phonons, in a solid. Some outstanding scientists (Heisenberg, Chew and others) disapprove of the hypothesis of quarks.

But in spite of everything the quark model is very attractive and even at the worst it is a very convenient mathematical tool for the formulation of unitary symmetry. The near future should give answer to one of the cardinal questions of the contemporary physics of elementary particles: are quarks real, and if so, in what sense, or are they a purely mathematical fiction?

§142. General appraisal of unitary symmetry

From the contents of the preceding sections it is seen that, owing to the hypothesis of the approximate invariance of the strong interaction with respect to the group SU(3), the physics of elementary particles has recently made much progress. The successes of unitary symmetry are numerous and impressive.

1. All stable hadrons and low-lying resonances are distributed over unitary multiplets whose members have one and the same spin, parity and baryonic number: octet of 0^- mesons, octet of $\frac{1}{2}^+$ baryons, decuplet $\frac{3}{2}^+$, nonets of 1^- and 2^+ meson resonances and others. No particle which in principle could not be placed in one of the unitary multiplets of not too high dimension has been discovered.

2. The quark model, which makes it possible to construct all hadrons from three fundamental particles and their antiparticles, is very attractive.

3. Different mass formulae have been obtained for isomultiplets within one unitary multiplet, which are in very good agreement with empirical data.

4. A number of relations between the coupling constants of baryons with mesons have been established, which for the most part need experimental verification.

5. The relations between the cross sections for different processes have been derived; when some subsidiary facts are accurately taken into account, none of them is in sharp disagreement with experiment.

6. Taking into account the electromagnetic interaction, mass formulae

have been obtained for individual members of the isomultiplets constituting a unitary multiplet. For example, on the basis of the theoretical relation

$$(m_n - m_p) - (m_{\Xi^0} - m_{\Xi^-}) = m_{\Sigma^-} - m_{\Sigma^+} \qquad (142.1)$$

the sign of the mass difference between the Ξ^0 and Ξ^--hyperons was predicted. This prediction was subsequently confirmed experimentally.

7. The relations between the magnetic moments of the baryons belonging to the same unitary multiplet were derived. In particular, the following equalities were shown to hold for the members of the octuplet $\frac{1}{2}^+$:

$$\mu_p = \mu_{\Sigma^+}, \qquad \mu_{\Xi^-} = \mu_{\Sigma^-}, \qquad \mu_n = 2\mu_\Lambda = -2\mu_{\Sigma^0}. \qquad (142.2)$$

8. Finally, Cabbibo succeeded in including the weak interaction also in the scheme of unitary symmetry, by which this theory acquired a certain harmony and completeness.

The first five points were discussed in §138—141. We shall not dwell on the remaining points, but refer the reader to the literature*.

On the other hand, the hypothesis of unitary symmetry has a number of essential shortcomings.

1. First of all, the group theoretical scheme does not contain any elements of dynamics, and it must only be a part of a theory of elementary particles which is still to come.

2. The question of the nature of unitary symmetry and its violation remains open. In connection with this different points of view have been put forward:

(a) Unitary symmetry is a fundamental property of the strong interaction, and is inherent to it in the same way as, say, its invariance with respect to transformations of the Lorentz group. In this case it is violated either by a moderately strong interaction whose coupling constant is of the order of 0.1 (we recall that the coupling constant of the strong interaction is equal to about 14), or owing to its interaction with the vacuum state which may have a complex structure (the so-called spontaneous violation of symmetry).

(b) Unitary symmetry is approximate in its very nature, 'because of a complex concurrence of different factors'**. In this case, one need not raise the question of the nature of the violation of SU(3) invariance, and the require-

---

* See, for example, the monograph of Nguyen van Hieu, *Lektsii po teorii unitarnoi simmetrii elementarnikh chastits* (*Lectures on the theory of elementary particles*) (Atomizdat, Moscow, 1967).

** G.Chew, *The analytic S-matrix* (Benjamin, New York, 1966).

ment that a unitary triplet of particles (quarks) necessarily exists makes no sense.

(c) There exists a strictly unitary symmetric superstrong interaction, and SU(3) invariance is violated as a result of switching on a real strong inter-action. Such a situation is analogous to that in the case of isospin symmetry.

3. The quark model in its initial version described in § 141 contains a number of logical difficulties. One of these lies in the fact that in trying to construct hadrons from three quarks the spatial-spin part of the wave func-tion turns out to be antisymmetric with respect to permutation of the quarks, which is unusual for the lowest state. In trying to overcome this difficulty the quark was assigned a new quantum number, which is equivalent to the consideration of not three but nine quarks, whence the harmony and elegance of the model are lost. There were even propositions to assume that quarks do not obey Fermi–Dirac statistics but a certain new type (so-called para-statistics) in which occupation numbers may take on, say, the values 0, 1 and 2.

4. The unitary symmetry scheme has a number of other shortcomings, which are not so fundamental but are still important. We shall only enumerate them:

(a) The situation concerning the distribution of higher resonance states over multiplets is not quite clear. For example, the problem as to where the particle Λ_{1405}, having spin $\frac{1}{2}^-$, is to be placed has not been elucidated.

(b) The problem of ω–φ mixing has not been finally solved.

(c) The question remains open as to why the mass formula for baryons in-volves the mass itself, while the mass formula for mesons involves the square of the mass.

(d) There is contradiction between the high accuracy of the mass formula for the decuplet $\frac{3}{2}^+$ and the poor relations for the probabilities of decay of these resonances, and so on.

It should also be noted that for a number of reasons the framework of the group SU(3) turns out to be too narrow:

1. It does not involve the baryonic number.

2. The parameters of ω–φ mixing are not predicted but are introduced into the theory from outside.

3. There are a number of intermultiplet relations pointing to the correlation of unitary quantum numbers with ordinary spin. Thus, for example, the parameters contained in the Okubo formula (140.10) for the octet and decuplet of baryons turn out to be almost equal; the following mass formula is valid

$$m_{K^*}^2 - m_\rho^2 = m_K^2 - m_\pi^2 \qquad (142.3)$$

and so on.

4. Some physicists are not content with the fact that the charge of quarks is a fractional number, which calls for a correction of the Gell-Mann—Nishijima relation by introducing into it a new quantum number, and this means increasing the rank of the basic symmetry group.

In trying to partially overcome the shortcomings mentioned a theoretical scheme was formulated invariant with respect to the group SU(6), which describes at the same time the unitary symmetry and spin properties of particles. Pseudoscalar and vector mesons fill the 35-dimensional multiplet of this group (every 0^- meson has one spin state, while the 1^- meson has three spin states, so that if in addition a unitary scalar 1^- particle is introduced, then we shall obtain in all $8 \times 1 + 8 \times 3 + 1 \times 3 = 35$ members of the multiplet), while $\frac{1}{2}^+$ and $\frac{3}{2}^+$ baryons fill the 56-plet ($8 \times 2 + 10 \times 4 = 56$). The most impressive result of the group SU(6) is the formula for the ratio of the magnetic moments of the proton and neutron:

$$\mu_n/\mu_p = -\tfrac{2}{3} \qquad (142.4)$$

(the experimental value being -0.68). But the group SU(6) is essentially non-relativistic and is unsuitable, for example, for a description of the processes of scattering of particles. However, in the attempt to make it relativistic insuperable difficulties have arisen associated with the probabilistic interpretation of the corresponding quantum-mechanical scheme.

The trend of further development of the theory is to be indicated by experiment.

SUBJECT INDEX